D1084495

*Optimization Using
Personal Computers*

OPTIMIZATION USING PERSONAL COMPUTERS

With Applications to Electrical Networks

THOMAS R. CUTHBERT, JR.

Director, Digital Signal Processing
Collins Transmission Systems Division
Rockwell International Corporation
Dallas, Texas

A Wiley-Interscience Publication

JOHN WILEY & SONS

New York   Chichester   Brisbane   Toronto   Singapore

Library of Congress Cataloging-in-Publication Data:

Cuthbert, Thomas R. (Thomas Remy), 1928–
    Optimization using personal computers.

    "A Wiley-Interscience publication."
    Includes index.
    1. Mathematical optimization—Data processing.
2. BASIC (Computer program language)   3. Electric
networks.   I. Title.

QA402.5.C88   1986   519.7′6   86-13319
ISBN 0-471-81863-1

Printed in the United States of America

10 9 8 7 6 5 4 3 2 1

To Ernestine,
for her understanding,
patience, and encouragement

Preface

Optimization is the adjustment of design variables to improve a result, an indispensable step in engineering and mathematical modeling. This book explains practical optimization, using personal computers interactively for learning and application. It was written for practicing engineers and scientists and for university students with at least senior class standing in mathematics, engineering, and other sciences. Anyone who has access to a BASIC-language computer and has been introduced to calculus and matrix algebra is prepared to master the fundamentals of practical optimization using this book.

Optimization, formally known as *nonlinear programming*, is the minimization of a scalar function that is nonlinearly related to a set of real variables, possibly constrained. Whether optimizing electrical circuit responses, structural design parameters, a system model, or curve fitting, there are usually free variables to choose so that an overall measure of goodness can be improved. The process of defining the objective function, selecting the variables and values, and considering their relationships often makes critical trade-offs and limitations evident. In many cases the process leads to a standard mathematical form so that a digital computer can adjust many variables automatically.

Early personal computers provided accessibility, responsiveness, autonomy, fixed cost. Recent models added large memory, high precision, and impressive speed, especially those with 16- or 32-bit microprocessors, numerical coprocessors, and software compilers. Although any computer can facilitate learning about iterative processes like optimization, recent personal computer models allow the addition of number-intensive optimization to their growing list of practical applications.

The first goal of this book is to explain the mathematical basis of optimization, using iterative algorithms on a personal computer to obtain key insights and to learn by performing the computations. The second goal is to acquaint the reader with the more successful gradient optimization techniques, especially Gauss–Newton and quasi-Newton methods with nonlinear constraints. The third goal is to help the reader develop the ability to read and comprehend the essential content of the vast amount of optimization literature. Many important topics in calculus and matrix algebra will be reinforced in that preparation. The last goal is to present programs and examples that illustrate the ease of obtaining exact gradients (first partial derivatives) for response functions of linear electrical networks and their analogues in the physical sciences.

Optimization is introduced in Chapter One by using fundamental mathematics and pictures of functions of one and two variables. Fortunately, these pictures apply to functions of many variables without loss of generality; the technique is employed throughout this book wherever possible. A general statement of the problem and some typical fields of application are provided. Issues involved in iterative process are discussed, such as number representation, numerical stability, illconditioning, and termination. Chapter One also includes comments concerning choices of programming languages and supporting software tools and gives some reassuring data concerning the speed of numerical operations using personal computers.

Chapters Two and Three furnish the essential background in linear and nonlinear matrix algebra for optimization. The approach is in agreement with Strang (1976:ix): linear algebra has a simplicity that is too valuable to be sacrificed to abstractness. Chapter Two reviews the elementary operations in matrix algebra, algorithms that are included in a general-purpose BASIC program called MATRIX for conveniently performing vector and matrix operations. The coverage of general matrix algebra topics is quite complete: notation, matrix addition, multiplication, inverse, elementary transformations, identities and inequalities, norms and condition numbers are defined and illustrated pictorially and by numerical examples. Matrix roles in space include linear independence, rank, basis, null space, and linear transformations, including rotation and Householder methods. Orthogonality and the Gram–Schmidt decomposition are described for later applications. The real matrix eigenproblem is defined and its significance is reviewed. The Gerchgorin theorem, diagonalization, and similarity are discussed and illustrated by example. Concepts concerning vector spaces are developed for hyperplanes and half-spaces, normals, projection, and the generalized (pseudo) inverse. Additions to program MATRIX are provided so that all the functions discussed may be evaluated numerically as well.

Chapter Three introduces linear and nonlinear functions of many variables. It begins with the LU and LDL^T factorization methods for solving square linear systems of full rank. Overdetermined systems that may be rank deficient are solved by singular value decomposition and generalized inverse using BASIC programs that are furnished. The mathematical and geometric properties of quadratic functions are described, including quadratic forms and exact linear (line) searches. Directional derivatives, conjugacy, and the conjugate gradient method for solution of linear systems are defined. Taylor series for many variables and the related Newton iteration based on the Jacobian matrix are reviewed, including applications for vector functions of vectors. Chapter

Three concludes with an overview of nonlinear constraints based on the implicit function theorem, Lagrange multipliers, and Kuhn–Tucker constraint qualifications.

Chapter Four describes the mathematics and algorithms for discrete Newton optimization and Gauss–Newton optimization. Both methods depend on first partial derivatives of the objective function being available, and BASIC programs NEWTON and LEASTP are furnished with numerical examples. Program NEWTON employs forward finite differences of the gradient vector to approximate the second partial derivatives in the Hessian matrix. Gauss–Newton program LEASTP is based on least-pth objective functions, the mathematical structure of which allows approximation of the Hessian matrix. The trust radius and Levenberg–Marquardt methods for limited line searches are developed in detail. Weighted least-pth objective functions are defined, and the concepts in numerical integration (quadrature) are developed as a strategy for accurate estimation of integral objective functions by discrete sampling.

Chapter Five covers quasi-Newton methods, using an iterative updating method to form successive approximations to the Hessian matrix of second partial derivatives while preserving a key Newton property. Program QNEWT is also based on availability of exact first partial derivatives. However, it is demonstrated that the BFGS search method from the Broyden family is sufficiently robust (hardy) to withstand errors in first derivatives obtained by forward differences, so much so that this quasi-Newton implementation is competitive with the best nongradient optimization algorithms available. Three kinds of line search are developed mathematically and compared numerically. The theory of projection methods for linear constraints is developed and applied in program QNEWT to furnish lower and upper bounds on problem variables. General nonlinear constraints are included in program QNEWT by one of the most successful penalty function methods due to Powell. The method and the algorithm are fully explained and illustrated by several examples. The theoretical considerations and limitations of most other methods for general nonlinear constraints are presented for completeness.

Chapter Six combines the most effective optimization method (Gauss–Newton) with projected bounds on variables and nonlinear penalty constraints on a least-pth objective function to optimize ladder networks, using program TWEAKNET. Fundamentals of electrical networks oscillating in the sinusoidal steady state are reviewed briefly, starting from the differential equations of an RLC circuit and the related exponential particular solution. The resulting complex frequency (Laplace) variable and its role in the impedance concept is reviewed so that readers having different backgrounds can appreciate the commonality of this subject with many other analogous physical systems. Network analysis methods for real steady-state frequency include an efficient algorithm for ladder networks and the more general nodal admittance analysis method for any network. The implementation for ladder networks in TWEAKNET includes approximate derivatives for dissipative networks and exact derivatives for lossless (inductor and capacitor) networks. The program utilizes optimization algorithms that were previously explained, and several numerical examples make clear the power and flexibility of nonlinear programming applications to electrical networks and analogous linear systems. Two methods for obtaining exact partial derivatives of any electrical network are explained, one based on Tellegen's theorem and adjoint networks and the other based on direct differentiation of the system matrix. Finally, the fundamental bilinear nature of sinusoidal responses of electrical networks is described in order to identify the best choice for network optimization. The underlying concept of sensitivity is defined in the context of robust response functions.

The text is augmented by six major programs and more than two dozen smaller ones, which may be merged into the larger programs to provide optional features. All programs are listed in Microsoft BASIC and may be converted to other BASICs and FORTRAN without serious difficulty. A floppy disk is available from the publisher with all the ASCII source code and data files included for convenience. A tenfold increase in computing speed is available by compiling the programs into machine code, which is desirable for optimization that involves more than five to ten variables. Compilers that link to the 8087 math coprocessor chip are necessary for very large problems to avoid overflow and to gain additional speed and precision. Programmed applications in this book demonstrate that current personal computers are adequate, so the reader is assured that future computers will allow even greater utilization of these algorithms and techniques.

This textbook is suitable for a one-semester course at the senior or graduate level, based on Chapters One through Five. The electrical network applications in Chapter Six might be included by limited presentation of material in Chapters Three and Five. The text has been used with excellent acceptance for a 32-hour industrial seminar for practicing engineers and scientists who desire an overview of the subjects. Its use in a university graduate course has also been arranged. Access to a BASIC-language computer is highly desirable and usually convenient at the present stage of the personal computer revolution. Closed-circuit network television or visible classroom monitors for the computer screen are optional but effective teaching aids that have been employed in the use of this material in industrial and university classrooms. Approximately 250 references are cited throughout the text for further study and additional algorithms.

I have been an avid student and user of nonlinear programming during the several decades that this subject has received the concentrated attention of researchers. My optimization programs have been applied in industry to obtain innovative results that are simply not available by closed-form analysis. I wish to express my sincere appreciation to my colleagues at Collins Radio Company, Texas Instruments, and Rockwell International who have made this possible. That certainly includes the outstanding librarians who have made new information readily available throughout those years.

I especially thank Dr. J. W. Cuthbert, Phil R. Geffe, John C. Johnson, and Karl R. Varian for their thorough reviews and comments on the manuscript. To Mr. Arthur A. Collins, whose support of my work in this field helped make optimization one of my professional trademarks, I extend my deepest gratitude.

THOMAS R. CUTHBERT, JR.

Plano, Texas
September 1986

Contents

3. Functions of Many Variables   96

4. Newton Methods   163

Optimization Using
Personal Computers

Chapter One ──────────────────────────

Introduction

This book describes the most effective methods of numerical optimization and their mathematical basis. It is expected that these techniques will be accomplished on IBM PC personal computers or comparable computers that run programs compatible with Microsoft BASIC. This chapter presents an overview of optimization and the unique approach made possible by modern personal computers.

Optimization is the adjustment of variables to obtain the best result in some process. This definition is stated much more clearly below, but it is true that everyone is optimizing something all the time. This book deals with optimization of those systems that can be described either by equations or more likely by mathematical algorithms that simulate some process. The field of electrical network design is one of many such applications. For example, an electrical network composed of inductors, capacitors, and resistors may be excited by a sinusoidal alternating current source, and the voltage at some point in that network will be a function of the source frequency and the values of the network elements. An objective function might be the required voltage behavior versus frequency. The optimization problem in that case is to adjust those network elements (the variables) to improve the fit of the *calculated* voltage-versus-frequency curve to the *required* voltage-versus-frequency curve over the frequency range of interest.

This chapter includes an overview of scalar functions of a vector (set) of variables, suggests some typical optimization problems from a number of technical fields, describes the nature of these iterative processes on computing machines, and justifies the choice of program language, computers, and tutorial approach that will be used to make these topics easier to understand.

1.1. Scalar Functions of a Vector

The scalar functions to be optimized by adjustment of the variables are described in greater mathematical detail in Sections 3.2 and 3.3. However, it is

1

important to see various geometrical representations that describe optimization before plunging into matrix algebra.

1.1.1. Surfaces Over Two Dimensions.   Consider the isometric representation of a function of two variables shown in Figure 1.1.1. The corresponding plot of its contours or level curves is shown in Figure 1.1.2. The equation for this function due to Himmelblau (1972) is

$$F(x, y) = -\left(x^2 + y - 11\right)^2 - \left(x + y^2 - 7\right)^2. \qquad (1.1.1)$$

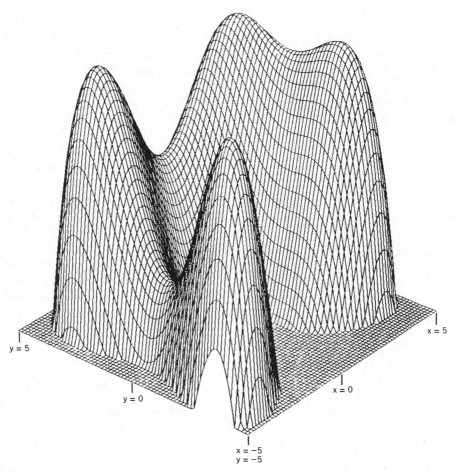

y = 5

y = 0

x = 5

x = 0

x = -5
y = -5

Figure 1.1.1.   A surface over two-dimensional space. The base is 150 units below the highest peaks.

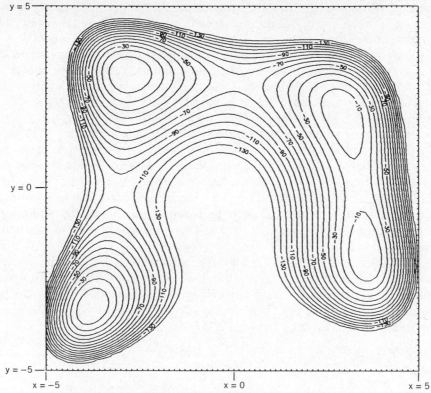

Figure 1.1.2. Contours of the surface depicted in Figure 1.1.1.

The function values are plotted below the x–y plane, in this case 2-space, where the variables are the coordinates in that space. In the general case, there will be n such variables in n space. The four peaks in Figure 1.1.1 each touch the x–y plane ($F = 0$), and the flat base represents a function value of $F = -150$. "As a rule, a theorem which can be proved for functions of two variables can be extended to functions of more than two variables without any essential change in argument," according to Courant (1936).

Now the optimization problem can be stated quite easily: Given some location on the surface in Figure 1.1.1, how can x and y be adjusted to find values corresponding to a maximum (or a minimum) of the function? This is the same task confronting the blind climber on a mountainside: What sequence of coordinate adjustments will carry the climber to a peak? The peak obtained depends on where the climber starts.

One can anticipate the use of slopes in various directions, especially that direction having the steepest slope (known as the gradient at that point). In

Figures 1.1.1 and 1.1.2 there are seven points in the $x-y$ space where the slopes in both the x and y directions are zero, yet three of those points are not at peaks. These three places are called saddle points for obvious reasons.

It is also assumed that there is at least one peak. Consider computation of compounding principal; if the x axis is the interest percentage, the y axis is the number of compounding periods, and the function surface is the principal for that interest rate and time, it is clear that the function grows without bound, so there is no peak at all.

Only static or parameter optimization will be discussed in this book and thus time is not involved. The surface in Figure 1.1.1 does not fluctuate in time, the variables all have the same status, and the solution is a set of numerical values, not a set of functions. Incidentally, the methods discussed could be extended to dynamic or time-dependent optimization including control functions (the optimal path problem). There would be additional mathematical principles involved, and the solution would be a set of functions of time.

Constraints on the variables, such as required relationships on and among them, are considered. For example, the variables might have upper and lower bounds, or they might be related by equalities or inequalities that could be linear or nonlinear functions. A constrained optimization problem of the latter type would be the maximization of (1.1.1) such that

$$(x - 3)^2 + (y - 2)^2 - 1 = 0. \tag{1.1.2}$$

This means that the solution must be the maximum point on the surface below the circle in the $x-y$ plane with unit radius and centered on $x = 3$, $y = 2$. Incidentally, a constraint such as (1.1.2), applied to an otherwise unbounded problem, would result in a unique, bounded solution.

Functions of two or more variables may be implicit because of an extensive algorithm involved in their computation. Therefore, explicit equations such as (1.1.1) are usually not available in practical problems such as a network voltage evaluation. These algorithms suggest another interpretation of optimization from the days when computer punch cards were in vogue. Then values for the specified variables were selected, and the formatted punched cards were submitted for computer execution. When the result was available, it was judged for acceptability, if not optimality. The usual case was a cycle of resubmittals to improve the result in some measurable way. Even though the result was obtained by some complicated sequence of computations instead of one or more explicit equations, that process is optimization in the sense of the illustration in Figure 1.1.1 for two variables.

In most practical cases, it will be assumed that the variables are continuous, real numbers (not integers) and that the functions involved are single-valued and usually smooth. These requirements will be refined in some detail later.

1.1.2.   Quadratic Approximations to Peaks.   The function described by (1.1.1) and shown in Figures 1.1.1 and 1.1.2 is of degree 4 in each variable. The four pairs of coordinates where the function has a maximum are given in Table 1.1.1. At each of these points, a necessary condition for a maximum or a minimum is that the first derivatives equal zero. Simple calculus yields these expressions from (1.1.1):

$$\frac{\partial F}{\partial x} = -4x(x^2 + y - 11) - 2(x + y^2 - 7), \qquad (1.1.3)$$

$$\frac{\partial F}{\partial y} = -2(x^2 + y - 11) - 4y(x + y^2 - 7). \qquad (1.1.4)$$

Substitution of each of the four pairs of values in Tables 1.1.1 will verify that these derivatives are zero at those points in 2-space. The program in Table 1.1.2 computes the function, its derivatives, and several other quantities of interest for any $x-y$ pair.

The peak nearest the viewer in Figure 1.1.1 is located at the $x-y$ values in the third column of Table 1.1.1. Running the program in Table 1.1.2 produces the results shown in Table 1.1.3.

Since this function will be used to illustrate a number of concepts, the program in Table 1.1.2 computes several other quantities, in particular the second derivatives:

$$\frac{\partial^2 F}{\partial x^2} = -12x^2 - 4y + 42, \qquad (1.1.5)$$

$$\frac{\partial^2 F}{\partial y^2} = -4x - 12y^2 + 26, \qquad (1.1.6)$$

$$\frac{\partial^2 F}{\partial x \, \partial y} = \frac{\partial^2 F}{\partial y \, \partial x} = -4(x + y). \qquad (1.1.7)$$

A matrix composed of these second derivatives is called the Hessian, and it is involved in many calculations explained later. Results from the program in Table 1.1.2 are usefully interpreted by comparison with the $x-y$ points in Figure 1.1.2.

Table 1.1.1.   The Four Maxima of the Function in Equation (1.1.1)

x	3.0000	3.5844	-3.7793	-2.8051
y	2.0000	-1.8481	-3.2832	3.1313

Table 1.1.2. A Program to Evaluate (1.1.1), Its Derivatives, and an Approximation

```
10 REM EVALUATE (1.1.1), DERIVS, & A QUAD FIT.
20 CLS
30 PRINT "INPUT X,Y=";:INPUT X,Y
40 T1=X*X+Y-11
50 P1=X+Y*Y-7
60 F=-T1*T1-P1*P1
70 G1=-4*X*T1-2*P1
80 G2=-2*T1-4*Y*P1
90 H1=-12*X*X-4*Y+42
100 H2=-4*X-12*Y*Y+26
110 H3=-4*(X+Y)
120 D=H1*H2-H3*H3
130 PRINT "X,Y="; X;Y
140 PRINT "EXACT F="; F
150 PRINT "1ST DERIVS WRT X,Y ="; G1;G2
160 PRINT "2ND DERIVS WRT X,Y ="; H1;H2
170 PRINT "2ND DERIV CROSS TERM ="; H3
180 PRINT "DET OF 2ND DERIV MATRIX ="; D
190 F1=-58.13*X*X-346.63*X+28.25*X*Y-182.95*Y-44.12*Y*Y-955.34
200 PRINT "APPROX F = "; F1
210 PRINT "ERROR=F-F1 ="; F-F1
220 PRINT
230 GOTO 30
240 END
```

The principle on which most efficient optimizers are based is that there exists a neighborhood of a maximum or minimum, where it is adequate to approximate the general function by one that is quadratic. For example, the peak at $x = -3.7793$ and $y = -3.2832$, which is nearest the reader in Figure 1.1.1, can be approximated by

$$F1(x, y) = ax^2 + bx + cxy + dy + ey^2 + k, \qquad (1.1.8)$$

where the constants are given in Table 1.1.4. The function is considered quadratic in variables x and y because the maximum degree is two (including the cross product involving xy). This is clearly not the case in (1.1.1), the function that is being approximated.

Table 1.1.3. Analysis of Equation (1.1.1) and an Approximation at $(-3.7793, -3.2832)$

```
INPUT X,Y=? -3.7793,-3.2832
X,Y=-3.7793 -3.2832
EXACT F=-1.916578E-08
1ST DERIVS WRT X,Y =-1.604432E-03  1.53765E-03
2ND DERIVS WRT X,Y =-116.2645 -88.23562
2ND DERIV CROSS TERM = 28.25
DET OF 2ND DERIV MATRIX = 9460.609
APPROX F =  7.56836E-03
ERROR=F-F1 =-7.568379E-03
```

Table 1.1.4.   Constants for the Quadratic Approximation Given in Equation (1.1.8)

$a = -58.13$	$c = 28.25$	$e = -44.12$
$b = -346.63$	$d = -182.95$	$k = -955.34$

The program in Table 1.1.2 also computes (1.1.8), and the results in Table 1.1.3 confirm that it is an excellent approximation at the peak. In fact, it is a good approximation in some small neighborhood of the peak, say within a radius of 0.3 units; the reader is urged to run the program using several trial values. The quadratic approximation in (1.1.8) is shown in the oblique illustration in Figure 1.1.3, comparable with Figure 1.1.1 for the original function. It appears that such an approximation of the other peaks in Figure 1.1.1 would be valid, but in smaller neighborhoods. The validity of this approximation is discussed in Chapter Three in connection with Taylor series for functions of many variables.

The important conclusions concerning quadratic approximations are (1) some informal scheme will be required to approach maxima or minima,

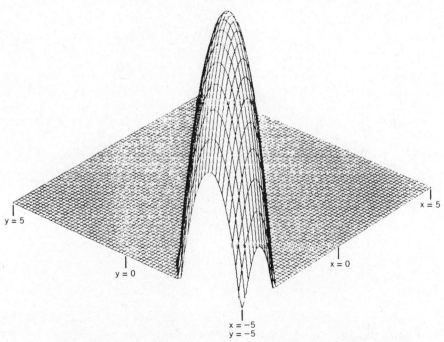

$x = 5$

$y = 5$

$y = 0$

$x = 0$

$x = -5$
$y = -5$

Figure 1.1.3.   A quadratic surface approximating a peak in Figure 1.1.1.

(2) a quadratic function will be the basis for optimization strategy near maxima and minima, and (3) the quadratic function makes the important connection between optimization theory and the solution of systems of linear equations. This last point is made clear by applying the necessary condition for a maximum or minimum to the quadratic approximation in (1.1.8); its derivatives are

$$\frac{\partial F1}{\partial x} = 2ax + b + cy, \tag{1.1.9}$$

$$\frac{\partial F1}{\partial y} = cx + d + 2ey. \tag{1.1.10}$$

The derivatives are equal to zero at the peak located at approximately $x = -3.7793$ and $y = -3.2832$. More important, equating the derivatives to zero produces a set of linear equations, and this is the connection between

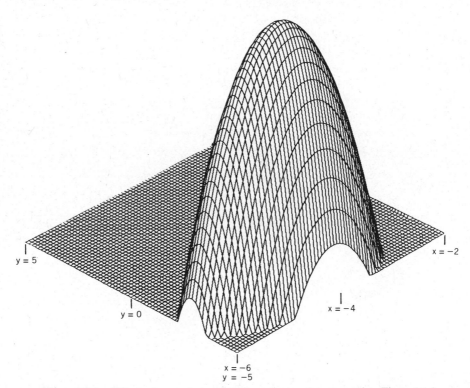

Figure 1.1.4.   Effects of poor choice of scale on the x axis compared to Figure 1.1.3.

optimization (and its quadratic behavior near peaks) and solution of systems of linear equations. This theme will be developed time and again as the methods for nonlinear optimization are explored in this book.

One last consideration in solving practical problems is the choice of scales for the variables. For example, Figure 1.1.3 is plotted over 10 units in x and y. Changing the scale for x to range over just four units produces the stretched surface shown in Figure 1.1.4; it is the same function. Finding peaks or solving the corresponding system of linear equations for severely stretched surfaces can be difficult. Mathematical description of these effects and how to deal with them are prime topics in subsequent chapters, since locally poor scaling is inevitable in practical optimization problems, especially in dimensions greater than 2.

1.2.   Types of Optimization Problems

Not much was known about optimization before 1940. For one thing, computers are necessary since applications require extensive numerical computation. However, there were some very early theoretical contributions; for example, in 1847 Cauchy described the method of steepest ascent (up a mountain) in connection with a system of equations (derivatives equated to zero). The field began to flourish in the 1940s and 1950s with linear programming—the case where all variables are involved linearly in both the main objective function and constraints. Successful algorithms for nonlinear unconstrained problems began with Davidon (1959). There has been steady progress since then, although optimization problems involving nonlinear constraints are often difficult to solve. The next section includes a more formal statement of optimization, consideration of some important objective functions, and mention of many fields where optimization is employed to great advantage, including the role of mathematical models.

1.2.1.   General Problem Statement.   The most comprehensive statement of the optimization problem considered in this book is to minimize or maximize some scalar function of a vector

$$F(\mathbf{x}) = F(x_1, x_2, \ldots, x_n), \tag{1.2.1}$$

subject to sets of constraints, expressed here as vector functions of a vector:

$$\mathbf{h}(\mathbf{x}) = \mathbf{0}, \quad \mathbf{h} \text{ a set } E \text{ containing } q \text{ functions,} \tag{1.2.2}$$

$$\mathbf{c}(\mathbf{x}) \geq \mathbf{0}, \quad \mathbf{c} \text{ a set } I \text{ containing } m - q \text{ functions.} \tag{1.2.3}$$

The notation introduced here covers the case where there are n variables, not just the two previously called x and y. In practice, n can be as high as 50 or more. The variables will henceforth be defined as a column vector, that is, the set

$$\mathbf{x} = (x_1, x_2, \ldots, x_n)^T. \tag{1.2.4}$$

The superscript T transposes the row vector into a column vector. There are q equality constraint functions in (1.2.2); a typical one might be that previously given in (1.1.2). For example, when $q = 3$, there would be $h_1(\mathbf{x})$, $h_2(\mathbf{x})$, and $h_3(\mathbf{x})$. There are $m - q$ inequality constraints in the vector $\mathbf{c}$ shown in (1.2.3), interpreted like the functions in $\mathbf{h}$.

All the variables and functions involved are continuous and smooth; that is, a certain number of derivatives exist and are continuous. This eliminates problems that have only integer variables or those with objective functions, $F(\mathbf{x})$, that jump from one value to another as the variables are changed. Linear programming allows only functions F, $\mathbf{h}$, and $\mathbf{c}$ that relate the variables in a linear way (first degree). There is also a classic subproblem known as quadratic programming, where $F(\mathbf{x})$ is a quadratic function, as was (1.1.8), and the constraint functions are linear. That case is analyzed in Section 5.4.1. This book emphasizes unconstrained nonlinear programming (optimization), with subsequent inclusion of linear constraints (especially upper and/or lower bounds on variables) and general nonlinear constraints.

1.2.2. Objective Functions.   So far optimization has been presented as maximizing a function, simply because it is easier to display surfaces with peaks as in Figure 1.1.1. Actually, there is a trivial difference mathematically between maximization and minimization of a function: Maximizing $F(\mathbf{x})$ is equivalent to minimizing $-F(\mathbf{x})$. In terms of Figure 1.1.1, plotting $-F(\mathbf{x})$ simply turns the surface upside down. Objective functions will be discussed in terms of minimizing some $F(\mathbf{x})$, especially the case where $F(\mathbf{x})$ can only be positive, thus the lowest possible minimum is zero.

A search scheme (iterative process) to find $\mathbf{x}$ such that $F(\mathbf{x}) = 0$ is called a one-point iteration function by Traub (1964). In many practical situations the optimization problem belongs to Traub's classification of multipoint iteration functions that are distinguished by also having an independent sample variable, say t. Thus, the problem is to attempt to obtain $F(\mathbf{x}, t) = 0$. One way to view this abstraction is the curve-fitting problem: consider Figure 1.2.1, which portrays a given data set that is to be approximated by some fitting function of five variables. To be specific, Sargeson provided data to be fit by the following function from Lootsma (1972:185):

$$f(\mathbf{x}, t) = x_1 + x_2\exp(-x_4 t) + x_3\exp(-x_5 t). \tag{1.2.5}$$

The data were provided in a table of 33 discrete data pairs, (t_i, d_i) evenly spaced along the t axis as plotted by the dots in Figure 1.2.1. Objective

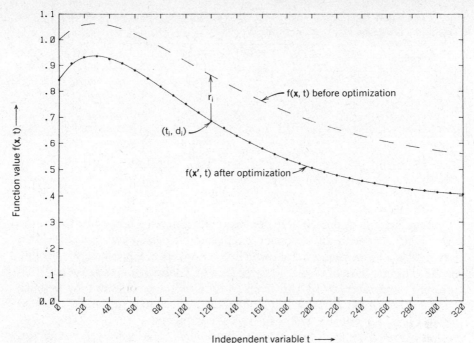

Figure 1.2.1. Sargeson's exponential fitting problem given 33 data samples and five variable parameters.

functions provide a measure of success for optimization. An important measure is the unconstrained least squares function used in this example:

$$F(\mathbf{x}, t) = \sum_{i=1}^{m} r_i^2, \qquad (1.2.6)$$

where there are m samples over the t space, and

$$r_i = f(\mathbf{x}, t_i) - d_i. \qquad (1.2.7)$$

The errors at each sample point, r_i in (1.2.7), are called *residuals* and are shown in Figure 1.2.1. Squaring each residual in (1.2.6) does away with the issue of sign.

The fitting function $f(\mathbf{x}, t)$ and values in the initial variables vector $\mathbf{x}$ are usually determined by some means peculiar to each problem. Practically, this must be a guess suitably close to a useful solution, as is the case shown in Figure 1.2.1. Denoting that initial choice as $\mathbf{x}^{(0)}$, the iterative search algorithm LEASTP from Chapter Four was employed for adjusting the set of five variables. Set number 9, $\mathbf{x}^{(9)}$, produced negligible improvement to a local minimum. Table 1.2.1 summarizes these results.

Table 1.2.1.   Initial and Final Values of Variables and Objective Function in the Sargeson Exponential Fitting Problem

Variable	Before	After	% Change
x_1	0.5000	0.3754	-24.92
x_2	1.5000	1.9358	29.05
x_3	-1.0000	-1.4647	-46.47
x_4	0.0100	0.01287	28.70
x_5	0.0200	0.02212	10.60
$F(\mathbf{x}, t)$	0.879E0	$0.546\mathrm{E}-4^a$	-99.994

[a] $\mathrm{E}-4$ denotes a factor of 0.0001.

A more general approach is to construct the objective function of residuals as in (1.2.6), but with the exponent 2 replaced by p, an even integer. This is known as least-pth minimization, which is described in Chapter Four, including the ordinary case of $p = 2$. Large values of p emphasize the residuals that represent the greater errors. This tends to equalize the errors but only as much as the mathematical model will tolerate. Also, values of p in excess of about 20 will cause numerical overflow in typical computers.

A measure of the error indicated in Figure 1.2.1 could very well be the area between the initial and desired curves. This suggests that principles of numerical integration might be relevant in the construction of an objective function. Indeed, Gaussian quadrature is one method of numerical integration that involves systematic selection of points t_i in the sample space as well as unique weighting factors for each sampled residual. Several variations of these methods are discussed in Section 4.4.4.

In some cases it is necessary to minimize the maximum residual in the sample space. One statement of this *minimax* objective is

$$\underset{\mathbf{x}}{\text{Min}} \ \underset{i}{\text{Max}} \ (r_i)^2 \qquad \text{for } i = 1 \text{ to } m. \tag{1.2.8}$$

In terms of Figure 1.2.1, adjustments to the variables are made only after scanning all the discrete points in the sample space (over t) to find the maximum residual. This sequence of adjustments usually results in different sample points being selected as optimization proceeds; therefore, the objective function is not continuous. An effective approach for dealing with these minimax problems was suggested by Vlach (1983); add an additional variable, x_{n+1}, and

$$\text{Minimize} \quad x_{n+1} \tag{1.2.9}$$

$$\text{such that} \quad x_{n+1} - (r_i)^2 \geq 0, \qquad \text{for all } i. \tag{1.2.10}$$

The iteration is started by selecting the largest residual; thereafter, the process is continuous.

Quite often, the least-squares solution is "close" to the minimax solution, and only a small additional effort is required to find it. However tempting it may be to use the absolute value of the residual, it is wise to avoid it because of its discontinuous effect on derivatives. Thus, these illustrations employ sums of squared residuals, which are smooth functions.

1.2.3. Some Fields of Application. The three classical fields of optimization, approximation, and boundary value problems for differential equations are closely related. Optimization per se is required in many problems that occur in the statistical and engineering sciences. Most statistical problems are essentially solutions to suitably formulated optimization problems. Resource allocation, economic portfolio selection, curve and surface fitting as illustrated previously, linear and nonlinear regression, signal processing algorithms, and solutions to systems of nonlinear equations are well-known applications of nonlinear optimization.

According to Dixon (1972b), the use of nonlinear optimization techniques is spreading to many areas of new application, as diverse as pharmacy, building construction, aerospace and ship design, diet control, nuclear power, and control of many production facilities. Fletcher (1980) provides a simple illustration of what is involved in the optimal design of a chemical distillation column to maximize output and minimize waste and cost. Bracken (1968) describes many of these applications and adds weapons assignment and bid evaluation. Nash (1979) describes optimal operation of a public lottery. Many more applications can be found in the proceedings of a conference on *Optimization in Action*, Dixon (1977), and scattered throughout the technical literature in great numbers. According to Rheinboldt (1974), "there is a growing trend toward consideration of specific classes of problems and the design of methods particularly suited for them."

All applications of optimization involve a model, which is essentially an objective function with suitable constraints, as in (1.2.1) through (1.2.3). These mathematical models are often used to study real-world systems, such as the human chest structure, including the lungs. In that case, pertinent mathematical expressions for the various mass, friction, and spring functions require that certain coefficients must be determined so that the model "fits" one human as opposed to another. In situations of this sort, physical experiments have been devised (an air-tight phone booth) to measure the physical system response (air volume, pressure, and breathing rate). Then the coefficients are determined in the mathematical model by optimization. In this case these coefficients often indicate the condition of the patient.

Bracken (1968) describes a rather elaborate model for the total cost of developing, building, and launching a three-stage launch vehicle used in space exploration. This is the only means for evaluating the results of alternative choices, since real-world experimentation is expensive, dangerous, and sometimes impossible. Models are not formulated as ends in themselves; rather they serve as means to evaluate free parameters that "fit" the system or to find those parameters that produce an optimum measure of "goodness," for example, minimum cost.

1.3.   Iterative Processes

Nonlinear optimization is an iterative process to improve some result. An iterative process is an application of a formula in a repetitive way (an algorithm) to generate a sequence of numbers from a starting initial value. The wisdom of relying on iterative processes for design as opposed to analytical, closed-form solutions is somewhat controversial. Acton (1970) went so far as to state that "minimum-seeking methods are often used when a modicum of thought would disclose more appropriate techniques. They are the first refuge of the computational scoundrel, and one feels at times that the world would be a better place if they were quietly abandoned. ... The unpleasant fact that the approach can well require 10 to 100 times as much computation as methods more specific to the problem is ignored—for who can tell what is being done by the computer?" Well, the owner/operator of a personal computer should certainly have a more balanced outlook, especially if he or she is aware of what Forsythe (1970) called "Pitfalls in computation, or why a math book isn't enough." This section discusses some of those issues.

1.3.1.   Iteration and Convergence.   Most optimization algorithms can be described by the simple iterative process illustrated in Figure 1.3.1. The initial estimate of the set of independent variables is $x^{(0)}$ and the corresponding scalar function value is $F^{(0)}$. The ancillary calculations noted in Figure 1.3.1 might include derivatives or other quantities that would support search or termination decisions.

The counter K is commonly referred to as the iteration number, that is, the number of times the process has been repeated in going around the outer loop shown in Figure 1.3.1. As in BASIC, FORTRAN, and many other programming languages, the statement $K = K + 1$ indicates a replacement operation; in this case the counter K is incremented by unity. The strategic part of the algorithm occurs in computing the next estimate, $x^{(k)}$. However chosen, it may not be satisfactory; for example, the corresponding function value $F^{(k)}$ may have increased when a minimum is desired. Other reasons for rejection include violation of certain constraints on the variables. In these events there may be a sequence of estimates for $x^{(k)}$ by some scheme, until a satisfactory estimate is obtained.

The decision to stop the algorithm, often called termination, can be surprisingly complicated and will be discussed further in Section 1.3.4. Somehow, if there is lack of progress or change in x and F, or the derivatives of F are approximately zero, or an upper limit in the number of iterations is

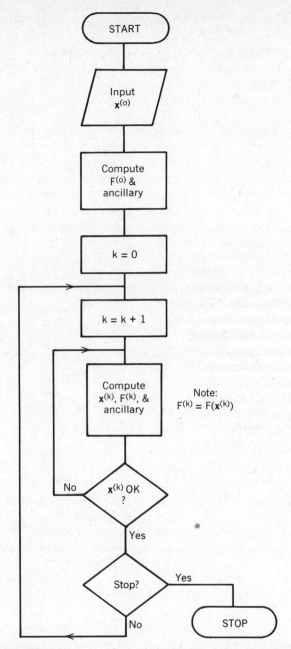

Figure 1.3.1. A typical iterative process for optimization or solution of nonlinear equations.

reached, these all may contribute to the decision to terminate the iterative process.

A graphical interpretation of a typical iterative process in one variable can be obtained by considering the classical *fixed-point problem* according to Traub (1964): Find the solution of

$$F(x) = x \tag{1.3.1}$$

by the so-called *repeated substitution* iteration

$$x^{(k+1)} = F(x^{(k)}). \tag{1.3.2}$$

If $x = a$ satisfies (1.3.1), then a is called a *fixed point* of F. Before showing the graphical solution of fixed-point problems, it is useful to relate them to minimization problems. Suppose that it is necessary to compute a zero of the function $f(x)$, or equivalently, a root of the equation $f(x) = 0$. Then the *fixed point* of the iteration function

$$F(x) = x - f(x) g(x) \tag{1.3.3}$$

coincides with the solution of $f(a) = 0$ if $g(a)$ is finite and nonzero. Two examples will illustrate these and other concepts.

Example 1.3.1.   Suppose that a root of $f(x) = 0$ is required, where

$$f(x) = x^2 - 1. \tag{1.3.4}$$

One such root is obviously $x = +1$. Referring to the iteration function in (1.3.3), choose $g(x) = 1/(2x)$, which meets the requirements placed on (1.3.3) and happens to be the Newton–Raphson iteration described in Section 5.1.1. Substitution of these choices for f and g into (1.3.3) yields an iteration function for this case that is

$$F(x) = \frac{x^2 + 1}{2x}. \tag{1.3.5}$$

This iteration function can be solved by the algorithm charted in Figure 1.3.1; the BASIC instructions and results are given in Table 1.3.1 as illustrated graphically in Figure 1.3.2. Most fixed-point problems are easily visualized because the $y = x$ line, a component of (1.3.3), always divides the first quadrant.

Example 1.3.2.   A second example of repeated substitution concerns finding a root of

$$f(x) = (x - 1)^2. \tag{1.3.6}$$

A root of multiplicity 2 is $x = +1$. The Newton formula requires that $g(x) = 0.5/(x - 1)$, which is the reciprocal of the first derivative of $f(x)$. This choice for $g(x)$ is satisfactory in the limit $x = 1$ by l'Hospital's rule from calculus. Substitution of these new choices for f and g into (1.3.3) yield the iteration function

$$F(x) = \frac{x + 1}{2}. \tag{1.3.7}$$

Table 1.3.1. **BASIC Program and Output to Find a Zero of $f(x) = x^2 - 1$ Using the Fixed-Point Iteration Function $F(x) = (x^2 + 1) / 2x$**

```
10 REM - FOR f=(X*X-1)
20 DEFDBL X,F
30 X=2
40 F=(X*X+1)/2/X
50 K=0
60 PRINT" K" TAB(15) "X" TAB(35) "F"
70 PRINT K TAB(5) X TAB(25) F
80 K=K+1
90 X=F
100 F=(X*X+1)/2/X
110 PRINT K TAB(5) X TAB(25) F
120 IF K=5 THEN STOP
130 GOTO 80
140 END
Ok
RUN
 K               X                        F
 0   2                          1.25
 1   1.25                       1.025
 2   1.025                      1.00030487804878
 3   1.00030487804878          1.000000046461147
 4   1.000000046461147         1.000000000000001
 5   1.000000000000001         1
```

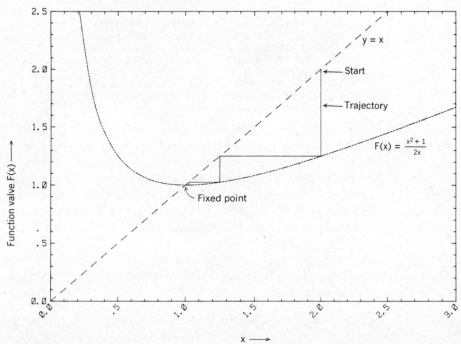

Figure 1.3.2. Repeated substitution of x in $F(x) = (x^2 + 1)/2x$, beginning with $x = 2$ and approaching the fixed point, $x = 1$.

Table 1.3.2.   BASIC Program and Output to Find a Zero of $f(x) = (x - 1)^2$

```
10 REM - FOR f=(X-1)^2
20 DEFDBL X,F
30 X=2
40 F=(X+1)/2
50 K=0
60 PRINT" K" TAB(15) "X" TAB(35) "F"
70 PRINT K TAB(5) X TAB(25) F
80 K=K+1
90 X=F
100 F=(X+1)/2
110 PRINT K TAB(5) X TAB(25) F
120 IF K=18 THEN STOP
130 GOTO 80
140 END
Ok
RUN
```

K	X	F
0	2	1.5
1	1.5	1.25
2	1.25	1.125
3	1.125	1.0625
4	1.0625	1.03125
5	1.03125	1.015625
6	1.015625	1.0078125
7	1.0078125	1.00390625
8	1.00390625	1.001953125
9	1.001953125	1.0009765625
10	1.0009765625	1.00048828125
11	1.00048828125	1.000244140625
12	1.000244140625	1.0001220703125
13	1.0001220703125	1.00006103515625
14	1.00006103515625	1.000030517578125
15	1.000030517578125	1.000015258789063
16	1.000015258789063	1.000007629394531
17	1.000007629394531	1.000003814697266
18	1.000003814697266	1.000001907348633

Again, the general algorithm in Figure 1.3.1 applies, and the BASIC instructions and results are shown in Table 1.3.2 as illustrated graphically in Figure 1.3.3. The trajectory or path of the repeated substitution generally appears to the right of the fixed point (as illustrated), or to the left, or encircles the fixed point. Furthermore, the trajectory may converge (as illustrated) or diverge. The reader is referred to Maron (1982:32) for illustration of all possible cases and to Traub (1964) for an exhaustive theoretical analysis.

Comparison of the data in Tables 1.3.1 and 1.3.2 shows a much slower convergence in the latter. To discuss rates of convergence, define the error in x at any iteration number k as

$$h^{(k)} = x^{(k)} - x^*, \tag{1.3.8}$$

where x^* is the fixed-point or optimal solution. Then convergence is said to be

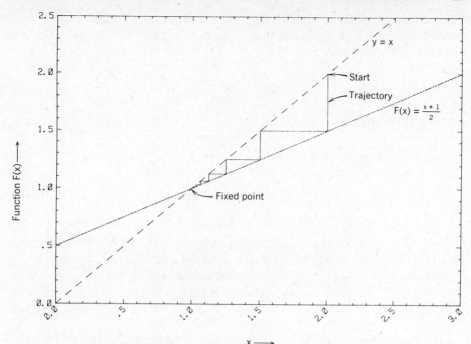

Figure 1.3.3. Repeated substitution of x in $F(x) = (x + 1)/2$, beginning with $x = 2$ and approaching the fixed point, $x = 1$.

linear if

$$\frac{|h^{(k+1)}|}{|h^{(k)}|} \to c_1 \tag{1.3.9}$$

for $c_1 \geq 0$, and *superlinear* for $c_1 = 0$. The arrow means "approaches." Convergence is said to be *quadratic* if

$$\frac{|h^{(k+1)}|}{|h^{(k)}|^2} \to c_2. \tag{1.3.10}$$

The data in Table 1.3.1 show that convergence is quadratic, satisfying (1.3.10) with $c_2 \doteq 0.5$. Roughly speaking, quadratic convergence means that the number of correct significant figures in $x^{(k)}$ doubles for each iteration. On the other hand, the data in Table 1.3.2 indicates linear convergence with $c_1 \doteq 0.5$.

This behavior of the repeated substitution algorithm can be predicted. Consider a Taylor series expansion of the iteration function about $x^{(k)} = x^*$:

$$F(x^{(k)}) = F(x^*) + F'(x^*)h^{(k)} + \frac{F''(x^*)(h^{(k)})^2}{2} + \cdots . \tag{1.3.11}$$

The error $h^{(k)}$ is that defined in (1.3.8), and F' and F'' are the first and second derivatives of F with respect to x, respectively. But the definition of the repeated substitution iteration in (1.3.2) enables restatement of (1.3.11) as

$$h^{(k+1)} = F'(x^*)h^{(k)} + \tfrac{1}{2}F''(x^*)(h^{(k)})^2 + \cdots. \tag{1.3.12}$$

Therefore, for very small errors, $h^{(k)}$,

$$h^{(k+1)} = F'(x^*)h^{(k)} \qquad \text{if } F'(x^*) \neq 0, \tag{1.3.13}$$

$$h^{(k+1)} = \frac{F''(x^*)(h^{(k)})^2}{2} \qquad \text{if } F'(x^*) = 0. \tag{1.3.14}$$

Considering the definitions of linear and quadratic convergence in (1.3.9) and (1.3.10), it can be concluded that the repeated substitution algorithm converges linearly if $0 < |F'(x^*)| < 1$ and quadratically if $F'(x^*) = 0$, but it diverges if $|F'(x^*)| > 1$. These conclusions explain the results in both Examples 1.3.1 and 1.3.2. The interested reader is again referred to Traub (1964).

1.3.2. Numbers and Stability. Computations in optimization algorithms are accomplished using floating-point numbers, namely, those in the form $x = ab^e$, where a is the mantissa, b is the base, and e the exponent. Though actual computing takes place in binary (base 2) arithmetic, the personal computer user's perception is that BASIC computes floating-point numbers in base 10 arithmetic. For these purposes it is quite adequate to note that IBM-PC BASIC provides six decimal digits of precision in the mantissa for single precision and 17 digits for double precision. Numbers can be represented in the range from 2.9E − 39 to 1.7E + 38. However, the optional 8087 math coprocessor integrated-circuit chip extends the number range from approximately 4.19E − 307 to 1.67E + 308. Furthermore, it has an internal format that extends the range from approximately 3.4E − 4932 to 1.2E + 4932. This data suggests that the user should simply be aware of some computational pitfalls that will be discussed. No exhaustive error analysis is required here; the interested reader is referred to Forsythe (1977) and Wilkinson (1963).

The troublesome phenomenon is simply that any digital computer provides only a finite set of points on the continuous real-number line. Values between these points are represented at an adjacent point; thus, there are *rounding errors*. These errors occur only in the mantissa and may accumulate to significant proportions during extended algorithms such as complicated iterative calculations. The computer user seldom observes the intermediate problems as they occur. When numbers on the real-number line exceed the largest numbers represented in the machine, then *overflow* occurs, usually as a result of multiplication. Similarly, multiplication of two nonzero numbers may have a nonzero product that falls between the two machine representable numbers adjacent to zero. This is called *underflow*, and the better software simply equates the result to zero without an error message.

Example 1.3.3.   Forsythe (1970) discusses an example of roundoff errors by Stegun and Abramowitz. One of the most common functions is e^x. An obvious (but dangerous) way to compute it is by its universally convergent infinite series

$$e^x = 1 + x + \frac{x^2}{2!} + \frac{x^3}{3!} + \cdots. \tag{1.3.15}$$

The program in Table 1.3.3 computes this power series in single-precision arithmetic. As noted, only six decimal digits are accurate, even though seven digits are stored and printed. Also, Table 1.3.3 contains the result of computing e^{-8}, the correct value being $3.3546262E - 4$ or 0.00033546262. The EXP function in IBM-PC BASIC gives the answer correctly to six significant figures, as expected. However, the power series method has only one correct significant figure!

This roundoff problem can be observed by removing "REM -" from lines 50 and 100 so that the Nth degree terms of (1.3.15) and the partial sum accumulated to that point are printed. This result is shown in Table 1.3.4. There is a lot of cancellation (subtraction) in forming the sum because of the alternating signs of the terms. Only six digits are accurate when using single

Table 1.3.3.   BASIC Program to Compute e^x by a Power Series with Optional Printing of Intermediate Terms and Sums

```
10 REM - COMPUTE EXPONENTIAL BASE E BY POWER SERIES
20 DEFSNG A,F,X
30 DEFINT I,N
40 PRINT "INPUT X:"; : INPUT X
50 REM - PRINT" N" TAB(7) "X^N/F" TAB(30) "SUM"
60 A=1
70 FOR N=1 TO 33
80 GOSUB 160
90 A=A+X^N/F
100 REM - PRINT N TAB(5) X^N/F TAB(28) A
110 NEXT N
120 PRINT "X,e^X = ";X,A
130 PRINT "e^X TO 6 FIG= ";EXP(X); "    % ERROR = "; (A-EXP(X))/EXP(X)*100
140 PRINT
150 GOTO 40
160 REM - COMPUTE F=N!
170 F=1
180 IF N=0 OR N=1 THEN RETURN
190 FOR I=2 TO N
200 F=F*I
210 NEXT I
220 RETURN
230 END
Ok
RUN
INPUT X:? -8
X,e^X = -8      3.865868E-04
e^X TO 6 FIG=   3.354627E-04    % ERROR =  15.23987
```

precision. However, the first significant digit in the answer occurs in the fourth decimal place. This means that the 9th term, -369.8681, contributes to the answer only by its last (and *inaccurate*!) digit. There are nine such terms that exceed 100, and the six accurate digits of each are lost.

Note that 33 terms other than unity have been computed; to go further than this results in overflows. However, 33 terms are adequate for $x = -8$, since it can be seen in Table 1.3.4 that the sum has stabilized in the fourth significant digit. Any remaining contribution from the remaining terms (number 34 onward) is called *truncation error*.

Another approach for the problem in this example is to change the single-precision declaration in line 20, Table 1.3.3, to double precision (DEFDBL). This gives an answer with at least three significant figures and perhaps more if more than 33 terms could be accumulated without overflow.

Table 1.3.4.   Intermediate Results for Each Term and the Partial Sum for the e^{-8} Power Series in Single Precision

```
INPUT X:? -8
  N      X^N/F                        SUM
  1   -8                       -7
  2    32                       25
  3   -85.33334                -60.33334
  4   170.6667                 110.3333
  5   -273.0667               -162.7333
  6   364.0889                 201.3556
  7   -416.1016               -214.746
  8   416.1016                 201.3556
  9   -369.8681               -168.5125
 10   295.8945                 127.382
 11  -215.196                  -87.81398
 12   143.464                   55.65
 13  -88.28552                 -32.63553
 14   50.44888                  17.81335
 15  -26.90607                  -9.09272
 16   13.45303                   4.360314
 17  -6.330839                  -1.970526
 18   2.813706                   .8431804
 19  -1.184718                  -.341538
 20    .4738874                  .1323494
 21  -.1805285                 -4.817912E-02
 22   6.564673E-02              1.746761E-02
 23  -2.283365E-02             -5.366035E-03
 24   7.611215E-03              2.24518E-03
 25  -2.435589E-03             -1.904089E-04
 26   7.49412E-04               5.590031E-04
 27  -2.22048E-04               3.369551E-04
 28   6.344228E-05              4.003974E-04
 29  -1.750132E-05              3.82896E-04
 30   4.667018E-06              3.875631E-04
 31  -1.204392E-06              3.863587E-04
 32   3.01098E-07               3.866598E-04
 33  -7.299345E-08              3.865868E-04
X,e^X = -8        3.865868E-04
e^X TO 6 FIG= 3.354627E-04       % ERROR =  15.23987
```

All three floating-point variables, A, F, and X, must contain more significant digits, not just the partial sum A.

There is another important lesson besides understanding roundoff error: Sometimes the problem can be formulated so as to *avoid cancellation*. In this case, simply compute e^{+8} and take the reciprocal. The power series calculation implied in Table 1.3.3 gives $e^8 = 2980.958$. The reciprocal is 3.3546263E–4, which happens to be the correct value of e^{-8} to eight significant figures. So one recurring theme in the methods that follow is that problems should be formulated to avoid numerical difficulties in the first place! Incidentally, there are much better ways to compute the function e^x than by power series. See Morris (1983).

According to Klema (1980), an algorithm is numerically stable if it does not introduce any more sensitivity to perturbation than is already inherent in the problem. *Stability* also ensures that the computed solution is "near" the solution of a problem slightly perturbed by floating-point arithmetic. An unstable algorithm can produce poor solutions even to a well-conditioned problem, as the preceding example showed. The next section deals with a problem inherent in optimization algorithms.

1.3.3. Illconditioned Linear Systems.

Linear systems of equations are of special interest in nonlinear optimization. Recall that a peak in the surface of Figure 1.1.1 was approximated by a quadratic mathematical model, the surface of which was shown in Figure 1.1.3. Approximated or not, the necessary condition for an extremum in a function is that the first derivatives must all vanish. For the quadratic function, it was shown in (1.1.9) and (1.1.10) that these derivatives, equated to zero, are in fact a set of linear equations. Thus, the central role of quadratic approximations in locating maxima and minima is synonymous with the solution of systems of linear equations.

Unfortunately, because linear systems are often badly conditioned, methods for solving them must account for that fact. "Until the late 1950's most computer experts inclined to paranoia in their assessments of the damage done to numerical computations by rounding errors. To justify their paranoia, they could cite published error analyses like the one from which a famous scientist concluded that matrices as large as 40×40 were almost certainly impossible to invert numerically in the face of roundoff. However, by the mid-1960s matrices as large as 100×100 were being inverted routinely, and nowadays equations with hundreds of thousands of unknowns are being solved during geodetic calculations worldwide. How can we reconcile these accomplishments with the fact that the famous scientist's mathematical analysis was quite correct? We understand better now than then why different formulas to calculate the same result might differ utterly in their degradation by rounding errors"—Hewlett-Packard (1982).

The symptoms of illconditioned linear systems and their corresponding quadratic functions are evident in the mild distortion seen by comparison of Figures 1.1.3 and 1.1.4; these showed the effects of changing the *x*-axis scale. Consider the contour plots in Figures 1.3.4 and 1.3.5, which differ in scale the same way. These contours are families of ellipses; the two figures obviously differ in their eccentricities. Although the ratio of major to minor axes is about 3 : 1 in Figure 1.3.5, ratios of 100 to 1000 or more are not uncommon in practice. Clearly, this eccentricity could create all kinds of havoc with algorithms that explore the surface of such shapes using preconceived finite steps. The point is that the corresponding system of linear equations is also illconditioned. This discussion and the following example serve to emphasize

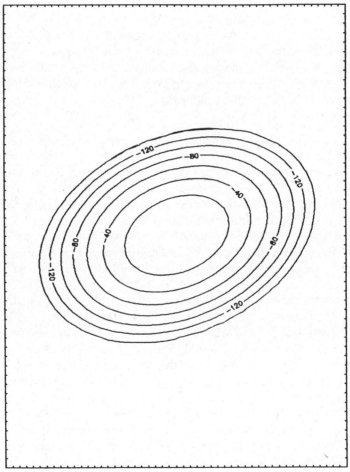

Figure 1.3.4. Contours of the quadratic peak shown in the surface plot of Figure 1.1.3. The coordinates have been shifted to center these contours.

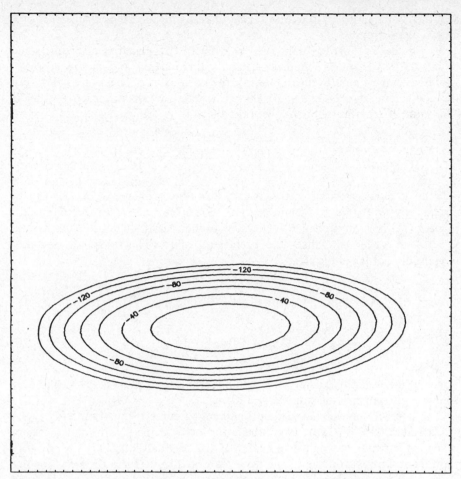

Figure 1.3.5. Contours of the illconditioned quadratic peak corresponding to Figure 1.1.4. The coordinates have been shifted to center these contours.

that the roundoff errors previously introduced can severely aggrevate the solution of illconditioned optimization problems.

Example 1.3.4. Consider the two linear equations treated by Forsythe (1970):

$$0.000100x + 1.00y = 1.00, \qquad (1.3.16)$$

$$1.00x + 1.00y = 2.00. \qquad (1.3.17)$$

The *Gauss–Jordan* elimination method (Cuthbert, 1983:9), solves this system by a series of equivalence operations that make the coefficient of y in (1.3.16) and the coefficient of x in (1.3.17) equal to zero and the other coefficients of x and y equal to unity.

In this case (1.3.16) is multiplied by 10,000; that result is also subtracted from (1.3.17). However, suppose that only three significant figures, correctly rounded, can be employed. In (1.3.17), the coefficient of x becomes zero, but the coefficient of y is $1.00 - 10,000 = -9,999.0$, which rounds to $-10,000$. The same effect occurs for the right-hand side of (1.3.17), so that it now is

$$-10,000y = -10,000. \tag{1.3.18}$$

The next Gauss–Jordan step requires that the coefficients of (1.3.18) be divided by $-10,000$ to obtain a unity coefficient of y. Furthermore, the revised (1.3.18) (multiplied by $-10,000$) is then subtracted from the revised (1.3.16) in order to cancel the coefficient of y. This yields a new but supposedly equivalent system

$$1.0x + 0.0y = 0.0, \tag{1.3.19}$$

$$0.0x + 1.0y = 1.0. \tag{1.3.20}$$

The solution computed using three significant figures is thus $(x, y) = (0.0, 1.0)$. That hardly satisfies the original (1.3.17).

The correct solution using nine significant figures throughout is $(x, y) = (1.00010001, 0.999899990)$. Even this solution obtained with much greater precision satisfies the right-hand sides of the original equations to only four significant figures (rounded). The reader is urged to perform the steps for this more accurate solution. The potential difficulties with rounding errors in an illconditioned calculation are thus experienced.

There are at least two ways to view the size of errors in solutions to problems discussed to this point. The *direct or forward error approach* asks the intuitive question, "How wrong is the computed solution for this problem?" This is the way that the results of the preceding example were viewed. There is a better way that is much more amenable to analysis. The *backward or inverse error analysis* technique asks, "How little change in the data (coefficients in linear systems of equations) would be necessary to make the computed solution be the exact solution to that slightly changed problem?" Backward error analysis has led to discovery of new and improved numerical procedures that are not obvious. Such analysis has made it possible to distinguish linear systems and related algorithms that are sensitive to rounding errors from those that are not. An excellent example of the remarkable results attributable to backward error analysis as applied to the solution of linear systems of equations is given by Noble (1969:270). These topics are treated quantitatively in Chapter Two.

1.3.4. Termination and Comparison of Algorithms. Referring to the flow chart in Figure 1.3.1 for a typical iterative process, one of the most confounding problems is when to stop or terminate the search procedure. This section discusses that problem and one that is even more subjective, namely, how to compare different algorithms applied to similar or identical problems.

Human ability to perceive trends and patterns far exceeds that of machines in most cases. For many purposes it is desirable to produce a nearly foolproof computer program, but an unwillingness to depend on human judgment should not force computer users to accept stupid decisions from the machine. No set of termination tests is suitable for all optimization problems. This issue is especially relevant for personal computer users, since they can see what is going on if the program is constructed to keep them properly informed and if they are sufficiently knowledgeable.

As Murray (1972:107) remarked, there are two kinds of algorithmic failures. The first is a miserable failure, which is discovered when an exasperated computer finally prints out a message of defeat. The second failure occurs when a trusting user mistakenly thinks that he and the computer have found the correct answer. Nash (1979:78) took an opposing view, that one of the most annoying aspects of perfecting numerical computation is that the foolhardy often get the right answer! Unfortunately, you may not get the right answer if the algorithm is stopped too soon; or the algorithm may never converge—not even to a wrong answer.

Again consider the surface in Figure 1.1.1 and a minimum-seeking algorithm. Even more specifically, suppose that a skier is descending by some strategy in the Swiss Alps. Since the function value will be his altitude, that could be a criterion for believing that a minimum has been reached. When the function fails to decrease adequately from iteration (search direction) to iteration, then it may be time to stop. On the other hand, if his speed is still high, the skier may be on a plateau and some of the variables (direction coordinate values) may be changing significantly, even if altitude isn't. Another explanation for this symptom is that the search direction may have been chosen to be nearly parallel to the contour lines (see Figure 1.1.2). However, the skier may not depend solely on changes in the variables from iteration to iteration. Consider that the skier may have gone over a cliff. In that case, there is little progress in the change of variables, but the altitude will decrease rapidly! Yet another criterion may be elements of the gradient, that is, the slope in each of the coordinate directions, say north and east. Near-zero gradient is a necessary condition for a minimum. Finally, the skier may have spent more time than allowed and thus stop because an allowable number of iterations (or minutes) has been reached.

The preceding ideas for stopping algorithms, such as that in Figure 1.3.1, can be described precisely. Let e be some small number, for example

$e = 0.0001$. Assuming minimization, the change in function value for termination after the kth iteration can be expressed by

$$F^{(k-1)} - F^{(k)} < e, \tag{1.3.21}$$

where $F^{(k)} = F(\mathbf{x}^{(k)})$. *Relative changes* in the function value would require division of the difference in (1.3.21) by $F^{(k-1)}$, which might approach zero in some cases, so relative function error is not recommended. The minimum function value is seldom known, and even if it is known it may not be attainable.

Because of roundoff errors and problem illconditioning, computers may fail to find the "exact" solutions to even simple problems. Computable criteria are not equivalent to exact mathematical properties, so convergence of the function value may not imply convergence of the variables. Gill (1981) has given a very flexible termination criterion for the variables; for the jth variable it is

$$\frac{|x_j^{(k)} - x_j^{(k-1)}|}{1 + |x_j^{(k-1)}|} < e. \tag{1.3.22}$$

This criterion is similar to *relative error* when the magnitude of x_j is large and is similar to *absolute error* when x_j is small. Suppose that $e = 0.001$. Then for very small x_j, the denominator in (1.3.22) is unity so that the changes between iterations for x_j must not occur in the first three *decimal places*. When $|x_j| = 1$, the effect is simply to double e. When x_j is very large, then the changes between iterations must both occur in the first three *significant figures*. Regrettably, most convergence tests on the variables are sensitive to scaling. Even if scaling is good at the start of the iterations, there is no assurance that it will remain that way, especially at a well-defined solution.

Good termination criteria require that the results pass the test in (1.3.22) for the objective function and for each variable. Many optimization algorithms allow different tolerances to be set for the function value and for each variable. The test for convergence of each component of the gradient (the slope in each coordinate direction) is discouraged since it is even more subject to roundoff "noise" than the other quantities. Another serious difficulty in that case is to decide what value is suitably near zero. Therefore, any tests on the gradient are usually made with a more forgiving tolerance value.

Instead of testing each component of the variable vector $\mathbf{x}$ as in (1.3.22), it is possible to test just the "length" or norm of $\mathbf{x}$ instead; see Gill (1981:306). Three such vector norms are described in Chapter Two. Others have suggested that termination should occur only when the computation has run the changes in quantities off the end of the computer word length, as in Table 1.3.1. That strategy often causes unnecessarily long execution time and the results may never converge because of roundoff noise. P. R. Geffe suggested that when two successive Newton iterates differ by half the machine precision, exactly two more iterations should be performed and the algorithm stopped. Engineers are often content with solution precision of about three significant figures, comparable to that attainable in the world of physical components. However, it is usually wise to compute to much higher precision than finally required, in order to detect blunders or other significant phenomena.

Personal computer users can observe the progress of iterative algorithms, so there is little reason to impose a fixed limit on the maximum number of iterations or elapsed time. However, it is undesirable to interrupt the algorithm without a planned ending, since certain data need to be summarized and the last vector of variables should be stored, perhaps on a permanent medium. For that purpose some machine-dependent scheme may be used, such as having the program check a special key that the user may press to cause an orderly termination of the program. A good procedure is to run algorithms with loose convergence requirements before imposing more stringent requirements and restarting at the previous termination point.

Finally, the reader certainly would like to select from this book or any other only those algorithms that are reliable or robust. *Reliability* may be indicated by the ability to solve a wide range of problems with little expectation of failure. A *robust algorithm* is hardy and efficiently utilizes all the information available for rapid convergence to a solution within a certain precision. Simple optimization algorithms are appealing, even if less efficient. However, usually in practice the much more complex and time-consuming computation occurs in evaluating the particular function being sampled, not in pursuing the optimization strategy that requires yet more such function evaluations. So it has become commonplace to measure optimization algorithms by the number of function evaluations required. For those algorithms that require gradient information, the calculation of derivatives sometimes requires as much work as for the function. Indeed, when the gradient is calculated by finite differences (small perturbations) the work required is increased by the factor n, where there are n variables in the problem! However, it is shown in Chapter Six that exact derivatives may be computed for many important problems with little additional work.

One measure of robustness is the convergence rate, previously discussed. This may be examined from data such as Tables 1.3.1 and 1.3.2 or by plotting the logarithms of function values versus iteration number. Comparison of algorithms is further complicated by their evaluation on different computers using programs written in different languages by a variety of programmers. As discussed in the next section, there has been some attempt to establish a standard timing unit (for a fixed amount of computational work) on various machines, but even those results vary. There are many well-known test problems; some sources are discussed in Appendix B. There is a strong tendency to use these problems to evaluate various changes in strategy, large and small. As unsatisfactory as this state of affairs remains, the effort to select superior methods where they can be identified is extremely important. As Nash (1979) remarked, "The real advantage of caution in computation is not, in my opinion, that one gets better answers but that the answers obtained are known not to be unnecessarily in error."

1.4.   Choices and Form

There are hundreds of books on optimization, but few are suitable for using personal computers to learn, reinforce, and then apply practical optimization techniques. The following sections explain the choices made for programming language, computers, and style in presenting this fascinating subject.

1.4.1.   Languages and Features.   The clear choice for programming language for this book is Microsoft BASIC, the standard for the IBM PC and compatible computers. It is furnished with almost every make and model personal computer in dialects that are trivially different. Its most common form is *interpreted*, as opposed to *compiled*, so programs can be run, modified, and rerun with an absolute minimum of effort. Nearly everyone having anything to do with computing can use it, notwithstanding its lack of sophistication and structure. The IBM (Microsoft) BASIC compiler is readily available and easy to use so that an order of magnitude increase in speed is available if required. Additional supportive opinion is available in Norton (1984:122, 123, 207). Equally important, compiled BASIC can be linked with the powerful Intel 8087 math coprocessor integrated-circuit chip, which costs about $100 and simply plugs into the PC computer, to provide numerical precision surpassing that available on many larger computers. This section will describe why and how Microsoft BASIC is implemented in this book.

In an interesting article on programming languages, Tesler (1984) said: "The great diversity of programming languages makes it impossible to rank them on a single scale. There is no best programming language any more than there is a best natural language." He also quoted Emperor Charles V: "I speak Spanish to God, Italian to women, French to men, and German to my horse." There is nothing to be gained by yet another debate on programming languages, but a few differences between BASIC and FORTRAN are worth mentioning. The author has more than two decades experience with FORTRAN, and there are valid reasons for its use for optimization programs, especially for its modularity. If one were to collect a number of standard routines (modules) for application to an ongoing series of unique problems, then FORTRAN would be the reasonable choice. Readers interested in this approach as a final outcome will benefit from reading Gill (1979a). However, FORTRAN is a compiled language, and the process of compiling and linking new modules is annoying, particularly when most failures occur with input and output formatting statements and other undistinguished pitfalls. Microsoft FORTRAN version 3.3 runs well on the IBM PC, meets and exceeds the FORTRAN 66 and 77 standards, and links to the 8087 coprocessor. The main point is that nearly anyone fluent in FORTRAN can easily translate from BASIC. The interested reader is referred to Wolf (1985).

Unlike FORTRAN, coding errors in BASIC can be repaired and tested almost as fast as one recognizes the problem. Program statements can be added to BASIC to display additional information at will, and the TRACE feature simplifies debugging. Any of several "cross-reference" software utility programs that tabulate program variable names versus line number occurrences are useful when working with BASIC programs. The listings in Appendix C for the major programs are followed by a list of variable names used; readers adding to the program or planned subroutines should be careful not to reuse variable names recklessly.

Names for variables in all BASIC programs in this book conform to the simple Dartmouth BASIC standard: They begin with any capital letter A through Z and are optionally followed by only one more digit from the numbers 0 through 9. The FORTRAN convention that integer variable names start with I, J, K, L, M, or N has been followed in these BASIC programs. To emphasize this practice, the BASIC type statements DEFINT, DEFSNG, and DEFDBL are used, sometimes redundantly. The lack of double precision on many computers that are not IBM compatible is not a fatal defect for purposes of learning the material in this book; however, some results may suffer from illconditioning. Similarly, a BASIC compiler is not mandatory, but the use of interpreted BASIC will limit most practical optimization algorithms to just a few variables or will cause them to run for many hours before solutions are obtained.

Many features available in the IBM-PC version of Microsoft BASIC have been avoided (e.g., the ELSE clause in IF statements). Simple screen menus have been employed where required instead of function keys, and no screen graphics have been used. Although PC-DOS (disk operating system) commands have been included in some programs to store and retrieve data, alternative means have been provided for extended data entry, mainly by using DATA and READ statements. Therefore, there should be little difficulty in adapting these programs to any conventional personal computer, even if it is not IBM PC compatible. Readers using computers that are not IBM PC compatible may find one appendix in the IBM BASIC manual especially useful; it describes the major differences between IBM and other versions of Microsoft BASIC. As mentioned, many of the incompatible features have been avoided, as well as several incompatibilities that exist between interpreted and compiled Microsoft BASIC.

A number of short programs are contained in tables in the main body of the text, but the larger and more important programs are listed in Appendix C. The pertinent sections of the text give explanations and test results to allow verification of programs entered manually. The index provides the page numbers where the references to each of the larger programs occur. Remarks (REM) have been used extensively to explain the use of variables or as titles for program sections. Any of these programs will run on a computer with fewer than 64 kilobytes of random-access memory (RAM), and considerable storage space can be saved by omitting the remarks embedded in the code. It has been assumed that the reader can perform the simple and conventional operations on his or her computer, especially relating to BASIC. For example, the program in Table 1.1.2 must be terminated using the ⟨Ctrl⟩⟨Break⟩ keys, and it is assumed that the user will realize that.

Several miscellaneous comments are provided to assist readers. BASIC programs in this book were written and run using IBM interpreted BASICA version A2.10, and many were compiled using IBM compiled BASIC version 1.00 and MicroWay 87BASIC versions 2.08 and 3.04. Interpreted BASIC was usually run without the /D switch option, which activates double-precision trancendental and trigonometric functions. Additional program segments are used throughout this book to be MERGE'd with major programs to add certain features. Users should be sure to merge the suggested program segments in the *order* stated. Numerous small data sets are required, especially vectors and matrices. Users of hard disks may wish to archive most of these on floppy disks, because the minimum file length on hard disks is usually about 4 kilobytes. These data sets may be created and modified without leaving program execution by using utility program Sidekick, which temporarily interrupts the ongoing program. There are many occasions when users will restart a program and type in the same data again. Utility programs, such as SuperKey, that assign macro files to specified keys to remember all the keystrokes required are great savers of time. The "cut-and-paste" feature also simplifies saving results from the computer screen for later reentry or storage. Many of the isometric and contour graphs in this book were plotted on a matrix printer, using program Plotcall by Golden Software.

1.4.2. *Personal Computers.*

Performance data on personal computers are often obsolete long before they can be published, but they will also be conservative for future equipment. Therefore, enough performance information is given to make the case for running optimization algorithms on IBM PC and compatible computers. This book and the included programs were written on an IBM PC-XT. (The XT designation originally was for the PC with a hard disk as well as a floppy disk drive.) It has an Intel 8088 microprocessor using a clock rate of 4.77 MHz. This is mentioned because higher clock speeds, other current microprocessors (Intel 8086, 80 X 86 series, and Motorola 68000 series) and software improvements are known to provide execution speeds many times faster than the PC-XT. Some IBM-PC data show that a compiler and 8087 math coprocessor chip provide the speed and accuracy necessary for practical optimization.

These data were obtained by averaging the times for 5,000 to 20,000 loops that included the indicated arithmetic operations. The data in Table 1.4.1 compare interpreted and compiled IBM (Microsoft) BASIC with and without an 8087 numeric coprocessor chip. The coprocessor works only with a modified BASIC compiler or some other compiled languages. These data show that

Table 1.4.1. **Milliseconds[a] for Mathematical Operations by IBM Interpreted BASICA and IBM Compiled BASIC With / Without 8087 Coprocessor[b]**

	Elementary Functions					
	SP ADD	DP ADD	SP MULT	DP MULT	SP SQR	DP SQR
BASICA	3.65	4.80	3.90	5.65	9.25	96.60[c]
Compiled	0.40	0.50	0.55	1.15	1.15	3.70
With 8087	0.15	0.20	0.20	0.20	0.15	0.20

	Trigonometric Functions					
	SP SIN	DP SIN	SP TAN	DP TAN	SP ATN	DP ATN
BASICA	17.40	39.80[c]	45.20	98.80[c]	10.40	30.80[c]
Compiled	3.40	12.80	7.20	27.00	4.00	16.00
With 8087	0.80	1.00	0.80	0.80	0.60	0.60

	Exponential Functions					
	SP EXP	DP EXP	SP Ln	DP Ln	SP Y^X	DP Y^X
BASICA	8.60	47.60[c]	9.60	62.80[c]	17.20	115.80[c]
Compiled	3.80	11.40	4.20	12.40	8.80	26.60
With 8087	0.60	0.60	0.40	0.60	0.80	0.80

[a] Loop overhead (nonmathematical operations) are included.
[b] SP = single precision, DP = double precision.
[c] BASICA switch/D set to obtain double precision.

the trigonometric functions are the slowest, and compiled BASIC is much faster than interpreted BASICA. Without the 8087, all double-precision calculations are slower than single precision, since they generally require about twice the work.

The Intel 8087 coprocessor greatly improves performance. It requires no more time to obtain double precision than single precision; in fact, it provides more precision than double precision by computing in 80-bit words. All elementary functions are computed at about the same speed (including subtraction and division, which are not shown). According to Fried (1984:204), the 8087 time for addition using Microsoft compiled BASIC is 0.134 milliseconds, which is close to the data in Table 1.4.1. All nonelementary functions are substantially faster than those without the coprocessor.

P. R. Geffe (private communication) has furnished a simple program that loops through 100,000 calculations of addition, subtraction, multiplication, division, and square roots. Some test results are shown in Table 1.4.2. In addition, it has been established that the IBM PC-AT computer is three times faster than the IBM PC computer, except when the math coprocessor is

Table 1.4.2. **Seconds for 100,000 Add, Subtract, Multiply, Divide, and Square Root Operations on IBM and HP Desktop Computers**

IBM PC (8088 and Microsoft interpreted BASICA)	2075
IBM PC (8088 and Microsoft compiled BASICA)	302
IBM PC (8088 & 8087 & Microway/IBM compiled BASICA)	138
IBM PC (8088 & 8087 & Microsoft compiled FORTRAN v.3.2)	87
Hewlett-Packard 9845 (HP BASIC)	740
Hewlett-Packard 9816 (HP BASIC)	300
Hewlett-Packard 9000 (HP BASIC)	24

utilized. For several technical reasons, the math coprocessor in the PC-AT does not increase speed very much, but it does provide increased precision, as previously described.

As previously mentioned, the 8087 coprocessor must be linked to the compiled BASIC in a special way. Several companies provide this capability, including Microway, Inc., which makes a software connection between the 8087 coprocessor and the IBM BASIC compiler. Although there are many different factors to be considered, the data in Table 1.4.2 show that the 8087 implementation compares favorably with the professional scientific computers from Hewlett-Packard, the HP-9000 being the newest of the models listed. Fried (1984) has provided many details concerning implementation of the 8087 coprocessor that are beyond the scope of this book. He mentions a PC user who discovered that his CRAY supercomputer executed one of his applications only 180 times faster than his PC with the 8087 installed. But since the CRAY was serving 100 users, the turn-around time for results was only twice as fast as that on the PC. This and other considerations led Fried to recommend the use of desktop computers to solve problems that involve up to 100 million floating-point operations, which includes all optimization applications contemplated in this book.

Finally, Fried (1984) noted that one matrix inversion algorithm now available on the PC executes at one-tenth the speed of an IBM 360 mainframe computer. Colville (1968) provided a FORTRAN program that inverts a 40×40 matrix of floating-point numbers 10 times. This was intended to provide a timing standard for the various kinds of computers on which testing of optimization algorithms would be accomplished. Although the data that have been published cannot be reproduced exactly, the results are useful as an indicator of scientific computing speed. Himmelblau (1972:368) notes that one of the faster mainframe computers, a CDC-6600, required 22 sec to invert the 40×40 test matrix 10 times in *single* precision. The Microsoft FORTRAN compiler version 3.2 required 510 sec (23 times as long) using the 8087 coprocessor working in excess of *double* precision. The conclusion that justifies the advocacy of this book is that number-intensive iterative processes are certainly feasible using mathematical coprocessors such as the Intel 8087, are often feasible using only compiled languages, but may easily be studied and appreciated using only interpreted languages.

1.4.3. Point of View.

1.4.3. Point of View. Of the hundreds of books on optimization, very few exemplify the relationship between personal computer users and the study and application of optimization. This is a pragmatic book that concentrates more on understanding than on rigor for its own sake. However, optimization is a pervasively mathematical subject involving matrix algebra and calculus, even though it is an intuitively appealing process. It can be understood by statement of requirements, behavior, techniques, and reduction to practice. Fortunately, visualization of the two-variable case generalizes to any number of variables. The linkage that this book provides the reader for that generalization is matrix algebra notation; it is not only unavoidable, but concise and attractive. Years of practice have convinced the author that experiencing computation occurring under the user's immediate control is an excellent learning medium that develops key insights. That is not to underestimate the theoretical preparation that must precede this experience. This book should motivate the reader to review and gain better theoretical underpinnings in the many fascinating topics from matrix algebra and calculus. This section furnishes some comments that differentiate this approach to optimization from the more conventional treatment.

Chapters Two and Three cover the numerous topics from matrix algebra that will be required throughout the remainder of the book. A program that performs the elementary matrix operations will be used to verify many important relationships. The major programs will be named. The one introduced in Chapter Two is called MATRIX. MATRIX eliminates one immediate obstacle for those studying matrix algebra, namely, the tedium of the extensive computations involved. For example, it is possible to observe convergence of a matrix process without even lifting a pencil by using the MATRIX program. The reasons for convergence will be described notationally so that it becomes apparent to the reader, but no proof of convergence or its rate will be offered. As Acton (1970) observed, "It is commonplace that numerical processes that are efficient usually cannot be proven to converge, while those amenable to proof are inefficient."

The reader may be aware of several sources for matrix software packages written in assembly language and callable from BASIC and other languages. Such a package is available from Microway, Inc., for example. These software utilities simplify and accelerate matrix algebra computations, but lack standardization and are thus avoided in this book. Also, as Acton (1970:329) noted, there are many instances where matrix notation, although very useful for derivations and proofs, can lead to very inefficient computation. Nevertheless, there is still an important trade-off to be made between a user's effort and time and that of a machine, so matrix algebra packages should be considered for any final implementation of optimization algorithms.

There are three important precautions worth mentioning in connection with optimization. First, partial derivatives of functions, implicit or explicit, need to be calculated for use in many algorithms. Mistakes made in programming those calculations are commonplace and disastrous, so almost every author emphasizes checking those calculations very carefully. In cases where these derivatives are programmed from explicit equations, it is strongly recommended that the initial results be confirmed by finite differences (perturbations). Second, the next likely difficulty to be encountered by the new optimization user is a bad choice of scales for the problem variables. If some variables have dimensions in inches so that a change of an inch or so makes a reasonable change in the outcome, then the scale is suitable. Expressing that same variable in a scale of miles would be totally disastrous, of course. Usually, different scales will be required for different variables. Third, many of the programs in this book employ double-precision calculations. They are often not necessary, and readers should not be too apprehensive if the computer used does not allow double precision. Readers should then expect some cases where single precision in BASIC is simply not adequate, and the symptoms often include sluggish convergence. The inner product of vectors is especially susceptible to cancellation because it is a sum of products. It is noted in passing, however, that Example 1.3.3 in this chapter requires that almost all variables (not just the accumulated result) be expressed in double precision in order to avoid excessive roundoff error.

Finally, the topics in this book employ a remarkably complete complement of the important facets of matrix algebra and a considerable number of concepts from the calculus. This is viewed as a significant reward for studying optimization. Beyond that, the role of the personal computer in the process of learning and application is vital. Personal computers are accessible, responsive, autonomous, and (usually) already paid for. These attributes, especially the last one, allow this book to differ in important ways from other means to discover and apply optimization.

Problems

1.1.   Sketch the contours of constant function value of

$$F(\mathbf{x}) = 2x_1 + x_2$$

for $F = 0$, 4, 8, and 12. Does this function have a finite minimum or maximum? Apply the constraint function

$$h(\mathbf{x}) = (x_1 - 3)^2 + (x_2 - 2)^2 - 1.$$

Sketch this constraint locus, and find x_1 and x_2 at the maximum of $F(\mathbf{x})$ subject to $h(\mathbf{x}) = 0$, if it exists.

1.2. Find the set of linear equations associated with the quadratic function

$$F(\mathbf{x}) = 5.5x_2^2 + 2x_1x_2 + 7x_1^2 - 94x_1 - 67x_2 + 500.$$

What are the values of the first partial derivatives of F with respect to x_1 and x_2 at the point $x_1 = 3$ and $x_2 = 7$? At what values of x_1 and x_2 will $F(\mathbf{x})$ have a minimum or maximum? What will be the values of the first derivatives at that point?

1.3. Evaluate equation (1.1.1) and its derivatives at the following points using the program in Table 1.1.2.

x	0.08668	3.38520	-3.07300
y	2.88430	0.07358	-0.08135

Locate these points on the representations of the surface in Figures 1.1.1 and 1.1.2. Any point $\mathbf{x} = \mathbf{a}$ such that the first derivatives of $F(\mathbf{x})$ vanish is called a *stationary point* of F. The three points given here are *saddle points* since they are neither local maxima or minima. What property of functions of a single variable distinguish a stationary point? How do stationary points of functions of a single variable differ from saddle points?

1.4. Rewrite the following constrained nonlinear programming problem in matrix notation as given in Section 1.2.1:

$$\text{Minimize} \quad 4x_1^3 + 3x_2 - 5x_3^2 + x_4$$
$$\text{such that} \quad x_1 = 3x_2^2,$$
$$x_3 - x_4 \le 0,$$
$$x_i > 0 \quad \text{for } i = 1 \text{ to } 4.$$

1.5. Check the first partial derivatives of equation (1.1.1) that are calculated in the program given in Table 1.1.2 by perturbing each variable in turn. First derivatives may be approximated by

$$\frac{\partial F}{\partial x_1} \doteq \frac{F(1.0001x_1, x_2) - F(x_1, x_2)}{0.0001x_1}$$

$$\frac{\partial F}{\partial x_2} \doteq \frac{F(x_1, 1.0001x_2) - F(x_1, x_2)}{0.0001x_2}.$$

Perform similar operations using the first derivatives to approximate the second derivatives. Note the accuracy of the approximation and the substantial possibility of roundoff error (cancellation).

1.6.   An important problem in financial analysis is the calculation of rate of interest in compounding problems. The function involved is

$$f(i) = PV + PMT\frac{1 - (1 + i)^{-n}}{i} + FV(1 + i)^{-n} = 0,$$

where PV is the present value, PMT is the payment amount for each of the n periods, i is the interest rate per period, and FV is the future value. In terms of the fixed-point iteration function (1.3.3), an algorithm to solve for the implicit variable i is Newton's method, where $g(x)$ in (1.3.3) is $1/f'(x)$, and $f'(x)$ is the derivative of f with respect to x. Alter the program in Table 1.3.2 to use the iteration function

$$i^{(k+1)} = i^{(k)} - \frac{f(i^{(k)})}{f'(i^{(k)})}$$

in line 100. Find i when $n = 60$ months, $PMT = \$100$ per month, $PV = \$10,000$ initial balance, and $FV = \$23,200$ is the desired balance 5 years hence. Begin the iteration with $i = 0.00833$ (10% per year).

1.7.   As in Problem 1.6, use Newton's method to find a zero of the function

$$f(x) = x^2(x - 1)^2,$$

starting from $x = 1.05$. Modify the program in Table 1.3.2 to perform and print the iterations. What is the rate of convergence?

1.8.   Sketch the following functions from Wright (1976):
(a) $y = (x - 2)^2(x + 1)^2$ for $-1.7 \le x \le 2.7$,
(b) $y = (e^x - 2)^2(e^x + 1)^2$ for $0.2 \le x \le 15$,
(c) $y = (x^2 - 2)^2(x^2 + 1)^2$ for $-1.7 \le x \le 1.7$.
Discuss the effects of scaling by transformation of variables (e.g., replace x with e^x or with x^2 in case a).

1.9.   Approximate the function in Problem 1.8(a) in the neighborhood of $x = 0.5$ using the first three terms of the Taylor series expansion about $x = a$:

$$y(x) = y(a) + y'(a)(x - a) + 0.5y''(a)(x - a)^2 + \cdots.$$

Tabulate the percentage error in this approximation over $0.0 \le x \le 1.0$ in steps of 0.1.

1.10. From Maron (1982), use repeated substitution to find the fixed points of

$$y(x) = \tfrac{1}{2}e^{\frac{1}{2}x}$$

to four significant figures. Sketch the iterative process. The search for one of the fixed points will always diverge.

1.11. Show that the partial derivative with respect to any x_j of the least-squares objective function in (1.2.6) is

$$\frac{\partial F}{\partial x_j} = 2 \sum_{i=1}^{m} r_i \frac{\partial r_i}{\partial x_j}.$$

Also, write the partial derivative of (1.2.5) with respect to all x_j, $j = 1$ to 5.

Suppose that the fitting function similar to (1.2.5) were linear instead of nonlinear. Then show why, in principle, a linear fitting function for minimum squared error could be found by solving one set of linear equations.

1.12. Run the following program due to Forsythe (1977) on your personal computer to determine the smallest floating-point number it can represent (in error at most by a factor of 2). This quantity is called the *machine precision* e_m. Also, try this in double precision if available.

```
10   REM FORSYTHE (1977) P.14 EPSILON TEST
20   E = 1
30   E = .5*E
40   E1 = E + 1
50   IF E1 > 1 THEN GOTO 30
60   PRINT "APPROXIMATE MACHINE PRECISION = ";E
70   END
```

1.13. Using the quadratic surface of (1.1.8), compute the constants of Table 1.1.4 for an approximation of surface (1.1.1) at point $x = 3$ and $y = 4$.

Chapter Two _____

Matrix Algebra and Algorithms

Vectors and matrices are the essential data structure for numerical applications such as optimization. The previous examples have treated the vector **x** as a set of independent variables or parameters subject to choice according to some objective. It is an easy and (one hopes) familiar step to extend that concept further to the matrix as a two-dimensional table of dependent or fixed coefficients. As Bellman (1960) has noted, the notation chosen to represent this arithmetic of higher mathematics is crucial to avoiding being swamped by a sea of arithmetical and algebraical detail. A well-designed notation expresses the underlying mathematics without obscuring or distracting the reader. Also, a great virtue of matrix notation is that the expressions do not depend on the number of variables involved.

Matrix computation is required in many fields, usually for solving systems of linear equations. Some examples are given by Jennings (1977:38–69) for electrical networks, surveying, heat transfer, and nonlinear cable analysis. Analysis of electrical network responses is described in Chapter Six. Since transformations lie at the heart of mathematics, it should be noted that matrices represent the most important of these, namely, the linear transformations.

It is assumed that the reader has been introduced to vectors and matrices, so the treatment in this chapter is limited to stating and illustrating the surprisingly complete set of these concepts required for optimization. One goal is to make it possible for the reader to absorb the analyses and results widely available in both new and old literature on optimization. Algorithms for computation are equally important to the personal computer user. For that purpose this chapter includes several BASIC programs, including a major one named MATRIX that simplifies ordinary matrix computation.

Chapter Two begins with the fundamental definitions and operations of matrix algebra and includes a description of the MATRIX program. Vector spaces, their geometry, and the concept of projection are described, and matrices with special structure are included. Pertinent matrix transformations and factorizations are explained, especially those related to real, symmetric matrices, their eigensystems, and illconditioning. With this specific background and its applications in Chapter Three, the reader should be able to understand and apply the several truly effective optimization algorithms that follow.

2.1. Definitions and Operations

Since notation is an important part of comprehending matrix algebra, this section elaborates on all aspects of notation. Program MATRIX, which simplifies fundamental matrix operations, is described for subsequent use. All the simple vector and matrix operations are defined, essentially the major four functions: addition, subtraction, multiplication, and inversion. Finally, the concepts and definitions of vector and matrix norms are introduced.

2.1.1. Vector and Matrix Notation.
Except for electrical network applications in Chapter Six, the notation describes sets of real numbers, as opposed to complex numbers that have both real and imaginary parts. Many excellent textbooks have been written for the more general case, that is, for sets of complex numbers. Although the matrix algebra is valid for the special case of sets of real numbers, some of the nomenclature is peculiar to the more general case. Since optimization per se involves only sets of real numbers, the reader should be aware when reading Noble (1969) and other classical texts that *unitary* translates to *orthogonal* and *hermitian* translates to *symmetric* between the complex and real cases, respectively.

Repeating the previous definition, a vector is a set of n numbers defined to be in *column* order:

$$\mathbf{x} = \begin{bmatrix} x_1 \\ x_2 \\ \vdots \\ x_n \end{bmatrix} \qquad \text{or} \qquad \mathbf{x} = (x_1\ x_2\ \cdots\ x_n)^T. \qquad (2.1.1)$$

A vector is denoted by a boldface, lower-case letter, and its elements or components are denoted by the same letter in ordinary lower-case type with subscripts that specify their order or position. The superscript T in the right-hand side of (2.1.1) denotes *transposition*, the interchanges of rows and columns. Thus, $\mathbf{x}^T$ is a *row vector*. The set of numbers in a vector represents the coordinates of a point in *Euclidean n-dimensional space* as shown in Figure 2.1.1 for $n = 3$. The *zero vector* **0** contains elements that are all zero. The *unit-direction vectors*, $\mathbf{e}_i$, $i = 1$, 2, and 3, shown in Figure 2.1.1, are associated with the direction and measurement along the three orthogonal axes. These

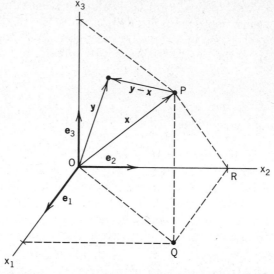

Figure 2.1.1. The vector $\mathbf{x} = (x_1\ x_2\ x_3)^T$ in Euclidean 3-space. The vector has the direction and length shown from point O to P. Its projection in the x_1–x_2 plane is the vector OQ. The elements of $\mathbf{x}$, x_i, are the units of its projections on the three orthogonal axes; for example, x_2 is the length OR.

unit vectors are defined by:

$$\mathbf{e}_1 = \begin{bmatrix} 1 \\ 0 \\ 0 \end{bmatrix}, \qquad \mathbf{e}_2 = \begin{bmatrix} 0 \\ 1 \\ 0 \end{bmatrix}, \qquad \mathbf{e}_3 = \begin{bmatrix} 0 \\ 0 \\ 1 \end{bmatrix}. \tag{2.1.2}$$

A matrix is a rectangular array of numbers:

$$\mathbf{A} = [a_{ij}] = \begin{bmatrix} a_{11} & a_{12} & \cdots & a_{1n} \\ a_{21} & a_{22} & \cdots & a_{2n} \\ a_{31} & a_{32} & \cdots & a_{3n} \\ & & \cdots & \\ a_{m1} & a_{m2} & \cdots & a_{mn} \end{bmatrix}. \tag{2.1.3}$$

A matrix is denoted by a boldface capital letter. The subscripts on each element or component of the matrix in (2.1.3) indicate row i and column j; there are m rows and n columns. Matrix size is often referred to as $m \times n$ (pronounced "m by n"), and sometimes it is useful to indicate the dimensions by subscripts, such as $\mathbf{A}_{m,\,n} = [a_{ij}]_{m,\,n}$. The *zero matrix* $\mathbf{0}$ contains elements that are all zero; it is distinguished from the zero vector only by its context.

Matrices may be rectangular, so it is not necessary that $n = m$, although that is a common case. For square matrices the elements a_{ii} constitute the *principal* (*main or leading*) *diagonal*. The *trace* of a matrix is the sum of the elements on the principal diagonal, that is, $\text{tr}(\mathbf{A}) = a_{11} + a_{22} + \cdots + a_{nn}$. A matrix having $a_{ij} = 0$ for $i \neq j$ is said to be a *diagonal matrix*. An *upper triangular matrix* $\mathbf{U}$ is a square matrix having all elements below the principal diagonal equal to zero. A similar definition applies for a *lower triangular matrix* $\mathbf{L}$.

The transposition operator applied to any matrix (square or not) swaps rows and columns, so that

$$\mathbf{A}^T = [a_{ji}]_{n,m}. \qquad (2.1.4)$$

If $\mathbf{A} = \mathbf{A}^T$, then elements $a_{ij} = a_{ji}$ and that *square* matrix is said to be *symmetric*. Many matrices required in optimization are symmetric and have other special properties. Another useful way to denote a matrix is by a set of column vectors:

$$\mathbf{A} = (\mathbf{a}_1 \ \mathbf{a}_2 \ \cdots \ \mathbf{a}_{n-1} \ \mathbf{a}_n), \qquad (2.1.5)$$

where the subscripted $\mathbf{a}_i$ are numbered *m*-vectors, each containing *m* elements. The *unit matrix* $\mathbf{I}$ is such a set of vectors; it is square, having all 1's on the principal diagonal and 0's elsewhere. In terms of (2.1.2) and (2.1.5), the unit matrix can be expressed as

$$\mathbf{I} = (\mathbf{e}_1 \ \mathbf{e}_2 \ \mathbf{e}_3) \qquad (2.1.6)$$

in 3-space. Of course, it is possible to make a vertical arrangement of a set of row vectors analogous to the column-vector arrangement in (2.1.5).

There is one more representation for a matrix, namely, as a set of submatrices:

$$\mathbf{T} = \begin{bmatrix} \mathbf{P} & \mathbf{Q} \\ \mathbf{R} & \mathbf{S} \end{bmatrix}, \qquad (2.1.7)$$

where $\mathbf{T}$, $\mathbf{P}$, $\mathbf{Q}$, $\mathbf{R}$, and $\mathbf{S}$ are all matrices of various dimensions (rows and columns). In fact, the designation of the submatrices corresponds to *partitioning* as indicated by the dashed lines in this particular matrix:

$$\mathbf{T} = \begin{bmatrix} 2 & 3 & 5 & 6 & 4 \\ -1 & 1 & -1 & 1 & 1 \\ -2 & 3 & -4 & 5 & -1 \\ 5 & 3 & -1 & 2 & -3 \\ -4 & 8 & -3 & 2 & 2 \end{bmatrix}. \qquad (2.1.8)$$

Superscripts in parentheses sometimes appear with both vectors and matrices to indicate a stage or iteration number, as mentioned in Chapter One. For example, a set of linear equations associated with the kth iteration of a process might appear as

$$\mathbf{H}^{(k)} \, d\mathbf{x}^{(k)} = \mathbf{g}^{(k)}, \qquad (2.1.9)$$

where **dx** and **g** are vectors and **H** is a matrix. Matrix multiplication and other operations indicated in (2.1.9) are discussed after the MATRIX utility program is introduced in the next section.

2.1.2.   Utility Program MATRIX.   Program C2-1 in Appendix C is the standard computation tool to verify and illustrate a large number of mathematical operations in matrix algebra in the rest of this book. Named MATRIX, it is the proving ground for algorithms that produce certain matrix results as well as some parts of those optimization algorithms that appear further on in a much more integrated form. The approximately 625 lines of BASIC instructions or code include numerous comments and were written in the most simple way to enhance reader comprehension and computer utilization. The code requires about 20,600 bytes in ASCII (American Standard Code for Information Interchange) format, 16,700 bytes in compressed binary (tokenized) format, and about 18,700 bytes in RAM during execution. A table of variable names used in MATRIX is appended to the program listing to aid in the addition of new subroutines. MATRIX is introduced at this point so that the reader can refer to relevant BASIC instructions and worked examples for the various matrix operations being described.

As presented in program C2-1, MATRIX enables the user to enter, view, print, and operate on as many as four matrices, **A**, **B**, **C**, and **D**, each as large as 6×6. Any or all of these four matrices might be a row or column vector, that is, with one dimension equal to unity. The command menu for MATRIX is shown in Table 2.1.1.

The data that describe the one to four matrices to be employed can be entered as BASIC data statements using the BASIC "MERGE" command or

Table 2.1.1.   Command Menu for Program C2-1, MATRIX

```
********* COMMAND MENU **********
0. DISPLAY A MATRIX IN FIXED FORMAT
1. SEE COMMAND HISTORY
2. TOGGLE PRINTER ON/OFF
3. EQUATE ONE MATRIX TO ANOTHER
4. TRANSPOSE
5. MATRIX TO/FROM DISK
6. SCALAR * (MATRIX)
7. A = B + C
8. A = D * C
9. D = (invB) & DETERMINANT(B); DESTROYS B!
10. SPARE
11. NORMS OF VECTOR OR MATRIX D
12. EXTREME ELEMENTS OF D
13. SPARE
    FORMAT:    FIXED   SCIENTIFIC   ALL COLUMNS
    PRINT =      14         16          18
    DISPLAY = 0 OR 15       17          19
20. EXIT (RESUME WITH 'GOTO 999')
************************************
```

Table 2.1.2.   **Two Data Entry Methods for Program C2-1, MATRIX**

Method 1: MERGE Lines 400–620	Method 2: ASCII File on Disk
400 N$ = "METHOD1Q"	"Method 2 for Q "
405 DATA 2, 3	2 3
410 DATA 5, 6, 4	5 6 4
415 DATA −1, 1, 1	−1 1 1
420 DATA 0, 0	

from ASCII files, using command 5 in Table 2.1.1, or both ways. Table 2.1.2 shows the formats for these two methods. Suppose that the five lines of BASIC instructions on the left side of Table 2.1.2 have been stored in an ordinary PC-DOS file named "METHOD1.BAS" and that the four lines of text have been placed into a file named "Q.MAT". Either of these files could be created by use of the EDLIN line editor that is a standard feature of the PC-DOS system. The file name extension ".MAT" is recommended for matrices and ".VEC" for vectors. The statements on the left could optionally have been created and saved in BASIC, using the standard BASIC program editor.

On the first page of the program C2-1 listing in Appendix C, lines 400 to 620 have been reserved to enter none, any, or all of the four matrices or vectors. The listing in Appendix C already contains BASIC lines 400, 410, and 420 for the case where no matrix data are entered by DATA statements. While operating in the BASIC environment with the MATRIX program already loaded into memory, issuing the DOS command [MERGE"METHOD1] (without the brackets) will load lines 400 to 420 as in Table 2.1.2. They will overwrite the original lines and will furnish data that are read by lines 660 to 730 *in row order*. Again referring to the left side of Table 2.1.2, the data set has the name "METHOD1Q" according to line 400. Line 405 gives the number of rows and columns in **A**; lines 410 and 415 give the first and second rows of **A**, respectively.

If matrix **B** were to be read in, then its numbers of rows and columns would appear in line 420; however, since **A** is the last matrix to be read into the program in this case, the two zeros are read by line 740 and the READ process is terminated by line 750. If all four matrices are to be read this way, then they must appear in the order **A**, **B**, **C**, and **D** *and no pair of zeros is required after the last row of* **D**. The first screen seen after the RUN command appears in Table 2.1.3 when data from the left side of Table 2.1.2 are not employed. The screen will freeze until the ⟨RETURN⟩ key is pressed so that the notes and status can be read. The notes are discussed below.

The ASCII file on the right side of Table 2.1.2 is structured the same way except that the delimiters are blanks instead of commas. Files written *to* disk by MATRIX command 5 are in one long column as opposed to the form on the right side of Table 2.1.2, which can be used equally well to give the appearance of a matrix. Command 5 first asks if the user needs to see the disk

Table 2.1.3.   First Screen from Program MATRIX Showing Notes and Data
Status, Having Executed the DATA Statements on the Left Side of Table 2.1.2.

```
**********ELEMENTARY VECTOR & MATRIX OPERATIONS***********

NOTES:
1. USE ONLY UPPER CASE LETTERS
2. MERGE VECTOR AND MATRIX DATA STATEMENTS
     INTO RESERVED LINE RANGE 400-620 (OPTIONAL)
3. IF 'BREAK' OCCURS, RESTART WITH 'GOTO 999 <RTN>'

WORKING WITH DATA SET: NONE
PRESS <RETURN> KEY TO CONTINUE -- READY?
```

directory one or more times. Then the user is asked what matrix is involved
and if the data are to be recalled or saved. *Caution*: Do not press the
⟨Ctrl⟩⟨Break⟩ keys while in the command 5 sequence. Since a disk file is open,
the disk directory can be damaged. If this should occur, use the DOS
command "CHKDSK   /F" to repair the damaged disk directory. Command
5 will have to be used once for each matrix to be read in. Upon receiving the
name of the file, in this case Q.MAT, the program reads the first line in
quotation marks on the right-hand side of Table 2.1.2 and then asks the user if
this operation should be aborted (the file may not be the one intended).

If a file is not found with the name given, the program will report that fact
and then "break" (cease execution). As indicated at command 20 in Table
2.1.1 and by note 3 in Table 2.1.3, the program can be continued *without loss
of data* by typing 'GOTO 999' and pressing the ⟨ENTER⟩ key. Another note
in Table 2.1.3 specifies that ⟨CAPS LOCK⟩ should be active (e.g., an answer
"Y" is expected, not "y"). Also, the impatient user can type up to 15
keystrokes ahead of the program INPUT commands, since the IBM PC has a
keyboard buffer with that capacity.

Screen results include command 1, which lists the first 100 commands that
have been issued to alter data. Command 2 successively toggles the printer to
an "on" or "off" state so that all matrix operations may be recorded (if the
printer has been made ready). The operation of command 5 to recall a disk file
is shown in Table 2.1.4. Commands 14 to 17 display the data conveniently on
the 80-column screen or printer, assuming that the matrices have maximum
dimensions of six columns (and rows). Commands 18 and 19 display all
available significant digits by matrix columns. This program employs double
precision in line 330 (not mandatory), so that 17 significant figures are
available. *To process matrices larger than* 6×6, *change the dimensions in line
380 and observe results with commands 18 and 19*. Comments on matrix
operational commands 3 to 9 and 11 to 12 are made where the mathematical
operations are introduced in the next section. SPARE commands 10 and 13
allow other matrix operations or sequences to be defined as indicated by line
6390. For example, additional program lines for the iterative power method
for finding eigenvalues are presented in Section 2.2.3 to execute under com-

Table 2.1.4. Screen Transactions While Entering the Disk File Q.MAT on the Right Side of Table 2.1.2.

```
SEE DIRECTORY (Y/N)?  Y
FILENAME SPECIFIER (LIKE *.* OR <RETURN>) = ? *.MAT
C:\ib
T    .MAT     P     .MAT     R     .MAT     S        .MAT
Q    .MAT
 782336 Bytes free

SEE DIRECTORY AGAIN (Y/N)? N
MATRIX INVOLVED IS A, B, C, OR D? A
RECALL OR SAVE MATRIX A (R/S)? R
FILE NAME IS? Q.MAT
READY TO READ FILE Q.MAT INTO MATRIX A TITLED:    Method 2 for Q
PRESS <RETURN> KEY IF OK, ELSE 'ABORT' <RETURN>?
   READ FILE Q.MAT INTO MATRIX A
PRESS <RETURN> KEY TO CONTINUE -- READY?
```

mand 10. The added instructions simply call a sequence of existing commands by calling subroutines and by making other calculations.

Program MATRIX is deliberately unsophisticated and has been limited in size (18,700 bytes in execution) for computers with limited memory. All functions are in subroutines beginning with the line numbers shown in Table 2.1.5. Program MATRIX operates according to the calculated GOSUB statement in line 1270, which reflects the line numbers in the third column of Table 2.1.5. As demonstrated by example, the SPARE commands can thus cause the program to visit new subroutines added by the user. These can in turn call existing subroutines after presetting the parameters that would normally be furnished by the user from the keyboard. For example, Table 2.1.5 shows that MATRIX command 6 normally causes the program to GOSUB 3870. The user could accomplish the same action by assigning string values to variables S7$ and S$ and then GOSUB 3920 as observed by noting program lines 3870 to 3920. This range of lines is shown in Table 2.1.5; they solicit the variable assignments normally made from the keyboard. Of course, the new subroutine(s) can also calculate new quantities as required. The names of the major variables and their usage are given in program lines 40 to 220, and all variables are listed after the code in Appendix C, program C2-1.

The reader will soon find that program C2-1, MATRIX, eliminates the tedium of computation while learning, shows how to program straightforward matrix computations, and enables clear display of all results from algorithms involving vectors and matrices.

2.1.3. Simple Vector and Matrix Operations. Simple matrix operations include three of the four major functions: addition, subtraction, multiplication, and inversion (comparable to division). The matrix inverse is discussed in the next section. Unlike multiplication of real numbers, matrix multiplication produces results that differ with the order of certain operations. This section presents the essential rules and illustrates them using program MATRIX.

Table 2.1.5.   MATRIX Command Subroutine Beginning and Preset-Parameter Entry Line Numbers

CMD No.	Command	Solicits Parameters	For Preset Parameters
0.	DISPLAY A MATRIX IN FIXED FORMAT	4180	4230
1.	SEE COMMAND HISTORY	1330	1330
2.	TOGGLE PRINTER ON/OFF	1410	1410
3.	EQUATE ONE MATRIX TO ANOTHER	1470	1560
4.	TRANSPOSE	2470	2510
5.	MATRIX TO DISK (SAVE)	4580	4770
5.	MATRIX FROM DISK (RECALL)	4580	5010
6.	SCALAR * (MATRIX)	3870	3920
7.	**A = B + C**	2890	2890
8.	**A = D * C**	3040	3040
9.	**D** = (inv **B**) & DETERMINANT (**B**)	3220	3220
10.	SPARE	—	—
11.	NORMS OF VECTOR OR MATRIX **D**	5360	5360
12.	EXTREME ELEMENTS OF **D**	5820	5820
13.	SPARE	—	—
14.	PRINT MATRIX IN FIXED FORMAT	4150	4230
15.	DISPLAY MATRIX IN FIXED FORMAT	4180	4230
16.	PRINT MATRIX IN SCIENTIFIC FORMAT	4070	4230
17.	DISPLAY MATRIX IN SCIENTIFIC FORMAT	4110	4230
18.	PRINT ALL COLUMNS DOUBLE PRECISION	6030	6080
19.	DISPLAY ALL COLUMNS DOUBLE PRECISION	6050	6080
20.	EXIT	6370	6370

Except for a few cases, only matrices are discussed, since a vector can often be treated as a special one-dimensional matrix.

Two matrices are equal if they have the same dimensions and each respective element is equal. For example, to require that $\mathbf{A}_{m,n} = \mathbf{B}_{p,q}$ means that $m = p$, $n = q$, and $a_{ij} = b_{ij}$. For 2×3 matrices,

$$\begin{bmatrix} a_{11} & a_{12} & a_{13} \\ a_{21} & a_{22} & a_{23} \end{bmatrix} = \begin{bmatrix} b_{11} & b_{12} & b_{13} \\ b_{21} & b_{22} & b_{23} \end{bmatrix}, \qquad (2.1.10)$$

so that $a_{11} = b_{11}$, $a_{12} = b_{12}$, etc. This process is accomplished in program MATRIX by command 3 (see Table 2.1.1) and lines 1450–2440 in the listing, Appendix C, program C2-1. If matrices **A**, **B**, **C**, or **D** are equated to the unit matrix **I**, then the dimension of that square matrix is requested and assigned by lines 1540 and 1600, respectively.

Multiplication of a matrix by a scalar (a nonvector, say h) is denoted by $\mathbf{A} = h\mathbf{B}$ and requires that $a_{ij} = hb_{ij}$ for all elements. Again using 2×3

matrices, for example, if $h = -1$, then

$$\begin{bmatrix} a_{11} & a_{12} & a_{13} \\ a_{21} & a_{22} & a_{23} \end{bmatrix} = \begin{bmatrix} -b_{11} & -b_{12} & -b_{13} \\ -b_{21} & -b_{22} & -b_{23} \end{bmatrix}. \qquad (2.1.11)$$

Scalars can be moved through products; for example, $(h\mathbf{A})\mathbf{B} = \mathbf{A}(h\mathbf{B}) = (\mathbf{AB})h$. Similarly, $(h\mathbf{A})^{-1} = (1/h)\mathbf{A}^{-1}$. A straight line in n-space, such as Figure 2.1.1 for $n = 3$, might begin at a point $\mathbf{x}$ and then proceed in direction $\mathbf{s}$. A scalar multiplier h enables points on that new straight-line vector, say $\mathbf{y}$, to be specified, namely, $\mathbf{y} = \mathbf{x} + h\mathbf{s}$. The last concept plays a major role in optimization.

Addition of two matrices, $\mathbf{A} = \mathbf{B} + \mathbf{C}$, is also on a respective element-by-element basis; thus, $a_{ij} = b_{ij} + c_{ij}$ for all nm elements. Command 7 in Table 2.1.1 accomplishes matrix addition. The names of BASIC variables containing the dimensions of each matrix are shown in program MATRIX line 40. If the dimensions of $\mathbf{B}$ and $\mathbf{C}$ do not agree, a warning is issued by program line 2900, and lines 2940 to 2950 set the dimension of $\mathbf{A}$ to the larger respective dimensions of the operands. Matrix subtraction is accomplished by performing the scalar multiplication on the subtrahend as shown in (2.1.11) and *then* performing matrix addition. Matrix addition and subtraction satisfy the same properties as the corresponding scalar operations:

$$\textit{Associativity}: \qquad \mathbf{A} + (\mathbf{B} + \mathbf{C}) = (\mathbf{A} + \mathbf{B}) + \mathbf{C},$$

$$(2.1.12)$$

$$\textit{Commutativity}: \qquad \mathbf{A} + \mathbf{B} = \mathbf{B} + \mathbf{A}.$$

Matrix multiplication is considerably more involved. It is performed by MATRIX command 8 in Table 2.1.1, namely $\mathbf{A} = \mathbf{D} * \mathbf{C}$. It is helpful to consider a particular case:

$$\begin{bmatrix} a_{11} & \boxed{a_{12}} \\ a_{21} & a_{22} \\ a_{31} & a_{32} \end{bmatrix} = \begin{bmatrix} \boxed{d_{11}} & d_{12} & d_{13} & d_{14} \\ d_{21} & d_{22} & d_{23} & d_{24} \\ d_{31} & d_{32} & d_{33} & d_{34} \end{bmatrix} \begin{bmatrix} c_{11} & \boxed{c_{12}} \\ c_{21} & c_{22} \\ c_{31} & c_{32} \\ c_{41} & \boxed{c_{42}} \end{bmatrix}. \qquad (2.1.13)$$

The rule is that each element of $\mathbf{C}_{mn}$ may be found by

$$a_{ij} = \sum_{k=1}^{p} d_{ik}c_{kj} \qquad \text{for } i = 1 \text{ to } m \text{ and } j = 1 \text{ to } n, \qquad (2.1.14)$$

where p is the column dimension of $\mathbf{D}$. The boxes superposed on (2.1.13) for computing a_{12} illuminate the procedure stated by Finkbeiner (1966): "In practice, we can perform this computation (for a_{ij}) easily by the technique of using the left index finger to run across the ith row of the left-hand matrix and simultaneously using the right index finger to run down the jth column of the right-hand matrix, multiplying elements in corresponding positions and adding successively the products obtained." From (2.1.13),

$$a_{12} = d_{11}c_{12} + d_{12}c_{22} + d_{13}c_{32} + d_{14}c_{42}. \tag{2.1.15}$$

This rule of thumb clarifies the fact that **D** and **C** must be *conformable*: the number of *columns* in **D** must equal the number of *rows* in **C**. Where the unit matrix is involved, $\mathbf{I}_m\mathbf{A} = \mathbf{AI}_n = \mathbf{A}$ when **A** is not symmetric ($m \neq n$), but $\mathbf{I}_m$ and $\mathbf{I}_n$ then have different dimensions so that the products are conformable.

To summarize, if $\mathbf{A}_{mn} = \mathbf{D}_{mp}\mathbf{C}_{qn}$, then it is required that $p = q$ (**D** and **C** are conformable), and the result has dimensions $m \times n$. Program MATRIX line 3050 warns the user if **D** and **C** are not conformable, but an operation is performed anyhow. Matrix multiplications can involve a great deal of computation; for the dimensions given above, a product requires $m \times p \times n$ scalar multiplications and nearly as many scalar additions.

Matrix multiplication does not satisfy all of the properties of the scalar case. For matrix multiplication:

$$\textit{Associativity}: \qquad\qquad (\mathbf{AB})\mathbf{C} = \mathbf{A}(\mathbf{BC})$$

$$\begin{aligned}&\textit{Distributivity} &&\tag{2.1.16}\\ &\textit{over matrix addition}: \qquad \mathbf{A}(\mathbf{B} + \mathbf{C}) = \mathbf{AB} + \mathbf{AC}.\end{aligned}$$

Generally, matrix multiplication is *not* commutative because of the roles of rows and columns as illustrated by the boxes in (2.1.13). Even if **D** and **C** were square and had the same dimensions, in general $\mathbf{DC} \neq \mathbf{CD}$. Since order matters, for $\mathbf{A} = \mathbf{DC}$ as in (2.1.13), **D** is said to *premultiply* **C**, and **C** is said to *postmultiply* **D**. As noted by Jennings (1977), even though the identity in (2.1.16) says that $\mathbf{A}(\mathbf{BC})$ has the same value as $(\mathbf{AB})\mathbf{C}$, the order of evaluation of the products may be very important in computation. For example, if **A** and **B** have dimension 100×100 and **C** has dimensions 100×1, then the total number of multiplications for $(\mathbf{AB})\mathbf{C}$ is 1,010,000 and for $\mathbf{A}(\mathbf{BC})$ is only 20,000!

There are several important rules and naming conventions of matrix algebra that involve transposition of matrices and vectors. One of these is the *reversal rule for transposed products*:

$$\text{If } \mathbf{A} = \mathbf{BCD}, \text{ then } \mathbf{A}^T = \mathbf{D}^T\mathbf{C}^T\mathbf{B}^T. \tag{2.1.17}$$

An important consequence is that

$$\mathbf{B} = \mathbf{A}^T\mathbf{A} \text{ is symmetric.} \tag{2.1.18}$$

This follows from $\mathbf{B}^T = (\mathbf{A}^T\mathbf{A})^T = \mathbf{A}^T\mathbf{A} = \mathbf{B}$. It follows in a similar fashion that

$$\mathbf{C} = \mathbf{A}^T\mathbf{BA} \text{ is symmetric if } \mathbf{B} \text{ is,} \tag{2.1.19}$$

even though the product of two symmetric matrices is in general *not* symmetric.

The *inner product* (or *scalar* or *dot* product) of two *n*-vectors is a *scalar*:

$$\mathbf{x}^T\mathbf{y} = x_1y_1 + x_2y_2 + \cdots + x_ny_n. \qquad (2.1.20)$$

Inner products were used to obtain the respective elements in the result of a matrix multiplication; the three boxes in (2.1.13) illustrate the inner product of the *i*th row vector and the *j*th column vector. The inner product does satisfy the rules of scalar multiplication:

Commutativity: $\qquad\qquad\qquad\qquad \mathbf{x}^T\mathbf{y} = \mathbf{y}^T\mathbf{x}$

Distributivity over vector $\qquad\qquad\qquad\qquad\qquad\qquad$ (2.1.21)
$\quad$ *addition:* $\qquad\qquad\qquad \mathbf{x}^T(\mathbf{y} + \mathbf{z}) = \mathbf{x}^T\mathbf{y} + \mathbf{x}^T\mathbf{z}.$

Figure 2.1.1 is an illustration of Euclidean *n*-dimensional space, specifically $n = 3$. Such space is characterized by a *distance function*: the distance between any two points $\mathbf{x}$ and $\mathbf{y}$ is z; its square is $z^2 = (\mathbf{x} - \mathbf{y})^T(\mathbf{x} - \mathbf{y})$. For example, the length of the vector in Figure 2.1.1 is the distance between the vector (i.e., the point in 3-space) and the null vector (i.e., the origin). Therefore, the length of the $\mathbf{x}$ vector in Figure 2.1.1 is $(\mathbf{x}^T\mathbf{x})^{1/2}$.

An *outer product* of two vectors is a *matrix*:

$$\mathbf{xy}^T = \begin{bmatrix} x_1 \\ x_2 \\ x_3 \end{bmatrix} [y_1 \ y_2] = \begin{bmatrix} x_1y_1 & x_1y_2 \\ x_2y_1 & x_2y_2 \\ x_3y_1 & x_3y_2 \end{bmatrix}. \qquad (2.1.22)$$

The outer product is sometimes called a *proportional matrix* because its rows are proportional to elements of $\mathbf{x}$, and a similar statement can be made about its columns. Outer products such as in (2.1.22) play an important role in optimization.

2.1.4. Inverse of a Square Matrix. This section will deal only with the inverse of square matrices, putting aside the case for rectangular matrices to be described in Section 3.1.2. The most familiar inverse of a matrix $\mathbf{B}$ is designated $\mathbf{B}^{-1}$ and satisfies

$$\mathbf{BB}^{-1} = \mathbf{B}^{-1}\mathbf{B} = \mathbf{I}, \qquad (2.1.23)$$

where $\mathbf{I}$ is the square unit matrix that is the same dimension as $\mathbf{B}$ and has 1's on its principal diagonal and 0's elsewhere. The inverse of a diagonal matrix is a diagonal matrix having elements that are the reciprocals of the given diagonal elements, that is, $1/b_{ii}$. The inverse of a matrix may not exist, analogous to division by a scalar zero. The matrix inverse of every square

$n \times n$ matrix **B** does exist if the determinant of **B** is nonzero; then the matrix is said to be *nonsingular*. The *determinant* of **B**, det(**B**), is a scalar quantity; for a 2×2 matrix, det(**B**) $= b_{11}b_{22} - b_{12}b_{21}$. In general, it may be computed by the recursive formula:

$$\det(\mathbf{B}) = b_{11}M_{11} - b_{12}M_{12} + b_{13}M_{13} - \cdots - (-1)^n b_{1n}M_{1n}, \quad (2.1.24)$$

where M_{ij} is a determinant of the matrix formed by removing row i and column j from matrix **B**. The determinant of an $n \times n$ matrix consists of a sum of $n!$ terms having alternating signs and each one a product of n matrix elements. Equation (2.1.24) is not an efficient way to compute the determinant; it can be found another way as a by-product of the inverse command 9 in program MATRIX.

Certain properties of determinants are useful in the following developments:

1. A determinant is zero if the matrix has two identical rows (columns).
2. The determinant of a matrix product equals the product of the individual determinants.
3. The determinant of a triangular matrix equals the product of the elements on the principal diagonal.
4. Multiplying each row (column) of a determinant of a matrix by a scalar factor increases its determinant by the same factor.
5. The determinant value is unchanged by adding a multiple of one row to another row.
6. Swapping two rows (columns) in a matrix reverses the sign of its determinant.

Both the determinant and the matrix inverse should be thought of as useful algebraic concepts rather than as aids to computation. This is made clear in Section 3.1. These two operations are included in MATRIX as a part of that program's use in the learning process. Two identities involving the matrix inverse that are important concern transposition,

$$(\mathbf{B}^{-1})^T = (\mathbf{B}^T)^{-1}, \quad (2.1.25)$$

and the *reversal rule for inverse products*:

$$\text{If } \mathbf{A} = \mathbf{BCD}, \text{ then } \mathbf{A}^{-1} = \mathbf{D}^{-1}\mathbf{C}^{-1}\mathbf{B}^{-1}. \quad (2.1.26)$$

The former shows that the inverse of a symmetric matrix is also symmetric. The latter leads to the identity $(\mathbf{A}^p)^{-1} = (\mathbf{A}^{-1})^p$.

The explanation of the implementation of MATRIX command 9 for $\mathbf{D} = \mathbf{B}^{-1}$ and det(**B**) requires introduction of *elementary transformation* *matrices,* a concept that is needed throughout this book. For purposes of the Gauss–Jordan method for matrix inversion that follows, three elementary *row* transformations are required as described by McCalla (1967). These are illustrated for the $n = 3$ case. First is the transformation matrix $E_i(h)$ that is the unit matrix except for row i multiplied by h:

$$E_3(h) = \begin{bmatrix} 1 & 0 & 0 \\ 0 & 1 & 0 \\ 0 & 0 & h \end{bmatrix}. \tag{2.1.27}$$

Premultiplying a matrix by $E_i(h)$ multiplies each element in the ith row of that matrix by h, as seen from (2.1.27) and the definition of matrix multiplication in (2.1.13) and (2.1.14). Second is the transformation matrix E_{ik} that is the unit matrix with rows i and k interchanged:

$$E_{12} = \begin{bmatrix} 0 & 1 & 0 \\ 1 & 0 & 0 \\ 0 & 0 & 1 \end{bmatrix}. \tag{2.1.28}$$

Matrix E_{ik} is a *permutation matrix*—one whose elements are either 0 or 1, with just one 1 in each row and column. Premultiplying a matrix by E_{ik} swaps rows i and k in that matrix; the reader is urged to write out the terms of $A = E_{12}B$ to see that rows 1 and 2 in B are exchanged in A. Third is the transformation $E_{ik}(h)$ that is the unit matrix that has h times row k added to row i:

$$E_{12}(h) = \begin{bmatrix} 1 & h & 0 \\ 0 & 1 & 0 \\ 0 & 0 & 1 \end{bmatrix}. \tag{2.1.29}$$

Premultiplying a matrix by $E_{ik}(h)$ adds h times row k to row i; the way to see that this is the case is to write down the result $A = E_{12}(h)B$, where B is a square 3×3 matrix.

The immediate application for a sequence of transformations is the Gauss–Jordan procedure previously applied in Example 1.3.4 for two equations in two unknowns. In the present context, suppose that matrix B is premultiplied by some transformation, say E_1, and that result is premultiplied by some other transformation, say E_2. Continuing that process, suppose that n of these transformations result in the unit matrix:

$$E_n E_{n-1} \cdots E_2 E_1 B = I. \tag{2.1.30}$$

*Post*multiplication of each side of (2.1.30) by B^{-1} and application of (2.1.23) yields

$$E_n E_{n-1} \cdots E_2 E_1 I = B^{-1}. \tag{2.1.31}$$

Table 2.1.6.   A Sequence of Gauss–Jordan Transformations to Compute a Matrix Inverse for Example 2.1.1

Evolution of (2.1.30)	Evolution of (2.1.31)	Determinant
$\mathbf{B} = \begin{bmatrix} 4 & 8 & 2 \\ 1 & 5 & 3 \\ 2 & 7 & 1 \end{bmatrix}$	$\mathbf{D} = \begin{bmatrix} 1 & 0 & 0 \\ 0 & 1 & 0 \\ 0 & 0 & 1 \end{bmatrix}$	$\begin{aligned} D1 &= b_{11}^{(0)} \\ &= 4 \end{aligned}$

Stage k = 1. Make $b_{11} = 1$ and zero rest of column 1:

| Do first: $\begin{bmatrix} 1 & 2 & \frac{1}{2} \\ 0 & 3 & \frac{5}{2} \\ 0 & 3 & 0 \end{bmatrix}$ | $\begin{bmatrix} \frac{1}{4} & 0 & 0 \\ -\frac{1}{4} & 1 & 0 \\ -\frac{1}{2} & 0 & 1 \end{bmatrix}$ | $\begin{aligned} D1 &= D1 * b_{22}^{(1)} \\ &= 4*3 = 12 \end{aligned}$ |

Stage k = 2. Make $b_{22} = 1$ and zero rest of column 2:

| Do first: $\begin{bmatrix} 1 & 0 & -\frac{7}{6} \\ 0 & 1 & \frac{5}{6} \\ 0 & 0 & -\frac{5}{2} \end{bmatrix}$ | $\begin{bmatrix} \frac{5}{12} & -\frac{2}{3} & 0 \\ -\frac{1}{12} & \frac{1}{3} & 0 \\ -\frac{1}{4} & -1 & 1 \end{bmatrix}$ | $\begin{aligned} D1 &= D1 * b_{33}^{(2)} \\ &= 12*\left(-\frac{5}{2}\right) \end{aligned}$ |

Determinant $= -30$.

Stage k = 3. Make $b_{33} = 1$ and zero rest of column 3:

| Do first: $\begin{bmatrix} 1 & 0 & 0 \\ 0 & 1 & 0 \\ 0 & 0 & 1 \end{bmatrix}$ | $\begin{bmatrix} \frac{8}{15} & -\frac{1}{5} & -\frac{7}{15} \\ -\frac{1}{6} & 0 & \frac{1}{3} \\ \frac{1}{10} & \frac{2}{5} & -\frac{2}{5} \end{bmatrix} = \mathbf{B}^{-1}$ | |

Of course, it turns out that there is a sequence of the three elementary transformation matrices just described that will produce a matrix inverse. It is emphasized that the sequence of transformations in (2.1.31) is simply a formalism; these elementary matrices are not actually stored in the computer. The easiest way to assimilate the sequence of Gauss–Jordan transformations is by example, and it is imperative that the reader work through each step. The process allows simple calculation of the determinant as well.

Example 2.1.1.   Consider 3×3 matrices **B** and **D** shown in Table 2.1.6. There are two steps for each of the stages $1, 2, \ldots, n$:

(*a*)   Divide row k by b_{kk}, leaving the new $b_{kk} = 1$.

(*b*)   Zero the remaining elements in column k by subtracting a suitable multiple of the new row k from row i, $i \neq k$.

Referring to stage 1 in Table 2.1.6, $b_{11} = 1$ was obtained by dividing row 1 of **B** by 4. *Each operation on* **B** *is applied to* **D** *as well.* Therefore, $d_{11} = 1/4$. The zeros in column 1 of **B** in stage 1 are obtained by: (1) observing that $b_{21} = 1$ in **B**, so subtraction of 1.0 times the new row 1 from row 2 will make the new $b_{21} = 0$, and (2) observing that $b_{31} = 2$ in **B** so that subtraction of two times the new row 1 from row 3 will make the new $b_{31} = 0$. Again, notice that the *same* row multiplication factor and subsequent subtraction from the corresponding rows in **D** were also accomplished.

Stage 2 transformations operate in behalf of column 2 in the same way as stage 1 except that the reference matrices are the new ones from stage 1 instead of the original ones given in stage 0. As for the determinant of **B**, McCalla (1967) shows that it is equal to the product of the pivots

$$\text{Det}(\mathbf{B}) = b_{11}^{(0)} b_{22}^{(1)} b_{33}^{(2)} \tag{2.1.32}$$

for the case in Table 2.1.6. As indicated, the answer is $\det(\mathbf{B}) = -30$.

McCalla (1967) shows that roundoff error can be minimized by adding a step. Before step 1, scan the current column on and below the principal diagonal for the largest coefficient magnitude; if it is not in the principal diagonal, the two rows are swapped in both **B** and **D**; see lines 3410 to 3590 in program C2-1. Each row interchange alternates the sign of the determinant in (2.1.32). Readers who understand the transformations in Table 2.1.6 should be able to follow the BASIC instruction in lines 3210–3840 that implement command 9 in MATRIX. The reader is urged to enter matrix **B** into program C2-1 and obtain $\mathbf{B}^{-1}$ and $\det(\mathbf{B})$ using command 9. The results are shown in Table 2.1.7.

There are several situations in optimization algorithms where it is useful to find a matrix inverse by partitioning. Suppose that the partitioned matrix in (2.1.7) is given and its inverse exists. Then

$$\mathbf{T}\mathbf{T}^{-1} = \begin{bmatrix} \mathbf{P} & \mathbf{Q} \\ \mathbf{R} & \mathbf{S} \end{bmatrix} \begin{bmatrix} \mathbf{A} & \mathbf{B} \\ \mathbf{C} & \mathbf{D} \end{bmatrix} = \begin{bmatrix} \mathbf{I}_p & \mathbf{0} \\ \mathbf{0} & \mathbf{I}_s \end{bmatrix}, \tag{2.1.33}$$

Table 2.1.7. Program MATRIX Matrix Inverse Command 9 Screen Display for the Matrix in Table 2.1.6.

```
      READ FILE B.MAT INTO MATRIX B
MATRIX B( 3 , 3 )  -
          4.00000          8.00000          2.00000
          1.00000          5.00000          3.00000
          2.00000          7.00000          1.00000
      D=inv(B)
      DETERMINANT(B)  = -30 . NOW B=I.
MATRIX D( 3 , 3 )  -
          0.53333         -0.20000         -0.46667
         -0.16667          0.00000          0.33333
          0.10000          0.40000         -0.40000
```

where $\mathbf{P}$ is $p \times p$ and $\mathbf{S}$ is $s \times s$. Then the unit matrices in (2.1.33) have those dimensions, and the zero *matrix*, $\mathbf{0}$, fills the remainder of the right-hand side. Multiplication of conformable, partitioned matrices proceeds according to the ordinary matrix rules illustrated by (2.1.13). Therefore,

$$\mathbf{PA} + \mathbf{QC} = \mathbf{I}_p,$$

$$\mathbf{PB} + \mathbf{QD} = \mathbf{0},$$

$$\mathbf{RA} + \mathbf{SC} = \mathbf{0}, \tag{2.1.34}$$

$$\mathbf{RB} + \mathbf{SD} = \mathbf{I}_s.$$

Equation (2.1.34) may be solved for $\mathbf{A}$, $\mathbf{B}$, $\mathbf{C}$, and $\mathbf{D}$ *under the assumption that* $\mathbf{S}$ *is nonsingular*; thus, (2.1.34) yields

$$\mathbf{A} = (\mathbf{P} - \mathbf{QS}^{-1}\mathbf{R})^{-1},$$

$$\mathbf{B} = -\mathbf{AQS}^{-1},$$

$$\mathbf{C} = -\mathbf{S}^{-1}\mathbf{RA}, \tag{2.1.35}$$

$$\mathbf{D} = \mathbf{S}^{-1} - \mathbf{S}^{-1}\mathbf{RB}.$$

The reader is urged to use program MATRIX command 9 to find the inverse of (2.1.8) as a whole and by (2.1.35).

An entirely different way to obtain a matrix inverse is to make a change to a known inverse. According to Fiacco (1968:179), the *rank annihilation method* or *Sherman–Morrison–Woodbury* formula for inverting matrices is summarized as follows:

$$(\mathbf{A} + \mathbf{QPR}^T)^{-1} = \mathbf{A}^{-1} - \mathbf{A}^{-1}\mathbf{Q}(\mathbf{P}^{-1} + \mathbf{R}^T\mathbf{A}^{-1}\mathbf{Q})^{-1}\mathbf{R}^T\mathbf{A}^{-1}. \tag{2.1.36}$$

The matrices have dimensions $\mathbf{A}_{nn}$, $\mathbf{Q}_{np}$, $\mathbf{P}_{pp}$, and $\mathbf{R}_{np}$. Equation (2.1.36) is a powerful way to obtain an inverse by modifying an existing one by adding a "rank p" term $\mathbf{QPR}^T$; rank is discussed in Section 2.2.1. See Problem 2.15. To practice the rules of matrix algebra, the reader is urged to verify (2.1.36) by *post*multiplying the right-hand side by the matrix in the parentheses on the left-hand side of (2.1.36). It will also be necessary to insert the unit matrix, $\mathbf{I} = \mathbf{P}^{-1}\mathbf{P}$ to reduce (2.1.36) to $\mathbf{0} = \mathbf{0}$.

2.1.5.   *Vector and Matrix Norms and Condition Number.*   It is well known that the length of the vector in Figure 2.1.1 is

$$\|\mathbf{x}\| = (\mathbf{x}^T\mathbf{x})^{1/2} = \left(x_1^2 + x_2^2 + x_3^2\right)^{1/2}. \tag{2.1.37}$$

The symbol on the left side of (2.1.37) denotes a *norm* of the vector **x**. Norms are non-negative scalars that are important because they measure length, size, or distance, depending on the context. For example, norms of matrices are comparable to their "magnitude"; also, the norm $\|\mathbf{A} - \mathbf{B}\|$ indicates how "close" one matrix is to another. Norms are the basis for quantitatively describing matrix illconditioning as well.

Not just any old equation will do for a norm; *vector norms* must satisfy the following three conditions:

1. $\|\mathbf{x}\| > 0$, for $\mathbf{x} \neq \mathbf{0}$, and $\|\mathbf{x}\| = 0$ implies $\mathbf{x} = \mathbf{0}$.
2. $\|h\mathbf{x}\| = |h| \, \|\mathbf{x}\|$ for any scalar h.   (2.1.38)
3. $\|\mathbf{x} + \mathbf{y}\| \leq \|\mathbf{x}\| + \|\mathbf{y}\|$ (the triangle inequality).

The *p-norm* of an *n*-vector satisfies all three conditions in (2.1.38):

$$\|\mathbf{x}\|_p = \left[\sum_{i=1}^{n} |x_i|^p \right]^{1/p}. \qquad (2.1.39)$$

The three ordinary values of p are $p = 1$, 2, and ∞, and the corresponding norms are called the *one-*, *two-*, *and infinity-norms*. The ordinary vector length in (2.1.37) is clearly the two-norm, often called the *Euclidean vector norm*. For very large p, $\|\mathbf{x}\|_\infty = \max_i |x_i|$, the element in **x** with maximum modulus. All three of these vector norms are computed by MATRIX command 11 for the norms of vector **d** as implemented in lines 5650 to 5790. Figure 2.1.2 depicts the locus of the vector point **x** in 2-space for fixed values for all three of these vector norms.

Recalling the definition of the inner product in (2.1.20), the *Schwarz inequality*

$$|\mathbf{x}^T \mathbf{y}| \leq \|\mathbf{x}\|_2 \|\mathbf{y}\|_2 \qquad (2.1.40)$$

can be proved using the Law of Cosines from trigonometry. Consider the three

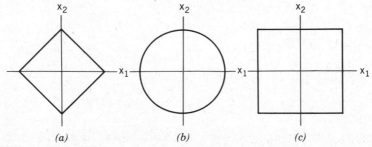

$$(a) \qquad\qquad\qquad (b) \qquad\qquad\qquad (c)$$

Figure 2.1.2. The locus of the **x** point in 2-space for fixed values of three vector norms: (*a*) the one-norm, (*b*) the two-norm, and (*c*) the infinity norm.

vectors **x**, **y**, and **y**–**x**; a special case is the illustration of **x** in Figure 2.1.1. In that space there exists a triangle having these three vectors as sides. The Law of Cosines relates the angle θ between **x** and **y** in terms of the lengths of the three sides:

$$\|\mathbf{y} - \mathbf{x}\|_2^2 = \|\mathbf{y}\|_2^2 + \|\mathbf{x}\|_2^2 - 2\|\mathbf{x}\|_2\|\mathbf{y}\|_2 \cos \theta. \qquad (2.1.41)$$

Replacing the left-hand side of (2.1.41) with $(\mathbf{y} - \mathbf{x})^T(\mathbf{y} - \mathbf{x})$ yields

$$\mathbf{x}^T\mathbf{y} = \|\mathbf{x}\|_2 \|\mathbf{y}\|_2 \cos \theta. \qquad (2.1.42)$$

But $-1 \le \cos \theta \le +1$, so (2.1.40) is obtained. Equation (2.1.42) is useful in its own right, since the angle between directions (vectors) is often an important issue in optimization.

Matrices usually occur in conjunction with vectors as in the system of linear equations

$$\mathbf{A}\mathbf{x} = \mathbf{b}, \qquad (2.1.43)$$

which is just matrix notation for

$$\begin{aligned}
a_{11}x_1 + a_{12}x_2 + \cdots + a_{1n}x_n &= b_1 \\
a_{21}x_1 + a_{22}x_2 + \cdots + a_{2n}x_n &= b_2 \\
&\cdots \\
a_{n1}x_1 + a_{n2}x_2 + \cdots + a_{nn}x_n &= b_n.
\end{aligned} \qquad (2.1.44)$$

Matrix norms are defined so that they are *compatible* with vector norms in the sense that

$$\|\mathbf{A}\mathbf{x}\| \le \|\mathbf{A}\| \|\mathbf{x}\|, \qquad (2.1.45)$$

where the first and last terms are vector norms and $\|\mathbf{A}\|$ is a matrix norm. Matrix norms satisfy the following four conditions:

1. $\|\mathbf{A}\| > 0$ for $\mathbf{A} \ne \mathbf{0}$, and $\|\mathbf{A}\| = 0$ implies $\mathbf{A} = \mathbf{0}$.
2. $\|h\mathbf{A}\| = |h| \|\mathbf{A}\|$ for any scalar h. $\qquad\qquad (2.1.46)$
3. $\|\mathbf{A} + \mathbf{B}\| \le \|\mathbf{A}\| + \|\mathbf{B}\|$.
4. $\|\mathbf{A}\mathbf{B}\| \le \|\mathbf{A}\| \|\mathbf{B}\|$.

Considering (2.1.45), the matrix norm is defined by

$$\|\mathbf{A}\| = \max \|\mathbf{A}\mathbf{x}\| \qquad \text{for all } \|\mathbf{x}\| = 1, \qquad (2.1.47)$$

where again it is important to note that norms $\|\mathbf{A}\mathbf{x}\|$ are *vector* norms. For example, Figures 2.1.1 and 2.1.2b suggest that $\|\mathbf{x}\|_2 = 1$ requires a search over the unit *hypersphere* to find the point on that surface that maximizes the vector $y = Ax$. Another way to look at matrix norms comes from (2.1.45): x can be "stretched" at most by the matrix norm, $\|A\|$, when x is multiplied by A.

Thus, it is said that the three vector norms, defined by (2.1.39) for $p = 1, 2$, and ∞, *induce* three norms for the matrix $A = [a_{ij}]$ according to (2.1.47):

$$\|A\|_1 = \max_j \sum_{i=1}^{m} |a_{ij}|, \text{ the maximum absolute column sum,}$$

$$\|A\|_2 = (w_{max})^{1/2}, \quad \begin{array}{l} \text{the square root of the maximum} \\ \text{eigenvalue of } A^T A, \end{array} \tag{2.1.48}$$

$$\|A\|_\infty = \max_i \sum_{j=1}^{n} |a_{ij}|, \text{ the maximum absolute row sum.}$$

If A is symmetric, then $\|A\|_\infty = \|A\|_1$. Readers interested in how these three "natural" norms are derived from (2.1.47) and (2.1.39) are referred to Noble (1969:429). The $\|A\|_2$ *spectral matrix norm* requires the set of n scalar numbers (possibly complex) called eigenvalues that are associated with any real matrix. The eigenvalues are often denoted by λ_i, but are denoted by w_i here; they are described in Section 2.2.3 and are mentioned here only for completeness. The presence of the transposed product $A^T A$ in (2.1.48) is not trivial; it plays a significant role in optimization.

One other matrix norm is well known; it is "compatible" according to (2.1.45) but is not "induced" by a vector norm according to (2.1.47). The *Frobenius matrix norm* is

$$\|A\|_F = \left(\sum_{i=1}^{m} \sum_{j=1}^{n} a_{ij}^2 \right)^{1/2}, \tag{2.1.49}$$

which is just the square root of the sum of the squares of all elements in the matrix. It is also equal to the square root of $\mathrm{tr}(A^T A)$, the trace of the transposed product. Table 2.1.8 shows the values of three of the preceding matrix norms for the 5×5 matrix in (2.1.8).

The solution of a system of linear equations when illconditioning occurs was described in Chapter One. The illconditioning can be quantified in the sense that each matrix has a defined *condition number*:

$$k(A) = \|A\| \|A^{-1}\|. \tag{2.1.50}$$

No matter which matrix norm is used, the condition number $k(A) \geq 1$, since (2.1.46) shows that $k(A) \geq \|AA^{-1}\| = \|I\| = 1$. *A matrix is said to be illconditioned if its condition number is much greater than unity.* Consider the matrix

Table 2.1.8.   Three Program MATRIX Matrix Norms for T in (2.1.8) and the Condition Number, $k(T)$

Norm	$\|\cdot\|_F$	$\|\cdot\|_1$	$\|\cdot\|_\infty$
T	17.1756	18.0000	20.0000
$\mathbf{T}^{-1}$	1.6723	3.0140	1.4793
$k(\mathbf{T})$	28.7228	54.2520	29.5860

product $\mathbf{AB}$, where $\mathbf{A}$ is a square matrix. Errors in the elements of $\mathbf{A}$ because of roundoff or for other reasons will propagate into the result where they may be magnified significantly. Suppose that matrix $\mathbf{A}$ is perturbed by the addition of matrix $\mathbf{dA}$ so that $\mathbf{A}$ becomes $\mathbf{A} + \mathbf{dA}$. The relative size of the perturbation in the product is

$$\frac{\|\mathbf{dA}\,\mathbf{B}\|}{\|\mathbf{AB}\|} = \frac{\|(\mathbf{dA}\,\mathbf{A}^{-1})\mathbf{AB}\|}{\|\mathbf{AB}\|}$$

$$\leq \|\mathbf{dA}\,\mathbf{A}^{-1}\|$$

$$\leq \|\mathbf{dA}\|\,\|\mathbf{A}^{-1}\| = k(\mathbf{A})\frac{\|\mathbf{dA}\|}{\|\mathbf{A}\|}. \tag{2.1.51}$$

Therefore, *condition number $k(\mathbf{A})$ measures how much the relative uncertainty of a matrix may be magnified in a matrix product*. See Table 2.1.8 for an example.

Suppose that there is a set of n right-hand vectors in the linear system described by (2.1.43). Then there are n solution vectors $\mathbf{x}$. Collecting the former in columns of $\mathbf{B} = (\mathbf{b}_1\ \mathbf{b}_2\ \ldots\ \mathbf{b}_n)$ and the latter in $\mathbf{X} = (\mathbf{x}_1\ \mathbf{x}_2\ \ldots\ \mathbf{x}_n)$, those *sets* of linear systems can be written

$$\mathbf{AX} = \mathbf{B}. \tag{2.1.52}$$

Clearly, uncertainties in the square system matrix $\mathbf{A}$ or in the right-hand side matrix $\mathbf{B}$ will also propagate into the solution vectors in $\mathbf{X}$. For small relative uncertainties $\mathbf{dA}$ in $\mathbf{A}$, say $\|\mathbf{dA}\|/\|\mathbf{A}\| \ll 1/k(\mathbf{A})$, the condition number closely approximates how much the relative uncertainty in $\mathbf{A}$ and/or in $\mathbf{B}$ can be magnified in the solutions contained in the columns of $\mathbf{X}$.

2.2.   Relationships in Vector Space

This section introduces some spatial concepts in matrix algebra that are relevant to optimization. These concepts have a geometry that is readily observed in two and three dimensions and remain valid in any number of dimensions. No attempt is made to generalize the approach to include abstract spaces, such as the space of functions. In fact, this presentation is intended to be quite straightforward so that the reader can form and retain a mental image of the principles on which optimization is based.

2.2.1.   The Matrix Role in Vector Space.   Figure 2.2.1 represents the same Euclidean 3-space (called E^3) shown in Figure 2.1.1. The three vectors, **x**, **y**, and **d** lie in an oblique plane that passes through the origin. Consider how one might locate any point in that plane; one way would be to locate any point **q** in that plane by forming a linear combination of two of the vectors shown, say

$$\mathbf{q} = r\mathbf{y} + s\mathbf{d}, \qquad (2.2.1)$$

where r and s are positive or negative scalars. The first point to be made is that there are always choices for r and s that will reach any point in the plane containing **y** and **d**, as illustrated in Figure 2.2.1, but no point outside that plane is available with (2.2.1). The plane described by (2.2.1) is called a *subspace* of the three-dimensional space represented by Figure 2.2.1 (not necessarily through the origin). Lines in that three-dimensional space, for example, the x_2 axis or the line coincident with vector **d**, are also subspaces. Clearly, (2.2.1) would not describe every point in the oblique plane if **y** and **d** were *collinear*, which corresponds to being parallel in this illustration. If they were collinear, then $\mathbf{y} = t\mathbf{d}$ for some nonzero scalar t, and it is said that **y** and **d** are *linearly dependent*. It is conventional also to write that condition:

$$r\mathbf{y} + s\mathbf{d} = \mathbf{0}, \qquad (2.2.2)$$

which implies that **y** and **d** are *linearly independent* if no nonzero r and s can

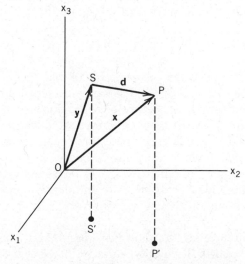

Figure 2.2.1.   Three vectors in Euclidean 3-space that form a linearly dependent set.

be found to make (2.2.2) true. In other words, $r\mathbf{y}$ and $s\mathbf{d}$ cannot go out and return to the origin. The *dimension* of a subspace is equal to the maximum number of vectors in that subspace that can be linearly independent.

So much for subspaces; clearly, three vectors are required to locate any point in E^3, as illustrated in Figure 2.2.1. Will $\mathbf{y}$, $\mathbf{d}$, and $\mathbf{x}$ do? No, because they are linearly dependent; that is, there exist three nonzero scalars such that

$$r\mathbf{y} + s\mathbf{d} + t\mathbf{x} = \mathbf{0}. \tag{2.2.3}$$

Again, the problem with those three vectors is that they form a closed loop and are consequently coplanar. The three unit vectors previously mentioned as having unit length and lying along the three axes are linearly independent and will locate any point $\mathbf{x}$ in E^3:

$$\mathbf{x} = \mathbf{I}\mathbf{x} = \begin{bmatrix} 1 \\ 0 \\ 0 \end{bmatrix} x_1 + \begin{bmatrix} 0 \\ 1 \\ 0 \end{bmatrix} x_2 + \begin{bmatrix} 0 \\ 0 \\ 1 \end{bmatrix} x_3. \tag{2.2.4}$$

Recall that $\mathbf{I} = (\mathbf{e}_1 \ \mathbf{e}_2 \ \mathbf{e}_3)$, where $\mathbf{e}_i$ is a column vector of the unit matrix $\mathbf{I}$. A set of vectors from E^n is said to *span* E^n if every vector in that space can be represented by a linear combination of that set. A *basis for* E^n is a linearly independent subset of the least number of vectors in E^n that spans the entire space. Clearly, the columns of $\mathbf{I}_3$ span E^3 and can be chosen as a basis for that space.

The preceding illustration with the identity matrix is too easy and may cause the reader to overlook a major concept. *Any matrix-vector product is equivalent to a linear combination of the columns of the matrix, the coefficients of each column being the respective elements of the vector.* It is not especially obvious, so the reader should verify this important identity using, say, $n = 3$:

$$\mathbf{y} = \mathbf{A}\mathbf{x} = (\mathbf{a}_1 \ \mathbf{a}_2 \ \cdots \ \mathbf{a}_n)\mathbf{x} = x_1\mathbf{a}_1 + x_2\mathbf{a}_2 + \cdots + x_n\mathbf{a}_n. \tag{2.2.5}$$

The vector $\mathbf{y}$ is said to be in the *column space* of matrix $\mathbf{A}$.

Now it should be clear that $\mathbf{A}\mathbf{x} = \mathbf{0}$ can only be satisfied if the columns of $\mathbf{A}$ are linearly dependent. The *null space* of $\mathbf{A}$ has dimension $n - r$, where r is the rank of $\mathbf{A}$. A square matrix whose columns are linearly dependent is said to be *singular* and its determinant is equal to zero. In general, the n columns of a *nonsingular* matrix span E^n and can be chosen as a basis for that space. The *range* or *column space* of a matrix $\mathbf{A}_{m,n}$ is the span of its column vectors. Such matrices are not unique, and there are an infinity of bases for any space; however, the representation of any vector in terms of a given set of basis vectors *is* unique.

Suppose that the three vectors shown in Figure 2.2.1 constituted the three columns of a 3×3 matrix. Then that matrix would be said to have rank equal to 2. The *rank* of a matrix, square or not, is the maximum number of linearly

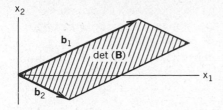

Figure 2.2.2. Geometric interpretation of a determinant in E^2 as the area of a parallelogram.

independent columns (or rows) and is equal to the dimension of the subspace that those vectors span. The row and column spaces of a matrix $A_{m,n}$ have the same dimensions. The determinant of a matrix was defined by (2.1.24). The order of the nonzero determinant of highest order contained in a matrix (square or not) is equal to the rank of that matrix. It is noted that the rank of a matrix is not changed by multiplication by a nonsingular matrix. Also, the matrix product $A^T A$ has the same rank as A.

Since determinants play a role in matrix singularity and rank, it is useful to note a geometric interpretation for determinants. Consider the 2×2 matrix

$$B = (b_1 \ b_2), \tag{2.2.6}$$

and the representation of its column vectors in E^2 as shown in Figure 2.2.2. The cosine of the angle between any two vectors was given by (2.1.42). The area of a parallelogram is equal to the product of the lengths of its two sides and the sine of the angle between them. Using the identity $\sin^2 \theta = 1 - \cos^2 \theta$ and the Euclidean ($p = 2$) vector norm for the vector lengths, routine algebra reveals that the area of the parallelegram in E^2 is equal to the magnitude of the determinant of B. In E^3 the *volume* of the parallelepiped with sides defined by matrix column vectors can be shown equal to the matrix determinant, and this generalizes to E^n. This leads to *Hadamard's inequality*,

$$\det(B) \le \|b_1\| \ \|b_2\| \ \cdots \ \|b_n\|, \tag{2.2.7}$$

since the volume of a parallelepiped cannot exceed the product of its sides.

As also seen from the left side of (2.2.5), a matrix can be viewed as a way to transform one vector into another, for example, x into y. The vector space containing x is called the domain of A; y is in the range of A. These are *linear transformations* because of the properties of matrix multiplication previously described. Therefore, the transformation of a linear combination of vectors is the same linear combination of the transformed vectors:

$$A(ry + sd) = A(ry) + A(sd) = r(Ay) + s(Ad). \tag{2.2.8}$$

If A transforms x into y, then the matrix inverse, A^{-1}, transforms y back into x.

2.2.2. Orthogonal Relationships.

The basis vectors found as columns of the identity matrix are orthogonal, that is, at right angles. Of course, orthogonal vectors need not be aligned with the major axes, so this section begins with some important remarks concerning orthogonality conditions. Equation (2.1.42) provided an expression for the angle between vectors. The relationship of a set of *orthogonal vectors* is

$$\mathbf{q}_i^T \mathbf{q}_j = 0, \qquad i \neq j. \tag{2.2.9}$$

Furthermore, the vectors $\mathbf{q}_i$ are said to be *orthonormal* if an additional property holds:

$$\mathbf{q}_i^T \mathbf{q}_i = 1. \tag{2.2.10}$$

Consider a matrix composed of these m-vectors in its n columns:

$$\mathbf{Q}_{m,n} = (\mathbf{q}_1 \ \mathbf{q}_2 \ \cdots \ \mathbf{q}_n). \tag{2.2.11}$$

A matrix is said to be an *orthogonal* (*orthonormal*) *matrix* if its columns are orthogonal (orthonormal). Simple algebraic multiplication of the case for $m = 4$ and $n = 3$ will verify that if $\mathbf{Q}$ is orthonormal, then

$$\mathbf{Q}^T \mathbf{Q} = \mathbf{I}_n, \tag{2.2.12}$$

which implies that $\mathbf{Q}^{-1} = \mathbf{Q}^T$ when $\mathbf{Q}$ is square.

When the basis vectors are orthogonal, the linear combination of those that define a given vector is easy to compute. Suppose that the basis vectors are the columns of $\mathbf{Q}$ in (2.2.11). Then a given vector $\mathbf{y}$ may be expressed as

$$\mathbf{y} = r_1 \mathbf{q}_1 + r_2 \mathbf{q}_2 + \cdots + r_n \mathbf{q}_n. \tag{2.2.13}$$

The typical ith inner product is

$$\mathbf{q}_i^T \mathbf{y} = r_1 \mathbf{q}_i^T \mathbf{q}_1 + r_2 \mathbf{q}_i^T \mathbf{q}_2 + \cdots + r_n \mathbf{q}_i^T \mathbf{q}_n = r_i \mathbf{q}_i^T \mathbf{q}_i, \tag{2.2.14}$$

since all but one term vanishes according to the vector orthogonality condition (2.2.9). Thus the ith coefficient in the linear combination shown on the right-hand side of (2.2.13) is

$$r_i = \frac{\mathbf{q}_i^T \mathbf{y}}{\mathbf{q}_i^T \mathbf{q}_i}, \tag{2.2.15}$$

where the denominator is unity when the basis vectors $\mathbf{q}_i$ are orthonormal.

The concepts introduced by (2.2.13) to (2.2.15) lead directly to the well known *Gram–Schmidt* procedure for finding a set of orthonormal vectors from a given set of linearly independent vectors. Specifically, given a set of n linearly independent m vectors, $(\mathbf{a}_1 \ \mathbf{a}_2 \ \cdots \ \mathbf{a}_n)$, construct an orthonormal set of vectors, $(\mathbf{q}_1 \ \mathbf{q}_2 \ \cdots \ \mathbf{q}_n)$, where the first j vectors, $\mathbf{q}_j$, are linearly related to $\mathbf{a}_j$. Therefore, in addition to mutual orthogonality among the $\mathbf{q}_i$ vectors, it is required that

$$\mathbf{a}_j = u_{jj}\mathbf{q}_j + u_{j-1,j}\mathbf{q}_{j-1} + \cdots + u_{1j}\mathbf{q}_1, \qquad (2.2.16)$$

where the u_{ij} are scalar coefficients of a linear combination of vectors. The pertinent equations are worked out below for $j = 1$, 2, and 3, then general expressions and a computer subroutine that runs in the MATRIX environment are presented.

The Gram–Schmidt procedure begins with $j = 1$ in (2.2.16):

$$\mathbf{a}_1 = u_{11}\mathbf{q}_1. \qquad (2.2.17)$$

The orthonormal requirement means that $\|\mathbf{q}_1\| = 1$, so $u_{11} = \|\mathbf{a}_1\|$, the length of the vector $\mathbf{a}_1$. (It is assumed that the two-norm is used throughout the Gram–Schmidt procedure.) Therefore, for $j = 1$ in (2.2.16), it is concluded that

$$\mathbf{q}_1 = \frac{\mathbf{a}_1}{u_{11}} \quad \text{and} \quad u_{11} = \|\mathbf{a}_1\|. \qquad (2.2.18)$$

For $j = 2$, (2.2.16) yields

$$\mathbf{a}_2 = u_{22}\mathbf{q}_2 + u_{12}\mathbf{q}_1. \qquad (2.2.19)$$

Premultiplying by $\mathbf{q}_1^T$ yields $u_{12} = \mathbf{q}_1^T\mathbf{a}_2$ due to (2.2.9) and (2.2.10). It is *not* possible to perform the same premultiplication to obtain u_{22} because $\mathbf{q}_2$ has not yet been found. However, (2.2.19) can be solved for $\mathbf{q}_2$:

$$\mathbf{q}_2 = \frac{\mathbf{a}_2 - u_{12}\mathbf{q}_1}{u_{22}}. \qquad (2.2.20)$$

The crucial step is to observe that (2.2.20) has the same form as (2.2.18), especially that u_{22} must cause $\mathbf{q}_2$ to have unit length. Therefore,

$$u_{12} = \mathbf{q}_1^T\mathbf{a}_2, \quad \mathbf{q}_2' = (\mathbf{a}_2 - u_{12}\mathbf{q}_1), \quad u_{22} = \|\mathbf{q}_2'\|, \quad \mathbf{q}_2 = \frac{\mathbf{q}_2'}{u_{22}}, \qquad (2.2.21)$$

where $\mathbf{q}_2'$ is the unnormalized orthogonal vector $\mathbf{q}_2$.

This procedure is illustrated for one more step before writing the general expressions. The $j = 3$ case for (2.2.16) yields

$$\mathbf{a}_3 = u_{33}\mathbf{q}_3 + u_{23}\mathbf{q}_2 + u_{13}\mathbf{q}_1. \qquad (2.2.22)$$

Premultiplying (2.2.22) by q_1^T yields u_{13} and premultiplying (2.2.22) by q_2^T yields u_{23}:

$$u_{13} = q_1^T a_3, \qquad u_{23} = q_2^T a_3. \qquad (2.2.23)$$

Therefore,

$$q_3' = (a_3 - u_{23}q_2 - u_{13}q_1), \qquad u_{33} = \|q_3'\|, \qquad q_3 = \frac{q_3'}{u_{33}}. \quad (2.2.24)$$

It is now possible to write the general expressions for the Gram–Schmidt orthonormalization procedure:

$$u_{ij} = q_i^T a_j, \qquad j > i,\ i = 1 \text{ to } n,$$

$$q_j' = (a_j - u_{j-1,j}q_{j-1} - u_{j-2,j}q_{j-2} - \cdots - u_{1j}q_1), \qquad (2.2.25)$$

$$u_{jj} = \|q_j'\|, \qquad q_j = \frac{q_j'}{u_{jj}}.$$

For both programming and subsequent solution of systems of linear equations, note that the equations written for $j = 1, 2, 3, \ldots, n$ from (2.2.16) are just the jth equation of the set expressed by the matrix equation $A_{m,n} = Q_{m,n}U_{n,n}$, or

$$(a_1\ a_2\ \cdots\ a_n) = (q_1\ q_2\ \cdots\ q_n)\begin{bmatrix} u_{11} & u_{12} & \cdots & u_{1n} \\ 0 & u_{22} & \cdots & u_{2n} \\ & & \cdots & \\ 0 & 0 & & u_{nn} \end{bmatrix}. \quad (2.2.26)$$

Subprogram C2-2, GSDECOMP, is a 60-line subroutine that performs the Gram–Schmidt orthonormalization procedure with respect to matrix **A**, placing the orthonormal vectors in the columns of program matrix **D** (comparable to **Q**) and the coefficients in the upper-right triangular program matrix **C** (comparable to **U**). Subprogram C2-2 also computes and displays the inner products of all combinations of orthonormal columns of **D** as a check on roundoff error.

Example 2.2.1. An example of the Gram–Schmidt process given by Noble (1969:315) involves the three column vectors in the 4×3 matrix

$$A = \begin{bmatrix} 1 & 2 & -1 \\ 1 & -1 & 2 \\ 1 & -1 & 2 \\ -1 & 1 & 1 \end{bmatrix}. \qquad (2.2.27)$$

Table 2.2.1. Results for the Gram–Schmidt Procedure Applied to the 4 × 3 Matrix in Example 2.2.1

```
READ FILE GSEX1 INTO MATRIX A

  D=orthodecomp(A), & C triangular
  D=orthodecomp(A), & C triangular
INNER PRODUCT OF COLUMNS 1  1  1
INNER PRODUCT OF COLUMNS 1  2  0
INNER PRODUCT OF COLUMNS 1  3 -6.938893903907229D-18
INNER PRODUCT OF COLUMNS 2  2  .999999989824544
INNER PRODUCT OF COLUMNS 2  3 -7.195134012205617D-08
INNER PRODUCT OF COLUMNS 3  3  1.000000121884437
PRESS <RETURN> KEY TO CONTINUE -- READY? MATRIX D( 4 , 3 ) -
       0.50000        0.86603       -0.00000
       0.50000       -0.28868        0.40825
       0.50000       -0.28868        0.40825
      -0.50000        0.28868        0.81650
```

Matrix **A** was stored in file "GSEX1". Then, in the BASIC environment, the commands LOAD"MATRIX followed by MERGE"GSDECOMP were executed. The composite program was then run with the results shown in Table 2.2.1. The columns of the matrix shown in Table 2.2.1 are orthonormal as verified by the combinations of all possible inner products shown there. Using program MATRIX, it is a simple matter to verify (2.2.12) numerically, in this case obtaining the 3 × 3 identity matrix.

Figure 2.2.3 is a geometric interpretation of two of the three Gram–Schmidt steps in E^3. The angle θ between two of the given vectors, $\mathbf{a}_1$ and $\mathbf{a}_2$, obeys

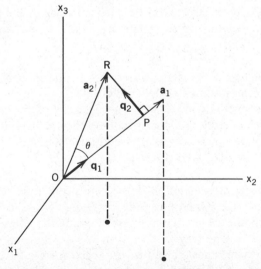

Figure 2.2.3. An illustration of equation (2.2.19) after the second step in a Gram–Schmidt orthonormalization. Unit vector $\mathbf{q}_1$ is a submultiple of the first given vector $\mathbf{a}_1$.

Table 2.2.2.   Orthogonalization of the 5 × 5 Matrix in (2.1.8) by the Gram–Schmidt Procedure

```
READ FILE T.MAT INTO MATRIX A

   D=orthodecomp(A), & C triangular
   D=orthodecomp(A), & C triangular
INNER PRODUCT OF COLUMNS 1   1   1.000000000511063
INNER PRODUCT OF COLUMNS 1   2   1.406787256130748D-10
INNER PRODUCT OF COLUMNS 1   3  -2.749956438778678D-10
INNER PRODUCT OF COLUMNS 1   4  -1.170160565666656D-10
INNER PRODUCT OF COLUMNS 1   5   2.541167451811099D-09
INNER PRODUCT OF COLUMNS 2   2   1.000000167059906
INNER PRODUCT OF COLUMNS 2   3   4.725349924544764D-08
INNER PRODUCT OF COLUMNS 2   4  -2.060781495727415D-07
INNER PRODUCT OF COLUMNS 2   5  -3.918210795704133D-07
INNER PRODUCT OF COLUMNS 3   3   .9999999982116635
INNER PRODUCT OF COLUMNS 3   4  -5.360577944816397D-08
INNER PRODUCT OF COLUMNS 3   5  -2.952260079973476D-08
INNER PRODUCT OF COLUMNS 4   4   .9999999927836797
INNER PRODUCT OF COLUMNS 4   5   6.048025740969937D-07
INNER PRODUCT OF COLUMNS 5   5   1.00000001904006
PRESS <RETURN> KEY TO CONTINUE -- READY? MATRIX D( 5 , 5 ) -
   0.28284        0.40226        0.77750        0.39196        0.00619
  -0.14142        0.06921       -0.06081        0.13625        0.97619
  -0.28284        0.24655       -0.42609        0.79939       -0.19657
   0.70711        0.51905       -0.45611       -0.13909        0.05664
  -0.56569        0.70937        0.04686       -0.41163       -0.07187
```

the inner product relationship in (2.1.42). The vector between points O and P in Figure 2.2.3 is the *projection* of a_2 on a_1. From (2.2.19), that projection is a multiple of the unit vector, namely, $u_{12}q_1$. Orthonormal vectors q_1 and q_2 are linearly independent and thus define the subspace that is the plane containing the three points O, P, and R. The third orthonormal vector q_3 is not shown in Figure 2.2.3, but it is constructed to be perpendicular to the plane defined by q_1 and q_2. Then the third given vector, a_3, could be shown in Figure 2.2.3 according to (2.2.22).

Example 2.2.2.   The 5 × 5 matrix in (2.1.8) was placed in file "T.MAT" and orthogonalized using program GSDECOMP merged into program MATRIX. The results are shown in Table 2.2.2. The columns of the matrix shown are orthonormal as verified by the inner products shown there. This orthonormal matrix D, since it is square, obeys the relationship that $D^{-1} = D^T$, as easily verified by using the MATRIX program.

Notice the increasing roundoff error evident in the inner products shown in Table 2.2.2. Since the Gram–Schmidt procedure worked from column 1 through column 5, the accumulation of errors made the approximation to zero become about 10^{-7}. One way to deal with cases where the process is more illconditioned is to reorthogonalize. In this case the reader can use program MATRIX to equate $A = D$ and rerun command 10. Then the approximation to zero is about 10^{-14} and will not improve with subsequent reorthogonalizations. Of course, the second orthonormal matrix is very nearly the same as the one first obtained.

Another method that attempts to minimize roundoff error is attributed to Rice and described by Morris (1983:165–167). Observe the angle θ between given vectors $\mathbf{a}_1$ and $\mathbf{a}_2$ in Figure 2.2.3. Since θ may be found using (2.1.42), one might choose some vector other than $\mathbf{a}_2$ that is more nearly orthogonal to $\mathbf{a}_1$, that is, is more linearly independent. This simply amounts to swapping the order of the given column vectors. At the third step the choice would involve comparing the angle between the vector that is normal to the $\mathbf{q}_1$–$\mathbf{q}_2$ plane and each of the remaining given column vectors. There is a recursive relationship for the entire process; it lengthens the computer code and can only minimize, not eliminate, roundoff error.

Golub (1970:238) noted that if $\mathbf{A}$ is at all illconditioned, the algorithm in program C2-2 would never be used without reorthogonalization, whereas consistently excellent results have been obtained by the Rice method. The interested reader is also referred to Lawson (1974:129–132) for additional analysis.

2.2.3. The Matrix Eigenproblem.

For every square matrix $\mathbf{A}_{n,n}$ there exist vectors $\mathbf{v}_i$, $i = 1$ to n, such that $\mathbf{A}$ transforms $\mathbf{v}_i$ into a vector proportional to itself:

$$\mathbf{A}\mathbf{v}_i = w_i\mathbf{v}_i. \qquad (2.2.28)$$

The scalar factor w_i is variously known as an eigenvalue, characteristic value, proper value, or latent root. As Bellman (1960) noted, the hybrid word *eigenvalue* is derived from *eigenwerte*, which means "characteristic value" in German. "Despite its ugliness, it seems to be too firmly entrenched to dislodge." Each vector $\mathbf{v}_i$ in (2.2.28) is called an *eigenvector*. The n eigenvectors are the basis for an *invariant subspace* in the sense that $\mathbf{x}$ and $\mathbf{A}\mathbf{x}$ are both linear combinations of the same basis vectors. If q of the n eigenvalues are equal to the same number, say $\bar{w}$, then the q-dimensional subspace spanned by the eigenvectors associated with them is called the *eigenspace* associated with $\bar{w}$.

The eigenproblem has a lot to do with the solution of linear equations, and thus with optimization as well. In vibration and similar instances of simple harmonic motion, eigenvalues are usually called the *normal modes* because they define the extreme positions between which the system oscillates when vibrating at a single "natural" frequency. As Noble remarked, "the computation of eigenvalues and eigenvectors...is a vast and technical subject. Many of the more obvious methods are computationally unstable and/or inefficient." In this section the discussion is directed toward clarity of concepts, and the computations here are performed with less than maximum efficiency. Techniques for improving algorithmic efficiency are discussed in the next section.

The matrix involved, say $\mathbf{A}$, is assumed to be real for purposes of optimization procedures. It may be unsymmetric, although the vast majority of applications in optimization involve only symmetric matrices. An equivalent statement of (2.2.28) is

$$(\mathbf{A} - w_i\mathbf{I})\mathbf{v}_i = \mathbf{0}, \qquad i = 1 \text{ to } n \text{ for } \mathbf{A}_{n,n}, \tag{2.2.29}$$

and w_i and $\mathbf{v}_i$ represent any corresponding pair. A nontrivial solution for $\mathbf{v}$ (to within a multiplicative constant) requires that

$$\det(\mathbf{A} - w\mathbf{I}) = 0, \tag{2.2.30}$$

which may be expanded to appear as

$$\det \begin{bmatrix} a_{11} - w & a_{12} & \cdots & a_{1n} \\ a_{21} & a_{22} - w & \cdots & a_{2n} \\ & & \cdots & \\ a_{n1} & a_{n2} & \cdots & a_{nn} - w \end{bmatrix} = 0. \tag{2.2.31}$$

The determinant for $n = 2$ yields

$$w^2 - (a_{11} + a_{22})w + \det(\mathbf{A}) = 0. \tag{2.2.32}$$

In general, (2.2.31) produces an nth-degree polynomial with real coefficients called the *characteristic equation*. It can always be factored to appear as

$$(w - w_1)(w - w_2) \cdots (w - w_i) \cdots (w - w_n) = 0, \tag{2.2.33}$$

Table 2.2.3.   Some Important Properties of the Eigenproblem

(1) $\text{tr}(\mathbf{A}) = \sum_{i=1}^{n} w_i$, i.e., $a_{11} + a_{22} + \cdots + a_{nn} = w_1 + w_2 + \cdots + w_n$.
(2) $\det(\mathbf{A}) = \Pi_{i=1}^{n} w_i$, i.e., $\det(\mathbf{A}) = w_1 w_2 \cdots w_n$, where Π is the product operator.
(3) The w_i of a triangular matrix are equal to the elements on the principal diagonal.
(4) The w_i are real or in complex conjugate pairs for real $\mathbf{A}$.
(5) The w_i of a real, symmetric matrix are *real* and the $\mathbf{v}_i$ are *orthogonal*.
(6) A real, symmetric matrix with orthonormal eigenvectors may be expanded as

$$\mathbf{A} = \sum_{i=1}^{n} w_i \mathbf{v}_i \mathbf{v}_i^T, \tag{2.2.34}$$

the *spectral decomposition* of $\mathbf{A}$, and if $\mathbf{A}$ is nonsingular, then

$$\mathbf{A}^{-1} = \sum_{i=1}^{n} w_i^{-1} \mathbf{v}_i \mathbf{v}_i^T. \tag{2.2.35}$$

(7) Eigenvectors of an arbitrary matrix corresponding to distinct eigenvalues are linearly independent.
(8) The eigenvalues of $(\mathbf{A} - h\mathbf{I})$ are each greater than those of $\mathbf{A}$ exactly by the amount h, and the eigenvectors are not changed.

where the w_i may not be distinct, that is, the eigenvalues may represent multiple roots. Root-finding algorithms are readily available [Cuthbert (1983:40)], so one means for computing eigenvalues is to solve the characteristic equation. However, finding the coefficients of the characteristic equation involves roughly n^4 multiplications, and root finding is often an illconditioned process. Better methods are described, but the characteristic equation is of great value in establishing many important properties of the eigenproblem.

The most important properties of the eigenproblem in the context of optimization are listed in Table 2.2.3 for subsequent discussion. The first two properties are a consequence of (2.2.31) through (2.2.33) and the well-known algebra of polynomials. From property (2) it is seen that a singular matrix has at least one eigenvalue equal to zero. Property (3) follows from the fact that the determinant of a triangular matrix is the product of the elements on the principal diagonal; thus, (2.2.31) and (2.2.33) lead to property (3). This is easily confirmed for $n = 2$ when $a_{21} = 0$ and $a_{12} \neq 0$; see (2.2.31).

It is common practice to normalize eigenvectors to the two-norm so that they have unit length, although it is clear from (2.2.28) that their lengths are completely arbitrary. Properties (4) and (5) are very important, because of the predominance and tractability of the situation for real, symmetric matrices. The transpose of any matrix has the same eigenvalues as the matrix, since the determinant of a matrix is equal to the determinant of its transpose. Property (6) involves sums of outer products and is discussed in the following. Property (8) is evident by inspection of (2.2.31).

Finally, it is noted that for every eigenvalue w associated with an eigenvector $\mathbf{v}$, there is also an eigenvector, $\mathbf{p}$, associated with $\mathbf{A}^T$:

$$\mathbf{A}^T\mathbf{p} = w\mathbf{p} \qquad \text{or} \qquad \mathbf{p}^T\mathbf{A} = w\mathbf{p}^T. \tag{2.2.36}$$

The equation on the right-hand side was obtained by transposition and associates $\mathbf{p}$ as the *left eigenvector* of $\mathbf{A}$. It can be shown that if $\mathbf{p}_k$ and $\mathbf{v}_j$ are, respectively, left and right eigenvectors corresponding to distinct eigenvalues, then $\mathbf{p}_k$ and $\mathbf{v}_j$ are orthogonal. Since a symmetric matrix is its own transpose, its left and right eigenvectors coincide.

Bounds on the magnitudes of eigenvalues are available from *Gerschgorin's theorem*. Consider (2.2.28) under the assumption that the eigenvector $\mathbf{v}$ has been scaled by the infinity norm, where v_k is its element with maximum modulus. Then

$$\begin{bmatrix} a_{11} & a_{12} & \cdots & a_{1k} & \cdots & a_{1n} \\ a_{21} & a_{22} & \cdots & a_{2k} & \cdots & a_{2n} \\ & & \cdots & & & \\ a_{k1} & a_{k2} & \cdots & a_{kk} & \cdots & a_{kn} \\ & & \cdots & & & \\ a_{n1} & a_{n2} & \cdots & a_{nk} & \cdots & a_{nn} \end{bmatrix} \begin{bmatrix} v_1 \\ v_2 \\ \vdots \\ 1 \\ \vdots \\ v_n \end{bmatrix} = w \begin{bmatrix} v_1 \\ v_2 \\ \vdots \\ 1 \\ \vdots \\ v_n \end{bmatrix}. \tag{2.2.37}$$

The kth equation from (2.2.37) is

$$w - a_{kk} = \sum_{j \neq k} a_{kj} v_j. \tag{2.2.38}$$

However, the infinity normalization provides that $|v_j| \leq 1$, so the result is that

$$|w - a_{kk}| \leq \sum_{j \neq k} |a_{kj}|. \tag{2.2.39}$$

In general, the eigenvalue w may be complex, so (2.2.39) states that on an Argand diagram (real and imaginary axes) the eigenvalue must lie within a circle of center a_{kk} with a radius given by the summation in (2.2.39). The most general statement possible, according to Gerschgorin's theorem, is that every eigenvalue must lie within the union of n such disks constructed from the rows of **A** according to (2.2.39). Since the major emphasis is on symmetric matrices, observe that all of those eigenvalues must be real, so that the intercepts of the disks lie on the real axis. Jennings (1977:36) shows that bounds on eigenvalues for real, symmetric matrices are:

$$(a_{ii})_{\max} \leq w_1 \leq \left(a_{kk} + \sum_{j \neq k} |a_{jk}| \right)_{\max},$$

$$\left(a_{kk} - \sum_{j \neq k} |a_{jk}| \right)_{\min} \leq w_n \leq (a_{ii})_{\min}, \tag{2.2.40}$$

where w_n is the minimum eigenvalue and w_1 is the maximum eigenvalue. Program C2-3, SYMBNDS, can be merged with program C2-1, MATRIX, so that command number 13 computes the bounds of eigenvalues of symmetric matrices. The reader interested in a more complete discussion of Gerschgorin's theorem is referred to Jennings (1977:35)

Example 2.2.3. One way to obtain a symmetric matrix is to form the transposed product of another matrix. Let $\mathbf{Z} = \mathbf{T}^T \mathbf{T}$, where **T** was given in (2.1.8). Table 2.2.4 shows that result. While in the BASIC environment, LOAD"MATRIX and then MERGE"SYMBNDS. Run the composite program, create matrix **Z** in program matrix **D**, and execute command number 13.

Table 2.2.4. A 5 × 5 Symmetric Matrix: The Transposed Product of (2.1.8)

```
MATRIX D( 5 , 5 ) -
     50.00000    -18.00000     26.00000      3.00000    -14.00000
    -18.00000     92.00000    -25.00000     56.00000     17.00000
     26.00000    -25.00000     52.00000      1.00000     20.00000
      3.00000     56.00000      1.00000     70.00000     18.00000
    -14.00000     17.00000     20.00000     18.00000     31.00000
```

The results from equations (2.2.40) place the maximum eigenvalue between 50 and 208, and the minimum eigenvalue between -38 and 31. The actual eigenvalues are calculated in the following development; the maximum and minimum eigenvalues are 149.42 and 0.3719, respectively.

Consider a real, *symmetric* matrix $\mathbf{A}$ and define a matrix $\mathbf{V}$ whose columns are the normalized eigenvectors of $\mathbf{A}$:

$$\mathbf{V} = (\mathbf{v}_1 \ \mathbf{v}_2 \ \cdots \ \mathbf{v}_n). \tag{2.2.41}$$

Further define a diagonal matrix $\mathbf{W}$, whose elements are the corresponding eigenvalues of $\mathbf{A}$:

$$\mathbf{W} = \begin{bmatrix} w_1 & 0 & \cdots & 0 \\ 0 & w_2 & \cdots & 0 \\ 0 & & \cdots & 0 \\ 0 & \cdots & 0 & w_n \end{bmatrix} = \mathrm{diag}(w_1 \ w_2 \ \cdots \ w_n). \tag{2.2.42}$$

Then all n equations represented by the eigenproblem in (2.2.28) can be stated by

$$\mathbf{AV} = \mathbf{VW}. \tag{2.2.43}$$

Since $\mathbf{V}$ is orthonormal, $\mathbf{VV}^T = \mathbf{I}$ and thus (2.2.43) yields

$$\mathbf{A} = \mathbf{VWV}^T, \tag{2.2.44}$$

which is equivalent to the summation in (2.2.34). Another arrangement of (2.2.43) is

$$\mathbf{V}^T\mathbf{AV} = \mathbf{W}, \tag{2.2.45}$$

a special transformation. By making one orthogonal change of coordinates in the *domain* of this transformation $\mathbf{A}$ and another orthogonal change in the *range*, the representation $\mathbf{W}$ becomes diagonal.

If there exists a matrix $\mathbf{P}$ such that

$$\mathbf{P}^{-1}\mathbf{AP} = \mathbf{B}, \tag{2.2.46}$$

then $\mathbf{B}$ is said to be similar to $\mathbf{A}$, and $\mathbf{B}$ is the result of a *similarity transformation*. Whenever $\mathbf{A}$ is symmetric, the similarity transformation in (2.2.45) results in a *diagonalization* of $\mathbf{A}$, and the eigenvalues of $\mathbf{A}$ are the elements on the principal diagonal of $\mathbf{W}$. (Note that $\mathbf{V}^{-1} = \mathbf{V}^T$.)

A similar result can occur for any arbitrary real matrix. According to Ralston (1965:471), there always exists a particular similarity matrix $\mathbf{P}$ such that (2.2.46) produces a triangular matrix, say $\mathbf{R}$:

$$\mathbf{P}^{-1}\mathbf{AP} = \mathbf{R}. \tag{2.2.47}$$

Recall that the principal diagonal of the triangular matrix $\mathbf{R}$ will contain the eigenvalues of $\mathbf{R}$ (some may be complex), according to property (3) in Table 2.2.3. But similar matrices have the same characteristic equation and, therefore, *the same eigenvalues*. To prove that $\mathbf{A}$ and $\mathbf{R}$ in (2.2.47) have the same eigenvalues, note that $\det(\mathbf{P}^{-1})\det(\mathbf{P}) = \det(\mathbf{P}^{-1}\mathbf{P}) = \det(\mathbf{I}) = 1$, so that

$$\det(\mathbf{R} - w\mathbf{I}) = \det[\mathbf{P}^{-1}(\mathbf{A} - w\mathbf{I})\mathbf{P}]$$

$$= \det(\mathbf{P}^{-1})\det(\mathbf{A} - w\mathbf{I})\det(\mathbf{P}) \qquad (2.2.48)$$

$$= \det(\mathbf{A} - w\mathbf{I}) = 0.$$

It so happens that the Gram–Schmidt orthogonalization algorithm that produced (2.2.26) is the basis for finding eigenvalues from (2.2.47) in every case and from (2.2.45) when all eigenvalues of an arbitrary $\mathbf{A}$ happen to be real. Orthogonal matrices are special cases of similarity matrices, that is, $\mathbf{P}^{-1} = \mathbf{P}^T$. Consider the following infinite iteration based on Gram–Schmidt orthonormalization as described in Section 2.2.2:

$$\mathbf{A}^{(1)} = \mathbf{A}, \qquad \mathbf{A}^{(k)} = \mathbf{Q}^{(k)}\mathbf{U}^{(k)}, \qquad (2.2.49)$$

$$\mathbf{A}^{(k+1)} = \mathbf{U}^{(k)}\mathbf{Q}^{(k)}, \qquad k = 1, 2, \ldots. \qquad (2.2.50)$$

This procedure calls for an orthonormalization of $\mathbf{A}$ as before. The next step is to obtain a new matrix $\mathbf{A}$ as the product of the *reversed* factors, namely, $\mathbf{UQ}$. Then that new $\mathbf{A}$ is decomposed into two factors using the same Gram–Schmidt procedure, those factors are reversed and multiplied, and so on. Note that (2.2.49) can be solved for $\mathbf{U}^{(k)} = \mathbf{Q}^{(k)T}\mathbf{A}^{(k)}$. When that is substituted into (2.2.50), it is found that

$$\mathbf{A}^{(k+1)} = \mathbf{Q}^{(k)T}\mathbf{A}^{(k)}\mathbf{Q}^{(k)}, \qquad (2.2.51)$$

which is a similarity transformation in the form of (2.2.47). After k steps, the algorithm described by (2.2.49) and (2.2.50) produces

$$\mathbf{A}^{(k+1)} = \left(\mathbf{Q}^{(1)}\mathbf{Q}^{(2)} \cdots \mathbf{Q}^{(k)}\right)^T \mathbf{A}\left(\mathbf{Q}^{(1)}\mathbf{A}^{(2)} \cdots \mathbf{Q}^{(k)}\right). \qquad (2.2.52)$$

As was the case for the elementary transformations employed in Section 2.1.4 in the Gauss–Jordan matrix inversion algorithm, the sequence of orthogonal transformations, $\mathbf{Q}$, are not explicitly collected to form similarity matrix $\mathbf{P}$ in (2.2.47). This procedure is called the *QR algorithm* and was first described by Francis (1961). He noted that the algorithm can be expected to be numerically stable because it employs orthogonal transformations. He proved that the matrix $\mathbf{A}^{(k+1)}$ tends to an upper triangular (or diagonal) matrix in which the elements on the principal diagonal are the eigenvalues in order of modulus,

Table 2.2.5. The Diagonalized Matrix from Table 2.2.4 After 16 Iterations of the QR Algorithm

```
MATRIX A( 5 , 5 ) -
    149.41766      -0.01060      -0.00000       0.00001      -0.00000
     -0.01060      81.27141       0.00968      -0.00000       0.00021
      0.00000       0.00969      48.00166       0.00000       0.00061
      0.00000       0.00000      -0.00000      15.93808       0.00007
      0.00000      -0.00000       0.00000      -0.00000       0.37119
COMPLETED ITERATION #   16
```

the first ($a_{11} = w_1$) being the largest. Francis also proved that the rate of convergence of the element $a_{ij}^{(k)}$ below the principal diagonal of $\mathbf{A}^{(k)}$ is proportional to $(w_i/w_j)^k$, which indicates that the rate of convergence to triangular or diagonal form is slowest when some eigenvalues are closely spaced.

Program C2-4, QRITER, in Appendix C implements the QR algorithm in conjunction with program C2-2, GSDECOMP, the Gram–Schmidt decomposition. The composite program is obtained by issuing the following three commands in the BASIC environment: LOAD"MATRIX, MERGE"GSDECOMP, and MERGE"QRITER. The Gram–Schmidt orthogonalization procedure appears as command 10. It is called automatically by the QR algorithm, which also *operates on matrix* **A** and is initiated by command 13. Because command 13 asks the user if the algorithm should continue after each iteration, the user may find that the IBM-PC keyboard buffer is useful in that it will remember as many as 15 ⟨RETURN⟩ keystrokes to keep the iterations going.

Example 2.2.4. Consider the symmetric matrix **Z** shown in Table 2.2.4. It was generated by the transposed product of the 5×5 matrix **T** in (2.1.8). The diagonal matrix in Table 2.2.5 was obtained after 16 iterations of the QR algorithm. Nearly all the nondiagonal elements have converged to zero, and the elements on the principal diagonal are the eigenvalues of the original matrix, ordered according to modulus. According to (2.1.48), the *spectral norm* of **T** is the square root of the largest eigenvalue of $\mathbf{T}^T\mathbf{T}$, namely, 149.42. Thus, $\|\mathbf{T}\|_2 = 12.22$, which is the same order of magnitude as the other matrix norms of **T** shown in Table 2.1.8.

Example 2.2.5. Consider the 3×3 nonsymmetric matrix shown in Table 2.2.6 The result obtained after seven iterations of the QR algorithm are also shown there. Observe that nonsymmetric matrices with all *real eigenvalues* are triangularized by the QR algorithm. In this special case the three real eigenvalues are shown on the principal diagonal (846.50, 20.46, and 3.03). Notice that properties (1) and (2) in Table 2.2.3 verify the accuracy of this decomposition (the determinant being available from command 9).

Example 2.2.6. Now copy matrix **T** from (2.1.8) into program matrix **A** and execute command 13 for the QR algorithm. The resulting matrix shown in

Table 2.2.6.   A 3 × 3 Nonsymmetric Real Matrix Having Real Eigenvalues Before and After Seven Iterations of the QR Algorithm

```
MATRIX A( 3 , 3 ) -
      530.00000       550.00000       150.00000
      275.00000       300.00000       100.00000
       80.00000        90.00000        40.00000
*****************************************************
MATRIX A( 3 , 3 ) -
      846.50580       266.74814       -96.24244
        0.00000        20.46343        14.29877
        0.00000         0.00003         3.03078
COMPLETED ITERATION #  7
```

Table 2.2.7.   Matrix from (2.1.8) after 20 Iterations of the QR Algorithm

```
MATRIX A( 5 , 5 ) -
    7.48306       0.91139      -2.25749      -1.79853      -0.99830
    0.50558      -3.22727      -6.58359      -2.15423       1.97081
   -0.00192       4.67517      -3.08647       7.27831       5.03873
    0.00000      -0.00004      -0.00005       3.56256       6.39289
   -0.00000       0.00000       0.00000      -0.00000      -1.73188
COMPLETED ITERATION #  20
```

Table 2.2.7 is almost upper triangular. Evidently, there is one pair of conjugate-complex eigenvalues as indicated by the *principal* 2×2 *submatrix* composed of a_{22}, a_{23}, a_{32}, and a_{33}. As Francis (1961) explained, the most likely occurrence of eigenvalues of equal modulus would be those in conjugate-complex pairs: "Then the nearly-triangular matrix **A** becomes split into independent principal submatrices coupled only in so far as the eigenvectors are concerned." The extraction of complex pairs of eigenvalues is beyond the scope of this treatment; the interested reader is referred to Francis (1962), Morris (1983:359), and Golub (1983).

Several important improvements can be made to make the QR algorithm more efficient and accurate. The major inefficiency in the QR algorithm implemented here is that the Gram–Schmidt decomposition procedure requires on the order of n^3 multiplications; Householder elementary transformations that will triangularize a matrix using about n^2 multiplications are described in the next section. There is a way to accelerate convergence of the a_{nn} element; it will usually converge first because in general the rate of convergence of subdiagonal elements to zero is governed by the ratio of adjacent eigenvalues, especially $a_{n,\,n-1} \to 0$ at a rate proportional to (w_{n-1}/w_n). According to property (8) in Table 2.2.3, the a_{nn} element in Table 2.2.7 can be made quite small by adding 1.73**I** to **A** before employing the QR algorithm. The before and after results are shown in Table 2.2.8. Notice that the last row converged in only three iterations. The conventional procedure is to (1) estimate w_n at each iteration and shift **A** accordingly, (2) *deflate* the matrix when a_{nn} has converged by eliminating the last row and column, and (3)

Table 2.2.8. **Matrix from (2.1.8) shifted by 1.7I for Rapid Convergence to the Eigenvalue of Smallest Modulus**

```
READ FILE T.MAT INTO MATRIX B
   C=I
   5X5 UNIT MATRIX
   1.7*C
   A=B+C
MATRIX A( 5 , 5 ) -
     3.70000       3.00000       5.00000       6.00000       4.00000
    -1.00000       2.70000      -1.00000       1.00000       1.00000
    -2.00000       3.00000      -2.30000       5.00000      -1.00000
     5.00000       3.00000      -1.00000       3.70000      -3.00000
    -4.00000       8.00000      -3.00000       2.00000       3.70000
 ***********************************************
MATRIX A( 5 , 5 ) -
     9.26277       0.53595       2.38470      -3.07325       1.94178
     1.86793      -1.46890      -3.19527      -3.66087      -0.52142
     0.97351       3.07693      -1.57100       4.58315       2.08605
     2.07081      -3.06019      -6.50438       5.30901      -7.92110
    -0.00000       0.00000       0.00000      -0.00000      -0.03188
COMPLETED ITERATION # 3
```

recommence QR iterations for the reduced $(n - 1, n - 1)$ matrix. Interested readers will find details in Morris (1983:353).

Finally, it is noted that any real matrix can be transformed into a Hessenberg, tridiagonal, or bidiagonal form using noniterative processes that preserve the eigenvalues. An *upper Hessenberg matrix* is upper triangular with one more nonzero subdiagonal adjacent to the principal diagonal (i.e., $a_{ij} = 0$ for $i > j + 1$). The reason for doing so is to apply the *iterative* QR algorithm on a matrix containing many zeros and known to converge much more rapidly than from a general matrix. See Acton (1970:317).

The remaining problem is finding the eigenvectors that correspond to the approximate eigenvalues that are now known. From (2.2.28), one approach is to solve that set of n equations for the unknown v_i. However, that system is actually of rank $n - 1$, since the eigenvector can be determined only to within an arbitrary factor as seen from (2.2.28). The *power method* successfully determines the eigenvalue of largest modulus and its corresponding eigenvector. Since it leads to a general scheme for finding all eigenvectors, it is discussed first. It assumes that the eigenvalues are ordered according to modulus, so that $|w_1|$ is the largest and $|w_n|$ is the smallest. Property (7) in Table 2.2.3 states that an arbitrary vector may be expressed as a linear combination of distinct eigenvectors of a matrix. Thus, suppose that

$$\mathbf{c}^{(0)} = p_1 \mathbf{v}_1 + p_2 \mathbf{v}_2 + \cdots + p_n \mathbf{v}_n, \qquad (2.2.53)$$

where the $\mathbf{v}_i$ are eigenvectors of matrix $\mathbf{B}$. Now premultiply $\mathbf{c}^{(0)}$ by $\mathbf{B}$; observing (2.2.28), the result can be expressed as

$$\mathbf{c}^{(1)} = \mathbf{B}\mathbf{c}^{(0)} = \sum_i p_i \mathbf{B}\mathbf{v}_i = \sum_i w_i p_i \mathbf{v}_i. \qquad (2.2.54)$$

Continuing to premultiply these results by **B**, after k times, the vector obtained is

$$\mathbf{c}^{(k)} = \mathbf{B}^k \mathbf{c}^{(0)} = w_1^k p_1 \mathbf{v}_1 + w_2^k p_2 \mathbf{v}_2 + \cdots + w_n^k p_n \mathbf{v}_n. \qquad (2.2.55)$$

Since the eigenvalues are distinct, $|w_1| > |w_2|$ and for $p_1 \neq 0$ and k sufficiently large, a limiting relationship is that

$$\mathbf{c}^{(k)} \doteq w_1^k p_1 \mathbf{v}_1. \qquad (2.2.56)$$

Therefore, $\mathbf{c}^{(k)}$ approaches being proportional to the *dominant eigenvector*, $\mathbf{v}_1$.

Example 2.2.7. Program C2-5, SHINVP, finds any eigenvector, but it is easily modified to accomplish the power method for finding the dominant eigenvector. LOAD"MATRIX, then MERGE"SHINVP. Then change the following two lines:

$$7000 \quad M9 = K9 : N9 = 1 : GOTO\ 7120$$
$$7380 \quad RETURN$$

This program assumes an initial vector $\mathbf{c}^{(0)}$ composed of all 1's; see (2.2.55). There are cases where coefficient p_1 is so small that convergence is to the second eigenvector instead of the first. A common implementation employs randomly selected normalized starting vectors. The infinity norm is used to scale each new estimated eigenvalue. Running this modified program with the top matrix in Table 2.2.6 *placed in program matrix* **B** results in the following calculations:

$$\begin{bmatrix} 530 & 550 & 150 \\ 275 & 300 & 100 \\ 80 & 90 & 40 \end{bmatrix} \begin{bmatrix} 1 \\ 1 \\ 1 \end{bmatrix} = \overset{c^{(0)}}{1230.0} \begin{bmatrix} 1 \\ .54878 \\ .17073 \end{bmatrix},$$

$$\begin{bmatrix} 530 & 550 & 150 \\ 275 & 300 & 100 \\ 80 & 90 & 40 \end{bmatrix} \begin{bmatrix} 1 \\ .54878 \\ .17073 \end{bmatrix} = \overset{c^{(1)}}{857.43} \begin{bmatrix} 1 \\ .53264 \\ .15887 \end{bmatrix},$$

$$\begin{bmatrix} 530 & 550 & 150 \\ 275 & 300 & 100 \\ 80 & 90 & 40 \end{bmatrix} \begin{bmatrix} 1 \\ .53264 \\ .15887 \end{bmatrix} = \overset{c^{(2)}}{846.78} \begin{bmatrix} 1 \\ .53223 \\ .15859 \end{bmatrix},$$

$$\begin{bmatrix} 530 & 550 & 150 \\ 275 & 300 & 100 \\ 80 & 90 & 40 \end{bmatrix} \begin{bmatrix} 1 \\ .53223 \\ .15859 \end{bmatrix} = \overset{c^{(3)}}{846.51} \begin{bmatrix} 1 \\ .53222 \\ .15858 \end{bmatrix}.$$

Therefore, the eigenvalue is 846.51, which agrees with Table 2.2.6, and the corresponding eigenvector is $(1 \quad 0.53222 \quad 0.15858)^T$. The convergence criterion in (1.3.22) was applied to the estimated eigenvalue (reported as the "trial factor" by the program) with a tolerance value of 0.0001.

The eigenvalue having the *least modulus* can be found by using $\mathbf{B}^{-1}$ in place of $\mathbf{B}$ and inverting the resulting eigenvalue (the eigenvector is the same). In other words, if $(w, \mathbf{v})$ is the dominant eigenpair for $\mathbf{B}$, then $(1/w, \mathbf{v})$ is the dominant eigenpair for $\mathbf{B}^{-1}$. This is called the *inverse power method*. The reader is urged to confirm this result for the matrix in Example 2.2.7. This and the shifting property (8) in Table 2.2.3 are the basis of the universally accepted method for finding eigenpairs: the *shifted inverse power method*.

Note that (2.2.28) can be modified to be

$$(\mathbf{B} - h\mathbf{I})\mathbf{v} = (w - h)\mathbf{v}. \tag{2.2.57}$$

Suppose that w_j is the eigenvalue nearest h. Then the least dominant eigenpair for $(\mathbf{B} - h\mathbf{I})$ is $(w_j - h, \mathbf{v}_j)$, and the dominant eigenpair for $(\mathbf{B} - h\mathbf{I})^{-1}$ is $[(w_j - h)^{-1}, \mathbf{v}_j]$. Therefore, given an estimated eigenvalue h, the eigenvalue nearest h is

$$w_j = \left[\text{dominant eigenvalue of } (\mathbf{B} - h\mathbf{I})^{-1}\right]^{-1} + h. \tag{2.2.58}$$

Program C2-5, SHINVP (unmodified), performs the calculations just described for the shifted inverse power method, including renormalization of the eigenvector to the two-norm. As before, the matrix (uninverted) is entered into program matrix $\mathbf{B}$; then command 13 requests an estimate for the eigenvalue and performs all other calculations. These unit-length eigenvectors are contained in matrix (vector) $\mathbf{A}$ and can be stored as obtained, one at a time. Then program C2-6, VECTOCOL, can be merged as a new command 13 to store each of the eigenvectors in the appropriate column of program matrix $\mathbf{D}$. *Caution*: Since $\mathbf{B}$ is inverted each time, the matrix must be recalled into $\mathbf{B}$ for each eigenvalue–eigenvector calculation.

Example 2.2.8. The 5×5 symmetric matrix in Table 2.2.4 has the approximate eigenvalues shown in Table 2.2.5. These can be used one at a time in program C2-5, SHINVP, to find their respective eigenvectors. These were placed into columns of program matrix $\mathbf{D}$ by program C2-6, VECTOCOL, as shown in Table 2.2.9. Use of MATRIX command 8 to form the transpose product $\mathbf{D}^T\mathbf{D} = \mathbf{I}$ verifies that $\mathbf{D}$ is in fact an orthonormal matrix.

Example 2.2.9. Program SHINVP was employed to characterize the 3×3 nonsymmetric matrix in the top of Table 2.2.6, using the eigenvalues in the bottom of Table 2.2.6. The resulting columns shown in Table 2.2.10 were stored using program VECTOCOL. In this case it is easily verified that the resulting matrix of eigenvectors is *not* orthogonal.

As Jacobs (1977) noted, "In the last decade inverse iteration has established

Table 2.2.9. An Orthonormal Matrix of Eigenvectors Characterizing the Symmetric Matrix in Table 2.2.4

MATRIX D(5 , 5) -

-0.19844	0.52573	0.65297	-0.26471	0.43333
0.75412	-0.04281	0.11400	-0.62599	-0.15690
-0.20313	0.69700	-0.33871	-0.21396	-0.55895
0.56318	0.40860	0.16550	0.69617	-0.06193
0.18302	0.26271	-0.64693	-0.08738	0.68654

Table 2.2.10. A Matrix of Eigenvectors Characterizing the Nonsymmetric Matrix at the Top of Table 2.2.6

MATRIX D(3 , 3) -

0.87424	0.70600	-0.63882
0.46528	-0.52418	0.69946
0.13864	-0.47623	-0.32041

itself as the standard algorithm for computing eigenvectors corresponding to specific eigenvalues." Finally, the *generalized eigenproblem* is mentioned for completeness. It is

$$\mathbf{Av} = w\mathbf{Bv}, \tag{2.2.59}$$

or

$$(\mathbf{A} - w\mathbf{B})\mathbf{v} = \mathbf{0}, \tag{2.2.60}$$

analogous to (2.2.29). The $\mathbf{v}$ is referred to as the eigenvector of $\mathbf{A} - w\mathbf{B}$. There are specialized algorithms for solution of the generalized eigenproblem, depending on the nature of matrix $\mathbf{B}$. Highly reliable FORTRAN routines are available for solutions to the eigenproblem, especially the LINPACK series (Dongarra, 1979), and the EISPACK series (Garbow, 1977; Smith, 1976).

2.2.4. Special Matrix Transformations. Several other important orthogonal matrices are introduced in this section. An orthonormal matrix is called *proper* if $\det(\mathbf{A}) = 1$ and *improper* if $\det(\mathbf{A}) = -1$. Consider the *plane rotation matrix* in E^2:

$$\mathbf{Q} = \begin{bmatrix} \cos\theta & -\sin\theta \\ \sin\theta & \cos\theta \end{bmatrix}. \tag{2.2.61}$$

The transformation $\mathbf{y} = \mathbf{Qx}$ using (2.2.61) represents two equations in two independent variables, x_1 and x_2, and two dependent variables, y_1 and y_2. It is not difficult to verify that this is the rotation of axes shown in Figure 2.2.4. It is also easy to verify that $\mathbf{Q}$ in (2.2.61) is orthonormal, that is, $\mathbf{Q}^T\mathbf{Q} = \mathbf{I}$. Any proper 2×2 or 3×3 matrix represents a rotation in Euclidean space; see Noble (1969:421). In general, orthonormal matrices employed in transformations like $\mathbf{y} = \mathbf{Qx}$ leave angles and lengths unchanged between ranges and

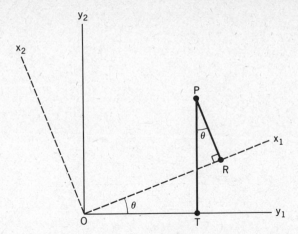

Figure 2.2.4. A rotation of axes according to $\mathbf{y} = \mathbf{Qx}$, where $\mathbf{Q}$ is an orthogonal rotation matrix in E^2.

domains. It is especially easy to see that lengths are preserved by orthonormal transformations,

$$\|\mathbf{Qx}\|_2 = \|\mathbf{x}\|_2, \tag{2.2.62}$$

because $\|\mathbf{Q}x\|_2^2 = (\mathbf{Qx})^T\mathbf{Qx} = \mathbf{x}^T\mathbf{x} = \|\mathbf{x}\|_2^2$.

It was remarked in the last section that there were means to transform general real matrices to an upper Hessenberg form and symmetric real matrices to a tridiagonal form. A bidiagonal form is also required in Section 3.1.2. These forms are illustrated in Figure 2.2.5. The process of arranging for zeros in matrices is called *element annihilation*. A very efficient way to perform those transformations is based on the *Householder transformation*:

$$\mathbf{Q} = \mathbf{I} - 2\mathbf{uu}^T, \qquad \|\mathbf{u}\|_2 = 1. \tag{2.2.63}$$

The Householder transformation of $\mathbf{x}$ into $\mathbf{y}$ is

$$\mathbf{y} = \mathbf{Qx} = \mathbf{x} - \mathbf{u}(2\mathbf{u}^T\mathbf{x}). \tag{2.2.64}$$

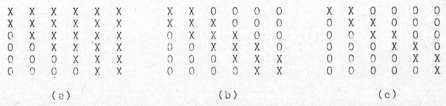

Figure 2.2.5. Three matrix forms that can be obtained by Householder similarity transformations: (a) Upper Hessenberg, corresponding to any real matrix; (b) tridiagonal, corresponding to any real, symmetric matrix; and (c) bidiagonal, for use in singular value decomposition.

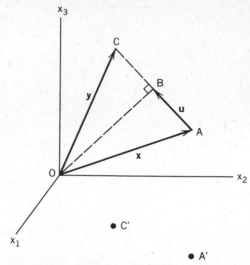

Figure 2.2.6. A geometric interpretation of the Householder transformation. B is the midpoint of AC, where OB is perpendicular. Therefore, $\mathbf{u}$ is the projection of $-\mathbf{x}$ on AC.

This shows that (1) $\mathbf{u}$ is in the $(\mathbf{x} - \mathbf{y})$ direction, (2) $y_i = x_i$ if and only if $u_i = 0$, and (3) $\mathbf{y} = \mathbf{x}$ only if $\mathbf{u}^T\mathbf{x} = 0$, the orthogonality condition discussed previously. In fact, $\mathbf{u}^T\mathbf{x}$ is the *projection* of $\mathbf{x}$ on $\mathbf{u}$ discussed in connection with the Gram–Schmidt procedure and illustrated in Figure 2.2.3. A similar geo-metrical interpretation for the Householder transformation is shown in Figure 2.2.6.

It is easily verified that the Householder matrix in (2.2.63) is symmetric $(\mathbf{Q}^T = \mathbf{Q})$ and orthonormal $(\mathbf{Q}^T\mathbf{Q} = \mathbf{I})$, and thus preserves Euclidean lengths according to (2.2.62). It will now be shown that the vector $\mathbf{u}$ can be chosen so that use of the Householder transformation in a similarity transformation, as in (2.2.46), can reduce general and symmetric matrices to Hessenberg and tridiagonal forms, respectively. This is accomplished in a finite number of steps, and eigenvalues are not affected. The principal application of that result is to simplify and accelerate solutions of the eigenproblem, including those described in the last section.

Consider the similarity transformation according to (2.2.46), using the Householder matrix:

$$\mathbf{B} = \mathbf{QAQ} = \mathbf{A}(\mathbf{I} - 2\mathbf{uu}^T) + 2\mathbf{uu}^T(2c\mathbf{I} - \mathbf{A}), \qquad (2.2.65)$$

where c is a scalar equal to the *quadratic form*

$$c = \mathbf{u}^T\mathbf{Au}. \qquad (2.2.66)$$

Carrying out the substitution of (2.2.63) into (2.2.65) will verify the latter. The fact that c is a scalar can be seen from the dimensions of its components; its actual value is not used in this analysis. As encountered earlier in (2.1.22), the outer product is

$$\mathbf{u}\mathbf{u}^T = \begin{bmatrix} u_1^2 & u_1 u_2 & \cdots & u_1 u_n \\ u_2 u_1 & u_2^2 & \cdots & u_2 u_n \\ & & \cdots & \\ u_n u_1 & u_n u_2 & \cdots & u_n^2 \end{bmatrix}. \tag{2.2.67}$$

Choose the vector $\mathbf{u}$ to have $u_1 = 0$ and unit length; then

$$\mathbf{u} = \begin{pmatrix} 0 & u_2 & u_3 & \cdots & u_n \end{pmatrix}^T, \qquad \|\mathbf{u}\|_2 = 1. \tag{2.2.68}$$

Also, the two parenthetical expressions in (2.2.65) are

$$(\mathbf{I} - 2\mathbf{u}\mathbf{u}^T) = \begin{bmatrix} 1 & 0 & \cdots & 0 \\ 0 & 1 - 2u_2^2 & \cdots & -2u_2 u_n \\ & & \cdots & \\ 0 & -2u_n u_2 & \cdots & 1 - 2u_n^2 \end{bmatrix}, \tag{2.2.69}$$

$$(2c\mathbf{I} - \mathbf{A}) = \begin{bmatrix} 2c - a_{11} & -a_{12} & \cdots & -a_{1n} \\ -a_{21} & 2c - a_{22} & \cdots & -a_{2n} \\ & & \cdots & \\ -a_{n1} & -a_{n2} & \cdots & 2c - a_{nn} \end{bmatrix}. \tag{2.2.70}$$

It is now easier to find expressions for the elements in the first column of $\mathbf{B}$ as defined by (2.2.65). Recall that the elements in the result of a matrix product are located by the row of the multiplier and the column of the multiplicand. For convenience, define a term

$$h = \sum_i a_{i1} u_i = a_{21} u_2 + a_{31} u_3 + \cdots + a_{n1} u_n. \tag{2.2.71}$$

Then the first column of $\mathbf{B} = [b_{ij}]$ is $b_{11} = a_{11}$, and

$$b_{21} = a_{21} - 2u_2 h = r$$

$$b_{31} = a_{31} - 2u_3 h = 0$$

$$\cdots$$

$$b_{i1} = a_{i1} - 2u_i h = 0 \tag{2.2.72}$$

$$\cdots$$

$$b_{n1} = a_{n1} - 2u_n h = 0.$$

If **B** is to be in upper Hessenberg form, then b_{11} and b_{21} are not zero; they are equal to a_{11} and some constant, r, respectively. Also, b_{31} through b_{n1} are to be zero. Two different operations on the set of equations in (2.2.72) will yield values for r and h, and thus for u_i. First, square each equation and then add them all together; the result is

$$\sum_i a_{i1}^2 - 4h\sum_i a_{i1}u_i + 4h^2\sum_i u_i^2 = r^2. \qquad (2.2.73)$$

But the second summation is equal to h according to (2.2.71), and the third summation is equal to unity because **u** has unit length. Thus an expression for r^2 is

$$r^2 = \sum_i a_{i1}^2 \qquad (2.2.74)$$

The second operation on the equations in (2.2.72) is to multiply each one by a_{i1} and then add them all together; the result is

$$\sum_i a_{i1}^2 - 2h\sum_i a_{i1}u_i = a_{21}r. \qquad (2.2.75)$$

But the first summation can be replaced by (2.2.74) and the second summation is equal to h. Therefore,

$$2h^2 = r^2 - a_{21}r. \qquad (2.2.76)$$

Since r is determined with an arbitrary sign by (2.2.74), its sign is chosen as opposite that of a_{21} so that there is no cancellation in (2.2.76). Then, referring to (2.2.72), the vector **u** can be written in terms of **A** using parameters r and h:

$$\mathbf{u}' = \begin{bmatrix} 0 & (a_{21} - r) & a_{31} & a_{41} & \cdots & a_{n1} \end{bmatrix}^T, \qquad (2.2.77)$$

and the outer product $\mathbf{u}\mathbf{u}^T$ must be divided by $2h^2$ so that **u** has unit length, that is, $\mathbf{u} = \mathbf{u}'/h/(2)^{1/2}$.

So much for the first column of **B**, a matrix in Hessenberg form. A little thought will show that making u_1 *and* u_2 zero in (2.2.68) will lead to a second similarity transformation that will annihilate the appropriate elements in the *second* column of **B**. A Hessenberg matrix as in Figure 2.2.5 has a lower triangular structure with exactly $n - 2$ columns defined equal to zero. Therefore, any real matrix may be reduced to upper Hessenberg form by the sequence of $n - 2$ similarity transformations starting with $\mathbf{A}^{(1)} = \mathbf{A}$, and

$$\mathbf{A}^{(n-1)} = \mathbf{Q}^{(n-2)} \cdots \mathbf{Q}^{(2)}\mathbf{Q}^{(1)}\mathbf{A}^{(1)}\mathbf{Q}^{(1)}\mathbf{Q}^{(2)} \cdots \mathbf{Q}^{(n-2)}, \qquad (2.2.78)$$

where **Q** is defined by (2.2.63), and

$$r = -(\text{sign } a_{j+1,j})(a_{j+1,j}^2 + a_{j+2,j}^2 + \cdots + a_{n,j}^2)^{1/2}, \quad (2.2.79)$$

$$2h^2 = r^2 - ra_{j+1,j}, \quad (2.2.80)$$

$$\mathbf{u}^{(j)} = \frac{[0 \cdots 0 \,(a_{j+1,j} - r)\, a_{j+2,j} \cdots a_{n,j}]^T}{h(2)^{1/2}}, \quad (2.2.81)$$

for $j = 1, 2, \ldots, n - 2$. The first j elements in the vector $\mathbf{u}^{(j)}$ are zero.

These equations have been programmed as MATRIX command 13; the additional instructions are contained in program HOUSE, C2-7 in Appendix C. This calculation may be run by invoking the BASIC environment, and then executing LOAD"MATRIX and then MERGE"HOUSE. Command 13 shows the effects of this particular Householder transformation on each successive column of the matrix placed in **A** and returns to the menu when the Hessenberg form has been obtained.

Example 2.2.10. The transformations to upper Hessenberg and tridiagonal forms were obtained for the 5×5 nonsymmetric matrix in (2.1.8) and its symmetric transposed product, namely, $\mathbf{Z} = \mathbf{T}^T\mathbf{T}$. The final results are shown in Tables 2.2.11 and 2.2.12, respectively.

Some reasons for wanting to obtain these forms have been given in connection with an increased rate of convergence of the QR process for finding eigenvalues, especially when matrix deflation is employed. Some other methods for finding the eigenvalues and eigenvectors of matrices with these

Table 2.2.11.   The Real Nonsymmetric Matrix in (2.1.8) After Transformation to Upper Hessenberg Form

```
MATRIX A( 5 , 5 ) -
    2.00000      0.14744     -7.59203     -5.00257     -1.82032
    6.78233      0.76087     -1.04183      4.48833      2.92576
   -0.00000     -5.85025     -1.27637      3.44000      5.25292
   -0.00000      0.00000     -0.51077      2.47215      6.76662
    0.00000      0.00000      0.00000      0.52385     -0.95665
```

Table 2.2.12.   The Real Symmetric Matrix Transposed Product of (2.1.8) After Transformation to Tridiagonal Form

```
MATRIX A( 5 , 5 ) -
   50.00000     34.71311     -0.00000     -0.00000      0.00000
   34.71311     67.77595     53.98535      0.00000     -0.00000
   -0.00000     53.98535     92.06268     35.62140     -0.00000
   -0.00000     -0.00000     35.62140     65.35001    -15.49466
    0.00000     -0.00000      0.00000    -15.49466     19.81138
```

special forms have not been discussed. The reader is referred to Wilkinson (1963), Ralston (1965), and Acton (1970). The reader is cautioned that program C2-7 has been deliberately left in a very inefficient state, since clarity is much more important for these purposes. However, Jennings (1977) and many others have shown that the full Householder matrix need not be constructed at all, thus conserving memory space and avoiding spurious multiplications by zero. When optimally programmed, the Householder method requires as many as $\frac{4}{3}n^3$ multiplications. As Acton (1970:329) noted, "We have here one of many examples where matrix notation, though very useful for derivations and proof, can lead the unwary computor astray."

The remainder of this section describes projection matrices. In order to treat that subject, it is necessary to formalize the concept of the hyperplane. A *hyperplane H* is the subspace described by vectors **x** in E^n that satisfy

$$\mathbf{n}^T\mathbf{x} = h, \tag{2.2.82}$$

where h is a scalar. The nonzero vector **n** is called the *normal* to the hyperplane, which is especially apparent when $h = 0$. In E^2, consider the case when $\mathbf{n} = (\frac{1}{2}\ 1)^T$ and $h = \frac{4}{5}$. Then (2.2.82) yields

$$\tfrac{1}{2}x_1 + 1x_2 = \tfrac{4}{5}. \tag{2.2.83}$$

This particular normal vector **n** and the hyperplane H are shown in Figure 2.2.7. Clearly, if $h = 0$, the hyperplane passes through the origin. Apparently, the scalar h causes a shift from point O to point O'. This particular hyperplane is a line having the coordinate y_1. Note that the **x** vectors that describe the line are not in the line, only their tips. Similarly, Figure 2.2.8 shows a two-dimensional hyperplane in a three-dimensional space. Any vector $\mathbf{y} = c_1\mathbf{y}_1 + c_2\mathbf{y}_2$ lies in the hyperplane.

When $h = 0$, the hyperplane passes through the origin in E^n. Then $\mathbf{n}^T\mathbf{y} = 0$ forms a vector subspace of dimension $n - 1$ which has orthonormal basis

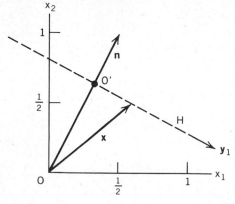

Figure 2.2.7. A hyperplane H in E^2. The normal vector is $\mathbf{n} = (\frac{1}{2}\ 1)^T$, and the hyperplane is one-dimensional (a line).

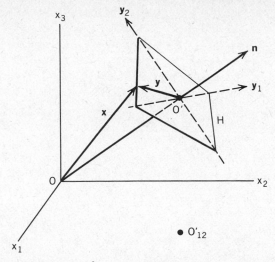

Figure 2.2.8. A hyperplane H in E^3. The hyperplane is two-dimensional and has coordinate vectors y_1 and y_2.

vectors $y_1, y_2, \ldots, y_{n-1}$. But normal vector n is linearly independent of these, so the $n - 1$ basis vectors in the hyperplane and the normal vector n form a basis for E^n. Therefore, every vector x in E^n that satisfies (2.2.82) can be written as the linear combination

$$x = c_1 y_1 + c_2 y_2 + \cdots + c_{n-1} y_{n-1} + c_n n. \tag{2.2.84}$$

Premultiplying (2.2.84) by n^T and substituting (2.2.82) yields

$$n^T x = c_n n^T n = h, \tag{2.2.85}$$

so that the last coefficient can be found equal to

$$c_n = \frac{h}{n^T n} = \frac{h}{\|n\|_2^2}. \tag{2.2.86}$$

But the squared length of any vector x from the origin to a point in the hyperplane is the sum of squared coefficients related to the basis vectors in (2.2.84):

$$x^T x = c_1^2 + c_2^2 + \cdots + c_{n-1}^2 + c_n^2 (n^T n). \tag{2.2.87}$$

Note that the last term accounts for the fact that n is not necessarily of unit length as are the y_i vectors in H. From (2.2.87) and the Pythagorean theorem, the minimum x occurs where x is collinear with n, that is, along O–O' in Figure 2.2.8. Therefore, the hyperplane is displaced $|h|/\|n\|_2$ from the origin in E^n, that is, from point O. This displacement is $\left(\frac{4}{5}\right)^{3/2}$ in Figure 2.2.7.

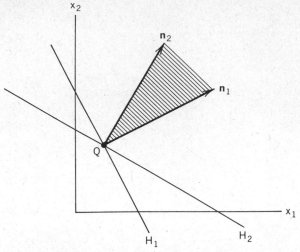

Figure 2.2.9. A polyhedral cone having a vertex at the intersection of two hyperplanes, H_1 and H_2. The cone is included in the half-spaces of the two hyperplanes and is generated by their respective normal vectors, $\mathbf{n}_1$ and $\mathbf{n}_2$. In this case the manifold Q is simply a point in E^2.

A *half-space* is the set of vectors lying on one side of a hyperplane, usually taken to be that vector space in the direction of the related normal vector. A *polyhedron* is the intersection of a finite number of half-spaces. If there are a number of hyperplanes, H_i, associated with normal vectors $\mathbf{n}_i$, then the hyperplanes are said to be linearly independent if the $\mathbf{n}_i$ are linearly independent. The intersection of q hyperplanes is called a *manifold*, Q. A manifold is sometimes called an *affine subspace* as opposed to a *linear subspace*, the technical distinction being that the latter passes through the origin.

The intersection of any two linearly independent hyperplanes is an $(n - 2)$-dimensional manifold in E^n. In the same way, the intersection of any $(n - 1)$ such hyperplanes determines a line, and n such intersecting hyperplanes determine a point in E^n, called the *vertex* of a polyhedron. A *polyhedral cone* is "generated" by two or more normal vectors at a vertex; see Figure 2.2.9.

The following analysis is simplified by two assumptions: (1) all normal vectors $\mathbf{n}_i$ that generate hyperplanes H_i have unit length, and (2) all hyperplanes contain the origin, that is, $h = 0$ in (2.2.82). After Rosen (1960), use q of these linearly independent normals to define the $\mathbf{N}_{m,q}$ matrix, $m \geq q$:

$$\mathbf{N} = (\mathbf{n}_1 \ \mathbf{n}_2 \ \cdots \ \mathbf{n}_j \ \cdots \ \mathbf{n}_q), \qquad (2.2.88)$$

which has rank q. Therefore, $\mathbf{N}^T\mathbf{N}$ is nonsingular and its inverse exists. The concepts of Figure 2.2.9 in E^2 can be extended to the E^3 space in Figure 2.2.8 by constructing two intersecting hyperplanes as illustrated in Figure 2.2.10. Q is the subspace or affine manifold representing the intersection of the hyperplanes; Q is generally of dimension $(m \times q)$ when there are q intersecting

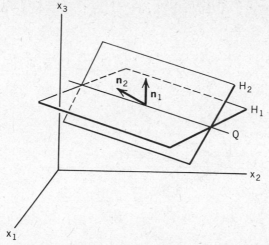

Figure 2.2.10. Two intersecting hyperplanes, H_1 and H_2, generated in E^3 by their respective normal vectors, $\mathbf{n}_1$ and $\mathbf{n}_2$. Q is the $(m - 2)$-dimensional manifold representing their intersection. A q-dimensional subspace is that spanned by the two normal vectors; it is called $\tilde{Q}$.

hyperplanes. If $\tilde{Q}$ is the subspace spanned (generated) by $\mathbf{n}_1$ and $\mathbf{n}_2$, then clearly E^m consists entirely of Q and $\tilde{Q}$ (their *direct sum*). Also, any vector $\mathbf{r}$ in Q is perpendicular to any vector $\mathbf{d}$ in $\tilde{Q}$, that is, $\mathbf{r}^T\mathbf{d} = 0$, so that Q and $\tilde{Q}$ are *orthogonal complements* of each other.

Now define a matrix

$$\tilde{\mathbf{P}} = \mathbf{N}(\mathbf{N}^T\mathbf{N})^{-1}\mathbf{N}^T. \tag{2.2.89}$$

The matrix $\tilde{\mathbf{P}}$ is a *projection matrix* that takes any vector in E^m into subspace $\tilde{Q}$. As a consequence of the preceding direct-sum statement, proof of projections by (2.2.89) may be satisfied by showing that $\tilde{\mathbf{P}}$ takes any vector in Q into the zero vector and any vector in $\tilde{Q}$ into itself. Again, if any vector $\mathbf{r}$ is in Q, then $\mathbf{n}_j^T\mathbf{r} = 0$, for $j = 1$ to q. That is equivalent to $\mathbf{N}^T\mathbf{r} = \mathbf{0}$, so that

$$\tilde{\mathbf{P}}\mathbf{r} = \mathbf{N}(\mathbf{N}^T\mathbf{N})^{-1}(\mathbf{N}^T\mathbf{r}) = \mathbf{0}. \tag{2.2.90}$$

Again, let $\mathbf{d}$ be any vector in $\tilde{Q}$, so that $\mathbf{d}$ may be expressed as a linear combination of its basis vectors, namely,

$$\mathbf{d} = \sum_{i=1}^{q} t_i\mathbf{n}_i = \mathbf{N}\mathbf{t}, \tag{2.2.91}$$

where $\mathbf{t} = (t_1 \ t_2 \ \cdots \ t_q)^T$. Therefore,

$$\tilde{\mathbf{P}}\mathbf{d} = \mathbf{N}(\mathbf{N}^T\mathbf{N})^{-1}(\mathbf{N}^T\mathbf{N})\mathbf{t} = \mathbf{N}\mathbf{t} = \mathbf{d}. \tag{2.2.92}$$

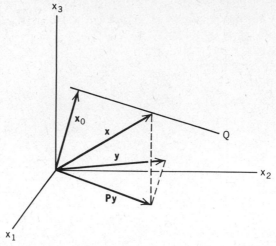

Figure 2.2.11. Projection of vector **y** from E^3 into the intersection (manifold) of the two hyperplanes in Figure 2.2.10. Notice that **Py** is "parallel" to the manifold.

Now a second projection matrix can be defined that takes any vector in E^m into the manifold Q, the intersection of q hyperplanes. Stated another way, this transformation projects the direction on which it operates into the orthogonal complement of the directions represented by the columns of **N**. It is

$$\mathbf{P} = \mathbf{I} - \tilde{\mathbf{P}} = \mathbf{I} - \mathbf{N}(\mathbf{N}^T\mathbf{N})^{-1}\mathbf{N}^T. \tag{2.2.93}$$

P has dimensions $m \times m$. Using the same vectors and properties of $\tilde{\mathbf{P}}$ as previously defined,

$$\mathbf{Pr} = \mathbf{r} - \tilde{\mathbf{P}}\mathbf{r} = \mathbf{r}, \tag{2.2.94}$$

$$\mathbf{Pd} = \mathbf{d} - \tilde{\mathbf{P}}\mathbf{d} = \mathbf{0}. \tag{2.2.95}$$

Figure 2.2.11 illustrates the projection of a vector in E^3 into a point on the intersection of the two hyperplanes shown in Figure 2.2.10. Vectors $\mathbf{x}_0$ and $\mathbf{x}$ terminate on the manifold Q. Vector **Py** is "parallel" to Q, and any vector **x** on Q may be expressed as

$$\mathbf{x} = \mathbf{x}_0 + \mathbf{Py}. \tag{2.2.96}$$

Projection matrices have a number of interesting properties. Their eigenvalues are either unity or zero. The rank of a projection matrix is equal to its trace, and the number of unity eigenvalues is equal to its rank. Projection matrices are *idempotent*, meaning that $\mathbf{P}^2 = \mathbf{P}$. The necessary conditions for a matrix to be a projection matrix are that it be both idempotent and symmetric.

There are several useful identities that involve projection matrices; these will be left to the problems at the end of this chapter.

Some of the terms on the right side of (2.2.93) have a pervasive importance in analysis of overdetermined systems of linear equations and other analysis related to optimization. At this time, the full-rank *generalized* (*pseudo*) *inverse* is defined as

$$\mathbf{A}^+ = (\mathbf{A}^T\mathbf{A})^{-1}\mathbf{A}^T, \tag{2.2.97}$$

where $\mathbf{A}_{m,n}$ is of rank n and $m \geq n$. $\mathbf{A}^+$ has dimension $n \times m$. When $m = n$, then $\mathbf{A}^+ = \mathbf{A}^{-1}$. Clearly, $\mathbf{P} = \mathbf{I} - \mathbf{N}\mathbf{N}^+$. Program GENINVP, C2-8 in Appendix C, can be merged into program MATRIX. Then command 10 computes the generalized inverse for the matrix in $\mathbf{A}$ according to (2.2.97). Also, command 13 computes the projection matrix defined by (2.2.93), using the data in matrix $\mathbf{A}$ which is automatically normalized by column as required for matrix $\mathbf{N}$ in (2.2.88). The answers are returned in matrix $\mathbf{A}$ in both cases.

Example 2.2.11.   Consider the 3×2 matrix

$$\mathbf{N} = \begin{bmatrix} 1 & 2 \\ 5 & 7 \\ 5 & 3 \end{bmatrix}. \tag{2.2.98}$$

To find its generalized inverse and the associated projection matrix, LOAD"MATRIX and then MERGE"GENINVP. The results are shown in Table 2.2.13. The upper 2×3 matrix is the generalized inverse of rank two; the reader is urged to confirm that $\mathbf{A}^+\mathbf{A} = \mathbf{I}_2$. The lower 3×3 matrix in Table 2.2.13 is the projection matrix, $\mathbf{P}$ defined in (2.2.93). The reader is urged to note that $\mathbf{P}$ is symmetric and its rank (equal to its trace) is 1, and to confirm that $\mathbf{P}^2 = \mathbf{P}$ ($\mathbf{P}$ is idempotent). The geometric significance of $\mathbf{P}$ can be observed by noting that the columns of matrix $\mathbf{N}$ in (2.2.98) define two hyperplanes through the origin in E^3; using (2.2.82), with $h = 0$,

$$x_1 + 5x_2 + 5x_3 = 0,$$
$$2x_1 + 7x_2 + 3x_3 = 0. \tag{2.2.99}$$

Table 2.2.13.   The Generalized Inverse and Projection Matrix $\mathbf{P} = \mathbf{I} - \mathbf{N}\mathbf{N}^+$ for the Matrix in (2.2.98)

```
MATRIX A( 2 , 3 ) -
          -0.09170        -0.11790         0.33624
           0.10917         0.21179        -0.23362
MATRIX A( 3 , 3 ) -
           0.87336        -0.30568         0.13100
          -0.30568         0.10699        -0.04585
           0.13100        -0.04585         0.01965
```

By substituting from one equation into the other, it is found that $x_2 = -7x_1/20$ and $x_3 = 3x_1/20$. Arbitrarily equating $x_1 = 4$, we find that $\mathbf{x} = (4 \ -\frac{7}{5} \ \frac{3}{5})^T$ is a vector lying in the intersection Q of the two hyperplanes, since they both pass through the origin. Each of the unit coordinate vectors $\mathbf{e}_i$ projects into this same direction, since $\mathbf{Pe}_i$, $i = 1$ to 3, represents each column of $\mathbf{P}$ shown in the lower part of Table 2.2.13, and each column is proportional to $\mathbf{x} = (4 \ -\frac{7}{5} \ \frac{3}{5})^T$. Furthermore, consider the arbitrary vector $\mathbf{y} = (2 \ 1 \ 3)^T$, having a length of $(14)^{1/2}$. Then $\mathbf{Py} = (1.82406 \ -0.64192 \ 0.27512)^T$, having a length of 1.96253. But by the formula for the angle between two vectors, (2.1.42), the angle between $\mathbf{y}$ and $\mathbf{Py}$ is 58.3648 degree, and the projection of length $(14)^{1/2}$ at that angle is 1.96253, verifying the relationship illustrated in Figure 2.2.11.

It was noted in Chapter One that inequality constraints to optimization problems could be expressed in the form

$$\mathbf{c(x)} \geq \mathbf{0}. \tag{2.2.100}$$

When $\mathbf{c(x)}$ is a set of q linear constraints, they may be expressed as

$$\mathbf{n}_j^T\mathbf{x} \geq \mathbf{0}, \qquad j = 1 \text{ to } q. \tag{2.2.101}$$

The question is: Given a point in space, $\mathbf{x}$, which constraints contain that point in their positive half-spaces? The answer lies in the sufficient condition

$$\mathbf{N}^T\mathbf{x} \geq 0, \tag{2.2.102}$$

where $\mathbf{N}$ is defined by (2.2.88). For example, when $\mathbf{N}$ is given by (2.2.98), then $\mathbf{x} = (2 \ 1 \ 3)^T$ is in the positive half-space of both hyperplanes since $\mathbf{N}^T\mathbf{x} = (22 \ 20)^T$. But $\mathbf{x} = (1 \ -1 \ 1)^T$ is in the positive half-space of H_1 but the negative half-space of H_2 since $\mathbf{N}^T\mathbf{x} = (1 \ -2)^T$.

Two final notes about projection matrices. The practical application of the projection of search direction vectors into applicable constraint hyperplanes requires that the columns of matrix $\mathbf{N}$ receive additions and deletions of one or more hyperplane normal vectors. Since $\mathbf{N}^{-1}$ is needed repeatedly for projection, the methods of Rosen (1960:189) enable the update of $\mathbf{N}^{-1}$ efficiently following additions and deletions of columns of $\mathbf{N}$. Finally, Willoughby (1973) noted that, "An understanding of the geometric significance of the equations associated with the projection process leads directly to an understanding of the calculations that are needed to apply the projection operator to any direct gradient search optimization method." As in previous methods, it is seldom necessary to explicitly compute and store the $\mathbf{P}$ matrix. There are a number of computational shortcuts that depend on the structure of the linear constraints; see Fiacco (1968:153).

Problems

2.1. A matrix defined so that $A^T = -A$, that is, $[a_{ji}] = [-a_{ij}]$, is called a *skew matrix*. Show that:

(a) A skew matrix is square.

(b) Only zero elements occur on the principal diagonal.

(c) For any square matrix A, $(A - A^T)$ is skew.

(d) Any square matrix can be uniquely decomposed into the sum of a symmetric and a skew matrix.

2.2. Show the form of the elements in the inverse of a real, diagonal matrix. Show that if a matrix A is symmetric, then A^{-1} is also symmetric. Show that the inverse of a nonsingular elementary Householder matrix $I - 2uu^T$ is also a Householder matrix involving the same two vectors.

2.3. If matrices A, B, and A + B all have inverses, show that

$$(A^{-1} + B^{-1})^{-1} = A(A + B)^{-1}B = B(A + B)^{-1}A.$$

2.4. Given the partitioned matrices

$$T = \begin{bmatrix} P & Q \\ R & S \end{bmatrix} \quad \text{and} \quad T^{-1} = \begin{bmatrix} A & B \\ C & D \end{bmatrix},$$

where T and P are nonsingular. Show that

$$A = P^{-1} + P^{-1}QC,$$
$$B = -P^{-1}QD,$$
$$C = DRP^{-1},$$
$$D = (S - RP^{-1}Q)^{-1}.$$

2.5. Consider the matrix product DA, where D is diagonal, that is, $D = \text{diag}(d_1 \ d_2 \ \cdots \ d_i \ \cdots \ d_n)$. Show that DA is equivalent to multiplying the ith row of A by d_i. Conversely, show that AD is equivalent to multiplying the ith column of A by d_i.

2.6. For orthogonal matrices, show that

(a) The product of two orthogonal matrices is also orthogonal.

(b) The inverse of an orthogonal matrix is orthogonal.

(c) All eigenvalues of an orthogonal matrix have modulus unity.

2.7. Show that the columns of a diagonal matrix are linearly independent if all of elements on the principal diagonal are nonzero.

2.8. A set of vectors forms a *vector space* if any two vectors, say x and y, are members and x + y and hx are also members, where h is a scalar.

Show that the eigenvectors $\mathbf{v}$ corresponding to a given eigenvalue w constitute a vector space, where $\mathbf{Av} = w\mathbf{v}$.

2.9.   Show that swapping any two rows (or columns) in a matrix $\mathbf{A}$ swaps the corresponding columns (or rows) of $\mathbf{A}^{-1}$.

2.10. Show that the $\det(\mathbf{A}_{3,3})$ is equal to

$$\det(\mathbf{A}) = a_{11}a_{22}a_{33} - a_{11}a_{32}a_{23} - a_{12}a_{21}a_{33}$$
$$+ a_{12}a_{31}a_{23} + a_{13}a_{21}a_{32} - a_{13}a_{31}a_{22}.$$

Note that no two elements in each product lie in the same row of $\mathbf{A}$, nor do any two elements lie in the same column.

2.11. Show that the product of two upper triangular matrices is also an upper triangular matrix. Also show that the elements on the principal diagonal of the product matrix are given by the product of the corresponding elements in the given matrices.

2.12. Use the result in Problem 2.10 to confirm the product of the eigenvalues 1, 3, and 4, of the matrix

$$\mathbf{A} = \begin{bmatrix} 3 & -1 & 0 \\ -1 & 2 & -1 \\ 0 & -1 & 3 \end{bmatrix}.$$

Verify that $\operatorname{tr}(\mathbf{A})$ equals the sum of the eigenvalues. Assume that the first element in each eigenvector is unity; then delete one equation from the defining set, $\mathbf{Av} = w\mathbf{v}$, and solve for all three eigenvectors $\mathbf{v}$. Since $\mathbf{A}$ is symmetric in this case, verify that the eigenvectors are orthogonal.

2.13. Prove that the eigenvalues of the matrix $c\mathbf{A}$ are cw_i, where w_i is an eigenvalue of $\mathbf{A}$ and c is a scalar. Similarly, prove that the eigenvalues of $\mathbf{A}^h$ are w_i^h, where h is a positive integer.

2.14. Show that every polynomial of degree n is the characteristic polynomial of the *companion matrix*:

$$A = \begin{bmatrix} -a_1 & -a_2 & -a_3 & \cdots & -a_{n-1} & -a_n \\ 1 & 0 & 0 & \cdots & 0 & 0 \\ 0 & 1 & 0 & \cdots & 0 & 0 \\ & & & \cdots & & \\ 0 & 0 & 0 & \cdots & 1 & 0 \end{bmatrix}.$$

There are several equivalent forms for the companion matrix.

2.15. Show that a special case of equation (2.1.36) is the rank 1 update equation:

$$\left(\mathbf{A} + \mathbf{xy}^T\right)^{-1} = \mathbf{A}^{-1} - \frac{\mathbf{A}^{-1}\mathbf{xy}^T\mathbf{A}^{-1}}{1 + \mathbf{y}^T\mathbf{A}^{-1}\mathbf{x}}.$$

Verify this identity numerically by using $\mathbf{T}_{5,5}$ in (2.1.8) and $\mathbf{x} = (1\ 2\ 3\ 2\ 1)^T$, $\mathbf{y} = (1\ 3\ 2\ 3\ 1)^T$. Check your answer by direct inversion, using MATRIX program command 9.

2.16.   Find an expression for the eigenvalues of a 2×2 general real matrix. Show that when it is symmetric the eigenvalues must be real; obtain expressions for the eigenvectors. What are the eigenvalues when a_{12} or a_{21} is zero?

2.17.   Show that the inner product of real vectors is linear (obeys superposition and scaling). Also, show that outer products are symmetric.

2.18.   Let matrix $\mathbf{B} = (\mathbf{b}_1\ \mathbf{b}_2\ \cdots\ \mathbf{b}_n)$. Then show that

$$\mathbf{AB} = (\mathbf{Ab}_1\ \mathbf{Ab}_2\ \cdots\ \mathbf{Ab}_n).$$

2.19.   For any two vectors in E^n, say $\mathbf{x}$ and $\mathbf{y}$, verify the parallelogram relation:

$$\|\mathbf{x} + \mathbf{y}\|^2 + \|\mathbf{x} - \mathbf{y}\|^2 = 2(\|\mathbf{x}\|^2 + \|\mathbf{y}\|^2).$$

2.20.   If matrices $\mathbf{A}$ and $\mathbf{B}$ are *similar*, prove that $\det(\mathbf{A}) = \det(\mathbf{B})$.

2.21.   In the *spectral decomposition* of real symmetric matrix $\mathbf{A}$, the rank 1 matrix $\mathbf{E}_i = \mathbf{v}_i \mathbf{v}_i^T$ is composed of the outer product of the orthonormal eigenvectors $\mathbf{v}_i$. Verify that $\mathbf{E}$ is idempotent ($\mathbf{E}^2 = \mathbf{E}$). Also, show that $\mathbf{A} = \mathbf{U(VU)}^{-1}\mathbf{V}$ is idempotent. Why are all vector outer products rank 1?

2.22.   Prove that if a *similarity transformation* $\mathbf{V}^T\mathbf{AV} = \mathbf{W}$ exists, where $\mathbf{W}$ is a diagonal matrix, then $\mathbf{A}^p = \mathbf{VW}^p\mathbf{V}^T$, for p a positive integer, or p a negative integer if $\mathbf{A}$ is nonsingular. Note that this is a simple expression for the pth power of a matrix.

2.23.   Find condition numbers using three different matrix norms for the matrix in Table 2.2.4 and compare them to those in Table 2.1.8. Compute the condition number of equation (2.1.8) using the $\|\mathbf{T}\|_2$ norm.

2.24.   For the projection matrix in equation (2.2.93), prove the following properties, where $\mathbf{x}$ and $\mathbf{y}$ are arbitrary vectors in E^n:

   (a) $\mathbf{P}^T = \mathbf{P}$.

   (b) $\mathbf{P(Px)} = \mathbf{Px}$.

   (c) $\mathbf{x} = \mathbf{Px} + \tilde{\mathbf{P}}\mathbf{x}$.

   (d) $\mathbf{y}^T(\mathbf{Px}) = (\mathbf{Py})^T(\mathbf{Px})$.

   (e) $\mathbf{y}^T(\mathbf{Px}) = (\mathbf{Py})^T\mathbf{x}$.

   (f) $\mathbf{x}^T(\mathbf{Px}) = \|\mathbf{Px}\|_2^2$.

Chapter Three _____

Functions of Many Variables

Linear functions provide the framework for analysis of general nonlinear functions, and this is as true for many variables as for just one variable. The point was made in Chapter One that quadratic functions would be used to approximate the neighborhood of extreme values in multivariable surfaces and, furthermore, that equating the derivatives of the general quadratic function to zero results in a set of linear equations. There is an even better reason for solving sets of linear equations in connection with nonlinear optimization: almost all descriptions of a promising search direction in n-dimensional space require the solution of a system of linear equations.

Chapter Three begins with pertinent methods for solving systems of linear equations, especially those that are symmetric and/or illconditioned. The most straightforward case involves systems having the same number of independent equations and unknowns. Certain matrix decompositions are introduced for the general and special cases useful in optimization. Then overdetermined systems having more equations than unknowns, but with rank equal to the number of unknowns, are considered. This latter class of systems is related to the weighted linear least squares problem, because of its close relationship to nonlinear least-pth problems to be discussed in Chapter Four. The generalized inverse and the singular value decomposition will provide important conceptual and computational tools, respectively.

Description of nonlinear functions begins with the quadratic function, which serves as a basis for introducing the Taylor series in many variables. The set of derivatives, or gradient vector, of the nonlinear function is crucially involved. Various related topics such as the implicit function theorem and convexity are introduced as required to set the stage for optimization algorithms.

Nearly all nonlinear optimization algorithms are based on the concept of choosing a sequence of promising search directions in multidimensional space. Then so-called line searches are accomplished in each of those directions until a minimum (or maximum) is reached. The mechanics of carrying out these line searches involves both theoretical and practical strategies. The most successful of these will be described for later use in optimization algorithms, beginning with some theoretically important practice on quadratic functions.

Finally, the application of constraints to optimization problems will be considered. For linear constraints, the projection matrices and line searches discussed previously will be involved in the theoretical and practical implementation, respectively. A natural extension of these concepts is the Lagrange multiplier method for dealing theoretically with general nonlinear constraints. After this chapter, the description of practical optimization algorithms can begin.

3.1. Systems of Linear Equations

The reader undoubtedly has been exposed to the solution of systems of linear equation from very early years. However, it is important to deal with the special cases related to optimization, using the methods that have proved most effective.

3.1.1. Square Linear Systems of Full Rank.
The solution of two equations in two unknowns is easily illustrated. Figure 3.1.1 shows the loci of the three linear equations $\mathbf{Nx} = (4\ 21\ 15)^T$, where $\mathbf{N}$ was given by (2.2.98). The intersections of all pairs are graphically obvious and are easily computed using a

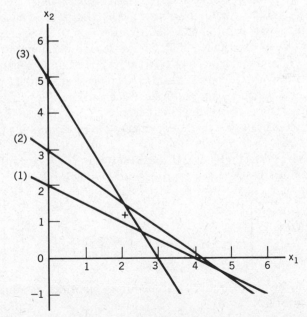

Figure 3.1.1. Three linear equations in two unknowns. (1) $(1\ 2)\mathbf{x} = 4$, (2) $(5\ 7)\mathbf{x} = 21$, (3) $(5\ 3)\mathbf{x} = 15$. The "+" marks the linear least-squares solution (LLS).

Table 3.1.1.   Data for The Three Linear Equations in Figure 3.1.1

Equation Pairs	x_1	x_2	$\|\mathbf{A}\|_F$	$\|\mathbf{A}^{-1}\|_F$	$k(\mathbf{A})$
(1)–(2)	4.6667	−0.3333	8.8882	2.9627	26.3333
(1)–(3)	2.5714	0.7143	6.2450	0.8921	5.5714
(2)–(3)	2.1000	1.5000	10.3923	0.5196	5.4000

succession of three $\mathbf{A}_{2,2}$ matrices in the equation $\mathbf{Ax} = \mathbf{b}$. Also the condition number (2.1.50) using the Frobenius matrix norm (2.1.49) can be computed by using MATRIX command 11. The results are summarized in Table 3.1.1. Clearly, the more nearly orthogonal the intersection of line pairs, the lower the condition number (that is, the solution is better conditioned). As previously indicated, the matrix inverse is not a computationally efficient means for obtaining the solution of linear equations. Two efficient methods are described, one for the general real matrix and a similar one for symmetrical real matrices; the latter situation is the case in nearly all applications of optimization.

The most general trouble-free case is when the set of linear equations $\mathbf{Ax} = \mathbf{b}$ involves a matrix $\mathbf{A}$ that is dimensioned $n \times n$ and is of rank n. In that case one of the most effective means for finding $\mathbf{x}$ given $\mathbf{A}$ and $\mathbf{b}$ is **LU** *factorization*. The method is very similar to the Gauss–Jordan procedure previously described in Section 2.1.4 for the matrix inverse. The Gauss–Jordan method employed a series of elementary transformations to reduce $\mathbf{A}$ to a diagonal form. Another well-known method is *Gaussian elimination*, which reduces $\mathbf{A}$ to a triangular form; **LU** factorization is entirely equivalent to that method. It has been determined that Gauss–Jordan elimination requires on the order of $n^3/2$ mathematical operations, while Gaussian elimination requires on the order of $n^3/3$ operations. (In keeping with standard practice, the operational count includes only multiplications and divisions.) Description of **LU** factorization here is worthwhile because of its pervasiveness in the literature and its relationship to another method more commonly employed in optimization.

As described by Vlach (1983:43), consider the $n = 4$ case, without loss of generality; suppose that $\mathbf{A}$ is decomposed into two factors:

$$\mathbf{A} = \mathbf{LU} = \begin{bmatrix} l_{11} & 0 & 0 & 0 \\ l_{21} & l_{22} & 0 & 0 \\ l_{31} & l_{32} & l_{33} & 0 \\ l_{41} & l_{42} & l_{43} & l_{44} \end{bmatrix} \begin{bmatrix} 1 & u_{12} & u_{13} & u_{14} \\ 0 & 1 & u_{23} & u_{24} \\ 0 & 0 & 1 & u_{34} \\ 0 & 0 & 0 & 1 \end{bmatrix}. \qquad (3.1.1)$$

Assuming that this factorization can be done, consider its application in solving $\mathbf{Ax} = \mathbf{b}$ for $\mathbf{x}$:

$$\mathbf{LUx} = \mathbf{b}. \qquad (3.1.2)$$

The well-known approach when two triangular matrices are involved is to define an auxiliary vector **y** such that

$$\mathbf{Ux} = \mathbf{y}. \tag{3.1.3}$$

Substituting (3.1.3) into (3.1.2), a second triangular system of linear equations is obtained:

$$\mathbf{Ly} = \mathbf{b}. \tag{3.1.4}$$

It is clear from the nature of **L** in (3.1.1) that the first equation in (3.1.4) is simply $l_{11}y_1 = b_1$, so that $y_1 = b_1/l_{11}$. The second equation in (3.1.4) involves just y_1 and y_2, so that the latter can now be determined, and so on. The process described for solving (3.1.4) is called *forward substitution*. By writing each equation represented by (3.1.4), it is easy to verify the general equations for forward substitution:

$$y_1 = \frac{b_1}{l_{11}}$$

$$y_i = \frac{b_i - \sum\limits_{j=1}^{i-1} l_{ij} y_j}{l_{ii}}, \qquad \text{for } i = 2 \text{ to } n. \tag{3.1.5}$$

At the completion of forward substitution, **y** is known so that (3.1.3) may be solved for **x**, the dependent variable of interest. However, the upper triangular structure of **U** evident in (3.1.1) makes it clear that the last equation in (3.1.3) is $x_4 = y_4$. Also, the next-to-last equation in (3.1.3) involves only x_4 and x_3, so that x_3 may be found. The process described for solving (3.1.3) is called *back substitution*. By writing each equation represented by (3.1.3), it is also easy to verify the general equations for back substitution:

$$x_n = y_n$$

$$x_i = y_i - \sum\limits_{j=i+1}^{n} u_{ij} x_j, \qquad \text{for } j = n-1, n-2, \ldots 1. \tag{3.1.6}$$

Approximately $n^2/2$ operations each are required for forward and for back substitution, as compared with $n^3/3$ for the **LU** factorization. There are at least three useful features of **LU** factorization:

1. The determinant is simply the product of all the l_{ii} as seen from (3.1.1).
2. The decomposition need not be recalculated for a sequence of different right-hand vectors, **b**, in solving the system $\mathbf{Ax} = \mathbf{b}$.
3. $\mathbf{A}^T\mathbf{x} = \mathbf{z}$ can be solved with the same **LU** factorization, a requirement in certain electrical network sensitivity calculations discussed in Section 6.3.4.

There are two remaining developments for **LU** factorization: (1) determination of the elements of **L** and **U** as defined in (3.1.1), and (2) economization of storage so that **L** and **U** are computed and stored in place of the given matrix **A**. The diagonal of 1s in **U** ($u_{ii} = 1$, $i = 1$ to n) as shown in (3.1.1) carries no unique information, so the conventional means for *storing* **L** and **U** in an $n \times n$ matrix ($n = 4$ for illustration) is to replace matrix **A** with

$$\begin{bmatrix} l_{11} & u_{12} & u_{13} & u_{14} \\ l_{21} & l_{22} & u_{23} & u_{24} \\ l_{31} & l_{32} & l_{33} & u_{34} \\ l_{41} & l_{42} & l_{43} & l_{44} \end{bmatrix}.$$ (3.1.7)

Equation (3.1.1) can be written to show explicitly each term in the **LU** product:

$$\begin{bmatrix} l_{11} & l_{11}u_{12} & l_{11}u_{13} & l_{11}u_{14} \\ l_{21} & l_{21}u_{12} + l_{22} & l_{21}u_{13} + l_{22}u_{23} & l_{21}u_{14} + l_{22}u_{24} \\ l_{31} & l_{31}u_{12} + l_{32} & l_{31}u_{13} + l_{32}u_{23} + l_{33} & l_{31}u_{14} + l_{32}u_{24} + l_{33}u_{34} \\ l_{41} & l_{41}u_{12} + l_{42} & l_{41}u_{13} + l_{42}u_{23} + l_{43} & l_{41}u_{14} + l_{42}u_{24} + l_{43}u_{34} + l_{44} \end{bmatrix}.$$ (3.1.8)

Now (3.1.8) is **LU**, which is equal to **A**:

$$\mathbf{LU} = \mathbf{A} = \begin{bmatrix} a_{11} & a_{12} & a_{13} & a_{14} \\ a_{21} & a_{22} & a_{23} & a_{24} \\ a_{31} & a_{32} & a_{33} & a_{34} \\ a_{41} & a_{42} & a_{43} & a_{44} \end{bmatrix}.$$ (3.1.9)

The respective elements in (3.1.8) and (3.1.9) must be equal; clearly, the first columns are easily equated, so that $l_{i1} = a_{i1}$, for $i = 1$ to n. The first row in (3.1.7) is equally easy to determine: $u_{1j} = a_{1j}/l_{11}$, $j = 1$ to n. These and other results are shown to correspond with the respective elements in (3.1.7):

$$\begin{bmatrix} l_{11} & a_{12}/l_{11} & a_{13}/l_{11} & a_{14}/l_{11} \\ l_{21} & a_{22} - l_{21}u_{12} & (a_{23} - l_{21}u_{13})/l_{22} & (a_{24} - l_{21}u_{14})/l_{22} \\ l_{31} & a_{32} - l_{31}u_{12} & a_{33} - l_{31}u_{13} - l_{32}u_{23} & (a_{34} - l_{31}u_{14} - l_{32}u_{24})/l_{33} \\ l_{41} & a_{42} - l_{41}u_{12} & a_{43} - l_{41}u_{13} - l_{42}u_{23} & a_{44} - l_{41}u_{14} - l_{42}u_{24} - l_{43}u_{34} \end{bmatrix}.$$ (3.1.10)

There are at least three different methods for computing the elements according to the relationships in (3.1.10) as discussed by Vlach (1983); the main constraint is that the a_{ij} elements in (3.1.9) cannot be overwritten until they are no longer required. The method that is equivalent to Gaussian elimination computes terms such as l_{33} in (3.1.7) by first computing $a_{33} - l_{31}u_{13}$ and later subtracting the remaining term $l_{32}u_{23}$. Similarly, l_{44} in (3.1.7) is computed by first computing $a_{44} - l_{41}u_{14}$ and later subtracting the remaining terms $l_{42}u_{24}$ and $l_{43}u_{34}$, and so on. The BASIC instructions for that process are contained in subprogram C3-1 called LUFAC in Appendix C. When it is

Table 3.1.2. LU Factorization Matrix for Matrix T in (2.1.8)

```
MATRIX D( 5 , 5 )  -
      2.00000         1.50000         2.50000         3.00000         2.00000
     -1.00000         2.50000         0.60000         1.60000         1.20000
     -2.00000         6.00000        -2.60000        -0.53846         1.61538
      5.00000        -4.50000       -10.80000       -11.61538        -0.84768
     -4.00000        14.00000        -1.40000        -9.15385       -12.29801
```

merged into the MATRIX program, command 10 performs LU factorization on matrix **D**, and command 13 performs forward and back substitution using vector **C**, returning the solution vector in **C**. There is one issue not addressed in program LUFAC: The "pivoting" strategy demonstrated in Section 2.1.4 with the Gauss–Jordan procedure and also employed in Gaussian elimination has not been incorporated in subprogram LUFAC. For example, it is clear in (3.1.10) that $a_{11} = l_{11} = 0$ would cause a divide-by-zero operation. In general, it is desirable to swap equations so that divisors with the largest modulus are always in position to minimize buildup of roundoff errors. That additional complication is not implemented in program LUFAC because nearly all systems of linear equations involved in optimization are symmetric and have other properties that eliminate that risk.

Example 3.1.1. To perform LU factorization on a matrix, first create a data file containing the vector $\mathbf{b} = (53 \ -1 \ -1 \ 35 \ 9)^T$; call it TEST.VEC. In the BASIC environment, LOAD"MATRIX and MERGE"LUFAC. Then run the combined program and recall the 5×5 matrix **T** as defined in (2.1.8) into program matrix **D**, using command 5 and the appropriate data file name. Also recall vector **b** from file TEST.VEC into program matrix **C**. Then command 10 can be used to produce the LU factorization of **T** in the format of (3.1.7). That result is shown in Table 3.1.2. Then command 13 solves for $\mathbf{x} = \mathbf{A}^{-1}\mathbf{b} = (5 \ 4 \ 3 \ 2 \ 1)^T$. By recalling any number of other right-hand vectors **b**, command 13 will effectively perform the solution without explicitly computing the inverse and without recalculation of the LU factorization. Of course, any of the results just demonstrated may be checked by using the matrix inverse command 9 as described in Chapter 2.

Nonlinear optimization involves quadratic functions, and it is seen that quadratic functions involve symmetric matrices that contain second partial derivatives of the function. Therefore, suppose that there is a real, symmetric matrix **A** and that it can be decomposed or factorized in the form:

$$\mathbf{A} = \mathbf{LDL}^T, \tag{3.1.11}$$

where $\mathbf{D} = \text{diag}(d_1 \ d_2 \ \cdots \ d_n)$, an $n \times n$ diagonal matrix, and

$$\mathbf{L} = \begin{bmatrix} 1 & 0 & 0 & 0 \\ l_{21} & 1 & 0 & 0 \\ l_{31} & l_{32} & 1 & 0 \\ l_{41} & l_{42} & l_{43} & 1 \end{bmatrix}. \tag{3.1.12}$$

Equation (3.1.12) and subsequent analysis will consider the $n = 4$ case without loss of generality. A matrix is said to be *positive definite* if all its eigenvalues are positive real; an equivalent condition is that the diagonal matrix $\mathbf{D}$ in the $\mathbf{LDL}^T$ decomposition has only positive elements. Since it is shown that positive definiteness is a crucial property of matrices that are related to quadratic functions, the $\mathbf{LDL}^T$ decomposition indicates that condition and provides a simple means to alter the matrix to make it positive definite.

The matrix factorization in (3.1.11) can be expressed as $\mathbf{A} = \mathbf{LD}^{1/2}\mathbf{D}^{1/2}\mathbf{L}^T = \mathbf{U}^T\mathbf{U}$, where $\mathbf{D}^{1/2}$ contains the square roots of positive diagonal elements in $\mathbf{D}$, and $\mathbf{U}$ is a general upper-triangular matrix. (Note that postmultiplication of $\mathbf{L}$ by a diagonal matrix scales each *column* of $\mathbf{L}$ by the respective diagonal element.) This form is found in the literature as *Cholesky factorization*, or the square-root method, which requires a positive-definite symmetrical matrix. In a sense, $\mathbf{U}$ is the "square root" of $\mathbf{A}$.

The decomposition in (3.1.11) can be derived by ignoring any elements above the main diagonal, since $\mathbf{LDL}^T$ is symmetric. The $\mathbf{LDL}^T$ product is equal to

$$
\begin{bmatrix}
d_1 & & & \\
d_1 l_{21} & d_1 l_{21}^2 + d_2 & & \\
d_1 l_{31} & d_1 l_{21}l_{31} + d_2 l_{32} & d_1 l_{31}^2 + d_2 l_{32}^2 + d_3 & \\
d_1 l_{41} & d_1 l_{21}l_{41} + d_2 l_{42} & d_1 l_{41}l_{31} + d_2 l_{42}l_{32} + d_3 l_{43} & d_1 l_{41}^2 + d_2 l_{42}^2 + d_3 l_{43}^2 + d_4
\end{bmatrix}.
$$

$$(3.1.13)$$

Symmetric matrices that are $n \times n$ actually contain only $n(n + 1)/2$ unique elements as opposed to the n^2 elements in the ordinary matrix format. Therefore, it is common practice to evolve a given symmetric matrix $\mathbf{A}$ as follows:

$$
\mathbf{LDL}^T =
\begin{bmatrix}
a_{11} & & & \\
a_{21} & a_{22} & & \\
a_{31} & a_{32} & a_{33} & \\
a_{41} & a_{42} & a_{43} & a_{44}
\end{bmatrix}
\rightarrow
\begin{bmatrix}
d_1 & & & \\
l_{21} & d_2 & & \\
l_{31} & l_{32} & d_3 & \\
l_{41} & l_{42} & l_{43} & d_4
\end{bmatrix}
$$

$$
\rightarrow
\begin{bmatrix}
H(1) & & & \\
H(2) & H(5) & & \\
H(3) & H(6) & H(8) & \\
H(4) & H(7) & H(9) & H(10)
\end{bmatrix}.
$$

$$(3.1.14)$$

The matrix $[a_{ij}]$ on the left in (3.1.14) is given; its decomposition is ordinarily represented as shown by the middle matrix. However, in computer programs, *both* the given matrix and later its decomposition are stored in a singly dimensioned array (vector) denoted here as $H(\cdot)$ and shown embedded in the lowest matrix in (3.1.14).

A sequence of equations may be obtained by equating the respective elements in (3.1.13) and the $[a_{ij}]$ matrix in (3.1.14). In the upper left-hand corner it is seen that $d_1 = a_{11}$. Equating elements in the first column of the second row, $a_{21} = d_1 l_{21}$, so that $l_{21} = a_{21}/d_1$. Since that is typical of the remainder of the first column, the relationships so far are:

$$d_1 = a_{11}, \quad l_{21} = a_{21}/a_{11}, \quad l_{31} = a_{31}/a_{11}, \quad \text{and} \quad l_{41} = a_{41}/a_{11}. \quad (3.1.15)$$

In the top of the second column, $a_{22} = d_1 l_{21}^2 + d_2$, so that

$$d_2 = a_{22} - \frac{a_{21}^2}{a_{11}}. \qquad (3.1.16)$$

Writing two more equations involving a_{32} and a_{42}, respectively, the results are

$$d_2 l_{32} = a_{32} - \frac{a_{21}a_{31}}{a_{11}}, \qquad (3.1.17)$$

$$d_2 l_{42} = a_{42} - \frac{a_{21}a_{41}}{a_{11}}. \qquad (3.1.18)$$

Continuing, the respective results for a_{33}, a_{43}, and a_{44} are:

$$d_3 = a_{33} - \frac{a_{31}^2}{a_{11}} - \frac{(d_2 l_{32})^2}{d_2}, \qquad (3.1.19)$$

$$d_3 l_{43} = a_{43} - \frac{a_{31}a_{41}}{a_{11}} - \frac{(d_2 l_{32})(d_2 l_{42})}{d_2}, \quad \text{and} \qquad (3.1.20)$$

$$d_4 = a_{44} - \frac{a_{41}^2}{a_{11}} - \frac{(d_2 l_{42})^2}{d_2} - \frac{(d_3 l_{43})^2}{d_3}. \qquad (3.1.21)$$

The sequence of computations is severely limited by the evolution of matrices in (3.1.14) in light of these last equations and the necessity to avoid "covering up" any data until they are no longer required. For example, it can be seen that (3.1.17) cannot be solved for l_{32} until it has been used in both (3.1.19) and (3.1.20). Nevertheless, it is possible to program these relationships so that the symmetric matrix **A** may be given by elements on and below the principal diagonal and then may be replaced by elements of the **LDL**^{T} factorization without requiring additional storage. Beyond that, the actual storage and all computed results can be realized in a single array $H(\cdot)$, having only $n(n + 1)/2$ elements. The $\mathbf{LDL}^T$ decomposition requires on the order of $n^3/6$ operations. The forward and back substitutions are similar to those for the $\mathbf{LU}$ factorization and require a total of about n^2 operations. The forward and back substitutions are described in detail in Section 4.1.3, where there is a direct application.

Program C3-2 called LDLTFAC in Appendix C accomplishes the $\mathbf{LDL}^T$ factorization with the same external indications of the preceding program LUFAC. Decomposition occurs as MATRIX command 10. Only the lower triangular part of the symmetric matrix must be placed in program matrix $\mathbf{D}$. Program lines 7040 to 7110 transfer $\mathbf{D}$ to a vector H where the actual computations occur (lines 7120 to 7310). If any diagonal element d_i is negative (the matrix is not positive definite), the program terminates and announces that situation. A symmetric matrix that is positive definite may be decomposed into the $\mathbf{LDL}^T$ format without pivoting; however, a symmetric matrix that is not positive definite has *none* of the numerical stability of the positive-definite case.

The final decomposition by command 10 in LDLTFAC is returned to matrix $\mathbf{D}$ for display or printing (lines 7690 to 7770); however, subsequent forward and back substitutions by command 13 employ array H, not $\mathbf{D}$, so the decomposition must precede any sequence of solutions using command 13. Following decomposition (command 10), the user is asked if an update is required. That part of the program is described next, so an answer of "N" is appropriate at this time.

Example 3.1.2. The positive-definite, symmetric matrix in Table 2.2.4 was obtained as $\mathbf{Z} = \mathbf{T}^T\mathbf{T}$, where $\mathbf{T}$ was given in equation (2.1.8). In the BASIC environment, LOAD"MATRIX, then MERGE"LDLTFAC and place matrix $\mathbf{Z}$ in program matrix $\mathbf{D}$. Command 10 performs the $\mathbf{LDL}^T$ factorization, leaving the results in vector $H(\cdot)$ and in matrix $\mathbf{D}$ for inspection in the format illustrated in the middle of equation (3.1.14). The matrix $\mathbf{Z}$ and its factors are shown in Table 3.1.3. For example, $d_2 = 85.52$ and $l_{43} = 0.27734$. If the vector $\mathbf{c} = (248\ 332\ 208\ 400\ 125)^T$ is placed in program matrix $\mathbf{C}$, then command 13 returns the solution to $\mathbf{Z}^{-1}\mathbf{c}$ in program matrix $\mathbf{C}$; the solution vector is $(5\ 4\ 3\ 2\ 1)^T$.

Table 3.1.3. The Symmetric Matrix from Table 2.2.4 and Its $\mathbf{LDL}^T$ Factorization

```
MATRIX D( 5 , 5 )  -
      50.00000      -18.00000       26.00000        3.00000      -14.00000
     -18.00000       92.00000      -25.00000       56.00000       17.00000
      26.00000      -25.00000       52.00000        1.00000       20.00000
       3.00000       56.00000        1.00000       70.00000       18.00000
     -14.00000       17.00000       20.00000       18.00000       31.00000
MATRIX D( 5 , 5 )  -
      50.00000        0.00000        0.00000        0.00000        0.00000
      -0.36000       85.52000        0.00000        0.00000        0.00000
       0.52000       -0.18288       35.61974        0.00000        0.00000
       0.06000        0.66745        0.27734       28.98235        0.00000
      -0.28000        0.13985        0.82727        0.09264        0.78120
```

Many important optimization algorithms require a sequence of rank 1 "updates" or additions to the inverse of a symmetric matrix. Problem 2.15 illustrated the "rank annihilation" method, a special case of equation (2.1.36), which is an identity that furnishes the inverse of the matrix that has been "updated" by addition of a matrix product. Gill and Murray (1972) have described a procedure for performing that operation in the context of $\mathbf{LDL}^T$ factorization; although it requires no fewer operations than the classical rank annihilation method $(3n^2/2)$, it does allow testing and correction when the result is not positive definite. In most cases, the theoretical result should be positive definite, but that condition may not be obtained because of roundoff errors or other practical difficulties. As previously noted, positive definiteness is established when all $d_i > 0$ in matrix $\mathbf{D}$.

The *rank 1 update* to a matrix $\mathbf{A}$ is

$$\mathbf{A}^* = \mathbf{A} + q\mathbf{z}\mathbf{z}^T, \tag{3.1.22}$$

where q is a scalar and $\mathbf{z}\mathbf{z}^T$ is a rank 1 outer product as defined by (2.1.22). The asterisk denotes the "updated" matrix. The Gill and Murray (1972) "method B" defines an equivalent relationship:

$$\mathbf{A}^* = \mathbf{L}(\mathbf{D} + q\mathbf{v}\mathbf{v}^T)\mathbf{L}^T = \mathbf{LDL}^T + q(\mathbf{Lv})(\mathbf{Lv})^T. \tag{3.1.23}$$

Comparison of (3.1.22) and (3.1.23) shows that

$$\mathbf{Lv} = \mathbf{z}, \tag{3.1.24}$$

so that forward substitution of the given $\mathbf{z}$ vector will provide the intermediate vector $\mathbf{v}$. The objective is to determine the factorization

$$\mathbf{A}^* = \mathbf{L}^*\mathbf{D}^*\mathbf{L}^{*T}. \tag{3.1.25}$$

The analysis requires that an $\mathbf{LDL}^T$ factorization be derived for

$$\mathbf{D} + q\mathbf{v}\mathbf{v}^T = \tilde{\mathbf{L}}\tilde{\mathbf{D}}\tilde{\mathbf{L}}^T. \tag{3.1.26}$$

When that is available, the left-hand side of (3.1.23) shows that

$$\mathbf{A}^* = \mathbf{L}\tilde{\mathbf{L}}\tilde{\mathbf{D}}\tilde{\mathbf{L}}^T\mathbf{L}^T, \tag{3.1.27}$$

so that $\mathbf{L}^* = \mathbf{L}\tilde{\mathbf{L}}$ and $\mathbf{D}^* = \tilde{\mathbf{D}}$.

The derivation begins by writing the expansion indicated in (3.1.26); the $n = 4$ case is shown without loss of generality:

$$\tilde{\mathbf{L}}\tilde{\mathbf{D}}\tilde{\mathbf{L}}^T = \begin{bmatrix} d_1 + qv_1^2 & & & \\ qv_2v_1 & d_2 + qv_2^2 & & \\ qv_3v_1 & qv_3v_2 & d_3 + qv_3^2 & \\ qv_4v_1 & qv_4v_2 & qv_4v_3 & d_4 + qv_4^2 \end{bmatrix}. \tag{3.1.28}$$

As in the previous case, (3.1.28) is compared with its triangular counterpart, $\tilde{L}\tilde{D}\tilde{L}^T$, which has exactly the form of (3.1.13). A series of equations results from equating corresponding elements. Beginning with respective elements in the first column as before, the first two relationships are

$$d_1^* = d_1 + qv_1^2, \tag{3.1.29}$$

$$d_1^* \tilde{l}_{21} = qv_2 v_1. \tag{3.1.30}$$

The next relationship comes from equating elements in the second row, second column:

$$d_2 + qv_2^2 = d_1^* \tilde{l}_{21}^2 + d_2^*, \tag{3.1.31}$$

which can be solved for d_2^*:

$$d_2^* = d_2 + qv_2^2 - \frac{q(v_2 v_1)^2}{d_1^*} = d_2 + v_2^2 (q - b_1^2 d_1^*). \tag{3.1.32}$$

The right side of (3.1.32) contains a new variable: $b_1 = qv_1/d_1^*$.

If the reader is persistent, it is possible to continue writing these equations in the manner illustrated, using the defined variable

$$b_j = \left(\frac{v_j}{d_j^*}\right)\left(q - \sum_{k=1}^{j-1} d_k^* b_k^2\right). \tag{3.1.33}$$

Then it can be found in general that

$$\tilde{l}_{ij} = v_i b_j, \qquad j < i, \tag{3.1.34}$$

$$d_i^* = d_i + v_i^2 \left(q - \sum_{j=1}^{i-1} b_j^2 d_j^*\right). \tag{3.1.35}$$

The rank 1 update defined by (3.1.22) has been appended as an option in the $\mathbf{LDL}^T$ factorization in program C3-2, LDLTFAC. The BASIC instructions for the update occur in lines 7350 to 7780, beginning with the choice for updating and the request for input of the scalar multiplier q in (3.1.22). The program assumes that the vector $\mathbf{z}$ has already been stored in program matrix $\mathbf{B}$.

Example 3.1.3.   LOAD"MATRIX and MERGE"LDLTFAC, then store matrix Z.MAT shown at the top of Table 3.1.3 in program matrix $\mathbf{D}$. Also store the update vector $\mathbf{z} = (1\ 2\ 3\ 4\ 5)^T$ in program matrix $\mathbf{B}$ and the right-hand-side vector $\mathbf{b} = (265.5\ 367\ 260.5\ 470\ 212.5)^T$ in program matrix $\mathbf{C}$. Use command 10 to factor $\mathbf{A}$; when the program asks if an update is desired, answer "Y" and furnish scalar $q = \frac{1}{2}$, that is, $\mathbf{A}^* = \mathbf{A} + \frac{1}{2}\mathbf{z}\mathbf{z}^T$. At this stage, the $\mathbf{LDL}^T$ factorization of $\mathbf{A}^*$ is stored in program vector $H(\cdot)$ and also program matrix $\mathbf{D}$. Command 13 then computes the solution vector $(\mathbf{A}^*)^{-1}\mathbf{b}$; the result in program matrix $\mathbf{A}$ should be $(5\ 4\ 3\ 2\ 1)^T$. Of course, any other right-hand-side vectors for $\mathbf{b}$ could be stored in program matrix $\mathbf{C}$ to obtain new solutions without refactoring the updated matrix. The reader is encouraged to verify this example numerically by the method in Problem 2.15 and by directly forming (3.1.22) and using matrix inverse command 9.

In summary, when there are square linear systems of full rank, solutions are obtained by methods that implicitly save the matrix inverse. The $\mathbf{LU}$ factorization and the $\mathbf{LDL}^T$ factorizations are efficient, requiring on the order of n^3 operations. The $\mathbf{LU}$ factorization is the same as Gaussian elimination, factors any real matrix, but requires pivoting to ensure numerical stability. The $\mathbf{LDL}^T$ factorization is similar to Cholesky factorization, factors only real symmetrical matrices, and is numerically stable without pivoting *only* if the matrix is positive definite, that is, has all positive real eigenvalues. The forward and back substitution operations required to obtain a solution vector from a right-hand-side vector require only on the order of n^2 operations. Rank 1 updating for a matrix can be performed by a well-known identity given in Problem 2.15, using on the order of n^2 operations. However, the Gill–Murray method performs that function on the $\mathbf{LDL}^T$ factorization and thereby enables detection and potential correction of a non-positive-definite matrix. This is a particular risk when the scalar multiplier q is negative in the rank 1 update, $q\mathbf{v}\mathbf{v}^T$. It is generally agreed that the $\mathbf{QR}$ (ne $\mathbf{QU}$), $\mathbf{LU}$, and $\mathbf{LDL}^T$ factorizations are the major ones in linear algebra.

3.1.2. Overdetermined Linear Systems of Full Rank.

Systems having more linear equations than unknowns are *overdetermined*. When there are m equations in n unknowns, the system is still represented by $\mathbf{Ax} = \mathbf{b}$, where $\mathbf{A}$ is of order $m \times n$, but the equations are said to be *inconsistent*. This section examines the case where $r = n < m$, and r is the rank of $\mathbf{A}$ (i.e., $\mathbf{A}$ has full column rank). This is the situation illustrated in Figure 3.1.1; those particular equations are:

$$\begin{bmatrix} 1 & 2 \\ 5 & 7 \\ 5 & 3 \end{bmatrix} \mathbf{x} = \begin{bmatrix} 4 \\ 21 \\ 15 \end{bmatrix}. \tag{3.1.36}$$

Since only pairs of the three equations can be satisfied (the three intersections in Figure 3.1.1), the crucial question for overdetermined, full-rank systems of linear equations is the nature of a compromise solution. As Noble (1969:142) described it, the classical compromise is to minimize the sum of the squared errors for each equation, namely, to minimize the square of the two-norm of the residual vector:

$$\mathbf{r} = \mathbf{b} - \mathbf{A}\mathbf{x}. \tag{3.1.37}$$

Labeling this error criterion $S = \|\mathbf{r}\|_2^2 = \mathbf{r}^T\mathbf{r}$ and using (3.1.37):

$$S = \mathbf{b}^T\mathbf{b} - \mathbf{x}^T\mathbf{A}^T\mathbf{b} - \mathbf{b}^T\mathbf{A}\mathbf{x} + \mathbf{x}^T\mathbf{A}^T\mathbf{A}\mathbf{x}. \tag{3.1.38}$$

A necessary condition for defining the minimum of the scalar error criterion S is that all partial derivatives of S with respect to each element of $\mathbf{x}$, namely, x_i, $i = 1$ to n, must be equal to zero.

For this immediate purpose and subsequent more general applications, it is necessary to examine the nature of each term in (3.1.38), especially the meaning of derivatives with respect to a vector. First, the *scalar "del"* operator

$$\nabla_i = \frac{\partial}{\partial x_i}, \tag{3.1.39}$$

and the *vector del* operator

$$\nabla = \begin{pmatrix} \nabla_1 & \nabla_2 & \cdots & \nabla_n \end{pmatrix}^T \tag{3.1.40}$$

are defined. They have no numerical significance by themselves. However, if h is a scalar function of $\mathbf{x}$, then the derivative of h with respect to $\mathbf{x}$ in E^n is called a *gradient vector* and is defined as the n vector whose ith element is the partial derivative of h with respect to x_i:

$$\nabla h = \begin{pmatrix} \nabla_1 h & \nabla_2 h & \cdots & \nabla_n h \end{pmatrix}^T. \tag{3.1.41}$$

The nature of the terms in (3.1.38) can be described by defining the *bilinear form* of two vectors, namely, the inner product $\mathbf{x}^T\mathbf{y}$. An inner product is a scalar quantity, and it is bilinear in the sense that it is a linear function of $\mathbf{x}$ for a fixed $\mathbf{y}$, and vice versa. In the more general case, suppose that $\mathbf{x}$ and $\mathbf{y}$ are in E^n and that $\mathbf{y} = \mathbf{C}\mathbf{z}$, where $\mathbf{C}$ is $n \times m$ and $\mathbf{z}$ is in E^m. Then $\mathbf{x}^T\mathbf{C}\mathbf{z}$ is also a bilinear form, which can be expressed according to (2.2.5) as

$$\begin{aligned}
\mathbf{x}^T\mathbf{C}\mathbf{z} &= \mathbf{x}^T\begin{pmatrix} \mathbf{c}_1 z_1 & \mathbf{c}_2 z_2 & \cdots & \mathbf{c}_m z_m \end{pmatrix} \\
&= \begin{pmatrix} z_1 \mathbf{x}^T\mathbf{c}_1 & z_2 \mathbf{x}^T\mathbf{c}_2 & \cdots & z_m \mathbf{x}^T\mathbf{c}_m \end{pmatrix} \\
&= \sum_{i=1}^{n}\sum_{j=1}^{m} c_{ij} x_i z_j.
\end{aligned} \tag{3.1.42}$$

An important special bilinear form is the *quadratic form*:

$$Q(\mathbf{x}) = \mathbf{x}^T \mathbf{C} \mathbf{x} = \sum_{i=1}^{n} \sum_{j=1}^{n} c_{ij} x_i x_j \tag{3.1.43}$$

$$= c_{11} x_1^2 + (c_{12} + c_{21}) x_1 x_2 + \cdots + c_{22} x_2^2 + \cdots + c_{nn} x_n^2.$$

The case when $\mathbf{C}$ is symmetric, $\mathbf{C} = \mathbf{C}^T$, is most often of interest. By writing the terms from (3.1.43) and taking the partial derivatives, it can be confirmed that the *partial derivative of a quadratic form* is

$$\nabla Q = 2\mathbf{C}\mathbf{x}, \quad \mathbf{C} = \mathbf{C}^T. \tag{3.1.44}$$

With that background concerning partial derivatives of a scalar with respect to a vector, the gradient of the sum of squared residuals in (3.1.38) is

$$\nabla S = -\mathbf{A}^T \mathbf{b} - (\mathbf{b}^T \mathbf{A})^T + 2\mathbf{A}^T \mathbf{A} \mathbf{x}, \tag{3.1.45}$$

and $\nabla S = \mathbf{0}$ is the necessary condition for a minimum of S. Note that $\mathbf{A}^T \mathbf{A}$ is always symmetric. Therefore, the solution to an overdetermined system of m equations in n unknowns, $\mathbf{A}\mathbf{x} = \mathbf{b}$, of rank n and $m \geq n$ is

$$\mathbf{x} = (\mathbf{A}^T \mathbf{A})^{-1} \mathbf{A}^T \mathbf{b} = \mathbf{A}^+ \mathbf{b}. \tag{3.1.46}$$

The *generalized inverse* matrix $\mathbf{A}^+$ has been defined previously in (2.2.97) in connection with projection matrices.

It will now be shown that the same solution for overdetermined systems of equations may be obtained by using orthogonal decomposition. Recall that the Gram–Schmidt decomposition $\mathbf{A}_{m,n} = \mathbf{Q}_{m,n} \mathbf{U}_{n,n}$ was described in (2.2.26), where $\mathbf{Q}$ is an orthonormal matrix and $\mathbf{U}$ is an upper triangular matrix. According to Jennings (1977:139), if $\mathbf{A}\mathbf{x} = \mathbf{b}$ and $\mathbf{A} = \mathbf{Q}\mathbf{U}$, then $\mathbf{Q}\mathbf{U}\mathbf{x} = \mathbf{b}$ and

$$\mathbf{U}\mathbf{x} = \mathbf{Q}^T \mathbf{b}, \tag{3.1.47}$$

since $\mathbf{Q}^T \mathbf{Q} = \mathbf{I}$. However, $\mathbf{U}$ is upper triangular, so the system of equations in (3.1.47) can be solved for $\mathbf{x}$ by back substitution. As Golub (1970:239) has noted, the matrix $\mathbf{A}^T \mathbf{A}$ is square, so that the condition numbers $k(\mathbf{A}^T \mathbf{A}) = k^2(\mathbf{A})$. Therefore, the loss of accuracy in solving (3.1.46) with an elimination method may well be twice that of solving (3.1.47) by a similar method, since $k(\mathbf{U}) = k(\mathbf{A})$.

Although (3.1.47) represents a practical way to solve systems of equations, overdetermined or not, further manipulation is useful to reveal a theoretical result. Premultiplying both sides of (3.1.47) by $\mathbf{U}^T$ and substituting $\mathbf{U}^T \mathbf{Q}^T \mathbf{Q} \mathbf{U}$ for $\mathbf{U}^T \mathbf{U}$ yields

$$\mathbf{U}^T \mathbf{Q}^T \mathbf{Q} \mathbf{U} \mathbf{x} = \mathbf{U}^T \mathbf{Q}^T \mathbf{b}. \tag{3.1.48}$$

Now substitution of $\mathbf{QU} = \mathbf{A}$ yields the same result as (3.1.46):

$$\mathbf{A}^T\mathbf{A}\mathbf{x} = \mathbf{A}^T\mathbf{b} \quad \text{or} \quad \mathbf{x} = (\mathbf{A}^T\mathbf{A})^{-1}\mathbf{A}^T\mathbf{b}. \tag{3.1.49}$$

The first set of equations in (3.1.49) is called the *normal equations*.

It is concluded that solution of an overdetermined system of linear equations by orthogonal decomposition is equivalent to minimizing the two-norm of the vector of residuals. This result is to be expected, since $\|\mathbf{Qr}\|_2 = \|\mathbf{r}\|_2$, where $\mathbf{Q}$ is an orthonormal matrix and $\mathbf{r}$ is the residuals vector defined by (3.1.37). Orthogonal matrices have desirable properties because they do not magnify errors and do not change Euclidean lengths. Henceforth, the minimum two-norm of the residuals corresponding to a set of overdetermined linear equations of any rank are called the *linear least-squares* problem, or LLS. It is also known as *linear regression* in statistical applications.

Example 3.1.4. The generalized inverse of the matrix in (3.1.36) and (2.2.98) was previously computed in Example 2.2.11, using program C2-8. The corresponding generalized inverse matrix was shown in the upper half of Table 2.2.13. Using the right-hand-side vector $\mathbf{b}$ from (3.1.36) in (3.1.46) with the generalized inverse yields the solution $\mathbf{x} = (2.20087 \ 1.37991)^T$. The same solution can be obtained by (3.1.47). This point is shown graphically by the "$+$" in Figure 3.1.1. The squared magnitude of the residual vector in (3.1.37) corresponding to this solution is 1.05677. The same measure applied to each of the three intersections in Figure 3.1.1 having the coordinates given in Table 3.1.1 produces 178.11, 9.87732, and 1.21, respectively. In fact, it is instructive to evaluate the sum of squared residuals at points in the immediate neighborhood of the least-squares solution to verify that it is indeed the minimum.

The overdetermined system $\mathbf{Ax} = \mathbf{b}$ is inconsistent because there is no exact solution. However, there is a related consistent system, $\mathbf{Ax}' = \mathbf{b}'$, where $\mathbf{b}'$ is the projection of the right-hand side $\mathbf{b}$ vector into the column space of $\mathbf{A}$. Consider the case in the preceding example, where $\mathbf{b}$ is in E^3 and the column space of $\mathbf{A} = (\mathbf{a}_1 \ \mathbf{a}_2)$ is spanned by $\mathbf{a}_1$ and $\mathbf{a}_2$, that is, a subspace having only two dimensions. This is pictured in Figure 3.1.2. According to (3.1.46), $\mathbf{x}' = \mathbf{A}^+\mathbf{b}$; premultiplying by $\mathbf{A}$,

$$\mathbf{Ax}' = \mathbf{b}' = (\mathbf{AA}^+)\mathbf{b}. \tag{3.1.50}$$

Even more explicit, $\mathbf{A}$ may be written in column-vector form to show the nature of $\mathbf{b}'$, as illustrated in Figure 3.1.2:

$$\mathbf{b}' = \mathbf{a}_1 x_1' + \mathbf{a}_2 x_2'. \tag{3.1.51}$$

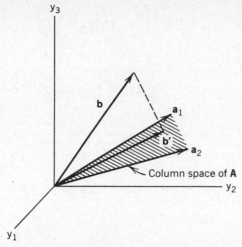

Figure 3.1.2. The 3-space containing the independent vector **b** in **Ax** = **b** also contains the subspace spanned by the two column vectors from **A**. The LLS solution is obtained by first projecting **b** onto that subspace using **b′** = **AA**$^+$**b** and then solving **Ax′** = **b′**.

However, (2.2.89) showed that $(\mathbf{AA}^+) = \tilde{\mathbf{P}}$ is a transformation that takes a vector into the subspace spanned by the columns of **A**, previously referred to as the normal vectors related to hyperplanes. This is the mathematical justification for the illustration in Figure 3.1.2. It is not to scale, since inevitably the angle between the **b** and **b′** vectors is quite small. For the system in Example 3.1.4, it is easily found from (2.1.42) that the angle in that case is only 2.26 degrees.

The major application of the LLS problem in classical mathematical and engineering fields is evaluating coefficients in a *linear mathematical model*. Assume that some function $d(t)$ of independent variable t is known only by m pairs of data points, say (t_i, d_i), $i = 1$ to m. Consider approximating the function $d(t)$ by a linear combination of n *basis functions*, $f_j(t)$:

$$f(\mathbf{x}, t) = x_1 f_1(t) + x_2 f_2(t) + \cdots + x_j f_j(t) + \cdots + x_n f_n(t). \quad (3.1.52)$$

This is the linear mathematical model because it is linear in the coefficients x_j, even though the functions f_j are nonlinear in the independent variable t. For example, $f(\mathbf{x}, t) = x_1 \cos(2t) + x_2 \exp(-3t) + x_3(t^2 - 1)^{1/2}$ is linear in **x**. This is in contrast to the nonlinear mathematical model described in (1.2.5), where some elements of **x** are also involved in the basis functions; that case is considered in a later chapter. Figure 3.1.3 shows the m data pairs at the "×" points in the *sample space t*. The linear mathematical model in (3.1.52) "fits" the data within some residual error, r_i, at each ith sample point as shown in Figure 3.1.3. Writing the equations for each sample point as if each residual

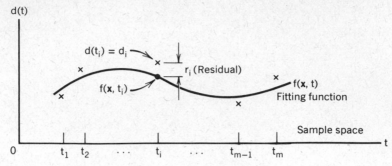

Figure 3.1.3. A curve-fitting example where a function $d(t)$ known by m given data pairs is to be approximated by a linear function $f(\mathbf{x}, t)$ that is the sum of given basis functions. The problem is to minimize the sum of the squared residuals r_i defined at each point in the sample space t.

were equal to zero (for $n = 3$ and $m = 5$ without loss of generality):

$$x_1 f_1(t_1) + x_2 f_2(t_1) + x_3 f_3(t_1) = d_1$$

$$x_1 f_1(t_2) + x_2 f_2(t_2) + x_3 f_3(t_2) = d_2$$

$$x_1 f_1(t_3) + x_2 f_2(t_3) + x_3 f_3(t_3) = d_3 \qquad (3.1.53)$$

$$x_1 f_1(t_4) + x_2 f_2(t_4) + x_3 f_3(t_4) = d_4$$

$$x_1 f_1(t_5) + x_2 f_2(t_5) + x_3 f_3(t_5) = d_5$$

These equations are overdetermined and therefore inconsistent.

It is seen from (3.1.53) in comparison to matrix equation $\mathbf{A}\mathbf{x} = \mathbf{b}$, where $\mathbf{A} = [a_{ij}]$, that

$$a_{ij} = f_j(t_i) \qquad \text{and} \qquad b_j = d_j, \qquad j = 1 \text{ to } n, \ i = 1 \text{ to } m. \quad (3.1.54)$$

Matrix $\mathbf{A}$ is called the *design matrix* and vector $\mathbf{b} = \mathbf{d}$ is called the *data vector*. Note that $\mathbf{A}$ depends only on the basis functions and the discrete sample space. To "fit" the curve $f(\mathbf{x}, t)$ to the discrete data pairs, it is only necessary to solve the LLS problem. Emphasis on particular residuals can be obtained by *weighted least squares*, where the ith row of $\mathbf{A}$ is multiplied by some positive scalar, called a weight. This amounts to scaling matrix $\mathbf{A}$, which can also be viewed as changing the column space and condition number, $k(\mathbf{A})$.

Example 3.1.5. Suppose that the piecewise-linear ramp shown by the two solid lines in Figure 3.1.4 is to be approximated by the linear mathematical model

$$f(\mathbf{x}, t) = x_1 + x_2 t + x_3 t^2 + x_4 t^3 + x_5 t^4. \qquad (3.1.55)$$

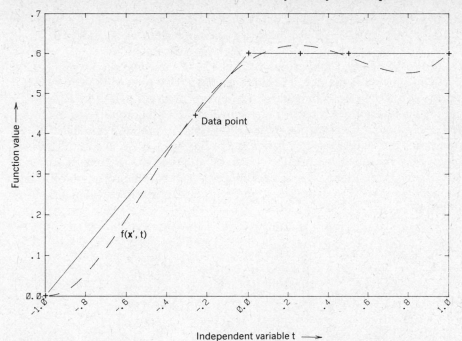

Figure 3.1.4. A 4th-degree power series LLS fit to a piecewise-linear ramp. In this case six unequally spaced data pairs are given as indicated by the "+" points. The approximating curve is shown by the dashed line.

The given data are shown in Table 3.1.4. Assuming unity weighting factors for each row (i.e., for each point in the sample space), the design matrix is:

$$
\mathbf{A} = \begin{bmatrix}
1 & t_1 & t_1^2 & t_1^3 & t_1^4 \\
1 & t_2 & t_2^2 & t_2^3 & t_2^4 \\
1 & t_3 & t_3^2 & t_3^3 & t_3^4 \\
1 & t_4 & t_4^2 & t_4^3 & t_4^4 \\
1 & t_5 & t_5^2 & t_5^3 & t_5^4 \\
1 & t_6 & t_6^2 & t_6^3 & t_6^4
\end{bmatrix}. \tag{3.1.56}
$$

Table 3.1.4.   Given Data Pairs for the Linear Least-Squares Curve Fitting Illustration in Example 3.1.5

i	1	2	3	4	5	6
t	−1.	−.25	0	.25	.5	1.
d_i	0.0	0.45	0.60	0.60	0.60	0.60

Matrix **A** in this case is the well-known *Vandermonde matrix*. The elements of vector **b**, the data vector, are the corresponding values of d_i shown in Table 3.1.4.

The MATRIX program can be used in any of three ways to solve this LLS problem: (1) solve the normal equations $A^TAx = A^Tb$, according to (3.1.46), using matrix inverse command 9; (2) merge program GENINVP and use command 10 to obtain the generalized inverse, according to (3.1.46), assuming full rank; and (3) merge program GSDECOMP and use command 10 to obtain the orthogonal decomposition, according to (3.1.47). The last method is highly recommended, because the system of equations that is actually solved is much better conditioned. The LLS solution **x′** and **b′**, the projection of **b** onto the column space of **A** as suggested in Figure 3.1.2 for different dimensions, is found to be:

$$\mathbf{x'} = \begin{bmatrix} 0.58105 \\ 0.32872 \\ -0.70796 \\ -0.02829 \\ 0.42712 \end{bmatrix}, \quad \mathbf{b'} = \begin{bmatrix} -0.00021 \\ 0.45674 \\ 0.58105 \\ 0.62021 \\ 0.59158 \\ 0.60063 \end{bmatrix}. \tag{3.1.57}$$

The angle between **b** and **b′** is 1.32 degrees. The dashed curve in Figure 3.1.4 represents the linear mathematical model in (3.1.55) with **x** defined in (3.1.57).

The basis functions in (3.1.55) in this example are the exponential terms from the power series; these usually result in a badly conditioned set of normal equations, (3.1.49). It is important to select basis functions so that the columns of **A** tend toward orthogonality, that is, the condition number of A^TA approaches unity. Better choices for basis functions include Chebyshev polynomials of the first kind; see Cuthbert (1983:21). Also, shifting the basis functions along the real axis and scaling their magnitudes are often necessary conditioning operations; see Forsythe (1977:200).

3.1.3.   Rank-Deficient Linear Systems.

As described by Lawson (1974:3), there are exactly six possible cases that may occur with sets of linear equations, depending on the number of equations (m), the number of unknowns (n), and the system rank (r). The two cases of interest in optimization concern square and overdetermined systems (i.e., $m = n$ and $m > n$, respectively). If **A** is *underdetermined* (i.e., $m < n$), then A^T can be analyzed instead without loss of generality (see Problem 3.14). Having dealt with the full rank case ($r = n$), this section provides a universal approach to systems of linear equations that are *rank-deficient*, ($r < n$). This corresponds with having some of the lines parallel in Figure 3.1.1.

It is possible to approach rank-deficient linear systems on the basis of orthogonal decompositions, as previously discussed; the interested reader is referred to Lawson (1974:13). However, Klema (1980) has called attention to the *singular value decomposition* (SVD): It now forms the cornerstone of modern linear algebra, it is the most elegant algorithm in numerical algebra for exposing quantitative information about the structure of a system of linear equations, and it is the only generally reliable method for numerical determination of rank. The SVD is useful for square matrices of full rank, but its greatest application is for solution and analysis of overdetermined systems, such as LLS, where there also may be rank deficiency. The singular value decomposition is defined, an algorithm in BASIC is provided, and several examples of rank-deficient systems of linear equations are used to illustrate some of its many applications.

The singular value decomposition of any matrix is

$$\mathbf{A} = \mathbf{USV}^T, \tag{3.1.58}$$

where the respective dimensions are $\mathbf{A}_{m,n}$, $\mathbf{U}_{m,m}$, $\mathbf{S}_{m,n}$, and $\mathbf{V}_{n,n}$. Matrices $\mathbf{U}$ and $\mathbf{V}$ are both orthonormal, and their columns are called the left and right *singular vectors* of $\mathbf{A}$, respectively. They are unique within a factor of ± 1. Rectangular matrix $\mathbf{S}$ has elements $s_{ij} = 0$ for $i \neq j$ and $s_{ii} = s_i \geq 0$ on the diagonal. The non-negative quantities s_i are called *singular values* of $\mathbf{A}$ and are equal to the positive square root of the eigenvalues of $\mathbf{A}^T\mathbf{A}$ or of $\mathbf{AA}^T$. In fact, the *matrix two-norm* of $\mathbf{A}$ described by (2.1.48) is equal to the largest singular value. It is called the *spectral norm*. Furthermore, the left and right singular vectors of $\mathbf{A}$ are particular choices of the eigenvectors of $\mathbf{AA}^T$ and $\mathbf{A}^T\mathbf{A}$, respectively. Reordering (3.1.58),

$$\mathbf{S} = \mathbf{U}^T\mathbf{AV}. \tag{3.1.59}$$

As Forsythe (1977:203) noted, the matrix transformation $\mathbf{A}$ may be modified by an orthogonal change of coordinates in its domain and by a second orthogonal change of coordinates in its range, so that the total transformation becomes diagonal. Since orthogonal transformations are of full rank, it is clear that the rank of a diagonal matrix is equal to the number of its nonzero diagonal elements.

The definition of the SVD given by (3.1.58) is equivalent to the summation of rank 1 matrices that are composed of the outer product of respective columns from $\mathbf{U}$ and $\mathbf{V}$:

$$\mathbf{A} = \sum_{i=1}^{n} s_i \mathbf{u}_i \mathbf{v}_i^T. \tag{3.1.60}$$

It is convenient to assume that the singular values are ordered from largest to smallest, so that $s_1 \geq s_2 \geq s_3 \geq \cdots \geq s_n \geq 0$. The smallest singular value s_n is a measure of how "close" $\mathbf{A}$ is to singularity. There are exactly r nonzero singular values, s_i, $i = 1$ to r, for a matrix of rank r. The rank r matrix in

(3.1.60) is thus the linear sum of r rank 1 matrices, each weighted by the respective nonzero singular value.

According to (3.1.59), the elements of S are

$$s_{ij} = \mathbf{u}_i^T \mathbf{A} \mathbf{v}_j. \tag{3.1.61}$$

The concept of singular value decomposition is that the respective columns of U and V can be chosen so that $s_{ij} = 0$. That process is the second step in the SVD computer algorithm due to Golub and Reinsch, which is a variant of Francis' iterative QR algorithm described in connection with (2.2.52), Section 2.2.3, to diagonalize a symmetric matrix. The first step in computing the SVD is to reduce the given matrix A to a matrix having only the principal diagonal and the one above it (the superdiagonal) using *Householder's bidiagonalization*. This transformation is similar to the Householder transformation previously described in Section 2.2.4, except that A is transformed by both pre- and postmultiplications, as indicated by the resultant U and V matrices in (3.1.58). This process is described in considerable detail in Golub (1983:171) and in Forsythe (1977:221). Program C3-3, SVD in Appendix C, follows the extremely robust program by Forsythe (1977:230), which in turn was based on the Golub and Reinsch Algol program by Golub in 1970. As the following example will make clear, there is no need to utilize the full dimensions given for each matrix shown in (3.1.58), since there can be no more than n nonzero singular values in S. Therefore, *in practice*, the SVD is computed using dimensions according to

$$\mathbf{A}_{m,n} = \mathbf{U}_{m,n} \mathbf{S}_{n,n} \mathbf{V}_{n,n}^T. \tag{3.1.62}$$

Example 3.1.6. Program SVD can be loaded and run in the BASIC environment after storing the matrix A given at the top of Table 3.1.5. The menu in program SVD is similar to that in program MATRIX. Command 1 can be used to load matrix A. On completion of command 2, command 0 can be used to see the results on the screen or command 3 prints any of the four matrices. Command 1 also enables storage on disk of any of the four matrices; this is useful for subsequent execution of program MATRIX for further processing of these results. For example, it is easy to confirm that $U^T U = I_3 = V^T V$, and that $A = USV^T$. Also, the fixed format in Table 3.1.5 indicates that element $s_{33} = 0$; command 19 in MATRIX shows that in fact $s_{33} = 0.00000022$. In this case, there is little uncertainty about the rank of S and, consequently, A (rank 2). Often it is useful to display the distribution of singular values, since the determination of a suitable approximation for zero is arbitrary. In this case A is rank 2 because its middle column is the average of the outer two; this is indicated by the third column of V, which has element magnitudes in the ratio $1 : 2 : 1$. In order that there be no subsequent confusion from showing matrices of reduced dimensions, the *actual* S matrix that appears in (3.1.58) is

Table 3.1.5.   A 6 × 3 Rank 2 Matrix and Its Singular Value Decomposition

```
MATRIX A( 6 , 3 ) -
        1.00000          7.00000         13.00000
        2.00000          8.00000         14.00000
        3.00000          9.00000         15.00000
        4.00000         10.00000         16.00000
        5.00000         11.00000         17.00000
        6.00000         12.00000         18.00000
MATRIX U( 6 , 3 ) -
       -0.31969         -0.64931         -0.34771
       -0.35348         -0.41267          0.76239
       -0.38726         -0.17602         -0.22093
       -0.42104          0.06063         -0.44084
       -0.45483          0.29727          0.23351
       -0.48861          0.53392          0.01359
MATRIX S( 3 , 3 ) -
       45.80601          0.00000          0.00000
        0.00000          3.28775          0.00000
        0.00000          0.00000          0.00000
MATRIX V( 3 , 3 ) -
       -0.19819          0.89110          0.40825
       -0.51582          0.25934         -0.81650
       -0.83345         -0.37241          0.40825
```

$$\mathbf{S} = \begin{bmatrix} 45.8 & 0 & 0 \\ 0 & 3.29 & 0 \\ 0 & 0 & 0 \\ 0 & 0 & 0 \\ 0 & 0 & 0 \\ 0 & 0 & 0 \end{bmatrix}.$$

The SVD is a reliable, well-conditioned tool for solution of systems of linear equations represented by $\mathbf{Ax} = \mathbf{b}$. Substitution of (3.1.58) yields

$$\mathbf{USV}^T\mathbf{x} = \mathbf{b} \qquad \text{or} \qquad \mathbf{SV}^T\mathbf{x} = \mathbf{U}^T\mathbf{b}. \tag{3.1.63}$$

Similar to the definition of an intermediate variable in the **LU** factorization procedure, define

$$\mathbf{z} = \mathbf{V}^T\mathbf{x}, \tag{3.1.64}$$

$$\mathbf{d} = \mathbf{U}^T\mathbf{b}. \tag{3.1.65}$$

These substitutions in (3.1.63) leave the diagonal system

$$\mathbf{Sz} = \mathbf{d}, \tag{3.1.66}$$

so that

$$z_i = \frac{d_i}{s_i}, \qquad i = 1 \text{ to } n, \, s_i \neq 0. \qquad (3.1.67)$$

Then the system solution from (3.1.64) is

$$\mathbf{x} = \mathbf{V}\mathbf{z}. \qquad (3.1.68)$$

Actually, (3.1.67) is misleading in the general case; Forsythe (1977:208) notes that there are exactly three cases to consider:

1. $s_i z_i = d_i$, if $i \leq n$ and $s_i \neq 0$.
2. $0 \cdot z_i = d_i$, if $i \leq n$ and $s_i = 0$.
3. $0 = d_i$, if $i > n$.

Case 1 is that described in (3.1.67). Note that $\mathbf{x}$ is determined by $\mathbf{z}$ in (3.1.68), $\mathbf{z}$ is determined by $\mathbf{d}$ in (3.1.66), and $\mathbf{d}$ is determined by $\mathbf{b}$ in (3.1.65); however, the *actual* dimensions of $\mathbf{U}^T$ are $m \times m$ and $\mathbf{b}$ are $m \times 1$. Case 2 cannot occur if the system has full rank ($r = n$), and case 3 cannot occur if the system is square ($m = n$). Otherwise, the system $\mathbf{A}\mathbf{x} = \mathbf{b}$ is inconsistent unless $d_i = 0$ whenever $s_i = 0$ or $i > n$ (overdetermined equations). In other words, (3.1.65) leads to the conclusion that $\mathbf{b}$ is not in the range of $\mathbf{A}$ in the sense previously portrayed in Figure 3.1.2.

The practical situation occurs in the LLS problem, where the norm of the residuals vector $\mathbf{r}$ can now be expressed in terms of the SVD:

$$\|\mathbf{r}\|_2 = \|\mathbf{A}\mathbf{x} - \mathbf{b}\| = \|\mathbf{U}^T(\mathbf{A}\mathbf{V}\mathbf{V}^T\mathbf{x} - \mathbf{b})\| = \|\mathbf{S}\mathbf{z} - \mathbf{d}\|. \qquad (3.1.69a)$$

Therefore, the SVD reduces the LLS problem to one involving the diagonal matrix $\mathbf{S}$, and clearly (3.1.67) does result in minimum $\|\mathbf{r}\|_2$ when $s_i \neq 0$. The remaining z_i and thus x_i are arbitrary when $s_i = 0$ (i between 1 and n) or when $i > n$. From (3.1.69a), the squared norm of the LLS residuals vector is simply Σd_i^2, summed over all i except those for which $s_i \neq 0$.

From Golub (1983:25), note that when $\mathbf{A} = \mathbf{U}\mathbf{S}\mathbf{V}^T$ has full rank, the solution vector $\mathbf{x}$ in the system $\mathbf{A}\mathbf{x} = \mathbf{b}$ may be expressed as

$$\mathbf{x} = \mathbf{A}^{-1}\mathbf{b} = (\mathbf{U}\mathbf{S}\mathbf{V}^T)^{-1}\mathbf{b} = \sum_{i=1}^{n} \frac{\mathbf{v}_i(\mathbf{u}_i^T\mathbf{b})}{s_i}. \qquad (3.1.69b)$$

In (3.1.69b), $\mathbf{v}_i$ is the ith column vector of $\mathbf{V}$; similarly, $\mathbf{u}_i$ is from $\mathbf{U}$, and s_i is the ith singular value. Note that the inner product $(\mathbf{u}^T\mathbf{b})$ in the summation is a scalar quantity. The expansion in (3.1.69b) shows that changes in $\mathbf{b}$ and changes in $\mathbf{A}$ reflected in $\mathbf{u}_i$ and $\mathbf{v}_i$ will drastically affect the solution $\mathbf{x}$ *when some singular values are relatively small.*

A condition number based on singular values has been defined; it is similar in concept, application, and numerical value as those based on matrix norms. The condition number using singular values is

$$k_2(\mathbf{A}) = \frac{s_n}{s_r}, \qquad (3.1.70)$$

where s_n is the largest singular value and s_r is the smallest nonzero singular value ($s_r = s_n$ when $\mathbf{A}$ is full rank). The condition number based on singular values is applicable to a rank-deficient matrix, a situation that is untenable for the norm-based condition number (2.1.50) that theoretically requires the matrix inverse. An example will illustrate the use of the singular-value condition number and serve to introduce the concept of *matrix scaling*. Recall that premultiplication of a matrix by a diagonal matrix multiplies each row of the multiplicand matrix by the corresponding element in the diagonal matrix. Similarly, postmultiplication of a matrix by a diagonal matrix multiplies each column by the corresponding diagonal element in the diagonal matrix.

Example 3.1.7.   Consider the matrix scaling operation $\mathbf{A} = \mathbf{BCD}$ on a matrix ($\mathbf{C}$) attributed to Bauer by Klema (1980):

$$
\mathbf{BCD} =
\begin{bmatrix}
1 & 0 & 0 & 0 & 0 & 0 \\
0 & 1 & 0 & 0 & 0 & 0 \\
0 & 0 & 1 & 0 & 0 & 0 \\
0 & 0 & 0 & 1 & 0 & 0 \\
0 & 0 & 0 & 0 & 8 & 0 \\
0 & 0 & 0 & 0 & 0 & 7
\end{bmatrix}
\begin{bmatrix}
-74 & 80 & 18 & -11 & -4 & -8 \\
14 & -69 & 21 & 28 & 0 & 7 \\
66 & -72 & -5 & 7 & 1 & 4 \\
-12 & 66 & -30 & -23 & 3 & -3 \\
3 & 8 & -7 & -4 & 1 & 0 \\
4 & -12 & 4 & 4 & 0 & 1
\end{bmatrix}
$$

$$
\times
\begin{bmatrix}
1 & 0 & 0 & 0 & 0 & 0 \\
0 & 1 & 0 & 0 & 0 & 0 \\
0 & 0 & 2 & 0 & 0 & 0 \\
0 & 0 & 0 & 3 & 0 & 0 \\
0 & 0 & 0 & 0 & 10 & 0 \\
0 & 0 & 0 & 0 & 0 & 10
\end{bmatrix}, \qquad (3.1.71)
$$

$$
\mathbf{A} = \mathbf{BCD} =
\begin{bmatrix}
-74 & 80 & 36 & -33 & -40 & -80 \\
14 & -69 & 42 & 84 & 0 & 70 \\
66 & -72 & -10 & 21 & 10 & 40 \\
-12 & 66 & -60 & -69 & 30 & -30 \\
24 & 64 & -112 & -96 & 80 & 0 \\
28 & -84 & 56 & 84 & 0 & 70
\end{bmatrix}.
$$

The multiplications of $\mathbf{C}$ by both $\mathbf{B}$ and $\mathbf{D}$ shown above may be verified using program MATRIX, although it is very useful to do this by hand to reinforce the concept of row scaling and column scaling by diagonal matrices. Further, program SVD shows that the singular values of $\mathbf{C}$ are 173.84, 64.86, 10.67, 1,

0.17525, and 0.00004744. Clearly, the last singular value can be taken as zero, so that $\mathbf{C}$ has rank 5 and condition number $k_2(\mathbf{C}) = 992$. Similarly, program SVD shows that the singular values of $\mathbf{A}$ are: 295.95, 181.66, 48.94, 12.88, 0.70960, and 0.001397. Again, the rank is 5 and the condition number $k_2(\mathbf{A}) = 417$.

For the system of linear equations $\mathbf{Cx} = \mathbf{b}$, a scaling of the rows (previously suggested in connection with weighted LLS problem) is achieved by transformation of $\mathbf{C}$ by premultiplying by a diagonal matrix $\mathbf{B}$. A scaling of the columns of $\mathbf{C}$ is achieved by postmultiplying by a diagonal matrix $\mathbf{D}$. This latter scaling amounts to a linear transformation of variables, $\mathbf{x} = \mathbf{Dy}$, since $\mathbf{Cx} = \mathbf{b}$ can be restated as $(\mathbf{CD})\mathbf{y} = \mathbf{b}$. For row and column scaling taken together, the scaled system of linear equations becomes $(\mathbf{BCD})\mathbf{y} = (\mathbf{Bb})$. The optimal choices of diagonal scaling matrices have been treated in literature attributed to Bauer (1963) and in his subsequent articles. A discussion that is easier to read appears in Noble (1969:438). The essence of *optimally scaled matrices* is that the absolute row sums should be about equal and the absolute column sums should be about equal. An absolute row sum is the sum of the absolute values of element in a particular row. For $\mathbf{C}$ in (3.1.71), the absolute row sums range from 23 to 195, and the absolute column sums range from 9 to 307. Scaling matrices $\mathbf{B}$ and $\mathbf{D}$ improved the matrix condition number from $k_2(\mathbf{C}) = 992$ to $k_2(\mathbf{A}) = 417$. This improvement is reflected in $\mathbf{A}$ in (3.1.71), where the absolute row sums range from 219 to 376 and the absolute column sums range from 160 to 435. Scaling of variables is also discussed in later chapters.

The SVD enables a more general statement of the *generalized* (*pseudo*) *inverse*. When matrix $\mathbf{A}$ has rank $r < n$, it was shown by (3.1.69a) that the squared residual $\|\mathbf{Ax} - \mathbf{b}\|_2^2$ is equal to

$$\|\mathbf{Sz} - \mathbf{d}\|_2^2 = (s_1 z_1 - d_1)^2 + \cdots + (s_r z_r - d_r)^2 + d_{r+1}^2 + \cdots + d_m^2. \quad (3.1.72)$$

The residual is thus independent of z_{r+1} through z_n; according to (3.1.68), the residual is likewise independent of x_{r+1} through x_n, and these are taken to be zero in order to minimize the length (norm) of $\mathbf{x}$. The remaining values of $\mathbf{z}$ (thus $\mathbf{x}$) are used to minimize (3.1.72); therefore, $z_i = d_i/s_i$, $i = 1$ to r, and the corresponding elements of $\mathbf{x}$ are determined by (3.1.68). A concise expression for $\mathbf{x}$ is available: from (3.1.64), $\mathbf{x} = \mathbf{Vz}$; from (3.1.65), $\mathbf{d} = \mathbf{U}^T\mathbf{b}$; and the linkage between $\mathbf{z}$ and $\mathbf{d}$ is indicated by (3.1.67) and the summation related to (3.1.72). Therefore, it is seen that the solution vector can be expressed as

$$\mathbf{x} = (\mathbf{VS}^+\mathbf{U}^T)\mathbf{b} = \mathbf{A}^+\mathbf{b}, \quad (3.1.73)$$

if the $n \times m$ $\mathbf{S}^+$ matrix is defined to be

$$\mathbf{S}^+ = \begin{bmatrix} s_1^{-1} & 0 & 0 & 0 & 0 & 0 \\ 0 & s_2^{-1} & 0 & 0 & 0 & 0 \\ 0 & 0 & 0 & 0 & 0 & 0 \end{bmatrix}. \quad (3.1.74)$$

The definition shown for S^+ in (3.1.74) has been given for a 3×6 matrix without loss of generality; the main specification is that the reciprocals of the r nonzero values of s_i are placed on the principal diagonal as shown for $r = 2$, and zeros occur elsewhere. The definition of $A^+ = VS^+U^T$, the pseudoinverse of A given in (3.1.73) and (3.1.74), is entirely consistent with the solution just reviewed for the rank-deficient LLS problem. There are a number of ways to compute the pseudoinverse, but the SVD is preferred because it provides a sound basis to determine the system rank. Like the ordinary inverse, there is seldom any reason to form the pseudoinverse explicitly, its main value being in the concept.

The pseudoinverse of matrix A satisfies four identities that are sometimes taken to define A^+:

$$AA^+A = A,$$

$$A^+AA^+ = A^+,$$

$$AA^+ = (AA^+)^T, \qquad (3.1.75)$$

$$A^+A = (A^+A)^T.$$

Note that the expression from (3.1.46) for the full-rank pseudoinverse, $A^+ = (A^TA)^{-1}A^T$, satisfies (3.1.75). In fact, it is easily verified that S^+ is the pseudoinverse of S; for example, $SS^+S = S$. Then it follows that the generalized inverse for any rank, $A^+ = VS^+U^T$, also satisfies (3.1.75).

The SVD has some interesting geometric properties. As previously noted, $\|Ax\|/\|x\|$ measures how much the transformation A "deforms" the unit hypersphere, $\|x\| = 1$. From Forsythe (1977:206), recall that $A = USV^T$, and let $z = V^Tx$. Since orthonormal matrices preserve Euclidean length, $\|z\| = \|x\|$, and

$$\|Ax\| = \|USV^Tx\| = \|Sz\|. \qquad (3.1.76)$$

But S is a diagonal matrix composed of the singular values $s_1 \geq s_2 \geq \cdots \geq s_r \geq 0$, $r \leq n$, so the length $\|Sz\|$ varies from $s_r\|z\|$ to $s_1\|z\|$, depending on direction. Therefore, the distortion introduced by the matrix transformation $b = Ax$ for x over the unit hypersphere is

$$s_{min} \leq \frac{\|Ax\|}{\|x\|} \leq s_{max}. \qquad (3.1.77)$$

The transformation $b = Ax$ actually maps an n-dimensional hypersphere into an r-dimensional hyperellipsoid embedded in m-dimensional space; the geometry is illustrated in Figure 3.1.5 for the case when $n = m = r = 2$. The lengths of the ellipsoidal axes correspond to the singular values, and the condition number k_2 is equal to its eccentricity. The matrix spectral norm,

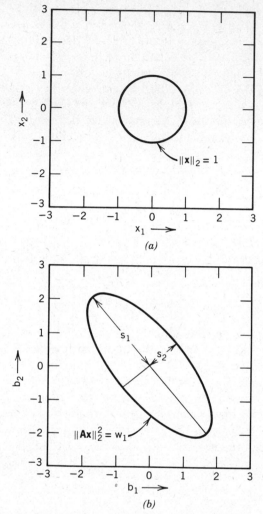

Figure 3.1.5. The matrix transformation $\mathbf{b} = \mathbf{Ax}$ in 2-space with full rank. The unit circle in (*a*) is mapped into an ellipse in (*b*) with axes corresponding to the singular values.

$\|\mathbf{A}\|_2 = s_1$, as defined earlier in (2.1.48). The matrix spectral norm is the length of the major axis in Figure (3.1.5(b)) and measures the maximum distortion because of the transformation $\mathbf{b} = \mathbf{Ax}$.

Now consider a *symmetric* matrix $\mathbf{B} = \mathbf{A}^T\mathbf{A}$. Since $\|\mathbf{Ax}\|_2^2 = \mathbf{x}^T\mathbf{A}^T\mathbf{Ax} = \mathbf{x}^T\mathbf{Bx}$, (3.1.77) leads directly to the well-known *Rayleigh quotient*:

$$w_r \le \frac{(\mathbf{x}^T\mathbf{Bx})}{(\mathbf{x}^T\mathbf{x})} \le w_1. \tag{3.1.78}$$

The eigenvalues of $\mathbf{B}$, $w_1 \ge w_2 \ge \cdots \ge w_r$ correspond to the squared singular values of $\mathbf{A}$.

Example 3.1.8.   Consider the nonsymmetrical matrix

$$A = \begin{bmatrix} 4 & 2 \\ 1 & 3 \end{bmatrix} \tag{3.1.79}$$

and its related symmetrical transposed product

$$B = A^T A = \begin{bmatrix} 17 & 11 \\ 11 & 13 \end{bmatrix}. \tag{3.1.80}$$

Program SVD computes the singular values of (3.1.79) as 5.11667 and 1.95440. Program MATRIX with GSDECOMP and QRITER merged with it computes the eigenvalues of (3.1.80) as 26.18034 and 3.81966. This confirms numerically that the singular values are equal to the positive square roots of the eigenvalues of the related symmetrical transposed product. According to (3.1.78), the maximum possible value of the quadratic form $Q = (Ax)^T(Ax) = x^T B x$ is $w_1 = 26.18034$ *for* x *on the unit circle* as in Figure 3.1.5(a). According to (3.1.43), when **B** is symmetric and given by (3.1.80), the quadratic form in E^2 is

$$Q = 17x_1^2 + 22x_1x_2 + 13x_2^2.$$

That is the locus plotted in Figure 3.1.5(b) when $Q = w_1 = 26.18034$.

Finally, the SVD easily confirms that a *similarity transformation*, as in (2.2.46), results in a diagonal matrix composed of positive semidefinite eigenvalues, $w_i \geq 0$, when applied to symmetrical matrices. Since

$$U^T A V = \text{diag}(s_1 \ s_2 \ \cdots \ s_n), \tag{3.1.81}$$

then

$$U^T(A^T A)V = \text{diag}(s_1^2 \ s_2^2 \ \cdots \ s_n^2). \tag{3.1.82}$$

While it is true that $w_i = s_i^2$, the eigenvalues should not be computed this way, since any small singular value might be reduced to rounding-error magnitude by the squaring process.

3.2.   Nonlinear Functions

The most elementary nonlinear function of many variables is the second-degree function centered at the origin and called the quadratic form; it was introduced in (3.1.43). A number of its important properties are described, including the various conic sections that the quadratic form may represent, before considering the more general second-degree function, the quadratic function. The first and second derivatives of the quadratic function are derived, and their use in locating the minimum function value is developed. Searches on a line in the variable space will be described, including first and second derivatives along the line, namely, the directional derivative and curvature. The step to the minimum along any line is applied to a sequence of several search directions that approach the quadratic function minimum.

The Taylor series in many variables is employed to describe a nonlinear function of any degree, and some properties of classical interest such as convexity and continuity are defined. The Jacobian and Hessian matrices will be introduced, especially as they apply to the solution of systems of nonlinear equations by the Newton–Raphson method. The general implications for line searches are developed for both elementary cutback methods and automatic methods based on interpolation on the line.

3.2.1. Quadratic and Line Functions.

The quadratic form introduced in (3.1.43) is only the numerator of the Rayleigh function in (3.1.78) and is thus a different function. The quadratic form is defined for a symmetric matrix **B** by

$$Q = \mathbf{x}^T \mathbf{B} \mathbf{x}, \tag{3.2.1}$$

where Q is a scalar quantity. A particular **B** matrix was considered in Example 3.1.8, and its quadratic form was expanded in the two components of **x** in (3.1.81). The *cross terms*, typically involving $x_i x_j$, had a nonzero coefficient so that the contour for a given value of Q was tilted at an angle to the axes as shown in Figure 3.1.5. To classify quadratic forms and for other important purposes, it is useful to employ a linear rotation of axes so that the cross terms do not appear in a new set of variables.

It was shown in Section 2.2.3 that a similarity transformation would diagonalize a symmetrical matrix. In particular, the orthonormal matrix whose columns are the eigenvectors of a symmetrical matrix transforms that matrix into a diagonal matrix with the eigenvalues as elements. From (2.2.45),

$$\mathbf{V}^T \mathbf{B} \mathbf{V} = \text{diag}(w_1 \ w_2 \ \cdots \ w_n) = \mathbf{W}, \tag{3.2.2}$$

where $\mathbf{V} = (\mathbf{v}_1 \ \mathbf{v}_2 \ \cdots \ \mathbf{v}_n)$, and the $\mathbf{v}_i$ are the eigenvectors of the corresponding w_i eigenvalues of **B**. Now define a new vector **y** that is also in E^n with **x**:

$$\mathbf{x} = \mathbf{V} \mathbf{y}. \tag{3.2.3}$$

Substituting (3.2.3) into (3.2.1),

$$Q = \mathbf{x}^T(\mathbf{B}\mathbf{x}) = (\mathbf{V}\mathbf{y})^T(\mathbf{B}\mathbf{V}\mathbf{y}) = (\mathbf{y})^T(\mathbf{V}^T\mathbf{B}\mathbf{V}\mathbf{y}) = \mathbf{y}^T\mathbf{W}\mathbf{y}. \tag{3.2.4}$$

The *canonical form* has been obtained:

$$Q = \mathbf{y}^T\mathbf{W}\mathbf{y} = w_1 y_1^2 + w_2 y_2^2 + \cdots + w_n y_n^2. \tag{3.2.5}$$

Table 3.2.1.   Canonical Forms in Three Variables for Positive Q

1. $Q = y_1^2/a^2 + y_2^2/b^2 + y_3^2/c^2$	Ellipsoid
2. $Q = y_1^2/a^2 + y_2^2/b^2 - y_3^2/c^2$	Hyperboloid, one sheet
3. $Q = y_1^2/a^2 - y_2^2/b^2 - y_3^2/c^2$	Hyperboloid, two sheets
4. $Q = y_1^2/a^2 + y_2^2/b^2$	Elliptic cylinder
5. $Q = y_1^2/a^2 - y_2^2/b^2$	Hyperbolic cylinder
6. $Q = y_1^2/a^2$	Pair of parallel planes

The canonical form enables simple classification of quadratic forms based on the conic forms shown in Table 3.2.1. The reduction of a quadratic form as in (3.2.1) to the canonical form (3.2.5) can also be accomplished by the ordinary method of algebra known as "completing the squares," beginning with (3.1.43). That method is known as *Lagrange's reduction* in this application (Ayres, 1962:132). It is mentioned only for completeness, since it lacks the notational and conceptual compactness inherent in diagonalization by similarity transformations.

The case of most interest is that involving a positive definite **B** (i.e., all the eigenvalues of **B** are strictly positive). Since a^2, b^2, and c^2 in case 1 of Table 3.2.1 are the lengths of the major axes of the ellipsoid when $Q = 1$, it follows that *the ellipsoidal axes are in the direction of the eigenvectors and have lengths that are inversely proportional to the square roots of the corresponding eigenvalues.* These relationships exist in higher dimensions but have no geometric significance in greater than three dimensions.

Example 3.2.1. Continuing the analysis of **B** in (3.1.80), program MATRIX with subprogram SHINVP merged into it can be used to find the eigenvectors corresponding to eigenvalues 26.15 and 3.82. They are $v_1 = (0.76775 \ 0.64075)^T$ and $v_2 = (-0.64075 \ 0.76775)^T$, respectively. The transformation of variables in (3.2.3) is thus $x_1 = 0.76775y_1 - 0.064075y_2$ and $x_2 = 0.64075y_1 + .76775y_2$, so that (3.2.5) yields $Q = 26.18y_1^2 + 3.82y_2^2$. These two axes are included in Figure 3.2.1, which shows three contours or level curves for $Q = 5$, 20, and 35.

Example 3.2.2. Consider the symmetrical matrix

$$\mathbf{C} = \begin{bmatrix} 17 & 21 \\ 21 & 11 \end{bmatrix}. \tag{3.2.6}$$

Program MATRIX with both GSDECOMP and QRITER merged into it can be used to find the two eigenvalues of **C**: $w_1 = 35.2132$ and $w_2 = -7.21320$. Since these two eigenvalues have opposite signs, the contours of constant values of $Q = x^T\mathbf{C}x$ form a hyperboloid as shown in Figure 3.2.2. Actually, what is shown in Figure 3.2.2 is a plot of the canonical equation $Q = 35.21y_1^2$

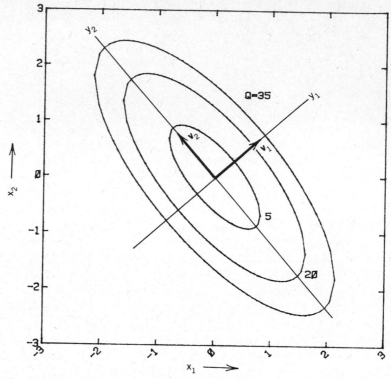

Figure 3.2.1. The y axes obtained by similarity transformation of the symmetric matrix in (3.1.80).

$- 7.21y_2^2$, where axes y_1 and y_2 are in the directions of the two eigenvectors. The saddle point at the origin is seen in both (a) the level curves and (b) the three-dimensional surface. At the saddle point, the first derivatives are zero, but the surface represents a minimum in the y_1 direction and a maximum in the y_2 direction.

There are several conclusions available from the preceding analysis. First, movement away from a given point x in the direction of the eigenvector corresponding to a very small eigenvalue effects very little change in the quadratic function value Q. Conversely, large changes in Q may be expected when movement is in the direction of an eigenvector corresponding to a large eigenvalue. Second, it should be apparent in Figure 3.2.1 that only two searches in the directions of the y_1 and y_2 axes are necessary to locate the minimum $Q = 0$, no matter where the starting point $x^{(0)}$ is located. The first search is in the direction of the y_1 axis that decreases Q as rapidly as possible; and the second search is accomplished by varying only y_2 in a similar way. Therefore, one search strategy for quadratic forms is to compute the eigenvectors and find the point of minimum value of Q by searching sequentially in those directions (or their negatives). That process is equivalent to minimizing

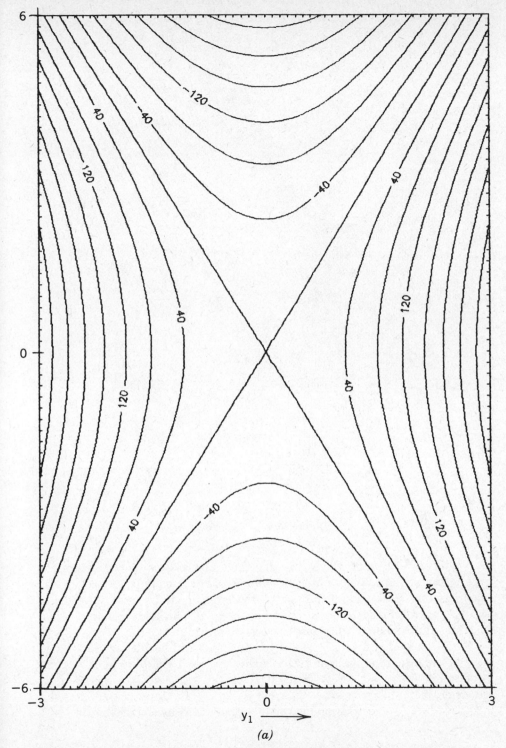

Figure 3.2.2. Hyperboloid in 2-space representing the quadratic form associated with matrix C in (3.2.6) for Example 3.2.2. (*a*) Level curves, (*b*) three-dimensional surface.

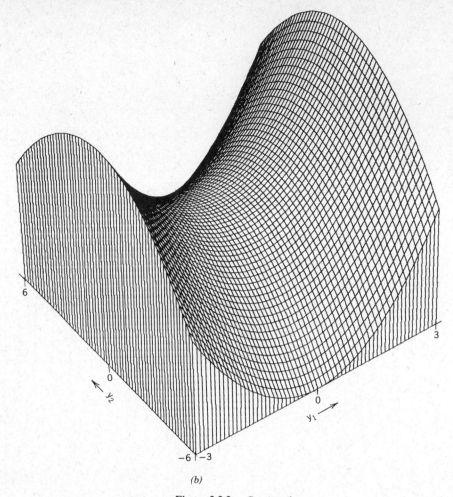

(b)

Figure 3.2.2. *Continued*

(3.2.5) by adjustment of each y_i, one at a time. The least one could conclude is that it is always possible to find the minimum of a quadratic form from any starting point in no more than n steps, where the quadratic form is defined in the E^n space. Of course, this is somewhat trivial as long as the quadratic form is centered at the origin; it is displaced as described next.

It turns out that a proper choice of directions will enable location of the point of minimum Q without diagonalizing the quadratic form at all. Before proceeding with that development, it is useful to define the more general quadratic function. A *quadratic function* is a quadratic form whose minimum has been both displaced from the origin and elevated by a constant:

$$q(\mathbf{x}) = c + \mathbf{b}^T\mathbf{x} + \tfrac{1}{2}\mathbf{x}^T\mathbf{B}\mathbf{x}. \qquad (3.2.7)$$

Both the value of the quadratic function q and the constant c are scalar quantities. The factor $\frac{1}{2}$ is traditionally used to simplify the *gradient* or vector of partial derivatives of q with respect to each component of $\mathbf{x}$:

$$\nabla q = \mathbf{g}(\mathbf{x}) = \mathbf{b} + \mathbf{B}\mathbf{x}. \tag{3.2.8}$$

The expression for the gradient vector, called ∇q or $\mathbf{g}$ interchangeably, is easily obtained from (3.1.41) and (3.1.44). Since $\mathbf{B}\mathbf{x} = (\mathbf{b}_1 x_1 \ \mathbf{b}_2 x_2 \ \cdots \ \mathbf{b}_n x_n)$, it is easy to see that the second derivative of q, called the *Hessian*, is

$$\nabla^2 q = \mathbf{B}, \tag{3.2.9}$$

where $\mathbf{B} = [b_{ij}] = [\partial^2 q/\partial x_i \, \partial x_j]$. Clearly, the Hessian of a quadratic function is a matrix constant. As noted previously, $\mathbf{B}$ must be positive definite for the quadratic form to have an unambiguous minimum (the hyperellipsoidal form). The necessary conditions for a minimum are that all first derivatives of q are equal to zero, that is, the gradient $\mathbf{g}(\mathbf{x}) = \mathbf{0}$. Setting (3.2.8) equal to zero yields

$$\mathbf{x}' = -\mathbf{B}^{-1}\mathbf{b}. \tag{3.2.10}$$

The vector $\mathbf{x}'$ denotes the point in E^n where a function is minimum.

Example 3.2.3. The quadratic form of Examples 3.1.8 and 3.2.1 is extended to the case of a quadratic function. Suppose that a quadratic function having the form of (3.2.7) is

$$q(\mathbf{x}) = 653 + (-202 \ -166)\mathbf{x} + \tfrac{1}{2}\mathbf{x}^T \begin{bmatrix} 34 & 22 \\ 22 & 26 \end{bmatrix} \mathbf{x}. \tag{3.2.11}$$

Note that the elements of matrix $\mathbf{B}$ have been doubled to account for the multiplier $\frac{1}{2}$ preceding the quadratic form in (3.2.11). Also, the scalar constant 653 has been chosen to make $Q = 0$ at the center of the quadratic form; see (3.2.33). Therefore, the family of level curves associated with that quadratic form is just that shown in Figure 3.2.1 except for a displacement from the origin. According to (3.2.10), that displacement vector is

$$\mathbf{x}' = -\mathbf{B}^{-1}\mathbf{b} = (4 \ 3)^T, \tag{3.2.12}$$

as easily computed by using program MATRIX. The displaced level curves are shown in Figure 3.2.3. According to the expansion for a quadratic form in (3.1.43), the ordinary algebraic notation for the surface $q(x_1, x_2)$ is

$$q = 653 - 202x_1 - 166x_2 + 17x_1^2 + 22x_1 x_2 + 13x_2^2. \tag{3.2.13}$$

Differentiating (3.2.13) by ordinary calculus, the elements of the gradient are

$$\nabla_1 q = -202 + 34x_1 + 22x_2, \tag{3.2.14}$$

$$\nabla_2 q = -166 + 22x_1 + 26x_2. \tag{3.2.15}$$

These last two equations are the same as those expressed in matrix notation by

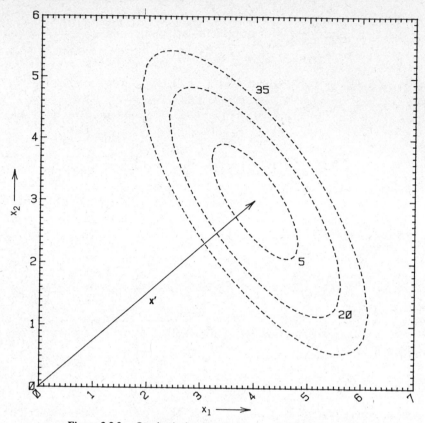

Figure 3.2.3.   Quadratic function in 2-space for Example 3.2.3.

(3.2.8), of course, and the values of $x_1 = 4$ and $x_2 = 3$ make both derivatives equal to zero as required for a minimum or a maximum. Additional differentiation of both (3.2.14) and (3.2.15) with respect to x_1 and with respect to x_2 result in the second partial derivatives of q; these are the elements of the matrix **B**, which is a constant matrix in the case of a quadratic function.

The concept of a *line* in E^n is important in optimization. Suppose that the initial guess at a point **x** for minimum $q(\mathbf{x})$ is $\mathbf{x}^{(0)} = (1.9 \ 4.5)^T$ in the x_1–x_2 plane. The next point, $\mathbf{x}^{(1)}$, is usually approached on a straight line, namely,

$$\mathbf{x}^{(k+1)} = \mathbf{x}^{(k)} + t_k \mathbf{s}^{(k)}. \tag{3.2.16}$$

The relationship in (3.2.16) is shown in Figure 3.2.4. Departure is from point $\mathbf{x}^{(k)}$ in direction $\mathbf{s}^{(k)}$ for distance t measured in units of $\|\mathbf{s}\|_2$. An alternative notation for a line in multivariable space (3.2.16) is less cluttered and is employed when it will not confuse the analysis:

$$\mathbf{x}^* = \mathbf{x} + t\mathbf{s}. \tag{3.2.17}$$

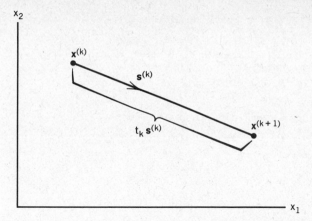

Figure 3.2.4. A straight line in 2-space illustrating vector search direction $s^{(k)}$ and scalar factor t_k.

There are several important concepts associated with the line function. First, it is a vector function of a scalar, namely, t. Notationally, $\mathbf{x}^* = \mathbf{x}^*(t)$. For any scalar function of a vector, say $F(\mathbf{x})$, the scalar function along the line is a function of only t, when given a starting point $\mathbf{x}$ and direction $\mathbf{s}$, that is, $F(\mathbf{x}^*) = F(t)$. A *line search* is the process of finding some t, say t^*, to minimize $F(t)$. For the general quadratic function, the value of t^* may be found by substituting (3.2.16) into (3.2.7):

$$F[\mathbf{x}^{(k+1)}] = c + \mathbf{b}^T[\mathbf{x}^{(k)} + t_k\mathbf{s}^{(k)}]$$
$$+ \tfrac{1}{2}[\mathbf{x}^{(k)} + t_k\mathbf{s}^{(k)}]^T\mathbf{B}[\mathbf{x}^{(k)} + t_k\mathbf{s}^{(k)}]. \tag{3.2.18}$$

Expanding the terms and setting the first derivative of (3.2.18) with respect to t equal to zero yields

$$t_k^* = -\frac{\mathbf{s}^{(k)T}\mathbf{g}^{(k)}}{\mathbf{s}^{(k)T}\mathbf{B}\mathbf{s}^{(k)}}. \tag{3.2.19}$$

When $\mathbf{s}^{(k)} = -\mathbf{g}^{(k)}$ and t_k^* is used in (3.2.16), $\mathbf{x}^{(k+1)}$ is called the *Cauchy point*.

Example 3.2.4. Again using the quadratic function in (3.2.11), start a line search at $x = (1.9 \ 4.5)^T$ in the direction $\mathbf{s} = (1.7 \ -0.6)^T$. This line is shown in Figure 3.2.5 from the starting point out to $x = (5.3 \ 3.3)^T$, where $t = 2$. Notice that the dash–dot–dash line has a point of tangency with the $q = 5$ contour. Use the short BASIC program in Table 3.2.2 to compute t^* and the function $F(t)$. According to (3.2.19) the minimum occurs at $t^* = 0.9716$, which coincides with the tangent point in Figure 3.2.5. It can be seen from

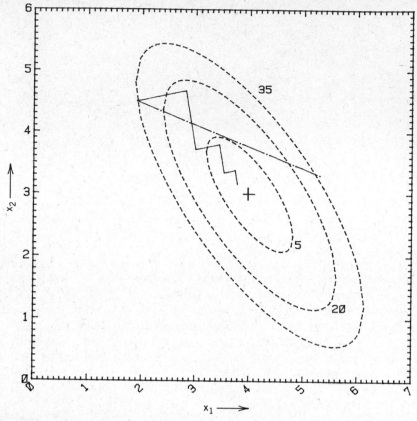

Figure 3.2.5.   Line searches on a quadratic function. The steepest-descent strategy usually causes zigzagging and slow convergence.

(3.2.16) that the range of t values depends entirely on the length of search direction **s**, which is arbitrary.

It is important to know both the first and second derivatives in a direction with respect to t, that is, the slope and its rate of change at any point on line functions that are not quadratic, especially for less well-behaved functions. After Fletcher (1980:6), recall the *chain rule* and apply some special notation as it relates to (3.2.17):

$$\frac{d}{dt} = \sum_i \frac{dx_i}{dt} \frac{\partial}{\partial x_i} = \sum_i s_i \frac{\partial}{\partial x_i} = \mathbf{s}^T \nabla . \tag{3.2.20}$$

The terms s_i and x_i are the ith components of **x** and **s**, respectively. Therefore, the slope at any point along the line $F'(t)$, called the *directional derivative*, is

$$F'(t) = \frac{dF}{dt} = \mathbf{s}^T \nabla F = \nabla F^T \mathbf{s}, \tag{3.2.21}$$

Table 3.2.2. A BASIC Program to Perform Searches on the Quadratic Function in (3.2.11)

```
10 REM - CALC (3.2.13)-(3.2.16) - LINE SEARCHES ON QUAD FNCN
20 CLS : PRINT"INPUT STARTING X1,X2"; : INPUT X1,X2
30 X3=X1 : X4=X2 : REM - SAVE STARTING POINT
40 GOSUB 180 : REM - COMPUTE & PRINT FUNCTION AND GRADIENT VALUES
50 PRINT
60 PRINT "INPUT DIRECTION COMPONENTS S1,S2"; : INPUT S1,S2
70 REM - CALC OPTIMAL STEP T=T* ON QUADRATIC SURFACE
80 T1 = (-S1*G1-S2*G2)/(34*S1^2+44*S1*S2+26*S2^2)
90 PRINT "OPTIMAL T=";T1
100 PRINT
110 REM - LOOP TO CALC POINT ON LINE GIVEN T
120 PRINT "T="; : INPUT T
130 X1=X3+T*S1
140 X2=X4+T*S2
150 GOSUB 180 : REM - COMPUTE & PRINT FUNCTION &  GRADIENT VALUES
160 GOTO 100
170 REM - SUBROUTINE TO CALC & PRINT FNCN & GRADIENT VALUES
180  Q=653-202*X1-166*X2+17*X1^2+22*X1*X2+13*X2^2
190 G1=-202+34*X1+22*X2
200 G2=-166+22*X1+26*X2
210 PRINT "X1,X2=";X1;X2
220 PRINT "FUNCTION Q=";Q
230 PRINT "GRADIENT COMPONENTS ARE:";G1;G2
240 RETURN
250 END
```

where the gradient vector ∇F and the direction vector $\mathbf{s}$ are evaluated at the particular $\mathbf{x}^*$ of interest on the line. Some authors refer to (3.2.21) as the *projected gradient*, since ∇F is "projected" onto $\mathbf{s}$. The second derivative of F with respect to t is the *curvature*. Applying the operator from (3.2.20) to (3.2.21), the curvature at any point on the line is the quadratic form

$$\frac{d^2 F}{dt^2} = \frac{d}{dt}\frac{dF}{dt} = \mathbf{s}^T \nabla(\nabla F^T \mathbf{s}) = \mathbf{s}^T \nabla^2 F \mathbf{s}. \qquad (3.2.22)$$

Since both the directional derivative and curvature depend on the length of the direction vector $\mathbf{s}$, it is common to require that $\|\mathbf{s}\|_2 = 1$. It can now be observed that the optimal step in parameter t to a minimum along a line as expressed in (3.2.19) is simply the negative of the ratio between the directional derivative and the curvature of the function with respect to the line.

Being at a point in variable space, say $\mathbf{x} = (1.9 \;\; 4.5)^T$ in Figure 3.2.5, the previous direction taken in Example 3.2.4 was not the one with the steepest slope at that point. From (2.1.42), the directional derivative $\mathbf{g}^T \mathbf{s} = \|\mathbf{g}\|_2 \|\mathbf{s}\|_2 \cos\theta$, where θ is the angle between vectors $\mathbf{g}$ and $\mathbf{s}$. In seeking a minimum, it might be supposed that the more negative slope the better; in that case, the direction $\mathbf{s}$ that provides the most negative directional derivative is seen to be $\mathbf{s} = -\mathbf{g}$. Thus the gradient vector is the direction of steepest ascent, and the negative gradient is the direction of steepest descent. As early as 1847,

Cauchy suggested the *method of steepest descent*: a sequence of line searches in the direction of the negative gradient, each line search terminating at the minimum on that line segment (called an *iteration*).

Example 3.2.5. It is easy to use the program in Table 3.2.2 to compute a sequence of steepest descents on the function portrayed in Figure 3.2.5. Starting from $\mathbf{x} = (1.9\ \ 4.5)^T$, the steepest-descent iterations always terminate at a point of tangency with a level curve or contour of constant function value. It is also seen that consecutive search directions are perpendicular, and this can be checked numerically without difficulty. Therefore, for quadratic functions there is substantially no difference in steepest-descent strategy and that of searching in the directions of principal axes. More important, the phenomena of *zigzagging* often associated with the steepest-descent method can be observed in Figure 3.2.5. Steepest descent is an inferior search strategy. The same can be said for searching in the directions of the coordinate axes, unless the function is in canonical form.

The difficulty encountered by the steepest-descent method relates to the eccentricity of the quadratic form. Crowder (1972:433) notes that the inequality

$$\frac{F[\mathbf{x}^{(k+1)}]}{F[\mathbf{x}^{(k)}]} \leq \left(\frac{k-1}{k+1}\right)^2 \tag{3.2.23}$$

holds for steepest descent on a quadratic function, where k is a condition number defined as the ratio of largest to smallest eigenvalues of matrix $\mathbf{B}$. The matrix related to Figure 3.2.5 was given in (3.2.11); in Example 3.1.8, its eigenvalues were found to be 26.18 and 3.82, so that $k = 6.85$ in this case. Then (3.2.23) predicts a sequential decrease in steepest-descent function values by a factor no greater than 0.56. Using the program in Table 3.2.2, it is found that every value of Q at the turning points of the zigzag descent decreased by a factor of 0.48. Convergence of the steepest-descent method is linear (large k), but it could converge in just n iterations ($k = 1$, circular contours).

The significance of a positive definite matrix has been developed in connection with ellipsoidal forms that have unambiguous minima. A family of direction vectors is now developed based on positive-definite matrices, that is, those having all positive eigenvalues. A more generally employed criterion for positive definiteness can be obtained from the spectral decomposition of a matrix according to (2.2.34). Such an expansion of a matrix is

$$\mathbf{B} = w_1\mathbf{v}_1\mathbf{v}_1^T + \cdots + w_k\mathbf{v}_k\mathbf{v}_k^T + \cdots + w_n\mathbf{v}_n\mathbf{v}_n^T, \tag{3.2.24}$$

where w_k is the kth eigenvalue and $\mathbf{v}_k$ is the kth eigenvector. Then a quadratic form for **B** involving a vector, say **s**, is

$$s^T B s = w_1 (v_1^T s)^T (v_1^T s) + \cdots + w_k (v_k^T s)^T (v_k^T s) + \cdots + (v_n^T s)^T (v_n^T s).$$

$$(3.2.25)$$

If **B** is positive definite, then all $w_i > 0$ so that each term in (3.2.25) is positive, and an alternate sufficient condition for a *positive-definite matrix* is

$$s^T B s > 0, \qquad \text{for all } s \neq 0. \tag{3.2.26}$$

Positive definiteness is a special property of a matrix with respect to a vector. A property of a positive definite $n \times n$ matrix with respect to n vectors is that of *conjugacy or B-conjugacy*:

$$s_i^T B s_j = \begin{cases} 0, & \text{for } i \neq j, \\ h_j > 0, & \text{for } i = j. \end{cases} \tag{3.2.27}$$

Furthermore, the n different s_j vectors are linearly independent, that is,

$$\sum_{j=1}^{n} r_j s_j = 0 \tag{3.2.28}$$

implies that all $r_j = 0$, $j = 1$ to n. This can be shown to be true by premultiplying (3.2.28) by $s_i^T B$. The result is equal to $h_j r_j$ according to (3.2.27); therefore, (3.2.28) implies that $r_j = 0$. But this condition is true for all $j = 1$ to n so that *conjugate vectors are linearly independent*. A trivial case of conjugacy is the orthogonal set of vectors in (3.2.27) when $B = I$, the identity matrix. Two examples are the unit vectors that define orthogonal axes and the eigenvectors of a positive-definite symmetrical matrix.

Fletcher (1980:25) has given a clear explanation of the important fact that a sequence of n optimal line searches in conjugate directions on a quadratic surface in E^n will terminate at the exact minimum. This property of *quadratic termination* is the basis of nearly all optimization algorithms that employ gradient information. Suppose that the n conjugate search directions (vectors) are placed in columns of a matrix:

$$S = [s^{(1)} \ s^{(2)} \ \cdots \ s^{(n)}], \tag{3.2.29}$$

where the superscripts now denote the different search directions. Since the $s^{(k)}$ are conjugate, they are also linearly independent so that any vector in E^n can be expressed as a linear combination of them. If the search is to begin at point $x^{(0)}$, then any point in E^n can be expressed in the form

$$x = x^{(0)} + \sum_{k=1}^{n} s^{(k)} t_k = x^{(0)} + S t, \tag{3.2.30}$$

where the coefficients t_k are collected in vector **t**:

$$\mathbf{t} = (t_1 \ t_2 \ \cdots \ t_n)^T. \tag{3.2.31}$$

The meaning of an optimal line search is that there is a value of t_k, say t_k^*, that minimizes $F[\mathbf{x}^{(k)} + t_k \mathbf{s}^{(k)}]$. It is assumed that the minimum value of the quadratic function is $F(\mathbf{x}')$ at point $\mathbf{x}'$ (see Figure 3.2.3) and that $\mathbf{x}'$ can be found after n optimal line searches, that is,

$$\mathbf{x}' = \mathbf{x}^{(0)} + \mathbf{S}\mathbf{t}^*. \tag{3.2.32}$$

It is noted that the standard quadratic form in (3.2.7) can be altered to read

$$q(\mathbf{x}) = c + \mathbf{b}^T\mathbf{x} + \tfrac{1}{2}\mathbf{x}^T\mathbf{B}\mathbf{x} = c' + \tfrac{1}{2}(\mathbf{x} - \mathbf{x}')^T\mathbf{B}(\mathbf{x} - \mathbf{x}'), \tag{3.2.33}$$

where $\mathbf{x}' = -\mathbf{B}^{-1}\mathbf{b}$ as in (3.2.12) and $c' = c - \tfrac{1}{2}\mathbf{x}'^T\mathbf{B}\mathbf{x}'$. Since c' is only a scalar elevation, it can be ignored in finding the minimum of

$$F(\mathbf{x}) = (\mathbf{x} - \mathbf{x}')^T\mathbf{B}(\mathbf{x} - \mathbf{x}'). \tag{3.2.34}$$

Upon substitution of (3.2.30) and (3.2.32) into (3.2.34), the $\mathbf{x}^{(0)}$ terms subtract out and the result is

$$F(t) = \tfrac{1}{2}(\mathbf{t} - \mathbf{t}^*)^T\mathbf{S}^T\mathbf{B}\mathbf{S}(\mathbf{t} - \mathbf{t}^*)$$

$$= \tfrac{1}{2}\sum_{k=1}^{n}(t_k - t_k^*)^2 h_k. \tag{3.2.35}$$

Note that the second expression in (3.2.35) is a result of B-conjugacy, (3.2.27). The very important conclusion to be drawn from (3.2.35) is that it is equivalent to making n line searches in n-space, and thus B-conjugacy implies a diagonalizing transformation $\mathbf{S}^T\mathbf{B}\mathbf{S}$ to a new coordinate system **t** that decouples the variables. This method is just as effective as the diagonalizing transformation that employs eigenvectors (Example 3.2.1).

Example 3.2.6. Again consider the quadratic function in (3.2.11) as illustrated in Figure 3.2.5. The properties of B-conjugacy state that two such search lines, each terminated at the minimum in that direction, must minimize a quadratic function of two variables. Consider two cases to verify that any two line segments joining an arbitrary point to the origin are conjugate *if their common point is a minimum point along the first line*. The program listed in Table 3.2.2 can make the following calculations. First, let $\mathbf{x}^{(0)} = (1.9 \ 4.5)^T$ and $\mathbf{s}^{(0)} = (1.7 \ -0.6)^T$; movement in that direction traces the dash–dot–dash line in Figure 3.2.5. The optimal step length for minimum function value according to (3.2.19) is $T_0^* = 0.9716289$. Then (3.2.16) yields the "turning point" $x^{(1)} =$ (3.551769 3.917023)^T. To verify conjugacy, suppose that the new search direction is $s^{(1)} = x' - x^{(1)}$, since the minimum of the quadratic function is known to be at $x' = (4\ 3)^T$, as shown in Figure 3.2.5. Therefore, $s^{(1)} = (0.44823\ -0.91702)^T$. To test for all cases of B-conjugacy according to (3.2.27), form the direction matrix $S = (s_1\ s_2)^T$ and then compute $S^T B S =$ diag(62.74 10.61)^T, confirming the B-conjugacy of s_1 and s_2. A second case might also start at point $x^{(0)} = (1.9\ 4.5)^T$ but proceed in another arbitrary direction, say $s^{(0)} = (4\ 5)^T$; the minimum along that line occurs at $t_0^* = 9.141753E\text{-}2$. Then the turning point is $x^{(1)} = (2.26567\ 4.95709)^T$. The direction from there to the origin is a multiple of $s^{(1)} = (1.73433\ -1.95709)^T$. Again, all cases of conjugacy are verified by computing $S^T B S$, equal to diag(2074.00 52.50) for this second case.

There are an infinite number of ways to select sets of n conjugate directions that lead to the minimum of a quadratic function, and later those with the most valuable additional properties are described. It is useful now to define one unique set of conjugate directions that constitute the *conjugate gradient method*. This method was originally designed by Hestenes and Stiefel to solve systems of linear equations (such as the set of linear gradient equations related to a quadratic function); a good description of its original application is given by Beckman (1960). The conjugate gradient method has since been used in optimization of nonlinear functions; see Cuthbert (1983:145) for a description and BASIC program.

The conjugate gradient algorithm involves the gradient of Q, namely, $\nabla Q = g = b + Bx$. The steps in this algorithm are shown in Table 3.2.3. For quadratic functions, t^* is found using (3.2.19) and the conjugate gradient algorithm will have found the minimum of Q (where the gradient is zero) after a total of n line searches. Note that each new direction is a linear combination of the gradient at the current and preceding turning points and thus accumulates information about the function. Unlike more sophisticated conjugate direction methods, only three vectors need to be stored, namely, the last x, g,

Table 3.2.3.   Steps in the Conjugate Gradient Algorithm for Quadratic Functions

1. $x^{(0)}$ is an arbitrary starting point.
2. $s^{(0)} = -g^{(0)}$ starts search in steepest-descent direction.
3. $x^{(k+1)} = x^{(k)} + t_k^* s^{(k)}$, where t^* determines minimum Q in direction $s^{(k)}$.
4. $g^{(k+1)} = g(x^{(k+1)})$ is the gradient at the turning point.
5. $h_k = \dfrac{\|g^{(k+1)}\|^2}{\|g^{(k)}\|^2}$ is a ratio of two-norms.
6. $s^{(k+1)} = -g^{(k+1)} + h_k s^{(k)}$.
7. Go to 3 if $k < n$, else stop.

and **s** vectors. The reader can use data from the following examples to confirm that each of the n defined directions is indeed B-conjugate according to (3.2.27). However, the conjugate gradient method *requires* the initial steepest-descent direction and line searches to the exact minimum for the conjugacy property to hold. Readers interested in the derivation or proof of the conjugate gradient method are referred to Fletcher in Murray (1972:79).

Example 3.2.7. In order to start with a positive definite symmetric matrix, form $\mathbf{D} = \mathbf{A}^T\mathbf{A}$, where $\mathbf{A}$ is the 3×3 matrix given in Table 2.1.6. (The eigenvalues of $\mathbf{D}$ are 167.6, 4.0820, and 1.3155.) That matrix is employed in the quadratic function $Q = \mathbf{b}^T\mathbf{x} + \frac{1}{2}\mathbf{x}^T\mathbf{D}\mathbf{x}$, where vector $\mathbf{b}$ is assigned the value $\mathbf{b} = (-200 \ -532 \ -160)^T$. It is left as an exercise to write a program that computes the conjugate gradient algorithm just described. The output of such a program that works in conjunction with program MATRIX is shown in Table 3.2.4. The first three data lines describe the given matrix and the next data line is the given vector. The program was written to start from $\mathbf{x} = (1 \ 1 \ 1)^T$. In iteration 1, the gradient $\mathbf{g} = \mathbf{b} + \mathbf{D}\mathbf{x}$ evaluated at the starting $\mathbf{x}$ is $\mathbf{g} = (-115 \ -305 \ -95)^T$. As shown in Table 3.2.4, the first search direction is the negative gradient vector; the coefficient to determine the minimum in that direction according to (3.2.19) is $t^* = 5.97115\text{E}{-3}$. The new "turning point" in the search is then computed, ending iteration 1. The remainder of the conjugate gradient algorithm proceeds in the same way, except that the search direction formula adds a certain proportion of the last search direction to the negative gradient at the turning point. The minimum value of Q occurs at $\mathbf{x}' = (4 \ 1 \ 5)^T$; it is easily verified that $\mathbf{g}(\mathbf{x}') = \mathbf{0}$.

There are several important points to make concerning this example. First, this problem was posed as minimization of a quadratic function, but it is exactly the same problem as solving the linear matrix equation $\mathbf{D}\mathbf{x} = -\mathbf{b}$. Second, the three search directions shown in Table 3.2.4 can form the columns of a matrix $\mathbf{S}$ as in (3.2.29) to compute $\mathbf{S}^T\mathbf{D}\mathbf{S} = \text{diag}(1.9305\text{E}7 \ 279.18 \ 25.398)$. That proves that the search directions are D-conjugate and that the process represents a diagonalization of a quadratic function according to (3.2.35). Finally, it is noted that direction matrix $\mathbf{S}$ is often illconditioned; using program SVD, the singular values of $\mathbf{S}$ are found to be 339.52, 9.5516, and 3.8029, corresponding to a condition number $k_2 = 89.28$.

Example 3.2.8. To illustrate how easily the conjugate gradient method can become illconditioned, consider the well-conditioned positive-definite symmetric 5×5 matrix given at the top of Table 3.1.3. (It has approximate eigenvalues of 150, 81, 48, 16, and 0.37.) The problem posed with that matrix, say $\mathbf{D}$, is equivalent to the linear matrix equations in Example 3.1.2. Now the constant matrix is $\mathbf{b} = (-248 \ -332 \ -208 \ -400 \ -125)^T$. The conjugate gradient algorithm working in double precision and starting at $\mathbf{x} = (1 \ 1 \ 1 \ 1 \ 1)^T$ finds the solution $\mathbf{x}' = (5 \ 4 \ 3 \ 2 \ 1)^T$ in five iterations as expected; the five search vectors placed as columns of the direction matrix $\mathbf{S}$ are shown in Table 3.2.5.

Table 3.2.4. Results from a Program That Performs the Conjugate Gradient Algorithm on a Given Positive-Definite 3 × 3 Matrix and Vector, Starting from a Given Solution Estimate, x = (1 1 1)^T

```
MATRIX D( 3 , 3 ) -
      21.00000      51.00000      13.00000
      51.00000     138.00000      38.00000
      13.00000      38.00000      14.00000
CONSTANT VECTOR B (TRANSPOSED) IS:
   -200.00000    -532.00000    -160.00000
VARIABLE VECTOR (TRANSPOSED) IS:
       1.00000       1.00000       1.00000
 COMPLETED ITERATION # 0   CONTINUE (Y/N)?

GRADIENT VECTOR TRANSPOSED IS:
    -115.00000    -305.00000     -95.00000
CURRENT SEARCH DIRECTION (TRANSPOSED VECTOR) IS:
     115.00000     305.00000      95.00000
OPTIMAL LINE SEARCH STEP T*= 5.9711504468326750D-03
NEW VARIABLE POINT IN X SPACE IS:
       1.68668       2.82120       1.56726
 COMPLETED ITERATION # 1   CONTINUE (Y/N)?

GRADIENT VECTOR TRANSPOSED IS:
      -0.32406       2.90237      -8.92587
CURRENT SEARCH DIRECTION (TRANSPOSED VECTOR) IS:
       0.41205      -2.66901       8.99855
OPTIMAL LINE SEARCH STEP T*= .31592637483586
NEW VARIABLE POINT IN X SPACE IS:
       1.81686       1.97799       4.41014
 COMPLETED ITERATION # 2   CONTINUE (Y/N)?

GRADIENT VECTOR TRANSPOSED IS:
      -3.63665       1.20775       0.52475
CURRENT SEARCH DIRECTION (TRANSPOSED VECTOR) IS:
       3.70654      -1.66043       1.00146
OPTIMAL LINE SEARCH STEP T*= .5889976917456501
NEW VARIABLE POINT IN X SPACE IS:
       4.00000       1.00000       5.00000
SOLUTION VECTOR (TRANSPOSED) IS JUST ABOVE
PRESS <RETURN> KEY TO CONTINUE -- READY?
```

Table 3.2.5. Search Direction Matrix S for the Conjugate Gradient Algorithm in Example 3.2.8

```
MATRIX A( 5 , 5 ) -
   201.00000    131.03000     22.31000      2.10500      0.01281
   210.00000    -31.85660     27.06400     13.62720     -0.00464
   134.00000     70.20460    -12.88581      8.29660     -0.01653
   252.00000    -21.04685    -21.14677    -17.04150     -0.00183
    53.00000    -34.67800    -20.24608      7.05850      0.02030
```

The solution was obtained with some luck; program SVD shows that the transformation $S^T D S = \text{diag}(1.716E7\ 2.166E6\ 7.915E4\ 1.134E4$ $-4.793E-4)$. The apparent rank deficiency is confirmed by using program SVD to compute the singular values of the S matrix given in Table 3.2.5; they are 414.8, 174.3, 46.47, 21.10, and 0.02957. The conclusion is that S is essentially of rank 4, not 5, even though the computer solution carried out in double precision survived that illconditioning.

3.2.2. General Nonlinear Functions.

The analysis of functions that are neither linear nor quadratic is based on their representation by a Taylor series that embodies those two well-behaved classes of functions. It is helpful first to review the elementary analysis of scalar functions in this context. For some function $y = f(x)$, it is commonly assumed that y is a continuous single-valued function of x, and at least the first and second derivatives exist so that they are "smooth" functions. In general, the class of functions having continuous derivatives through order k is often denoted by C^k. Two properties that disqualify such ideal functions are illustrated in Figure 3.2.6a and b. A function is said to be *convex* in some range if a line between two points (linear interpolation) in the range overestimates the function. See Figure 3.2.6c.

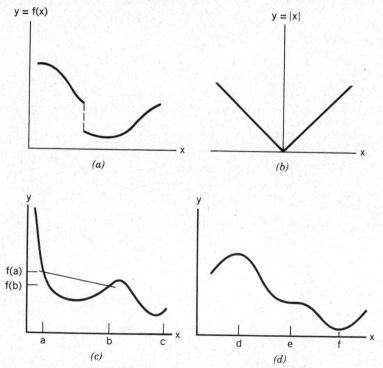

Figure 3.2.6. Some properties of functions. (a) A discontinuous function, (b) a continuous, nonsmooth function, (c) convex range $a \le x \le b$ on a nonunimodal function, and (d) a smooth function having a maximum, a point of inflection, and a minimum.

Mathematically, a function $y = f(x)$ is convex on an interval $h = b - a$ if

$$f(a + th) \leq tf(b) + (1 - t) f(a) \qquad \text{for } 0 \leq t \leq 1. \qquad (3.2.36)$$

Convexity implies that a minimum exists within the range of convexity, but it does not necessarily imply smoothness. Conversely, the function is said to be *concave* when linear interpolation between two points underestimates the function. If $f(x)$ is convex, then $f(-x)$ is concave.

A *global minimum* is said to exist if it is the only minimum over the entire range of the independent variable; such functions in space E^1 are called *unimodal*. The function shown in Figure 3.2.6*b* is unimodal. A minimum that is not unique is called a *local minimum*. The function in Figure 3.2.6*c* has two local minima and is not unimodal. The necessary condition for a minimum or a maximum value of a scalar function is that its first derivative equal zero, such as at points d, e, and f in Figure 3.2.6*d*. The maximum requires that the second derivative be negative, $f''(d) < 0$, and the minimum requires that the second derivative be positive, $f''(f) > 0$. A point of inflection occurs when the second derivative is zero, $f''(e) = 0$. Usually, the following analysis will deal with minima, since a minimum in $y = f(x)$ is a maximum in the function $z = -y = -f(x)$. All of these properties carry over into functions of many variables. For example, a quadratic function is convex everywhere and has a global minimum if its matrix is positive definite. If the matrix has both positive and negative eigenvalues, it is said to be indefinite, and the surface is a hyperboloid having a saddle point as in Figure 3.2.2.

Finally, for a scalar function of a scalar variable, recall that an *infinite Taylor series* expansion about $x = a$ is

$$y(x) = y(a) + y'(a)\, dx + \left(\tfrac{1}{2!}\right) y''(a)\, dx^2 + \left(\tfrac{1}{3!}\right) y'''(a)\, dx^3 + \cdots,$$

$$(3.2.37)$$

where the displacement from the expansion point a is

$$dx = (x - a). \qquad (3.2.38)$$

Example 3.2.9. Consider the function

$$y = 4x - x^2 - \ln(x) - 2, \qquad (3.2.39)$$

so that the first and second derivatives are

$$y'(x) = 4 - 2x - \frac{1}{x} \qquad \text{and} \qquad y''(x) = -2 + \frac{1}{x^2}. \qquad (3.2.40)$$

The function $y(x)$ is shown in Figure 3.2.7*a*. The minimum and maximum are found by solving $y'(x) = 0$ using the quadratic formula. The first three terms of the Taylor series in (3.2.37) are adequate in a small neighborhood of $x = a$ such that dx^3 is insignificant. Therefore, an approximate representation of

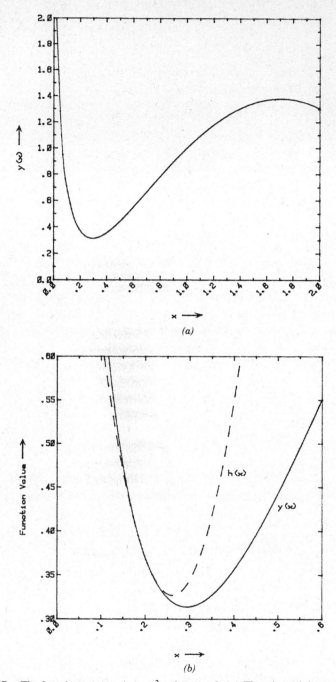

(a)

(b)

Figure 3.2.7. The function $y(x) = 4x - x^2 - \ln(x) - 2$. (a) There is a minimum at $x = 0.2929$ and a maximum at $x = 1.7071$. (b) Quadratic approximation $h(x)$ was obtained from a Taylor series expansion about point $x = 0.2$.

$y(x)$ in the neighborhood of $x = 0.2$ is

$$h(x) = 0.369438 - 1.4(x - 0.2) + 11.5(x - 0.2)^2. \qquad (3.2.41)$$

Both $y(x)$ and $h(x)$ are shown in Figure 3.2.7b.

The *multivariable Taylor series* for a scalar function is a generalization of the scalar variable case:

$$F(\mathbf{dx}) = F(\mathbf{p}) + \mathbf{g(p)}^T \mathbf{dx} + \tfrac{1}{2} \mathbf{dx}^T \mathbf{H(p)} \, \mathbf{dx} + \cdots . \qquad (3.2.42)$$

The higher-order terms (not shown) must be expressed as "tensors" and are seldom discussed in connection with optimization. The Taylor series in (3.2.42) is an expansion about the vector point $\mathbf{p}$, so that the multidimensional displacement about $\mathbf{p}$ is

$$\mathbf{dx} = (\mathbf{x} - \mathbf{p}) = (dx_1 \ dx_2 \ \cdots \ dx_n)^T. \qquad (3.2.43)$$

The gradient vector evaluated at point $\mathbf{p}$ is $\mathbf{g(p)}$, and the Hessian matrix of partial derivatives evaluated at point $\mathbf{p}$ is $\mathbf{H(p)}$. Since multidimensional Taylor series are seldom contemplated greater than degree 2, it is clear that the approximations of interest are simply the quadratic functions treated in the preceding section.

Example 3.2.10. In Chapter One, a surface over two dimensions was described mathematically and geometrically in Sections 1.1.1 and 1.1.2. Equations (1.1.1) through (1.1.8) gave the function, its first and second derivatives, and a quadratic approximating function. Figures 1.1.1 and 1.1.3 show the general surface and the quadratic approximation of a peak, respectively. In fact, the quadratic approximation employed was the Taylor series in (3.2.42) about the point $\mathbf{p} = (-3.7793 \ -3.2832)^T$, which is the exact location of a peak. Therefore, $F(\mathbf{p}) = 0$ and $\mathbf{g(p)} = \mathbf{0}$, so that the approximation is

$$F(\mathbf{dx}) = \tfrac{1}{2}(dx_1 \ dx_2)^T \begin{bmatrix} 116.2645 & -28.2500 \\ -28.2500 & 88.2356 \end{bmatrix} \begin{bmatrix} dx_1 \\ dx_2 \end{bmatrix}. \qquad (3.2.44)$$

Since (3.2.44) is a quadratic form, its expansion proceeds as given by (3.1.43). The approximation shown in (1.1.8) is $F(\mathbf{x})$, which required expansion of the first and second powers of the terms $dx_1 = x_1 - p_1$ and $dx_2 = x_2 - p_2$. Of course $F(\mathbf{dx})$ and $F(\mathbf{x})$ are equivalent except for the points of reference, namely, $\mathbf{x} = \mathbf{p}$ and $\mathbf{x} = \mathbf{0}$, respectively.

The first three terms in the Taylor series approximation given by (3.2.42) are all functions of $\mathbf{p}$, the point of reference on the scalar surface over the

n-dimensional variable space $\mathbf{x}$. If $F(\mathbf{x})$ is a quadratic function such as (3.2.7), then $\mathbf{g}(\mathbf{p})^T = \nabla F(\mathbf{p}) = \mathbf{b} + \mathbf{B}\mathbf{p}$ and $\mathbf{H}(\mathbf{p}) = \mathbf{B}$. However, it is emphasized that when $F(\mathbf{x})$ is a general nonlinear function, then

$$\mathbf{g}(\mathbf{p}) = [g_1(\mathbf{x}) \; g_2(\mathbf{x}) \; \cdots \; g_n(\mathbf{x})]^T, \tag{3.2.45}$$

where each $g_i(x)$ is a scalar nonlinear function of the vector $\mathbf{x}$. Furthermore, the Hessian matrix (in the 3×3 case, for example), is

$$\mathbf{H} = \begin{bmatrix} \dfrac{\partial^2 F}{\partial x_1^2} & \dfrac{\partial^2 F}{\partial x_2 \partial x_1} & \dfrac{\partial^2 F}{\partial x_3 \partial x_1} \\[2mm] \dfrac{\partial^2 F}{\partial x_1 \partial x_2} & \dfrac{\partial^2 F}{\partial x_2^2} & \dfrac{\partial^2 F}{\partial x_3 \partial x_2} \\[2mm] \dfrac{\partial^2 F}{\partial x_1 \partial x_3} & \dfrac{\partial^2 F}{\partial x_2 \partial x_3} & \dfrac{\partial^2 F}{\partial x_3^2} \end{bmatrix} = \begin{bmatrix} \nabla_1(\nabla_1 F) & \nabla_2(\nabla_1 F) & \nabla_3(\nabla_1 F) \\[2mm] \nabla_1(\nabla_2 F) & \nabla_2(\nabla_2 F) & \nabla_3(\nabla_2 F) \\[2mm] \nabla_1(\nabla_3 F) & \nabla_2(\nabla_3 F) & \nabla_3(\nabla_3 F) \end{bmatrix}.$$

$$\tag{3.2.46}$$

Because the Hessian matrix $\mathbf{H}$ is always symmetric, there are $n(n+1)/2$ nonlinear functions of $\mathbf{x}$ involved in E^n.

Since the gradient of a quadratic $F(\mathbf{x})$ with respect to $\mathbf{x}$ is $\mathbf{g}(\mathbf{x}) = \mathbf{b} + \mathbf{B}\mathbf{x}$ according to (3.2.8), the gradient of $F(\mathbf{dx})$ with respect to $\mathbf{dx}$, using the first three terms in (3.2.42), must be

$$\nabla F(\mathbf{dx}) = \mathbf{g}(\mathbf{p}) + \mathbf{H}(\mathbf{p})\,\mathbf{dx}. \tag{3.2.47}$$

Therefore, the step to the minimum where $\nabla F(\mathbf{dx}) = \mathbf{0}$ from any point $\mathbf{p}$ on a quadratic surface is

$$\mathbf{dx}' = -\mathbf{H}(\mathbf{p})^{-1}\mathbf{g}(\mathbf{p}). \tag{3.2.48}$$

This is the well-known *Newton step* in the *Newton–Raphson* search procedure, and $\mathbf{x} = \mathbf{p} + \mathbf{dx}'$ is called the *Newton point*. In practice, the set of linear equations $\mathbf{H}\,\mathbf{dx} = -\mathbf{g}$ would be solved by LU factorization as opposed to obtaining the matrix inverse shown in (3.2.48).

Example 3.2.11. A quadratic function is illustrated first to show how the Newton step in (3.2.48) works exactly. Again consider the quadratic function given by (3.2.11) that has the level curves shown in Figure 3.2.5. As in that figure, start from $\mathbf{p} = (1.9 \; 4.5)^T$; since $\mathbf{g}(\mathbf{p}) = \mathbf{b} + \mathbf{B}\mathbf{p}$ according to (3.2.8), the gradient at that point is $\mathbf{g} = (-38.4 \; -7.2)^T$. Using that and the matrix from (3.2.8) in (3.2.48), $\mathbf{dx} = (2.1 \; -1.5)^T$. From (3.2.43), $\mathbf{x} = \mathbf{p} + \mathbf{dx} = (4 \; 3)^T$, which is the minimum point and center of the quadratic form in Figure 3.2.5. Assuming that the Hessian matrix is available, a single Newton step minimizes a quadratic function.

Example 3.2.12. The scalar function of a vector of interest in this book is neither linear nor quadratic, and neither are its derivatives. Consider the nonlinear function

$$F = -x_1 x_2 + x_3 \left[2 - 4x_1 + x_1^2 + \ln(x_1) + x_2 \right], \qquad (3.2.49)$$

where $\ln(x_1)$ represents the natural logarithm of $x_1 > 0$. The first partial derivatives of F with respect to x_i, $i = 1$ to n, are the components of the gradient vector:

$$g_1 = \nabla_1 F = -x_2 + x_3(-4 + 2x_1 + 1/x_1)$$

$$g_2 = \nabla_2 F = -x_1 + x_3, \qquad (3.2.50)$$

$$g_3 = \nabla_3 F = 2 - 4x_1 + x_1^2 + \ln(x_1) + x_2.$$

The Hessian matrix $\mathbf{H} = [h_{ij}]$ is written in new notation (called the *Jacobian matrix*) of $\mathbf{g}$ that is still equivalent to (3.2.46) in this case:

$$\mathbf{J} = \begin{bmatrix} \nabla_1 g_1 & \nabla_2 g_1 & \nabla_3 g_1 \\ \nabla_1 g_2 & \nabla_2 g_2 & \nabla_3 g_2 \\ \nabla_1 g_3 & \nabla_2 g_3 & \nabla_3 g_3 \end{bmatrix} = \begin{bmatrix} (\nabla g_1)^T \\ (\nabla g_2)^T \\ (\nabla g_3)^T \end{bmatrix}. \qquad (3.2.51)$$

For instance, $\nabla_2 g_3$ is the first partial derivative of g_3 with respect to x_2, and that is identical to the second partial derivative of F with respect to x_2 and x_3. (In general situations, the Jacobian need not be a square matrix.) In this example, the matrix of second partial derivatives of $F(\mathbf{x})$ is

$$\mathbf{J} = \begin{bmatrix} \left[x_3(2 - 1/x_1^2) \right] & -1 & (-4 + 2x_1 + 1/x_1) \\ -1 & 0 & 1 \\ (-4 + 2x_1 + 1/x_1) & 1 & 0 \end{bmatrix}. \qquad (3.2.52)$$

The location of $\mathbf{x} = \mathbf{x}'$ such that $\mathbf{g}(\mathbf{x}') = \mathbf{0}$ is now an iterative procedure. If the initial guess for the starting $\mathbf{x}^{(0)}$ is sufficiently close to $\mathbf{x}'$, then a *finite* second-order Taylor series as in equation (3.2.42) represents the function reasonably well and convergence to $\mathbf{x}'$ can be expected. Program C3-4 (LAGRANGE) can be merged with program MATRIX to perform a sequence of calculations for $\mathbf{x}^{(k+1)} = \mathbf{x}^{(k)} + \mathbf{dx}^{(k)}$. This calculation is started by selecting menu command 13 and giving the requested initial guess for $\mathbf{x}^{(0)}$, say $(2\ 2\ 2)^T$. This Newton process has second-order convergence and in just two iterations the solution $\mathbf{x}' = (2.06545\ 1.27037\ 2.06545)^T$ is obtained. In this example, x_1 happens to equal x_3 because of the form of $g_2(\mathbf{x}) = 0$.

The reason for changing notation (**J** instead of **H**) in the preceding example is that it demonstrates a classical mathematical procedure entirely equivalent to iterative Newton methods that minimize some scalar function of a vector $F(\mathbf{x})$. Suppose that instead of $F(\mathbf{x})$ the nonlinear *vector function of a vector* $\mathbf{g}(\mathbf{x})$ had been given. As the use of the Jacobian matrix in (3.2.51) suggests, this solution of a set of nonlinear equations, $g_i(\mathbf{x}) = 0$, $i = 1$ to n, is obtained by a sequence of first-order (linear) approximations in E^n, namely, $\mathbf{dx} = -\mathbf{J}^{-1}\mathbf{g}$.

The Jacobian matrix plays an important role in the following paragraphs. Regardless of how problems are posed, the Hessian matrix of a scalar function $F(\mathbf{x})$ is equivalent to the Jacobian matrix of the related vector function $\mathbf{g}(\mathbf{x})$. However, note that a given set of equations $\mathbf{g}(\mathbf{x})$ might be neither symmetric nor positive definite. In those cases, the correspondence between Hessian and Jacobian matrices cannot be made. The methods of Chapter Four are relevant to that more complicated situation.

3.3. Constraints

In Section 1.2.1 it was noted that the most general optimization problem discussed in this book potentially involves constraints on the main objective function, that is, the minimization of $f(\mathbf{x})$ subject to $\mathbf{h}(\mathbf{x}) = \mathbf{0}$ and $\mathbf{c}(\mathbf{x}) \geq \mathbf{0}$. If some of the inequalities are not "binding," that is, some $c_i > 0$, then they can be ignored, since they are satisfied. The essential part of the constrained problem is to deal with the equality constraints. Constrained optimization is a difficult subject and there is so far no completely satisfactory method. However, it is clear that a student of that facet of optimization must have an understanding of the implicit function theorem, the Lagrange multiplier technique, and the Kuhn–Tucker constraint qualification concept.

3.3.1. Implicit Function Theorem.   Most of the functions to be minimized are implicit, being the result of an algorithmic computation. That fact and some fundamental theoretical principles require a brief description of the *implicit function theorem*. For example, suppose that there is an implicit function of two variables, say $h(x, y) = 0$. Further suppose that the derivative of y with respect to x is required, under the assumption that y is a function of x, written $y(x)$. The following two examples illustrate problems usually solved in an ordinary calculus course.

Example 3.3.1.   Given the implicit function

$$h(x, y) = 2xy + y^2 - 1 = 0,$$

it is well known from calculus that the derivative of y with respect to x is

$$\frac{dy}{dx} = -\frac{\nabla_x h}{\nabla_y h}, \qquad \nabla_y h \neq 0,$$

so that

$$\frac{dy}{dx} = -\frac{y}{x + y}.$$

It turns out that the sufficient condition that y is a function of x is that $\nabla_y h \neq 0$.

Consider the case of m implicit functions in n unknowns ($m < n$), say $\mathbf{h}(\mathbf{x}) = \mathbf{0}$, where $\mathbf{h}$ is in E^m and $\mathbf{x}$ is in E^n. Then *the number of dependent variables equals the number of functions*, that is, it is possible to solve for m of the variables in terms of the remaining $n - m$ variables under the condition that the related $m \times m$ Jacobian matrix is not singular:

$$\mathbf{J_u} = \begin{bmatrix} \nabla_1 h_1 & \nabla_2 h_1 & \cdots & \nabla_m h_1 \\ \nabla_1 h_2 & \nabla_2 h_2 & \cdots & \nabla_m h_2 \\ & & \cdots & \\ \nabla_1 h_m & \nabla_2 h_m & \cdots & \nabla_m h_m \end{bmatrix}. \tag{3.3.1}$$

It is convenient to partition the n variables into two subvectors: dependent $\mathbf{u}$ in E^m and independent $\mathbf{v}$ in E^{n-m}:

$$\mathbf{x} = \begin{bmatrix} \mathbf{u} \\ \mathbf{v} \end{bmatrix}. \tag{3.3.2}$$

Let each of the m dependent variables in $\mathbf{u}$ be functionally related to the $n - m$ independent variables in $\mathbf{v}$ by the single-valued and continuous functions

$$x_i = u_i = y_i(\mathbf{v}), \qquad i = 1 \text{ to } m. \tag{3.3.3}$$

Then the implicit function theorem states that the derivatives of these functions, $\nabla_{v_k} y_i$, are solutions to the $n - m$ sets of linear systems

$$\mathbf{J_u} \begin{bmatrix} \nabla_{v_k} y_1 \\ \nabla_{v_k} y_2 \\ \cdots \\ \nabla_{v_k} y_m \end{bmatrix} = - \begin{bmatrix} \nabla_{v_k} h_1 \\ \nabla_{v_k} h_2 \\ \cdots \\ \nabla_{v_k} h_m \end{bmatrix}, \qquad k = 1 \text{ to } n - m. \tag{3.3.4}$$

Example 3.3.2. Consider two functions in three variables ($m = 2$, $n = 3$):

$$h_1(\mathbf{x}) = x_1^2 + x_2^2 + x_3^2 - 27,$$

$$h_2(\mathbf{x}) = x_1^2 + 2x_2^2 - x_3 - 24.$$

The choice of which two of the three variables are dependent is arbitrary; in general there are $n!/m!(n - m)!$ choices. Level curves of h_1 are spheres about the origin, and level curves of h_2 are paraboloids of revolution about the x_3 axis. The solution level curves $h_1(\tilde{\mathbf{x}}) = 0 = h_2(\tilde{\mathbf{x}})$ pass through the point $\tilde{\mathbf{x}} = (3\ 3\ 3)^T$. In this case, $\mathbf{u} = (x_1\ x_2)^T$ and $\mathbf{v} = x_3$, so that (3.3.4) yields

$$\begin{bmatrix} 6 & 6 \\ 6 & 12 \end{bmatrix} \begin{bmatrix} \nabla_3 y_1 \\ \nabla_3 y_2 \end{bmatrix} = \begin{bmatrix} -6 \\ 1 \end{bmatrix}.$$

The determinant of the matrix is nonzero, so the matrix is nonsingular and the system has a solution—the derivative of y_1 with respect to x_3 is $-\frac{13}{6}$, and the derivative of y_2 with respect to x_3 is $\frac{7}{6}$. In this contrived example, functions $h_1 = 0$ and $h_2 = 0$ have been revealed explicit and can be solved for the relationships of dependent to independent variables. They are:

$$x_1 = y_1(x_3) = \left(30 - x_3 - 2x_3^2\right)^{1/2},$$

$$x_2 = y_2(x_3) = \left(x_3^2 + x_3 - 3\right)^{1/2}.$$

It is easy to obtain the derivatives of these last two equations and evaluate those at $\tilde{\mathbf{x}} = (3\ 3\ 3)^T$ to verify the solution obtained from (3.3.4). Also, the definitions of the functions y_1 and y_2 are valid only in the neighborhoods $-4.131 \leq x_3 \leq -2.3028$ and $1.3028 \leq x_3 \leq 3.6310$.

The important properties defined by the implicit function theorem are summarized in Table 3.3.1. Property (4c) in Table 3.3.1 deserves further explanation. The differential formula for changes in the dependent variables is

$$du_i = \nabla_{v_1} y_i\, dv_1 + \nabla_{v_2} y_i\, dv_2 + \cdots + \nabla_{v_{n-m}} y_i\, dv_{n-m}, \qquad i = 1 \text{ to } m. \quad (3.3.5)$$

However, the partial derivatives of the dependent variable functions, $u_i = y_i(\mathbf{v})$, may be obtained by solving (3.3.4). Therefore, the linear approximation of dependent variables according to (3.3.5) is always possible if $\mathbf{J_u}$ is nonsingular.

There is one additional interpretation of the implicit function theorem. Note that the ith row of the Jacobian matrix in (3.3.1) is the gradient vector in

Table 3.3.1. Properties of the Implicit Function Theorem

(1) The point $\tilde{\mathbf{x}}$ is defined to satisfy $\mathbf{h}(\tilde{\mathbf{x}}) = \mathbf{0}$, for $\mathbf{h}$ in E^m and $\mathbf{x}$ in E^n.

(2) There are m components of $\mathbf{x}$ in $\mathbf{u}$ that are dependent on the remaining $n - m$ components of $\mathbf{x}$ in $\mathbf{v}$ ($\mathbf{x}$ is partitioned into $\mathbf{u}$ and $\mathbf{v}$).

(3) Each function h_i belongs to C^1 (first derivatives exist), and the Jacobian (3.3.1) is nonsingular at $\mathbf{x} = \tilde{\mathbf{x}}$.

(4) There is a neighborhood about $\tilde{\mathbf{v}}$ where a set of m functions $u_i = y_i(\mathbf{v})$ are single-valued, are continuous, and have the additional properties that:

    a. $\tilde{u}_i = y_i(\tilde{\mathbf{v}})$, for some unknown function y_i.

    b. The various partial derivatives satisfy the $n - m$ systems of linear equations described by (3.3.4).

    c. For any $\mathbf{v}$ in this neighborhood of $\tilde{\mathbf{v}}$, the values of u_i may be approximated linearly.

E^m of equation $h_i(\mathbf{x})$. Furthermore, each $h_i(\mathbf{x}) = 0$ represents a level curve (a surface of dimension $n - 1$) in E^n, so according to (2.2.82) each level curve at point $\tilde{\mathbf{x}}$ has a tangent hyperplane

$$[\nabla h_i(\tilde{\mathbf{x}})]^T \mathbf{x} = [\nabla h_i(\tilde{\mathbf{x}})]^T \tilde{\mathbf{x}}, \qquad i = 1 \text{ to } m. \tag{3.3.6}$$

The solution of the m equations in (3.3.6) locates the point $\mathbf{x}$ that would lie in all m hyperplanes simultaneously; both $\mathbf{v}$ and the dependent $\mathbf{u}$ partitions of $\mathbf{x}$ would thus be found. The enabling condition that a solution exists is that $[\nabla h_i(\tilde{\mathbf{x}})]^T = \mathbf{J_u}$ is nonsingular, the same requirement that has already been assumed for (3.3.4) in the implicit function theorem.

Example 3.3.3. From Kaplan (1959:96), consider the two functions defined in E^4:

$$h_1 = x_1 + 2x_2 - x_3^2 + x_4^2 = 0,$$

$$h_2 = 2x_1 - x_2 - 2x_3x_4 = 0.$$

Since there are two equations ($m = 2$), there can be as many as two dependent variables; choose these to be $x_3(x_1, x_2)$ and $x_4(x_1, x_2)$. In other words, the dependent functions are $\mathbf{u}(\mathbf{v})$, where $\mathbf{u} = (x_3 \ x_4)^T$ and $\mathbf{v} = (x_1 \ x_2)^T$. The Jacobian for these choices according to (3.3.1) is

$$\mathbf{J_u} = \begin{bmatrix} -2x_3 & 2x_4 \\ -2x_4 & -2x_3 \end{bmatrix}.$$

Dispense with the functional notation y_i, so that $\nabla_{v_1} y_2 = \nabla_{x_1} x_4$, for example. Then the derivatives of x_3 and x_4 may be found according to (3.3.4):

$$\mathbf{J_u} \begin{bmatrix} \nabla_{x_1} x_3 \\ \nabla_{x_1} x_4 \end{bmatrix} = -\begin{bmatrix} 1 \\ 2 \end{bmatrix} \quad \text{and} \quad \mathbf{J_u} \begin{bmatrix} \nabla_{x_2} x_3 \\ \nabla_{x_2} x_4 \end{bmatrix} = -\begin{bmatrix} 2 \\ -1 \end{bmatrix}.$$

These two sets of linear equations may be solved by Cramer's rule; in any event, the reader can verify that the partial derivatives obtained are

$$\nabla_{x_1} x_3 = \frac{x_3 + 2x_4}{2x_3^2 + 2x_4^2},$$

$$\nabla_{x_1} x_4 = \frac{2x_3 - x_4}{2x_3^2 + 2x_4^2},$$

$$\nabla_{x_2} x_3 = \frac{2x_3 - x_4}{2x_3^2 + 2x_4^2},$$

$$\nabla_{x_2} x_4 = -\frac{x_3 + 2x_4}{2x_3^2 + 2x_4^2}.$$

The implicit functions $x_3(x_1, x_2)$ and $x_4(x_1, x_2)$ will exist as long as x_3 and x_4 are not simultaneously equal to zero.

3.3.2. Equality Constraints by Lagrange Multipliers.

The classical optimization problem is the minimization of $f(\mathbf{x})$ subject to the equality constraint $h(\mathbf{x}) = 0$, where one or both functions are implicitly related to the components of $\mathbf{x}$. The nature of this problem can be seen in Figure 3.3.1, which shows the locus of the implicit constraint function, $h(\mathbf{x}) = 0$ in E^2 (solid line). The dashed lines are level curves that represent constant values of some objective function, $f(\mathbf{x})$; these indicate decreasing function values toward the upper right corner. Because the minimum must occur on the constraint locus, it is clear that the constrained minimum is at $x_1 = 2.1$ and $x_2 = 1.3$.

In classical mathematics there are several important properties usually considered in developing solutions to the constrained problem, such as convexity, continuity, strong and weak extrema, and so on. Here it will suffice to

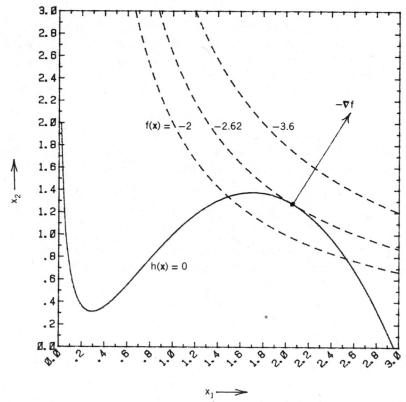

Figure 3.3.1. The minimum of an objective function $f(\mathbf{x})$ (dashed lines) constrained by an equality $h(\mathbf{x}) = 0$ (solid line) for Example 3.3.4. The negative gradient vector of the objective function, $-\nabla f$, is perpendicular to the tangents of both $f(\mathbf{x}') = -2.62$ and $h(\mathbf{x}') = 0$ at the optimal point.

follow the development by Hadley (1964:61), which proceeds rather directly to a usable but satisfying explanation. The development is in terms of two variables and one constraint function as illustrated in Figure 3.1.1. Suppose that Jacobian $\mathbf{J_u}$ (3.3.1) is nonsingular at some point, say $\mathbf{x}'$; in this simple development assume that the partial derivative of h with respect to, say, x_2 is not zero, that is, $\nabla_2 h \neq 0$. Then by the implicit function theorem there is a neighborhood about $\mathbf{x}'$ in which x_2 is some function of x_1, say, $x_2 = y(x_1)$. Therefore, the objective function is a function only of independent variable x_1: $f = f[x_1, y(x_1)]$ in a neighborhood of $\mathbf{x}'$. If f has a minimum near $\mathbf{x}'$, then $df(\mathbf{x}')/dx_1 = 0$ is a necessary condition. From the rules for differentiating compound functions (Kaplan 1959:86),

$$\frac{dh}{dx_1} = \nabla_1 f + \nabla_2 f \frac{dy}{dx_1} = \nabla_1 f - \nabla_2 f \frac{\nabla_1 h}{\nabla_2 h} = 0, \qquad (3.3.7)$$

where the implicit function theorem has enabled the substitution

$$\frac{dy}{dx_1} = -\frac{\nabla_1 h}{\nabla_2 h} \qquad (3.3.8)$$

according to (3.3.4).

The *Lagrange multiplier* is now defined:

$$p = \frac{\nabla_2 f}{\nabla_2 h} = \frac{\nabla_1 f}{\nabla_1 h}, \qquad (3.3.9)$$

evaluated at $\mathbf{x}'$. The identity in (3.3.9) comes from the chain relationships $\nabla_1 f = \nabla_2 f \nabla_1 y$ and $\nabla_1 h = \nabla_2 h \nabla_1 y$. Also note that this entire analysis can be carried out assuming x_1 is a function of x_2, which requires $\nabla_1 h \neq 0$. Therefore, for a minimum of function $f = f[x_1, y(x_1)]$ at $\mathbf{x}'$, it is necessary to satisfy three equations in this case:

$$\nabla_1 f - p \nabla_1 h = 0,$$

$$\nabla_2 f - p \nabla_2 h = 0, \qquad (3.3.10)$$

$$h = 0.$$

A convenient form for the necessary conditions in (3.3.10) is to define a *Lagrangian function*

$$L(\mathbf{x}, p) = f(\mathbf{x}) - p h(\mathbf{x}). \qquad (3.3.11)$$

Then the necessary conditions for minimizing $f(\mathbf{x})$ subject to $h(\mathbf{x}) = 0$ are the same for minimizing the unconstrained $L(\mathbf{x}, p)$ in the *three* variables x_1, x_2, and p.

Example 3.3.4. The functions illustrated in Figure 3.3.1 are the objective function $f(\mathbf{x}) = -x_1 x_2$ (dashed lines) and the constraint function $h(\mathbf{x}) = -[2 - 4x_1 + x_1^2 + \ln(x_1) + x_2] = 0$ (solid line). This problem involving the Lagrange multiplier p has already been solved in Example 3.2.12 using Newton iterations in the program C3-4, LAGRANGE. The Lagrangian function in (3.3.11) is (3.2.49) in this case, where p was replaced with x_3. The minimization of the Lagrangian function, which is equivalent to the three necessary conditions in (3.3.10), was obtained in Example 3.2.12, involving equations (3.2.49) to (3.2.52). The solution to the constrained problem shown in Figure 3.1.1 occurs at $\mathbf{x}' = (2.06545 \ 1.27037)^T$, and the Lagrange multiplier $p = 2.06545$ in this case (it is not usually equal to x_1). The reader can evaluate the Jacobian (Hessian) matrix in this case, equation (3.2.52), at the optimal solution. Using MATRIX, DECOMP, and QRITER, its eigenvalues are 3.94, -1.27, and 0.98, indicating a saddle point. That is typical of solutions of the Lagrangian function (3.3.11), because, at the optimum solution x_1', x_2', p', $F(\mathbf{x}, p')$ is a minimum of $\mathbf{x}$ and $F(\mathbf{x}', p)$ is a _maximum_ of p.

The case for $\mathbf{x}$ in E^2 and only one constraint, $h(\mathbf{x}) = 0$, is used to derive a meaningful interpretation of Lagrange multipliers. Suppose that instead of $h(\mathbf{x}) = 0$, the equality was changed to $h(\mathbf{x}) = e$ or $h(\mathbf{x}) - e = 0$, where e is some small number. Clearly, the solution to an optimization problem with this constraint is a function of e, that is, the optimal objective function $f(\mathbf{x}')$, x_1', and x_2' are all functions of e. According to the implicit function theorem, these functions are all differentiable with respect to e in some neighborhood about $\mathbf{x}'$. In that neighborhood the chain rule yields

$$\nabla_e f = \nabla_1 f \, \nabla_e x_1 + \nabla_2 f \, \nabla_e x_2, \tag{3.3.12}$$

$$\nabla_e h = \nabla_1 h \nabla_e x_1 + \nabla_2 h \nabla_e x_2 - 1 = 0, \tag{3.3.13}$$

where $h(\mathbf{x}) - e = 0$. Since (3.3.13) is zero, it can be multiplied by p and the result subtracted from (3.3.12):

$$\nabla_e f = p + [\nabla_1 f - p \nabla_1 h] \nabla_e x_1 + [\nabla_2 f - p \nabla_2 h] \nabla_e x_2. \tag{3.3.14}$$

However, both quantities in the brackets, $[\cdot]$, are zero according to the necessary conditions for an optimum (3.3.10). Therefore, _the Lagrange multiplier is equal to the partial derivative of the optimal constrained objective function with respect to constraint displacement_ e:

$$p = \nabla_e f(\mathbf{x}'), \tag{3.3.15}$$

where $\mathbf{x}'$ is the solution vector for the undisplaced constraint.

Example 3.3.5. The previous example is continued by computing the sensitivity of the optimal objective function with respect to the constraint according to (3.3.15). The constraint was $h(\mathbf{x}) = -[2 - 4x_1 + x_1^2 + \ln(x_1) + x_2] = 0$; suppose $h(\mathbf{x}) = 0.1$. This can be incorporated into program C3-4, LAGRANGE, by changing 2 to 2.1 in line 7030. Again merging LAGRANGE into MATRIX and starting at $\mathbf{x} = (2\ 2\ 2)^T$, the solution obtained is $\mathbf{x}' = (2.04469\ 1.18276\ 2.04469)^T$. Since the objective function was $f(\mathbf{x}) = -x_1 x_2$, this solution with the perturbed constraint is $f(\mathbf{x}') = -2.418378$ compared to the unperturbed constraint case of $f(\mathbf{x}) = -2.623886$. The sensitivity in (3.3.15) can be interpreted as an approximation for small differentials so that $df = p\,de$ to first order. In this case the actual df turned out to be $+0.2055$. An estimate of that change is $p\,de = 2.06545(0.1) = 0.2065$, where $e = 0.1$ represents a 5 percent change in the constant part of $h(\mathbf{x})$. The reader can verify that $e = 0.01$ (a 0.5 percent change in the constraint constant) causes approximation (3.3.15) to agree with the actual change in objective function to five significant figures.

Although the Lagrangian has been introduced in limited dimensions, its derivation for larger dimensions proceeds from the same principles but involves more general notation. The central concept in the use of Lagrange multipliers is to convert a constrained problem into an unconstrained problem. The resulting *classical Lagrangian function* is

$$L(\mathbf{x}, \mathbf{p}) = f(\mathbf{x}) - \mathbf{p}^T \mathbf{h}(\mathbf{x}). \qquad (3.3.16)$$

Again, $\mathbf{x}$ is in E^n; since there are m constraints, $\mathbf{h}$ and the Lagrange multipliers $\mathbf{p}$ are in E^m. The sensitivity of the optimal objective function is now with respect to perturbations of each of the m constraints, typically, $h_j(\mathbf{x}) = e_j$, where the perturbation to the jth constraint is e_j. Then the Lagrange multipliers may be interpreted as

$$p_j = \nabla_{e_j} f(\mathbf{x}'), \qquad j = 1 \text{ to } m. \qquad (3.3.17)$$

To interpret each of the Lagrange multipliers as sensitivity coefficients in this way, the constraint functions must have been scaled so that similar perturbations in $\mathbf{x}$ cause similar perturbations in each $h_j(\mathbf{x})$, $j = 1$ to m. In the linear programming art, these sensitivities are called *shadow costs* according to certain economic problems commonly encountered in that field. Readers interested in the complete derivation and details leading to (3.3.16) and (3.3.17) are referred to Hadley (1964:64).

3.3.3. Constraint Qualifications—The Kuhn–Tucker Conditions.

It is now appropriate to consider the case of the generally constrained optimization problem previously mentioned: Minimize $f(\mathbf{x})$ such that $\mathbf{h}(\mathbf{x}) = \mathbf{0}$ and $\mathbf{c}(\mathbf{x}) \geq \mathbf{0}$, where $\mathbf{x}$ is in E^n, $\mathbf{h}$ is in E^m, and $\mathbf{c}$ is in E^t. In other words, there are n

variables, m equality constraints, and t inequality constraints. Suppose that q of the inequality constraints are equal to zero at some point in $\mathbf{x}$ space, say $c_j(\mathbf{x}) = 0$ for $j = 1$ to $q < t$; then they are indistinguishable from the satisfied equality constraints. The central question asked in this section is: Under what conditions is that point $\mathbf{x}$ an optimal point for convex functions $f(\mathbf{x})$, $\mathbf{h}(\mathbf{x})$, and $\mathbf{c}(\mathbf{x})$? The answer depends on propositions known as Farka's lemma and the Kuhn–Tucker conditions. These concepts are rooted in the ideas presented in Section 2.2.4, especially those concerning normal vectors, hyperplanes, and polyhedral cones.

It is necessary to define more precisely a *convex polyhedral cone* generated by two or more vectors as previously illustrated in Figure 2.2.9. A vector $\mathbf{g}$ lies within a convex polyhedral cone if and only if it can be expressed as

$$\mathbf{g} = \mathbf{Nu} = \sum_{j=1}^{q} u_j \mathbf{n}_j, \qquad u_j \geq 0. \tag{3.3.18}$$

$\mathbf{N}$ is an $n \times q$ matrix composed of column vectors $\mathbf{N} = (\mathbf{n}_1 \ \mathbf{n}_2 \ \cdots \ \mathbf{n}_j \ \cdots \ \mathbf{n}_q)$ and vector $\mathbf{u} = (u_1 \ u_2 \ \cdots \ u_j \ \cdots \ u_q)^T$. Then (3.3.18) states that any vector $\mathbf{g}$ that is in the cone is expressible as a *positive* linear combination of the vectors $\mathbf{n}_j$ that "generate" or "span" the cone. For example, a convex cone similar to Figure 2.2.9 might be generated by vectors $\mathbf{n}_1 = (1, 2)^T$ and $\mathbf{n}_2 = (4, 3)^T$. Then $\mathbf{g} = 2\mathbf{n}_1 + 1\mathbf{n}_2$ lies within that convex cone; the reader is urged to plot this case.

Farka's lemma relates a matrix and vectors, and it is fundamental to the subject of constrained optimization. *Farka's lemma* is: Given vectors $\mathbf{n}_j$, the columns of $\mathbf{N} = [\mathbf{n}_j]$, and vector $\mathbf{g}$, there is *no* (direction) vector $\mathbf{s}$ that satisfies the conditions

$$\mathbf{g}^T \mathbf{s} < 0, \tag{3.3.19}$$

$$\mathbf{N}^T \mathbf{s} \geq 0, \qquad \text{i.e., } \mathbf{n}_j^T \mathbf{s} \geq 0, \qquad j = 1 \text{ to } q, \tag{3.3.20}$$

whenever $\mathbf{g}$ is in the convex cone generated by the vectors $\mathbf{n}_j$ as described by (3.3.18). It will become apparent that $\mathbf{s}$ is a line search direction and that $\mathbf{g}$ is a gradient vector, so that (3.3.19) is a negative (downhill) directional derivative. First, it is useful to interpret Farka's lemma geometrically.

Recall from (2.2.82) in Section 2.2.4 that a hyperplane is defined by $\mathbf{n}^T \mathbf{x} = h$. The hyperplane goes through the origin when $h = 0$ and is displaced from the origin in the direction $\mathbf{n}$ for $h > 0$ (see Figure 2.2.7). Thus, $\mathbf{n}^T \mathbf{x} \geq 0$ represents a *closed half-space* bounded by a hyperplane through the origin and extending infinitely in the direction $\mathbf{n}$. Figure 3.3.2 illustrates two closed half-spaces bounded by hyperplanes H_1 and H_2 and extending in the directions $\mathbf{n}_1$ and $\mathbf{n}_2$, respectively. In the context of set theory, the *intersection* of the two half-spaces is indicated by the wavy line in Figure 3.3.2. This

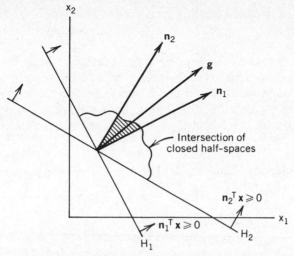

Figure 3.3.2. An illustration of Farka's lemma. Hyperplanes H_1 and H_2 define closed half-spaces that include a cone defined by their normal vectors. The directional derivative $\mathbf{g}^T\mathbf{s}$ cannot be negative for any direction $\mathbf{s}$ that is in the intersection of the closed half-spaces when $\mathbf{g}$ lies in the cone.

intersection is described mathematically by (3.3.20). Geometrically, Farka's lemma states that if direction vector $\mathbf{s}$ lies in that intersection, then the projection of vector $\mathbf{g}$ on $\mathbf{s}$ is positive, that is, the directional derivative (3.3.19) cannot be negative.

Farka's lemma is now applied to constraint qualification. The constraints of interest are those *q binding inequality constraints*, namely, $c_j(\mathbf{x}) = 0$, $j = 1$ to q. The other $t - q$ constraints are strictly greater than zero and are therefore satisfied by some margin. Define the gradients of the constraints as

$$\mathbf{n}_j = \nabla c_j(\mathbf{x}), \qquad j = 1 \text{ to } q, \qquad (3.3.21)$$

that is, the gradient vectors of the binding constraints are simply the column vectors comprising the $n \times q$ matrix $\mathbf{N} = [\mathbf{n}_j]$. Matrix $\mathbf{N}$ is assumed to be of full rank. The *Kuhn–Tucker constraint qualification condition* is: A necessary condition for $\mathbf{x}'$ to minimize $f(\mathbf{x})$ such that $\mathbf{c}(\mathbf{x}) \geq \mathbf{0}$ is that the gradient vector of the objective function, $\mathbf{g}(\mathbf{x}') = \nabla f(\mathbf{x}')$, lies within the cone generated by the gradients of the binding constraints. Mathematically, the necessary condition for a constrained minimum is

$$\nabla f(\mathbf{x}') = \mathbf{N}\mathbf{u} = \sum_{j=1}^{q} u_j \mathbf{n}_j, \qquad u_j \geq 0. \qquad (3.3.22)$$

In words, the objective function gradient must be a *positive* linear combination of the gradients of the binding constraints. The requirement for $u_j \geq 0$ stems from inequality constraints $\mathbf{c}(\mathbf{x}) \geq \mathbf{0}$. For equality constraints, $\mathbf{c}(\mathbf{x}) = \mathbf{0}$, u_j may have either sign.

Furthermore, the gradient operator with respect to $\mathbf{x}$ may be applied to both sides of the Lagrangian, (3.3.16). Since $\nabla_{\mathbf{x}} L(\mathbf{x}') = \mathbf{0}$ when $f(\mathbf{x}')$ is the constrained minimum where $\mathbf{c}(\mathbf{x}) = \mathbf{0}$, it is concluded that the u_j coefficients in (3.3.22) are in fact the Lagrange multipliers. Thus, $\nabla f_{\mathbf{x}}(\mathbf{x}') = \mathbf{Np}$ is the sum of the constraint normals, each scaled or weighted by the respective Lagrange multiplier.

Example 3.3.6. Suppose that it is desired to minimize the objective function

$$f(\mathbf{x}) = -x_1^2 x_2 \tag{3.3.23}$$

subject to

$$g_1(\mathbf{x}) = x_2 - x_1^2 \geq 0, \tag{3.3.24}$$

$$g_2(\mathbf{x}) = -x_1 - x_2 + 2 \geq 0. \tag{3.3.25}$$

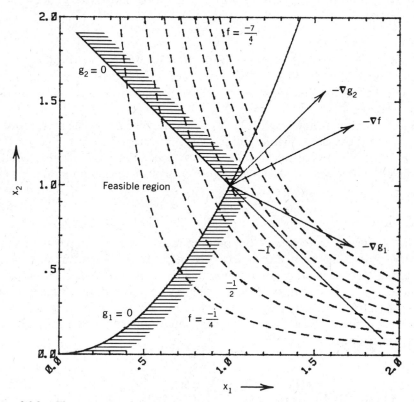

Figure 3.3.3. The optimal solution for the two binding constraints in Example 3.3.6. The Kuhn–Tucker conditions are satisfied, because the negative gradient of the objective function lies in the cone generated by the two negative gradients of the constraint functions.

This problem is illustrated in Figure 3.3.3. The feasible region to the left of the hatched borders represents the set of points **x** that satisfy both $g_1 \geq 0$ and $g_2 \geq 0$. Both these constraints are binding at $\mathbf{x}' = (1\ 1)^T$ where $f(\mathbf{x}') = -1$, the optimal constrained objective function value. It is more convenient and just as valid to employ the Kuhn–Tucker conditions in terms of negative gradients. Recall that the gradient of a nonlinear equality function ($g_1 = 0$) is simply the normal to a tangent line at that point. Clearly, the two constraint gradients form a cone that contains the gradient of the objective function. Therefore, $\mathbf{x}' = (1\ 1)^T$ fulfilled the necessary conditions for a constrained minimum.

The essence of the Kuhn–Tucker conditions is seen in Figure 3.3.3: There is no small step from the optimum into the feasible region that will further reduce the value of the objective function. Stated another way, if a direction **s** pointing into the feasible region such that $\mathbf{s}^T\mathbf{g} < 0$ can be found, then the point in question is not an optimal solution. It is easy to construct cases involving nonconvex functions where the Kuhn–Tucker conditions are satisfied, yet there are other, better minima that are unreachable once a search algorithm is "trapped" in a local constrained minimum. Readers interested in more complete analyses of these situations should consult Hadley (1964:197) and Powell in Gill (1974b:8).

Problems

3.1. Use forward substitution to solve the equation $\mathbf{L}\mathbf{y} = \mathbf{b}$, where

$$\mathbf{L} = \begin{bmatrix} 2 & 0 & 0 & 0 \\ -2 & -1 & 0 & 0 \\ -1 & 1 & -3 & 0 \\ 0 & 2 & 4 & 9 \end{bmatrix}, \qquad \mathbf{b} = \begin{bmatrix} -2 \\ 13 \\ -7 \\ 10 \end{bmatrix}.$$

3.2. Use back substitution to solve the equation $\mathbf{U}\mathbf{x} = \mathbf{y}$, where

$$\mathbf{U} = \begin{bmatrix} 1 & 3 & 0 & -1 \\ 0 & 1 & 3 & -2 \\ 0 & 0 & 1 & 0 \\ 0 & 0 & 0 & 1 \end{bmatrix}, \qquad \mathbf{y} = \begin{bmatrix} -1 \\ -11 \\ -1 \\ 4 \end{bmatrix}.$$

3.3. Suppose that matrix $\mathbf{Z} = \mathbf{T}^T\mathbf{T}$, and that $\mathbf{T}$ is given by (2.1.8).

   (a) Compute the *Cholesky factorization* $\mathbf{Z} = \mathbf{LDL}^T = \mathbf{U}^T\mathbf{U}$ where $\mathbf{U} = \mathbf{LD}^{1/2}$. Why is this always possible? Is $\mathbf{Z}$ positive-definite?

   (b) Compute the *Gram–Schmidt factorization* $\mathbf{T} = \mathbf{QU}$. Compare this matrix $\mathbf{U}$ with matrix $\mathbf{U}$ in part (a).

   (c) Discuss the corresponding $\mathbf{U}$ matrices in parts (a) and (b) using equation (3.1.47).

3.4. Verify the calculations for the rank 1 update of the $\mathbf{LDL}^T$ factorization in Example 3.1.3.

3.5. From the rule for differentiating a product, verify that

$$\nabla(\mathbf{u}^T\mathbf{v}) = (\nabla\mathbf{u}^T)\mathbf{v} + (\nabla\mathbf{v}^T)\mathbf{u},$$

where ∇ is defined by (3.1.40) and $\mathbf{u}$ and $\mathbf{v}$ depend on $\mathbf{x}$. For the quadratic form $Q = \mathbf{x}^T\mathbf{C}\mathbf{x}$, $\mathbf{C} = \mathbf{C}^T$, set $\mathbf{u} = \mathbf{x}$ and $\mathbf{v} = \mathbf{C}\mathbf{x}$ to prove (3.1.44), namely, $\nabla Q = 2\mathbf{C}\mathbf{x}$.

3.6. Convert the nonlinear least-squares fitting function in (1.2.5) to a linear mathematical model by assigning $x_4 = 0.01287$ and $x_5 = 0.02212$. Then

$$f(\mathbf{x}, t) = x_1 + x_2\exp(-0.01287t) + x_3\exp(-0.02212t).$$

Suppose that the given data are a representative subset of that shown in Figure 1.2.1:

i	1	2	3	4	5
t	0	20	60	120	320
d_i	0.844	0.932	0.881	0.685	0.406

Solve the linear least-squares (LLS) system in (3.1.53) using three methods:

(*a*) The normal equations (3.1.46) using program MATRIX command 9.

(*b*) The generalized inverse using programs MATRIX and GENINVP command 10.

(*c*) Orthogonal decomposition (3.1.47) using programs MATRIX and GSDECOMP command 10.

Compare your results with those in Table 1.2.1.

3.7. Prove the outer-product singular value decomposition

$$\mathbf{A} = \mathbf{USV}^T = \sum_{i=1}^{n} s_i\mathbf{u}_i\mathbf{v}_i^T, \tag{3.1.60}$$

where $\mathbf{U} = [\mathbf{u}_i]$, $\mathbf{V} = [\mathbf{v}_i]$, and the s_i are the singular values.

3.8. Using programs SVD and also MATRIX with GSDECOMP and QRITER, verify that the eigenvalues of matrix $\mathbf{Z} = \mathbf{T}^T\mathbf{T}$ are equal to the respective squared singular values of $\mathbf{Z}$ for $\mathbf{T}$ given by (2.1.8).

3.9. Are eigenvectors conjugate under any conditions? Prove your answer.

3.10. Using the matrix $Z = T^T T$, T in (2.1.8), numerically test the conjugacy of some column vectors in the $U = [u_i]$ and $V = [v_j]$ matrices associated with the singular value decomposition $A = USV^T$. Note from (3.1.61) that singular values $s_{ij} = 0$ for $i \neq j$. For computational convenience, also note that

$$s_{ij} = (Ue_i)^T A (Ve_j),$$

where e_i is a unit-direction vector such as (2.1.2); storage of the set of e_i on disk mass storage simplifies this computation. Are u_1 and v_2 conjugate?

3.11. Determine the rank of the following matrices:

$$A = \begin{bmatrix} -1 & 0 & 1 & 2 \\ -1 & 1 & 0 & -1 \\ 0 & -1 & 1 & 3 \\ 0 & 1 & -1 & -3 \\ 1 & -1 & 0 & 1 \\ 1 & 0 & -1 & -2 \end{bmatrix}.$$

$$B = \begin{bmatrix} 4 & -2 & 2 \\ -1 & 5 & 13 \\ -3 & -1 & -9 \\ 2 & -3 & -5 \end{bmatrix}.$$

3.12. Solve the linear system of equations in (3.1.36) using singular value decomposition in conjunction with

(a) Equations (3.1.65) through (3.1.68), and the alternative

(b) Equation (3.1.69b).

3.13. The Rayleigh quotient (3.1.78) gives the extreme values of $f(x) = x^T B x$ such that $x^T x = 1$, where B is symmetric. Form the Lagrangian function (3.3.16) for the constrained problem and show that the necessary conditions for a minimum or a maximum agree with (3.1.78).

3.14. Section 3.1.2 described overdetermined sets of linear equations, $Ax = b$, for $A_{m,n}$, $m > n$. Suppose that instead of too many equations, there are too few, that is, $m < n$. Form the Lagrangian function to minimize $x^T x$ such that $Ax = b$. Equating the derivatives of that Lagrangian function to zero and using the fact that $\nabla(p^T A x) = A^T p$, show that the solution for minimum $\|x\|_2$ when $m < n$ is

$$x = A^T(AA^T)^{-1}b.$$

Compare this result with (3.1.49). Explain why the case $m < n$ need not be treated separately from the $m > n$ case.

3.15.   Show that the column vectors of $\mathbf{U}$ are the eigenvectors of the symmetric matrix $\mathbf{AA}^T$, where the singular value decomposition of $\mathbf{A} = \mathbf{USV}^T$.

3.16.   Consider the quadratic form $Q = \mathbf{x}^T\mathbf{Bx}$, where $\mathbf{x}$ is in E^2 and $\mathbf{B}^T = \mathbf{B} = [b_{ij}]$. Show that

$$Q(\mathbf{x}) = b_{11}x_1^2 + 2b_{12}x_1x_2 + b_{22}x_2^2,$$

$$x_2 = \frac{-b_{12}x_1 \pm \left[b_{12}^2x_1^2 - b_{22}\left(b_{11}x_1^2 - Q\right)\right]^{1/2}}{b_{22}}.$$

3.17.   For the quadratic function $F(\mathbf{x}) = c + \mathbf{b}^T\mathbf{x} + \frac{1}{2}\mathbf{x}^T\mathbf{Bx}$ where $c = 500$, $\mathbf{b} = (-94 \ -67)^T$, and $\mathbf{B} = \begin{bmatrix} 14 & 2 \\ 2 & 11 \end{bmatrix}$,

(a)   Compute F, $\nabla_1 F$, $\nabla_2 F$, and the directional derivative in the direction $\mathbf{s} = (1 \ -2)^T$, all at the point $\mathbf{x} = (3 \ 7)^T$,

(b)   Find the minimum value, $F(\mathbf{x}')$ and $\mathbf{x}'$. Obtain the eigensolution for $\mathbf{B}$ and the canonical form of $Q = \mathbf{x}^T\mathbf{Bx}$ in new coordinates $\mathbf{y}$. Sketch the level curves for $F(\mathbf{x}) = 50.5, 60.5, 70.5,$ and $150.5,$

(c)   Write the Taylor series in $\mathbf{dx}$ about the point $\mathbf{x} = (3 \ 7)^T$ and find the $\mathbf{dx}$ step from that point to $\mathbf{x}'$.

(d)   Locate the minimum $F(\mathbf{x}')$ using the conjugate gradient algorithm in Table 3.2.3, starting from $\mathbf{x} = (3 \ 7)^T$. Show all values of $\mathbf{x}$, t_k^*, $\mathbf{s}$, and $\nabla F = \mathbf{g}$ involved. Show numerically that the search directions $\mathbf{s}^{(k)}$ are B-conjugate and that the gradient at each turning point is orthogonal to the last search direction.

3.18.   Change step 5 in the conjugate gradient algorithm, Table 3.2.3, to read

$$h_k = \frac{\left[\mathbf{g}^{(k+1)}\right]^T \left[\mathbf{g}^{(k+1)} - \mathbf{g}^{(k)}\right]}{\left[\mathbf{s}^{(k)}\right]^T \left[\mathbf{g}^{(k+1)} - \mathbf{g}^{(k)}\right]}.$$

Incorporate this change in your BASIC program that was suggested in Example 3.2.7 and rerun the program. It has been shown that this expression for h_k provides better protection against roundoff error and also ensures that $[\mathbf{s}^{(k+1)}]^T\mathbf{Ds}^{(k)} = 0$ even when the other steps have been computed inaccurately.

3.19.   Define $\mathbf{dg} = \mathbf{g}^{(k+1)} - \mathbf{g}^{(k)}$ and $\mathbf{dx} = \mathbf{x}^{(k+1)} - \mathbf{x}^{(k)}$. Then use (3.2.7) and (3.2.8) to show that there is an invariant mapping of corresponding differences in gradient and position on quadratic functions called the *secant condition*:

$$\mathbf{dg} = \mathbf{H}\,\mathbf{dx},$$

assuming constant Hessian matrix $\mathbf{H}$. Show that $\mathbf{dx}$ and $\mathbf{dg}$ are in the same direction only if **dx** is an eigenvector of the Hessian matrix. Also, use (3.2.48) and (3.2.42) to obtain the "altitude" $F(\mathbf{p})$ at a point **p** above that at the minimum $F(\mathbf{x}')$:

$$F(\mathbf{p}) - F(\mathbf{x}') = \tfrac{1}{2}[\mathbf{g}(\mathbf{p})]^T \mathbf{H}^{-1}[\mathbf{g}(\mathbf{p})].$$

Davidon noted that the inverse Hessian matrix is a *metric* (a measure of distance); this fact provides a family name for certain *variable metric search methods* discussed later.

3.20. Prove (3.2.35).

3.21. *Stationary points* of $F(\mathbf{x})$ are points where $\nabla F(\mathbf{x}) = \mathbf{0}$; they may represent maxima, minima, or neither (saddle points). From Fletcher (1980:30), find the stationary points of the following functions:
(a) $F(\mathbf{x}) = 2x_1^3 - 3x_1^2 - 6x_1x_2(x_1 - x_2 - 1)$.
(b) $F(\mathbf{x}) = (x_2 - x_1^2)^2 + x_1^5$.
(c) $F(\mathbf{x}) = 2x_1^2 + x_2^2 - 2x_1x_2 + 2x_1^3 + x_1^4$.
(d) $F(\mathbf{x}) = (x_1x_2)^2 - 4x_1^2x_2 + 4x_1^2 + 2x_1x_2^2 + x_2^2 - 8x_1x_2 + 8x_1 - 4x_2$.

3.22. Modify the program in Table 1.1.2 to compute Newton steps, **dx** in (3.2.48), so that the minima in Table 1.1.1 can be computed from a nearby starting point **x**. Note that which minimum is found depends entirely on the arbitrary starting point and behavior of the Newton-step algorithm.

3.23. Derive the gradient vectors in Example 3.3.6 and verify that they satisfy the Kuhn–Tucker conditions at the optimal point.

3.24. From Dixon (1972b:92), consider the constrained optimization problem:

$$\text{Minimize } f(\mathbf{x}) = 2x_1^2 - 2x_1x_2 + 2x_2^2 - 6x_1$$

such that

$$3x_1 + 4x_2 \le 6 \quad \text{and} \quad -x_1 + 4x_2 \le 2.$$

It is easy to verify that the unconstrained minimum is at $\mathbf{x} = (2\ 1)^T$, which satisfies the second constraint but not the first one. Convert this problem to the Lagrange format by adding squared *slack variables* to each constraint to make them equalities. By this classical ploy, solve the following problem by creating the Lagrangian function and using a suitably *modified* program C3-4, LAGRANGE:

$$\text{Minimize } f(\mathbf{x}) = 2x_1^2 - 2x_1x_2 + 2x_2^2 - 6x_1$$

such that

$$3x_1 + 4x_2 + x_3^2 - 6 = 0 \quad \text{and} \quad -x_1 + 4x_2 + x_4^2 - 2 = 0.$$

With the addition of slack variables x_3 and x_4 and two Lagrange multipliers, say x_5 and x_6, there are a total of six variables. The Newton iteration was started at $\mathbf{x} = (1\ 1\ 1\ 1\ 1\ 1)^T$ and converged on the eighth iteration to the following values:

i	1	2	3	4	5	6
x_i	1.45946	0.40541	−0.00000	1.35567	−0.32432	−0.00000
w_i	8.40516	5.51117	0.64865	−4.90579	−2.68335	1.67281.

The w_i are the eigenvalues of the *Hessian* matrix after convergence. What kind of stationary point is this solution? What is the significance of the slack variable and Lagrange multiplier values at the solution? Are any of the original constraints binding? Test the Kuhn–Tucker condition.

3.25. A circle of fixed diameter D has a rectangle of sides x and y inscribed within it. Use the Lagrangian function to find the values of x and y that will maximize the area of the rectangle.

Chapter Four

Newton Methods

The Newton–Raphson method was applied in Chapter Three for a vector function of a vector. Newton's method deals with a related scalar function of a vector and is based on a quadratic approximation to a multidimensional function. The Taylor series showed that this is reasonable as long as the starting vector of variables was suitably close to a minimum of the function. This chapter deals with three pragmatic questions about Newton's method:

1. How large is the neighborhood in which the quadratic model is valid?
2. How can Newton's method be made robust (hardy) when started far away from the minimum?
3. By what means can explicit expressions for the second derivatives be avoided?

The last question is answered in this chapter by assuming that exact first derivatives are available, so that either finite differences can approximate the second derivatives or the objective function is a sum of squares. The second question is answered in this chapter as it is for different methods described later—just start by steepest descent if necessary, and then change at some point to a method that is better nearer the minimum. The ingredient in the recipe that differs here from elsewhere is how and when the transition is made. The first question is of central importance to the subject of this chapter, namely, the concept of a "trust region" where the quadratic model of the function is valid. That leads to what Fletcher (1980) has called *restricted step methods*—the ones that make sense for a blind man on a mountain, who would hardly dare to take large, unrestricted steps without some assurance of avoiding disaster!

4.1. Obtaining and Using the Hessian Matrix

Usually, expressions for *second* derivatives are not available and creation of an algorithm for their calculation is complicated. However, it is possible to employ finite differences that use first derivative functions to approximate second derivatives. Some other alternatives are (1) to assume a useful structure in the objective function that allows a reasonable approximation of the Hessian or (2) to build an approximation to the Hessian by a succession of steps based on the idea of conjugate gradients introduced in Chapter Three. There is a small chance that a positive-definite Hessian might become indefinite in the approximation, but the real risk is that far from a minimum on a generally nonlinear function, the Hessian is negative definite or, more likely, indefinite. This problem must be faced when starting far from an unknown minimum, because all sophisticated search techniques assume a positive-definite Hessian matrix. This section describes the practical details necessary to approximate and to deal with a Hessian (symmetrical) matrix expressed in the factored $\mathbf{LDL}^T$ form and stored in a vector instead of the wasteful two-dimensional array.

4.1.1. Finite Differences for Second Derivatives.
Recall that the mathematically strict *definition* of a *derivative* of a scalar function $f(x)$ of a scalar variable x is

$$f'(x) = \frac{df}{dx} = \lim_{dx \to 0} \frac{f(x + dx) - f(x)}{dx}, \qquad (4.1.1)$$

where dx is an increment. The same concept applies for a partial derivative of a scalar function $F(\mathbf{x})$ of a vector $\mathbf{x}$:

$$\nabla_i F(\mathbf{x}) = \lim_{dx_i \to 0} \frac{F(\mathbf{x} + dx_i \, \mathbf{e}_i) - F(\mathbf{x})}{dx_i}, \qquad (4.1.2)$$

where dx_i is an increment in the ith element of $\mathbf{x}$ and $\mathbf{e}_i$ is the ith unit vector [e.g., see (2.1.2)]. The notation ∇_i is the del operator (3.1.39).

Not only can first derivatives be *computed* by the two preceding equations, but the second partial derivatives can be obtained in a similar way. Suppose that the gradient vector function $\mathbf{g}(\mathbf{x}) = (\nabla_1 F \ \nabla_2 F \ \cdots \ \nabla_i F \ \cdots \ \nabla_n F)^T$ is available, where the elements of $\mathbf{g}$ are defined by (4.1.2). Then *column vectors* that constitute the Hessian matrix of second partial derivatives, $\mathbf{H} = [\mathbf{h}_1 \ \mathbf{h}_2 \ \cdots \ \mathbf{h}_n]$, are defined by

$$\mathbf{h}_j = \lim_{dx_j \to 0} \frac{\mathbf{g}(\mathbf{x} + dx_j \, \mathbf{e}_j) - \mathbf{g}(\mathbf{x})}{dx_j}, \qquad j = 1 \text{ to } n. \qquad (4.1.3)$$

To see the details of why this is so, examine the first column of the Hessian matrix defined by (3.2.46). If $\mathbf{H} = [h_{ij}]$, then the elements in the first column are $h_{11} = \nabla_1(\nabla_1 F)$, $h_{21} = \nabla_1(\nabla_2 F), \ldots, h_{i1} = \nabla_1(\nabla_i F), \ldots,$ and $h_{n1} = \nabla_1(\nabla_n F)$. The idea is that the quantities in the parentheses, $\nabla_i F = g_i(\mathbf{x})$, are the elements of the gradient vector, which is presumed to be an available function. The same statement can be made for all other columns of **H**. Therefore, by perturbing just x_j, it is possible to finite difference each of the g_i elements in the spirit of (4.1.2) and thus obtain the entire jth column of the Hessian matrix, according to (4.1.3).

Since the Hessian matrix is symmetric, the method described for obtaining all n columns of the Hessian provides redundant elements that are off the main diagonal. To reduce the inherent errors described in the next paragraph, the symmetric approximation of the Hessian $\tilde{\mathbf{H}}$ is used:

$$\tilde{\mathbf{H}} = \frac{\mathbf{H} + \mathbf{H}^T}{2}. \tag{4.1.4}$$

Derivatives are limit operations that assume that the increment involved tends to zero. Consider the approximation of the gradient vector in light of the Taylor series in (3.2.42). The approximation that

$$F(\mathbf{dx}) - F(\mathbf{p}) = \mathbf{g}(\mathbf{p})^T \mathbf{dx} \tag{4.1.5}$$

is valid only if the quadratic form $\mathbf{dx}^T \mathbf{H}(\mathbf{p}) \, \mathbf{dx}$ and higher-order terms can be neglected. According to (3.1.43), quadratic forms contain only second-degree terms, in this case $(dx_i)^2 = (x_i - p_i)^2$, so the neglected higher-order terms do tend to zero faster than the linear $\mathbf{dx}$ term in (4.1.5) as $\mathbf{dx} \rightarrow \mathbf{0}$. The error in finite-differenced derivatives $(dx_i \neq 0)$ because of the neglected terms in the Taylor series representation of the function is called *truncation error*. That is just one of the two sources of error that is incurred by picking some small number, dx_i, and proceeding to calculate partial derivatives according to (4.1.2) and (4.1.3).

The total error incurred in computing derivatives by finite differences is the sum of truncation error and cancellation error. *Cancellation error* results from subtraction of two very nearly equal numbers on a computer having finite word length, especially in the mantissa. Table 4.1.1 shows the machine precision e_m as obtained by the short algorithm in problem 1.12 in Chapter 1. Table 4.1.1 shows that 1E–6 is about the smallest value of dx_i that would register a change in x_i when used in (4.1.2) on an IBM 370 in short precision.

Table 4.1.1. Several Machine Precision Constants in Base 10 and Base 2

Computer	Condition	e_m in Base 10	e_m in Base 2
IBM 370	Short Precision	9.50E–7	2^{-20}
IBM PC	BASICA DEFSNG	5.96E–8	2^{-24}
HP 85	HP BASIC	3.64E–12	2^{-38}
IBM PC	BASICA DEFDBL	1.39D–17	2^{-56}

According to Gill (1981:128, 345), the finite difference increment dx_i might reasonably be about 100–1000 times the machine precision e_m. Dennis (1984:1766) noted that a good rule of thumb in finite difference calculations is to perturb half as many digits in x_i as are accurate in the function, in this case g_j. However, the concern is with the number of digits in the mantissa and not the exponent, so a *relative increment* of, say, $dx_i = 0.0001x_i$ has always been used by the author, with good results, especially if the termination criterion has about the same relative magnitude. The case of $x_i = 0$ must be anticipated; in that event, $dx_i = 1D-6$ is set when working in double precision.

Finite differencing of either first or second partial derivatives requires the nominal function value without perturbed variables plus n additional evaluations for the set of perturbed variables. So in case of either (4.1.2) or (4.1.3), there are n extra function evaluations required for derivatives by finite differences. This can seriously lengthen the optimization program execution time. The effects of truncation error on first derivatives causes more problems in optimization algorithms than their effects on second derivatives employed in Newton's method in the case at hand. However, the idea of finite differencing *both* the first derivatives and then the second derivatives is especially bad, since the compounded truncation error has a severe effect on gradient algorithms. A part of the following program C4-1, NEWTON (lines 2420–2680), finite differences the gradient obtained by formula to provide approximate second derivatives that are accurate to four or five significant figures.

4.1.2. Forcing Positive-Definite Factorization.

It is useful to relate the steepest-descent method to Newton's method in a special way that ultimately relates to positive definiteness of the Hessian matrix. Consider a *linear model* of a function from the Taylor series in (3.2.42)

$$F^* = F(\mathbf{dx}) = F(\mathbf{y}) + \mathbf{g(y)}^T \mathbf{dx}, \qquad (4.1.6)$$

where $\mathbf{dx} = \mathbf{x} - \mathbf{y}$, $\mathbf{y}$ being the fixed point about which the expansion is constructed. The first scenario is to minimize F^* such that the step is restricted to the unit hypersphere $\|\mathbf{dx}\|_2 = 1$. Lagrangian function (3.3.11) for constraining the step size is

$$L(\mathbf{dx}, p) = F + \mathbf{g}^T\mathbf{dx} + p(\mathbf{dx}^T\mathbf{dx} - 1). \qquad (4.1.7)$$

Setting the gradient $\nabla L = \mathbf{0}$ yields $\mathbf{g} + 2p\,\mathbf{dx} = \mathbf{0}$, or

$$\mathbf{dx} = \frac{-\mathbf{g}}{2p}, \qquad (4.1.8)$$

and p is chosen so that $\|\mathbf{dx}\| = 1$.

It is seen that a linear function model constrained with a two-norm unit step is in the steepest-descent direction. There is an interesting geometric interpretation for this analysis. Comparing (4.1.6) with (2.2.82), it can be visualized for $\mathbf{x}$ in E^3 that fixed values of F^* define a set of planes, some of which cut the unit sphere centered at $\mathbf{x} = \mathbf{y}$. The constrained minimum value of F^* determines the unique plane that is tangent to the unit sphere. The gradient $\mathbf{g}(\mathbf{x})$ is normal to the planes.

A second scenario is to retain the constrained linear model but to change the norm of the unit step to an *elliptic norm* (see Figure 3.2.1):

$$\|\mathbf{dx}\|_H = \left(\mathbf{dx}^T \mathbf{H} \, \mathbf{dx}\right)^{1/2}, \tag{4.1.9}$$

where $\mathbf{H}$ is positive-definite. Thus, the Lagrangian function becomes

$$L(\mathbf{dx}, p) = F + \mathbf{g}^T \mathbf{dx} + p\left(\mathbf{dx}^T \mathbf{H} \, \mathbf{dx} - 1\right). \tag{4.1.10}$$

Again, setting the gradient $\nabla L = 0$ yields $\mathbf{g} + 2p\mathbf{H}\,\mathbf{dx} = 0$,

$$\mathbf{H}\,\mathbf{dx} = -\mathbf{g}, \tag{4.1.11}$$

where p has been ignored, because it affects only the step size. Of course, (4.1.11) is just Newton's step as previously obtained in (3.2.48), so it may be concluded that Newton's method is equivalent to steepest descent under the elliptic norm of (4.1.9). Put another way, under the elliptic norm the steepest descent method is obtained when $\mathbf{H} = \mathbf{I}$, the identity matrix. A geometric interpretation similar to that just given is that $\mathbf{H}^{-1}$ deflects the $\mathbf{dx}$ vector from the normal vector to the hyperplane, thus locating the point where that particular plane is tangent to the unit ellipsoid.

The descent property is paramount for an optimization algorithm, that is, each step must proceed downhill. It was shown in Section 3.2.1 that the directional derivative, $F' = \mathbf{g}^T \mathbf{s}$ evaluated at a point $\mathbf{x}$, is simply the slope of $F(\mathbf{x})$ in the $\mathbf{s}$ direction. In the Newton case, $\mathbf{g}$ can be replaced with (4.1.11) so that the directional derivative becomes

$$F' = -\mathbf{dx}^T \mathbf{H} \, \mathbf{dx}, \tag{4.1.12}$$

which is strictly negative (downhill) for any $\mathbf{dx}$ if and only if the Hessian $\mathbf{H}$ is positive-definite. If the Hessian is computed at some point not too close to a minimum function value, it may not be positive-definite. Then $\mathbf{H}$ *must* be altered in order to preserve the downhill property.

There are at least a half dozen ways that the Hessian can be forced positive-definite; two are discussed in this section and another in Section 4.2.2. According to Gill (1974b), Greenstadt proposed to employ the spectral decomposition of the Hessian as described in (2.2.34):

$$\mathbf{H} = \sum_{i=1}^{n} w_i \mathbf{v}_i \mathbf{v}_i^T, \tag{4.1.13}$$

where the w_i and v_i are the corresponding eigenvalues and eigenvectors of **H**, respectively. It has been reported that simply changing all negative eigenvalues, $w_i < 0$, to positive constants equal to about 10 times the machine precision constant in the summation of (4.1.13) produces a useful positive-definite matrix, say $\overline{\mathbf{H}}$. Then $\overline{\mathbf{H}} = \mathbf{LDL}^T$ can be used in (4.1.11) to find a **dx** which is a somewhat arbitrary step having a length and a downhill direction. A complete eigensystem analysis of the Hessian would be required for each step, and that requires between $2n^3$ and $4n^3$ operations, an exorbitant price to pay in run time and code. The interested reader is referred to Gill (1981:107).

Gill (1981:109) describes a means to detect and correct a negative definite symmetrical matrix during $\mathbf{H} = \mathbf{LDL}^T$ factorization. The diagonal matrix $\mathbf{D} = [d_1 \; d_2 \; \cdots \; d_n]$ will contain only $d_i > 0$ if **H** is positive-definite. [This is proved in (4.1.17).] If a negative d_i is encountered it may be set to some small positive constant, thus producing a positive-definite $\overline{\mathbf{H}} = \mathbf{H} + \mathbf{E}$, where **E** is a non-negative diagonal matrix. Gill describes a test to select the small positive constant so that **H** is disturbed as little as possible. In program C4-1, NEWTON, described in Section 4.3, the author simply sets any negative d_i to $+1\text{D}{-}6$; see lines 2730 and 2920. Otherwise, the $\mathbf{LDL}^T$ factorization in NEWTON (lines 2700 to 2960) is exactly that given in program C3-2, LDLTFAC.

When one or more eigenvalues of the Hessian matrix are negative, then that point is a saddle point. It is possible to compute a *negative curvature descent direction*, that is, a direction **s** such that when $\mathbf{s}^T\mathbf{Hs} < 0$ then $\mathbf{s}^T\mathbf{g} < 0$. Fletcher (1977) describes a successful method for this approach, based on a different $\mathbf{LDL}^T$ factorization. It seems to be less desirable than simply forcing positive definiteness, using the modified $\mathbf{LDL}^T$ factorization method just described.

4.1.3. Computing Quadratic Forms and Solutions.

As noted in Section 3.1.1, symmetric matrices that are $n \times n$ actually contain only $n(n + 1)/2$ unique elements, as opposed to the n^2 elements in general matrices. Therefore, it is common practice to evolve a given symmetric matrix **H** as previously given in (3.1.14):

$$\mathbf{LDL}^T = \begin{bmatrix} h_{11} & & & \\ h_{21} & h_{22} & & \\ h_{31} & h_{32} & h_{33} & \\ h_{41} & h_{42} & h_{43} & h_{44} \end{bmatrix} \rightarrow \begin{bmatrix} d_1 & & & \\ l_{21} & d_2 & & \\ l_{31} & l_{32} & d_3 & \\ l_{41} & l_{42} & l_{43} & d_4 \end{bmatrix}$$

$$\rightarrow \begin{bmatrix} H(1) & & & \\ H(2) & H(5) & & \\ H(3) & H(6) & H(8) & \\ H(4) & H(7) & H(9) & H(10) \end{bmatrix}. \tag{4.1.14}$$

In the application in this chapter where the Hessian matrix is computed by finite differencing the gradient vector, the resulting symmetrical matrix given by $\tilde{\mathbf{H}}$ in (4.1.4) is first stored as indicated by the top-left matrix in (4.1.14). After factorization, the original h_{ij} elements are replaced by elements from the $\mathbf{L}$ and $\mathbf{D}$ matrices as shown by the upper-right matrix in (4.1.14). The actual storage of these elements is in the single subscripted array, $H(\cdot)$ as indicated at the bottom of (4.1.14).

It is necessary to compute quadratic forms, such as $\mathbf{e}^T\mathbf{He}$ which occurred in (3.2.19), where $\mathbf{e} = \mathbf{g}$ in connection with the Cauchy point. Note that

$$Q = \mathbf{e}^T\mathbf{He} = \mathbf{e}^T\mathbf{LDL}^T\mathbf{e} = (\mathbf{L}^T\mathbf{e})^T\mathbf{D}(\mathbf{L}^T\mathbf{e}) = \mathbf{t}^T\mathbf{Dt}. \tag{4.1.15}$$

For the $n = 4$ case, the defined vector $\mathbf{t} = \mathbf{L}^T\mathbf{e}$ is

$$\mathbf{t} = \begin{bmatrix} 1 & l_{21} & l_{31} & l_{41} \\ 0 & 1 & l_{32} & l_{42} \\ 0 & 0 & 1 & l_{43} \\ 0 & 0 & 0 & 1 \end{bmatrix}\begin{bmatrix} e_1 \\ e_2 \\ e_3 \\ e_4 \end{bmatrix} = \begin{bmatrix} e_1 + l_{21}e_2 + l_{31}e_3 + l_{41}e_4 \\ e_2 + l_{32}e_3 + l_{42}e_4 \\ e_3 + l_{43}e_4 \\ e_4 \end{bmatrix}. \tag{4.1.16}$$

Note that $\mathbf{L}$ was previously defined by (3.1.12). Once $\mathbf{t}$ has been obtained, the desired quadratic form is

$$Q = \mathbf{e}^T\mathbf{He} = \mathbf{t}^T\mathbf{Dt} = \sum_{i=1}^n d_i t_i^2. \tag{4.1.17}$$

An important incidental conclusion from (4.1.17) is that $\mathbf{D}$ is positive definite if $\mathbf{H}$ is, thus all $d_i > 0$. These calculations have been coded in program C4-1, NEWTON, in lines 2100 to 2310.

It should be noted that if $\mathbf{e}$ in (4.1.17) is the Newton step, $\mathbf{e} = -\mathbf{H}^{-1}\mathbf{g}$, then

$$Q = (\mathbf{H}^{-1}\mathbf{g})^T\mathbf{H}(\mathbf{H}^{-1}\mathbf{g}) = \mathbf{g}^T\mathbf{H}^{-1}\mathbf{HH}^{-1}\mathbf{g} = \mathbf{g}^T\mathbf{H}^{-1}\mathbf{g}. \tag{4.1.18}$$

This quadratic form for the inverse Hessian is obtained in program C4-1, NEWTON, line 1920, to predict the expected decrease in function value (4.2.1).

It is timely to be specific about how the linear system $\mathbf{He} = \mathbf{u}$ is solved for $\mathbf{e} = \mathbf{H}^{-1}\mathbf{u}$, given $\mathbf{u}$ and $\mathbf{H} = \mathbf{LDL}^T$. (This topic was deferred from Section 3.1.1.) Temporary variables $\mathbf{t}$ and $\tilde{\mathbf{t}}$ can be defined so that

$$\mathbf{He} = \mathbf{LD}(\mathbf{L}^T\mathbf{e}) = \mathbf{L}(\mathbf{Dt}) = \mathbf{L}\tilde{\mathbf{t}} = \mathbf{u}. \tag{4.1.19}$$

The substitutions that comprise the steps for solution are given in Table 4.1.2. Step 1 in Table 4.1.2 involves the $\mathbf{L}$ matrix defined by (3.1.12). Step 2 is solved easily, since $\mathbf{D}$ is a diagonal matrix; the elements of its inverse are simply $1/d_i$. Step 3 is the solution of (4.1.16).

Table 4.1.2.   Steps for Forward and Back Substitution with LDL^T Factorization

1.   Solve $\mathbf{L}\tilde{\mathbf{t}} = \mathbf{u}$ for $\tilde{\mathbf{t}}$ given $\mathbf{u}$ (forward substitution).
2.   Solve $\mathbf{Dt} = \tilde{\mathbf{t}}$ for $\mathbf{t}$, that is, $t_i = \tilde{t}_i/d_i$, $i = 1$ to n.
3.   Solve $\mathbf{L}^T\mathbf{e} = \mathbf{t}$ for $\mathbf{e}$ (back substitution).

The code for the operations defined in Table 4.1.2 is contained in program C4-1, NEWTON, lines 2980 to 3200 (equivalent to lines 7800 to 8020 in C3-2, LDLTFAC). The interested reader can verify the code for the $n = 4$ case by observing (3.1.12), (4.1.14), and (4.1.16). Step 1 in Table 4.1.2 occurs in lines 2990 to 3070. Step 2 is accomplished in line 3080 and especially in line 3120. Step 3 is accomplished in lines 3090 to 3190.

4.2.   Trust Neighborhoods

Newton's method converges at a quadratic rate in the immediate vicinity of a local minimum, but without restrictions on its step size it is often unreliable elsewhere, even in the single-variable case. On the other hand, when the starting point is well removed from a minimum, it is usually true that good initial progress can be made in the direction of the negative gradient. As recently remarked by Dennis (1984:1767), "Someone once said that this local convergence property is not important for Newton's method because most of the work is expended in getting close enough to be able to take the full step (to the Newton point) $\mathbf{x}'$. This is somewhat like saying that jet travel is not an important part of a two-week trip to Europe because it occupies so little of the time."

   Newton's method belongs to the class of optimization strategies that are "without memory," because there is ordinarily no information carried from one iteration to the next. In this section, some reasonably simple trust neighborhoods (regions) are defined to establish when to use and then to switch from steepest-descent to Newton directions, and when to limit step lengths to a reasonable maximum. The implementation of these switching policies in second-order methods is often more important than how Hessian positive definiteness is forced. Also, line searches in methods employing second derivatives are less crucial than those methods that employ only first derivatives (see Chapter Five). Finally, an important classical method for constraining a Newton search to a neighborhood in which a quadratic model is valid is explained.

4.2.1.   Trust Radius.   Figure 4.2.1 is an enlarged view of the quadratic function previously shown in Figure 3.2.5. As in the latter case, the point where the search is started continues to be $\mathbf{x}^{(0)} = (1.9 \ 4.5)^T$. The first leg of a steepest-descent path from that point terminates at a minimum called the

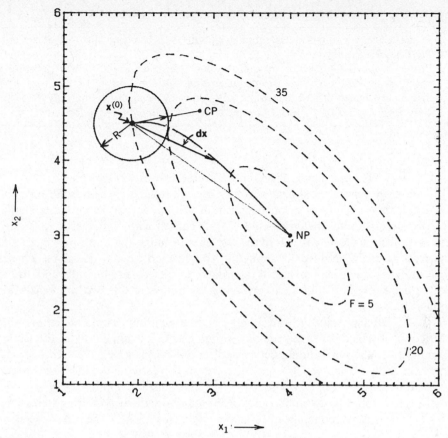

Figure 4.2.1. Level curves for the function in equation (3.2.11). A trust circle of radius $\frac{1}{2}$ is centered at $x = (1.9 \ 4.5)^T$, from which point two straight lines join the Cauchy and Newton points. The dash-dot-dash curve is the Levenberg trajectory.

Cauchy point, CP, which depends on $x^{(0)}$; this was described in Section 3.2.1. The *Newton point*, NP, does not depend on $x^{(0)}$; it was described in Section 3.2.2. A circular neighborhood of radius R about $x^{(0)}$ has been added to Figure 4.2.1; this might be used to inhibit the zigzagging inherent in the steepest-descent search.

For example, consider centering that neighborhood of radius R at every turning point, $x^{(k)}$, and employing the policy stated in Table 4.2.1. Consider the effects of this policy. The zigzagging characteristic of steepest descent (Figure 3.2.5) is avoided well before arriving in the neighborhood of the minimum. Once the search comes within radius R of the minimum, the rapid convergence of the Newton method is likely to prevail. In between those two states, the Newton step is limited to length R, presumably providing a reasonable rate of progress. By multiplying the step length by a factor as small

Table 4.2.1.   A Trust Neighborhood[a] Policy for Program C4-1, NEWTON

1. If the Cauchy point is outside the neighborhood, then step distance R in the steepest-descent direction.
2. If the Cauchy point is within the neighborhood, then step to the Newton point if it is within the neighborhood or to distance R in that direction.
3. If the chosen step fails to reduce the function value, sequentially multiply the step length by a factor of 4^{-1} until the step produces a function reduction.

[a]See Figure 4.2.1.

as $4^{-10} = 9.54\text{E}{-}6$, a downhill search is assured by (4.1.12).

Of course, general nonlinear surfaces are approximately quadratic only in the immediate vicinity of a local minimum. The policy in Table 4.2.1 is still reasonable, assuming that trust radius R is chosen with some wisdom. There are several indicators that will allow the user to judge the appropriateness of the trust radius. Certainly the algorithm should report major decisions. These include when the Hessian matrix is forced to be positive-definite and what policy decisions in Table 4.2.1 are being taken. There are two more helpful indicators.

First, Fletcher (1980:78) has suggested comparing the *actual* function reduction obtained on each step **dx** to that which is available from the same step on an *ideal* quadratic model (based on data from where the step began). Problem 3.19 in Chapter Three outlined means for deriving the ideal reduction in altitude $F(\mathbf{p}) - F(\mathbf{x}')$, where $\mathbf{p} = \mathbf{x}^{(k)}$, the starting point, and $\mathbf{x}'$ is the minimum point of a quadratic function based on gradient $\mathbf{g}[\mathbf{x}^{(k)}]$ and Hessian $\mathbf{H}[\mathbf{x}^{(k)}]$. That ideal reduction in function value *corresponding to a Newton step* using (4.1.18) is

$$F(\mathbf{p}) - F(\mathbf{x}') = \tfrac{1}{2}[\mathbf{g}(\mathbf{p})]^{T}[\mathbf{H}(\mathbf{p})]^{-1}[\mathbf{g}(\mathbf{p})]. \qquad (4.2.1)$$

Then a *quadratic factor r* is defined to be the ratio

$$r = \frac{F(\mathbf{p}) - F(\mathbf{p} + \mathbf{dx})}{F(\mathbf{p}) - F(\mathbf{x}')}. \qquad (4.2.2)$$

As defined in (4.2.2), this quadratic factor r is somewhat less general than Fletcher contemplated in that the reference step on the quadratic surface is not arbitrary. However, the trend is the same in the limit: $r \to 1$ as the minimum is approached. Fletcher proposed ways to expand and contract the trust radius R, based on the behavior of quadratic factor r from iteration to iteration. He also suggested using r in conjunction with the method in Section 4.2.2 that interpolates between the steepest-descent and Newton search directions.

A valuable second indicator for the user of second-order search methods is the angle between the negative gradient and the Newton search direction. That angle at $\mathbf{x} = (1.9\ 4.5)^T$ in Figure 4.2.1 is about 57 degrees. It is computed according to (2.1.42):

$$\theta = \cos^{-1}\left(\frac{\mathbf{g}^T\mathbf{dx}}{\|\mathbf{g}\|_2\|\mathbf{dx}\|_2}\right), \tag{4.2.3}$$

where $\mathbf{dx}$ is the Newton step in (4.1.11). It can be seen that if the Hessian condition number implicit in Figure 4.2.1 were much worse (larger), then the much more narrow elliptical level curves could cause values of θ approaching 90 degrees. In fact, there is a definite tendency for that angle to be between 70 and 90 degrees, which results in trajectories nearly tangent to the level curves (perpendicular to the gradient). Under those conditions the function value will show little decrease, tending to cause premature termination of the search.

All three indicators discussed in this section are reported by program C4-1, NEWTON:

1. Forced Hessian positive definiteness.
2. Quadratic factor r for Newton steps other than after report (1).
3. The angle between the negative gradient and the Newton step.

4.2.2. *Levenberg–Marquardt Methods.* This section provides a means for interpolating between a full Newton step and an infinitesimal steepest-descent step. The classical Levenberg–Marquardt (LM) method is often described in connection with nonlinear least-squares problems (Section 4.4). That unnecessarily complicates its exposition because those two topics are in fact separate subjects. The Levenberg–Marquardt method will not be employed in the program described next because the Hessian matrix of second partial derivatives is obtained by finite differences, an accurate but computationally expensive technique. However, the Levenberg–Marquardt method has everything to do with trust neighborhoods, so it is important to develop this subject at this time.

As opposed to the preceding considerations of constrained steps on a linear model (4.1.6), consider such a step on the *quadratic model* of a function from the Taylor series in (3.1.6)

$$F^* = F(\mathbf{y}) + \mathbf{g}(\mathbf{y})^T\mathbf{dx} + \tfrac{1}{2}\mathbf{dx}^T\mathbf{H}(\mathbf{y})\mathbf{dx}, \tag{4.2.4}$$

where $\mathbf{dx} = \mathbf{x} - \mathbf{y}$, $\mathbf{y}$ being the fixed point about which the expansion is constructed. As earlier, the goal is to minimize F^* such that the step is restricted to the unit hypersphere $\|\mathbf{dx}\|_2 = 1$. The related Lagrangian function is

$$L(\mathbf{dx}, p) = F + \mathbf{g}^T\mathbf{dx} + \tfrac{1}{2}\mathbf{dx}^T\mathbf{H}\,\mathbf{dx} + p(\mathbf{dx}^T\mathbf{dx} - 1). \tag{4.2.5}$$

Setting the gradient $\nabla L = 0$ yields $\mathbf{g} + \mathbf{H}\,\mathbf{dx} + 2p\,\mathbf{dx} = \mathbf{0}$, or

$$(\mathbf{H} + v\mathbf{I})\,\mathbf{dx} = -\mathbf{g}, \tag{4.2.6}$$

where the scalar parameter $v = 2p$ has been substituted for convenience.

Equation (4.2.6) is a very famous equation, having been introduced by Levenberg (1944) and amplified by Marquardt (1963) and countless others since then. If the parameter $v = 0$, then (4.2.6) reduces to the Newton equation in (4.1.11). If v is much larger than the elements of $\mathbf{H}$, then (4.2.6) describes a small steepest-descent step of length $\|\mathbf{g}\|/v$. The Lagrangian development aside, (4.2.6) specifies a means for dealing with non-positive-definite Hessian matrices, since property (8) in Table 2.2.3 guarantees that the eigenvalues of $\mathbf{H}$ will all be increased by amount v. It is also apparent from (4.2.6) that the vector $\mathbf{dx}$ is a function of v, notationally $\mathbf{dx}(v)$, since both $\mathbf{H}$ and $\mathbf{g}$ are evaluated at expansion point $\mathbf{y}$.

Example 4.2.1.   Again consider the quadratic function specified by (3.2.11) as pictured in Figures 3.2.5 and 4.2.1. Expanding that function about point $\mathbf{y} = (1.9 \;\; 4.5)^T$ according to (4.2.4) requires the data that

$$\mathbf{H} = \begin{bmatrix} 34 & 22 \\ 22 & 26 \end{bmatrix} \quad \text{and} \quad \mathbf{g}(\mathbf{y}) = \mathbf{b} + \mathbf{H}\mathbf{y} = \begin{bmatrix} -38.4 \\ -7.2 \end{bmatrix}. \tag{4.2.7}$$

Program MATRIX can be used to find the eigenvalues of $\mathbf{H}$: 52.36 and 7.64. From (4.2.6) and $\mathbf{x} = \mathbf{y} + \mathbf{dx}$, the trajectory $\mathbf{x}(v)$ is

$$\mathbf{x} = \mathbf{y} - (\mathbf{H} + v\mathbf{I})^{-1}\mathbf{g}. \tag{4.2.8}$$

For this simple system in E^2, the algebraic expression for the matrix inverse is

$$(\mathbf{H} + v\mathbf{I})^{-1} = \frac{1}{v^2 + 60v + 400} \begin{bmatrix} v + 26 & -22 \\ -22 & v + 34 \end{bmatrix}. \tag{4.2.9}$$

Notice that the roots of the polynomial are 52.36 and 7.64. Using (4.2.9) in (4.2.8) generates the dash-dot-dash curved trajectory shown in Figure 4.2.1. It begins at the Newton point (NP) for $v = 0$ and approaches $\mathbf{x}^{(0)}$, the center of the trust circle, as v increases without bound. The arrow in Figure 4.2.1 illustrates a typical step $\mathbf{dx}$ for some value of v.

Since $\mathbf{x}$ is a function of v, F is a function of v also. Both x_1 and F are shown as functions of v for this case in Figure 4.2.2. Notice that the sensitivity of these dependent variables on v is such that a *geometric progression* of values of v is required to scale the abscissa properly. That is a typical requirement for variations in parameter v. A value of v that corresponds to a given value for $R = \|\mathbf{dx}\|_2$ can be found by iteratively choosing values of v in (4.2.8), which is an implicit function $v(R)$. For example, the value $v = 51.2$ corresponds to $R = \frac{1}{2}$ as shown in Figure 4.2.1.

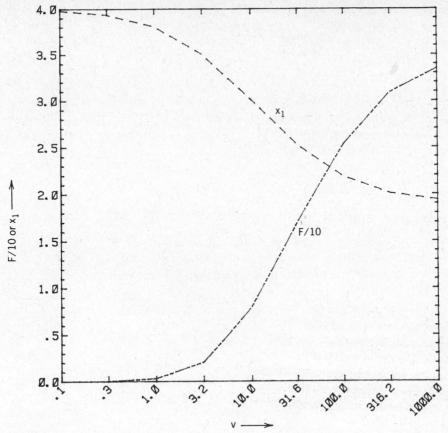

Figure 4.2.2. Function value and variable x_1 as functions of Levenberg parameter v in Example 4.2.1.

From the example and the corresponding Levenberg curved trajectory shown in Figure 4.2.1, it should be apparent that a given trust radius R specifies a point on the trajectory. Therefore, any given $R < \|\mathbf{x}' - \mathbf{y}\|$ determines a vector from $\mathbf{y} = \mathbf{x}^{(0)}$ that terminates on the trajectory. That vector, $\mathbf{dx}(v)$, is an interpolation between the Newton vector, $-\mathbf{H}^{-1}\mathbf{g}$, and the steepest descent vector, $-\mathbf{g}$; the interpolation parameter is v. Except for the $n = 2$ case, *the Levenberg interpolation vector is not in the Newton-gradient plane*, that is, points on the curve are not coplanar in the general case.

The function $\mathbf{dx}(v)$ is derived to provide the reader with a basis for understanding the numerous proposed algorithms involving the Levenberg trajectory. Rewrite (4.2.6):

$$\mathbf{dx} = -(\mathbf{H} + v\mathbf{I})^{-1}\mathbf{g}. \qquad (4.2.10)$$

Then use the spectral decomposition in (2.2.35) to express

$$(\mathbf{H} + v\mathbf{I})^{-1} = \sum_{i=1}^{n} (w_i + v)^{-1} \mathbf{v}_i \mathbf{v}_i^T. \tag{4.2.11}$$

Matrix $(\mathbf{H} + v\mathbf{I})$ is assumed to be of full rank since it is invertible and therefore nonsingular. Therefore, its eigenvalues are linearly independent so that the orthogonal expansion in (2.2.13) can be used to express gradient $\mathbf{g}$ as a linear combination of the eigenvectors:

$$\mathbf{g} = \sum_{i=1}^{n} a_i \mathbf{v}_i, \tag{4.2.12}$$

where the scalar coefficient $a_i = \mathbf{v}_i^T \mathbf{g}$ according to (2.2.15). Since matrix $(\mathbf{H} + v\mathbf{I})$ is symmetric, its eigenvectors are orthogonal, that is, $\mathbf{v}_i^T \mathbf{v}_j = 0$, $i \neq j$. If the $\mathbf{v}_i$ are also orthonormal, that is, orthogonal and with unit length, then $\mathbf{v}_i^T \mathbf{v}_i = 1$, and

$$\mathbf{dx} = \sum_{i=1}^{n} \frac{-a_i}{w_i + v} \mathbf{v}_i. \tag{4.2.13}$$

It may not be obvious why (4.2.13) is the product of (4.2.11) and (4.2.12) until the reader writes down and multiplies the two products for the $n = 2$ case. The simplification occurs precisely because of the orthonormal properties of the eigenvectors.

Equation (4.2.13) more clearly displays the explicit dependency of $\mathbf{dx}$ on v than does the alternative expression (4.2.6). It is a rational function with poles at $v = -w_i$, $i = 1$ to n. Notice in connection with (4.2.11) that there is an eigensystem $(\mathbf{H} + w\mathbf{I})\mathbf{v} = \mathbf{0}$ for specific eigenvalues w_i and corresponding eigenvectors $\mathbf{v}_i$. Thus, it should be clear why the function $(\mathbf{H} + v\mathbf{I})^{-1}$ has poles (singularities) at $v = -w_i$, $i = 1$ to n.

The reader is urged to compute the eigenvalues w_1 and w_2 of $\mathbf{H}$ in (4.2.7) using command 13 in program MATRIX with merged additions GSDECOMP and QRITER. Those results correspond to the polynomial appearing in the denominator of (4.2.9). In fact, $\mathbf{dx}(v)$ is a finite, continuous function on the domain $-w_n \leq v \leq \infty$, where w_n is the smallest eigenvalue. (All eigenvalues are positive for positive definite $\mathbf{H}$.)

The most important task in the Levenberg problem is to select v so that $(\mathbf{H} + v\mathbf{I})$ is positive definite, in case it is not. That fact would be discovered during $\mathbf{LDL}^T$ factorization (e.g., with program C3-2, LDLTFAC). Other than the method described in Section 4.1.2 for forcing positive definiteness, the inverse power method described in Section 3.2.2 could be used to compute the smallest eigenvalue of $\mathbf{H}$, say w_n. Then any value of v greater than the magnitude of w_n would make $(\mathbf{H} + v\mathbf{I})$ positive definite. Alternatively, a conservative upper bound for w_n is available from Gerschgorin's theorem (2.2.40), as noted by Dennis (1983:60).

Suppose that a step $\mathbf{dx}$ according to (4.2.10) produces an increase in function value $F[\mathbf{x}^{(k)} + \mathbf{dx}] > F[\mathbf{x}^{(k)}]$. Then an increase in v would shorten the step and produce an interpolation as previously discussed. An increase in $\mathbf{H}$ of $v\mathbf{I}$ also requires an *update* to the existing matrix inverse, a task that would be accomplished using (4.2.11) when the eigensolution was available. However, the Sherman–Morrison equation (2.1.36) is a more efficient method.

Several ways to select a value of v that is greater than that needed to make $(\mathbf{H} + v\mathbf{I})$ positive definite have been mentioned. When a step taken with a positive-definite Hessian produces an increase in function value, then one common method is to *increase v* in steps in search of a minimum $F(v)$. That amounts to a kind of line search from the Newton point (NP) along the curved Levenberg trajectory toward the current turning point $\mathbf{x}^{(0)}$; see Figure 4.2.1. In contrast to Figure 4.2.2 the function *actually* encountered may have a larger value at $v = 0$ and a smaller value at $v = 1000$, with an anticipated minimum in between.

Trial values of v should increase in geometrical progression. There are several means for making good use of such samples, for instance, polynomial approximation or systematic placement of samples at strategic intervals. Regarding interpolation or function fitting, the rational functional form evident in (4.2.13) has provided a number of ways to locate the minimum of an actual nonlinear function along the Levenberg trajectory. The interested reader is referred to Hebden (1973). Techniques for line searches are developed in Chapter Five.

Many users undoubtedly feel more comfortable selecting a value for trust radius R than for parameter v. Unfortunately, the solution of (4.2.10) for a value of v given a step length $\|\mathbf{dx}\| = R$ is a highly nonlinear problem. Powell (1970) suggested a piecewise-linear approximation to the Levenberg trajectory as illustrated in Figure 4.2.3. Figure 4.2.3 is similar to Figure 4.2.1 except that the trust neighborhood radius is $R = 1.2$ so that the Cauchy point (CP) is within the neighborhood and the Newton point is not. It can be shown for quadratic models that the Cauchy point is always closer to the turning point, $\mathbf{x}^{(0)}$, than is the Newton point (NP). Therefore, the two line segments ("dogleg") in Figure 4.2.3 that connect the point $\mathbf{x}^{(0)}$ to the Cauchy point and thence to the Newton point approximate the curved Levenberg trajectory shown by dash-dot-dashed line. As noted by Dennis (1984:1771), the dogleg is exactly the conjugate gradient algorithm (Section 3.2.1) applied to solve $\mathbf{H}\,\mathbf{dx} = -\mathbf{g}$ in the subspace defined by the Cauchy step vector and the Newton step vector.

It is not difficult to find the intersection of the circle of radius R with the line segment between CP and NP. The step $\mathbf{dx}$ shown in Figure 4.2.3 *approximately* solves the quadratic minimization problem in (4.2.4) such that $\|\mathbf{dx}\|_2 = R$; the *exact* solution is the intersection of the circle with the curved Levenberg trajectory. Dennis (1979a:456 and 1983:139) proposed a "double

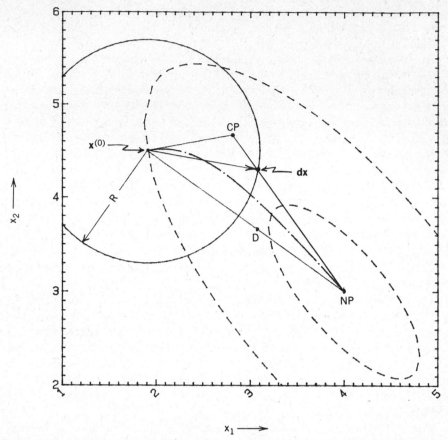

Figure 4.2.3. An illustration of the "dog-leg" approximate solution for a step of given length R on the Levenberg trajectory. This case is valid only when the Cauchy point (CP) is within the trust neighborhood (R).

dogleg" whereby the line segment from CP is brought first to some point such as D in Figure 4.2.3 on the Newton vector and thence to NP. It was his opinion that the single dogleg has too much bias toward steepest descent.

The reader should keep firmly in mind that the quadratic basis for all these analyses is highly suspect at points far from the minimum of a general nonlinear function. However, there are some who conjecture that the Levenberg trajectory in some way approximates the curving valleys that exist in n space. An interesting analysis and graph that suggests that conclusion is given in Wilde (1967:300). There is abundant evidence that the Levenberg–Marquardt concepts have contributed to more efficient algorithms. These vary from being reasonably straightforward such as Powell's dogleg and a spiral solution (Jones 1970) to a graphical display of normalized actual functions along the Levenberg trajectory (Antreich 1984).

4.3. Program NEWTON

Program C4-1, NEWTON, provides an illustration of Newton's method using the trust neighborhood policy stated in Table 4.2.1. The program starts with a short menu to make certain initial choices of parameters and variables. The objective function and its gradient are calculated in subroutines 5000 and 7000, respectively; these must be merged into NEWTON. One two-variable and one four-variable problem are furnished as examples. Finally, a simple means for enforcing upper and lower bounds on optimizer NEWTON is included for merging with the main program.

4.3.1. The Algorithm and Its Implementation. The listing for program C4-1, NEWTON, is contained in Appendix C. As noted by the remarks in lines 130 to 230, there are certain major programming names associated with the mathematics described in this chapter. The vector of variables, $\mathbf{x}$, is contained in array $X(\ )$. The objective function $F(\mathbf{x})$ is named F when newly computed and $F1$ when saved as a preceding value. The gradient vector of first derivatives, $\nabla F = \mathbf{g}$, is contained in $G(\ )$; that is finite differenced to obtain the Hessian matrix of second derivatives which is stored in vector form in $H(\ )$. The last major variable is $E(\ )$ which contains step vector $\mathbf{dx}$. A complete list of variable names employed in NEWTON is appended to the program listing so that the user will not violate previous naming assignments when supplying subroutines 5000 and 7000. In general, subroutines 5000 and 7000 are not called from loops or involved in current use of integer variables I, J, K, or L, so that these may be employed by the user. Notice that the BASIC function FNACS( ) is defined on line 280 for computing the inverse cosine.

The dimension of arrays of program variables is set to 30 in line 320, so that up to 30 optimization variables can be accommodated in the merged objective and gradient subroutines. The exception to this dimension is for the vector $H(\ )$ that stores the symmetric Hessian matrix in vector form. The

Table 4.3.1. Major Subroutines in Optimizer Program C4-1, NEWTON

Name	Lines
Enter Number & Value of Variables & Trust R	1200–1280
Enter/Revise Control Parameters	1300–1390
Main Optimization Algorithm (Figure 4.3.1)	1440–2080
Compute Quadratic Form Using $\mathbf{H} = \mathbf{LDL}^T$	2100–2310
Display Function, Gradient, and Variables	2330–2400
Compute Hessian from Gradient by Differences	2420–2680
$\mathbf{LDL}^T$ Factorization of Hessian in situ in $H(\ )$	2700–2960
Solution for $\mathbf{dx}$ in $\mathbf{H}\,\mathbf{dx} = -\mathbf{g}$ (Newton step)	2980–3200
Objective Function $F(\mathbf{x})$ (user supplied)	5000–6999
Gradient Vector $\nabla F = \mathbf{g}$ (user supplied)	7000–8999

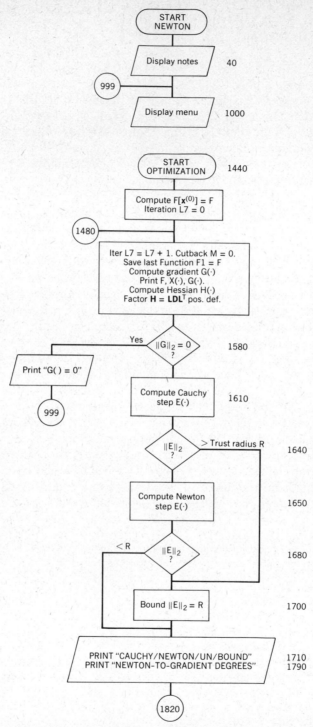

Figure 4.3.1. Flow chart for optimizer program C4-1, NEWTON.

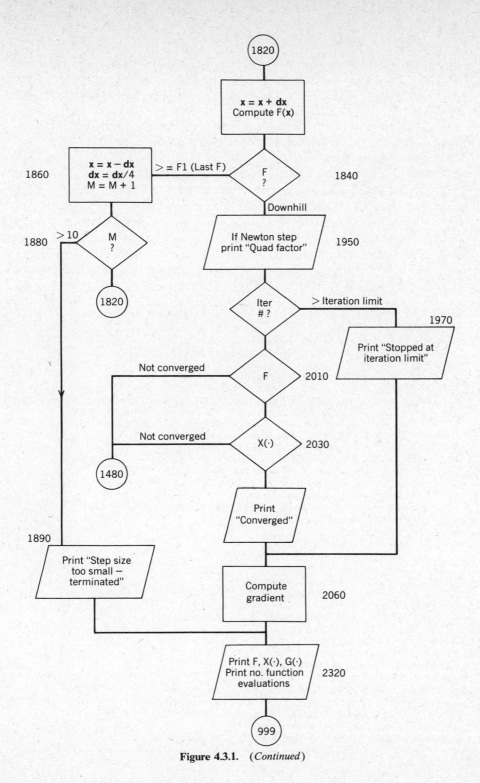

Figure 4.3.1. (*Continued*)

dimension of $H(\ )$ must be equal to $n(n + 1)/2$; when $n = 30$, the dimension of $H(\ )$ is 465 as shown in line 320. Users can reduce the memory required for execution of NEWTON by reducing these dimensions to fit their particular application.

A list of major subroutines and their line numbers in program NEWTON is given in Table 4.3.1. In addition to the preceding structure, program NEWTON has a menu scheme similar to that in program C2-1, MATRIX.

Table 4.3.2 shows the information displayed on the screen initially and during menu choices 1 and 2. The "NOTES" are similar to those used in program MATRIX, especially the recovery method after a ⟨Ctrl⟩ ⟨Break⟩ or EXIT, when ⟨GOTO 999⟩ ⟨Rtn⟩ will place the program back into menu selection without resetting any program variables. Menu choice 1 sets the number and value of optimization variables for the corresponding subroutines 5000 and 7000 supplied by the user. A trust radius value of zero is converted to 1E6, which means that only NEWTON steps are employed. Menu choice 2

Table 4.3.2.   Screen Displays for Notes and Menu Operation for Program NEWTON

```
********* NEWTON OPTIMIZER *************
NOTES:
1.   USE ONLY UPPER CASE LETTERS
2.   IF 'BREAK' OCCURS, RESTART WITH 'GOTO 999'
3.   USER MUST PROVIDE SUBROUTINE 5000 FOR FUNCTION EVALUATION
           AND SUBROUTINE 7000 FOR GRADIENT EVALUATION
4.   ENTER DEFAULT ANSWERS TO QUESTIONS BY <RETURN>.
PRESS <RETURN> KEY TO CONTINUE -- READY?
************* COMMAND MENU *************
1. ENTER STARTING VARIABLES (AT LEAST ONCE)
2. REVISE CONTROL PARAMETERS (OPTIONAL)'
3. START OPTIMIZATION
4. EXIT (RESUME WITH 'GOTO 999')
**********************************
INPUT COMMAND NUMBER:? 1
NUMBER OF VARIABLES = ? 2
ENTER STARTING VARIABLES X(I):
    X( 1 )=? -1.2
    X( 2 )=? 1
TRUST REGION RADIUS =? .5
PRESS <RETURN> KEY TO CONTINUE -- READY?
************* COMMAND MENU *************
1. ENTER STARTING VARIABLES (AT LEAST ONCE)
2. REVISE CONTROL PARAMETERS (OPTIONAL)'
3. START OPTIMIZATION
4. EXIT (RESUME WITH 'GOTO 999')
**********************************
INPUT COMMAND NUMBER:? 2
MAXIMUM # OF ITERATIONS (DEFAULT=50):? 60
STOPPING CRITERION (DEFAULT=.0001):? .001
ENTER FINITE DIFF FACTOR (DEFAULT=.0001):? .00001
PRINT EVERY Ith ITERATION (DEFAULT=1):? 3
PRESS <RETURN> KEY TO CONTINUE -- READY?
```

sets the four program parameters: (1) maximum number of iterations, (2) stopping criterion, (3) Hessian finite difference factor, and (4) screen printing interval for iteration results. All four parameters have default values as shown in the lower lines of Table 4.3.2; these have been set in line 290 so that menu choice 2 need not be exercised unless changes are desired. Also, after optimization (menu choice 3), menu choices 1 and 2 are undisturbed. This can be useful when selecting choice 3 again to continue optimization (with a reset iteration count).

The flow chart for program NEWTON in Figure 4.3.1 has a structure similar to the flow chart in Figure 1.3.1 for a generic iterative process. The four-digit numbers in Figure 4.3.1 correspond to the BASIC line numbers in the program C4-1 (NEWTON) listing in Appendix C. Reentry balloon 1480 is the starting point for each iteration or step in a search direction. At that time, the iteration and cutback ($\mathbf{dx} = \mathbf{dx}/4$) counters are incremented or set to zero, respectively. Also, the function value is saved at the beginning of each iteration for later comparison to be sure that each iteration actually produces a decrease in the function value.

There are a number of tests of vector lengths (two-norm). The test at line 1580 is the necessary condition for a minimum, but the real purpose is to avoid division by zero in subsequent calculations. Lines 1640 to 1700 implement the trust neighborhood policy defined in items 1 and 2 in Table 4.2.1. Item 3 in Table 4.2.1 is implemented by lines 1860 to 1880; if the attempt to obtain a reduced function value by decreasing step size exceeds a $10^6 : 1$ reduction, then the optimization is terminated (line 1890).

The normal directional step for each iteration is taken at balloon (line) 1820. If that results in a reduced function value, then the function value and each of the variables are subjected to the termination tests of (1.3.22). The algorithm continues if any one component fails.

4.3.2. *Some Examples Using Program NEWTON.* Investigators of nonlinear optimization have collected a large set of standard test problems over the years. A number of important references are discussed in Appendix B. In this section two of the most popular problems have been chosen for illustration of the NEWTON optimizer: Rosenbrock's function and Wood's function. The reader should be aware that these results are significant only in the most general sense, because the myriad of programming decisions, parameter choices, and computer characteristics will defeat precise comparisons. Also, it is always possible to construct problems that will cause a particular algorithm to fail. The criteria employed for the purposes of this book include the number of function evaluations and freedom from false convergence. Program NEWTON will not be judged on the number of gradient calculations, especially those numerous recalculations required to obtain the Hessian matrix by finite differences. Of course, the user may wish to evaluate analytical expressions for the second derivatives if they are available; usually, they are not.

Example 4.3.1. Probably the most frequently used test function in nonlinear optimization is Rosenbrock's function in E^2:

$$F(\mathbf{x}) = 100\left(x_2 - x_1^2\right)^2 + \left(1 - x_1\right)^2. \qquad (4.3.1)$$

The first derivatives are:

$$\nabla_1 F = g_1 = -400\left(x_1 x_2 - x_1^3\right) - 2\left(1 - x_1\right), \qquad (4.3.2)$$

$$\nabla_2 F = g_2 = 200\left(x_2 - x_1^2\right). \qquad (4.3.3)$$

These three equations are contained in Appendix C, program C4-2, ROSEN, which should be merged with NEWTON. Since these must be compatible with optimizer NEWTON, (4.3.1) is programmed in subroutine 5000, and (4.3.2) and (4.3.3) are programmed in subroutine 7000.

The main feature of the Rosenbrock function is the long curving valley shown in Figure 4.3.2. The standard starting point is $\mathbf{x}^{(0)} = (-1.2 \; 1)^T$, and the global (only) optimum point is at $\mathbf{x}' = (1 \; 1)^T$ for $F(\mathbf{x}') = 0$ as can be seen by inspection of (4.3.1). The two trajectories shown in Figure 4.3.2 display the variations in $\mathbf{x}$ made by NEWTON for $R = 1$ (solid line) and $R = 0.1$ (line with longer dashes). Although the dashed-line trajectory appears smoother, it

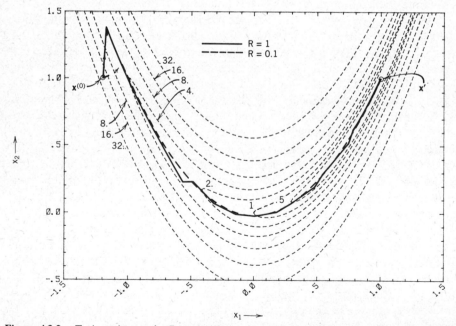

Figure 4.3.2. Trajectories on the Rosenbrock function. The solid trajectory is for trust radius $R = 1$ and the dashed trajectory is for $R = 0.1$.

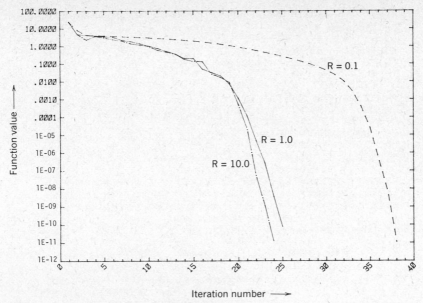

Figure 4.3.3. Descent on Rosenbrock's function in E^2 at $x = (-1.2 \ 1)^T$ from $F = 24.2$ to nearly 1E–12 for three values of trust radius R.

required 38 iterations, whereas the solid-line trajectory required only 24. Notice that the smaller trust radius caused the initial steps to proceed in the direction of the negative gradient, whereas the Newton steps started out a longer way around.

The effect of choosing different values for the trust radius can be observed better in Figure 4.3.3, which corresponds to the preceding trajectories. If the user will run these problems, it is apparent that, barring instability, the unbounded Newton method ($R = 10$) requires fewer iterations and fewer function values. Typical information displayed on the screen during optimization is shown in Table 4.3.3 for a starting point near the minimum point. Notice that the printing parameter was set for reporting every third iteration.

These results compare favorably with those for similar algorithms used on the Rosenbrock $n = 2$ problem. Some typical results for a discrete Newton algorithm and a description of the Rosenbrock problem for $n > 2$ are contained in O'Leary (1982).

Example 4.3.2. A second popular test for optimizers is Wood's function in E^4:

$$
\begin{aligned}
F(\mathbf{x}) = {} & 100\left(x_2 - x_1^2\right)^2 + (1 - x_1)^2 + 90\left(x_4 - x_3^2\right)^2 \\
& + (1 - x_3)^2 + 10.1\left[(x_2 - 1)^2 + (x_4 - 1)^2\right] \\
& + 19.9(x_2 - 1)(x_4 - 1).
\end{aligned}
\tag{4.3.4}
$$

Table 4.3.3. Screen Display for the Rosenbrock Function Starting Near the Minimum

```
NUMBER OF VARIABLES = ? 2
ENTER STARTING VARIABLES X(I):
   X( 1 )=? 1.06
   X( 2 )=? .96
TRUST REGION RADIUS =? .02
PRESS <RETURN> KEY TO CONTINUE -- READY?
AT START OF ITERATION NUMBER 1
   FUNCTION VALUE = 2.680096
  I           X(I)              G(I)
  1          1.060000         69.486336
  2          0.960000        -32.720000
****************************** CAUCHY (BOUNDED)
****************************** CAUCHY (BOUNDED)
****************************** CAUCHY (BOUNDED)
AT START OF ITERATION NUMBER 4
   FUNCTION VALUE = 6.707251E-02
  I           X(I)              G(I)
  1          1.005891         10.429429
  2          0.985926         -5.178332
****************************** NEWTON (BOUNDED)
                NEWTON-TO-GRADIENT DEGREES= 61.4
                QUADRATIC BEHAVIOR FACTOR R= 1.408021
****************************** NEWTON (UNBOUNDED)
                NEWTON-TO-GRADIENT DEGREES= 54.1
                QUADRATIC BEHAVIOR FACTOR R= 1.003378
****************************** NEWTON (UNBOUNDED)
                NEWTON-TO-GRADIENT DEGREES= 76.0
                QUADRATIC BEHAVIOR FACTOR R= 1.158249
AT START OF ITERATION NUMBER 7
   FUNCTION VALUE = 1.284383E-07
  I           X(I)              G(I)
  1          1.000357          0.002123
  2          1.000710         -0.000683
****************************** NEWTON (UNBOUNDED)
                NEWTON-TO-GRADIENT DEGREES= 81.2
****************************** NEWTON (UNBOUNDED)
                NEWTON-TO-GRADIENT DEGREES= 71.9
    ###### CUT BACK STEP SIZE BY FACTOR OF 4 #####
CONVERGED; SOLUTION IS:
AT START OF ITERATION NUMBER 9
   FUNCTION VALUE = 3.159744E-10
  I           X(I)              G(I)
  1          0.999982         -0.000013
  2          0.999964         -0.000019
TOTAL NUMBER OF FUNCTION EVALUATIONS = 10
PRESS <RETURN> KEY TO CONTINUE -- READY?
```

The components of the gradient are:

$$g_1 = -400x_1(x_2 - x_1^2) - 2(1 - x_1)$$

$$g_2 = 200(x_2 - x_1^2) + 20.2(x_2 - 1) + 19.8(x_4 - 1)$$

$$g_3 = -360x_3(x_4 - x_3^2) - 2(1 - x_3)$$

$$g_4 = 180(x_4 - x_3^2) + 20.1(x_4 - 1) + 19.8(x_2 - 1).$$

$$(4.3.5)$$

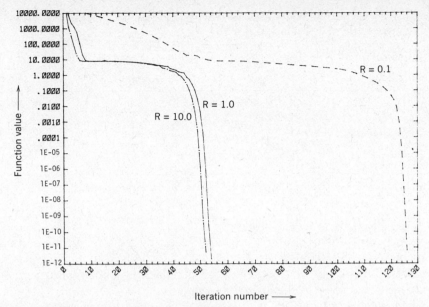

Figure 4.3.4. Descent on Wood's function in E^4 at $\mathbf{x} = (-3 \ -1 \ -3 \ -1)^T$ from $F = 19192$ to nearly 1E–12 for three values of trust radius R.

These equations are programmed in C4-3, WOODS, for merging with NEWTON. Figure 4.3.4 again shows that unbounded Newton steps give more rapid convergence when they are stable.

Wood's function is distinguished by a part of the trajectory where the Hessian is indefinite; that accounts for the flat region between iterations 10 and 40. The standard starting point is $\mathbf{x}^{(0)} = (-3 \ -1 \ -3 \ -1)^T$ and the global minimum is at $\mathbf{x}' = (1 \ 1 \ 1 \ 1)^T$ as seen by inspection of (4.3.4). The reader is urged to run program NEWTON with WOODS (C4-3) merged, using the standard starting point, trust radius $R = 1$, and stopping criterion 0.000001. At about iteration 10 in the vicinity of $\mathbf{x} = (-1 \ 1 \ -1 \ 1)^T$, a series of forced positive definiteness and cutbacks from uphill steps begin. That lasts until about iteration 40 in the vicinity of $\mathbf{x} = (1.4 \ 2 \ -0.15 \ 0)^T$. These obstacles should remind designers of algorithms that the implementation of an optimizer is often as important as the underlying convergence theory that is valid only in the neighborhood of a minimum.

Unfortunately, numerical tests are not totally reliable indicators of robustness. Table 4.3.4 shows the number of function evaluations for four starting points for both Rosenbrock's and Wood's functions. A trust radius of unity was chosen. These data compare favorably with some published earlier [e.g., Dennis (1979a)].

4.3.3. Simple Lower and Upper Bounds on Variables.
Most constraints encountered in practice are lower and upper bounds on variables; of those,

Table 4.3.4.   Number of Iterations and Function Evaluations for Four Starting Points Using Program NEWTON[a]

Rosenbrock's Function		Wood's Function	
Starting Point	#Iter/Feval	Starting Point	#Iter/Feval
$(-1.2 \ 1)$	24/31	$(-3 \ -1 \ -3 \ -1)$	54/79
$(2 \ -2)$	22/28	$(-1.2 \ 1 \ 1.2 \ 1)$	24/28
$(-3.635 \ 5.621)$	43/48	$(-3 \ 1 \ -3 \ 1)$	52/75
$(6.39 \ -0.221)$	26/40	$(-1.2 \ 1 \ -1.2 \ 1)$	44/64
$(1.489 \ -2.547)$	16/20		

[a] Number of function evaluations does not include those for finite differences.

bounds to keep variables positive are most frequently required. In this section, program C4-4, NBOUNDS, is described as an addition to the NEWTON optimizer. The purpose is to start the reader thinking about the constraint requirement and some of the difficulties that arise.

When merged with NEWTON, NBOUNDS causes a new menu item to appear: "5. SEE &/OR RESET LOWER/UPPER BOUNDS ON VARI-ABLES". As seen in the listing in Appendix C for C4-4, line 325 dimensions an integer array $L5(30)$ to contain 0s or 1s that indicate which variables have no bounds or one or both bounds set, respectively. Array $P5(30, 2)$ contains any pairs of bounds defined; $P5(I, 1)$ and $P5(I, 2)$ are the lower and upper bounds for the Ith variable, respectively. Default values selected by command 5 are $-10,000$ to $+10,000$.

Referring to the flow chart for NEWTON in Figure 4.3.1, balloon (circle) 1820 is the point where step $\mathbf{dx}$ in array $E(\ )$ has been determined. Before that step is taken, NBOUNDS adds a test to see if that step would violate any bounds. If so, the offending components of $\mathbf{dx} = (dx_1 \ dx_2 \ \cdots \ dx_n)^T$ are set so that the subsequent step will terminate on that boundary (see subroutine 3510 in NBOUNDS). Also, "flag" variable $L6 = 1$ is set to force the next step to be steepest descent (Cauchy). Flag variable $L6$ is tested in the new line 1635 of NBOUNDS; if $L6 = 1$ the Newton step is bypassed.

A typical screen display is shown in Table 4.3.5. Level curves of quadratic function (3.2.11) are shown in Figure 4.3.5 where the unconstrained global minimum is at $\mathbf{x}' = (4 \ 3)^T$. Table 4.3.5 shows that a lower bound of $x_2 \geq 3.9$ was set and that the optimizer was started at $\mathbf{x}^{(0)} = (1.9 \ 4.5)^T$ with a trust radius of $R = 1$. Results from every second iteration were printed. Note that iteration 3 started with x_2 binding at its lower limit, and thereafter only Cauchy steps were taken. The reason for this strategy can be seen in Figure 4.3.5. At the start of iteration 3, the function value was approximately on the $F = 5$ level curve, and a normalized negative gradient vector is shown at that point in Figure 4.3.5. The Newton step to the unconstrained minimum is also shown.

Table 4.3.5.   Screen Display for NEWTON with NBOUNDS on Quadratic Function (3.2.11)

```
BOUNDS NOW SET ARE:
I          LOWER              UPPER
NONE.   SET OR RESET ANY BOUNDS (Y/N)? Y
ENTER O TO RETURN TO MENU, ELSE ENTER VARIABLE # =? 2
PRESS <RETURN> IF NO BOUND DESIRED
  LOWER BOUND =? 3.9
  UPPER BOUND =?
ENTER O TO RETURN TO MENU, ELSE ENTER VARIABLE # =? O
PRESS <RETURN> KEY TO CONTINUE -- READY?
NUMBER OF VARIABLES = ? 2
ENTER STARTING VARIABLES X(I):
   X( 1 )=? 1.9
   X( 2 )=? 4.5
TRUST REGION RADIUS =? 1
PRESS <RETURN> KEY TO CONTINUE -- READY?
AT START OF ITERATION NUMBER 1
   FUNCTION VALUE = 34.92
  I          X(I)            G(I)
  1        1.900000     -38.400000
  2        4.500000      -7.200000
******************************** NEWTON (BOUNDED)
               NEWTON-TO-GRADIENT DEGREES= 46.2
               QUADRATIC BEHAVIOR FACTOR R= 4.161398
******************************** NEWTON (BOUNDED)
               NEWTON-TO-GRADIENT DEGREES= 46.2
               QUADRATIC BEHAVIOR FACTOR R= .7442283
AT START OF ITERATION NUMBER 3
   FUNCTION VALUE = 4.969744
  I          X(I)            G(I)
  1        3.527467       3.733879
  2        3.900000      13.004276
******************************** CAUCHY (BOUNDED)
   ###### CUT BACK STEP SIZE BY FACTOR OF 4 #####
******************************** CAUCHY (BOUNDED)
   ###### CUT BACK STEP SIZE BY FACTOR OF 4 #####
AT START OF ITERATION NUMBER 5
   FUNCTION VALUE = 4.766693
  I          X(I)            G(I)
  1        3.428480       0.368313
  2        3.900000      10.826556
******************************** CAUCHY (BOUNDED)
   ###### CUT BACK STEP SIZE BY FACTOR OF 4 #####
******************************** CAUCHY (BOUNDED)
   ###### CUT BACK STEP SIZE BY FACTOR OF 4 #####
AT START OF ITERATION NUMBER 7
   FUNCTION VALUE = 4.764701
  I          X(I)            G(I)
  1        3.418116       0.015952
  2        3.900000      10.598558
******************************** CAUCHY (BOUNDED)
   ###### CUT BACK STEP SIZE BY FACTOR OF 4 #####
CONVERGED; SOLUTION IS:
AT START OF ITERATION NUMBER 8
   FUNCTION VALUE = 4.764692
  I          X(I)            G(I)
  1        3.417740       0.003159
  2        3.900000      10.590280
TOTAL NUMBER OF FUNCTION EVALUATIONS = 13
PRESS <RETURN> KEY TO CONTINUE -- READY?
```

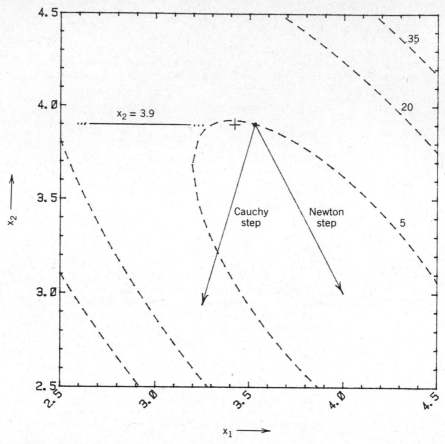

Figure 4.3.5. Level curves for (3.2.11) and two search directions at $\mathbf{x} = (3.527 \;\; 3.900)^T$.

If downhill progress is to be made at a point where one or more constraints are binding, then the projection of the unconstrained step vector onto the subspace of the unconstrained coordinates must be downhill. In this case, the *unconstrained subspace* is simply the x_1 coordinate direction, since x_2 is binding. In Figure 4.3.5 the projection of the Newton step onto the line $x_2 = 3.9$ goes off to the right, but the projection of the Cauchy step (negative gradient) onto $x_2 = 3.9$ goes off to the left. The constrained optimum is to the left at the point marked "$+$" in Figure 4.3.5 at $\mathbf{x}' = (3.4177 \;\; 3.9)^T$ according to Table 4.3.5. Note that $\nabla_1 F(\mathbf{x}') \doteq 0$. The important point to be observed is that not just any satisfactory unconstrained descent direction is suitable for projection onto the unconstrained subspace. Figure 4.3.5 clearly shows that no fraction of the projected Newton direction would be downhill, no matter how small. As described in Section 3.3.3, the *Kuhn–Tucker conditions* require that the gradient of the function must lie within the polyhedral cone formed by the constraint normal vectors. In this simple case the gradients of the function and the constraint coincide. *At points in a neighborhood of the constrained optimum, the negative gradient projection onto the unconstrained subspace always points toward the constrained optimum.*

One well-known method for incorporating linear constraints (not just bounds) is the *gradient projection method* by Rosen (1960); also see Hadley (1964:315). However, there are much more effective search strategies than steepest descent, especially some that retain the descent property in unconstrained subspaces. These more effective methods are described in Section 5.4.1. Program NBOUNDS will not be effective in cases where steepest descent is not effective, namely, where the function has curved valleys in the unconstrained subspace.

4.4. Gauss–Newton Methods

Second derivatives of functions subject to optimization are relatively difficult to obtain. It has been demonstrated that computed first derivatives (the gradient) may be perturbed in order to obtain reasonably accurate approximations of the second derivatives, but at a high price: n additional evaluations of the gradient, where there are n variables.

Remarkably, objective functions with the least-squares structure enable a positive-definite approximation to second derivatives that constitute the Hessian matrix. Least-squares objective functions are ideally suited to curve-fitting applications, including optimization of electrical networks and analogous scientific problems (Aaron 1956). Happily, there are extremely efficient means for computing the response functions and their first derivatives for electrical networks.

The positive-definite approximation of the Hessian matrix associated with optimization of least-squares objective functions is the essence of the *Gauss–Newton* method. This section describes nonlinear least squares (NLLS) and how it differs from linear least squares (LLS). Weighting coefficients may be employed to emphasize certain residual sampled errors in the objective function, and it is shown that the *least-pth objective function* that generalizes the LLS formulation ($p = 2$) is a way to emphasize automatically the larger residual errors. Also, in the limit as $p \to \infty$, minimization of the least-pth objective objective function approaches the infinity or minimax norm.

The formulation of the positive-definite approximation of the Hessian matrix is derived, and the additional considerations necessary to apply the Levenberg–Marquardt (LM) techniques are developed for effective use of approximate Newton optimization steps.

4.4.1. Nonlinear Least-pth Objective and Gradient Functions. The development begins with a precise basis for forming the nonlinear least-squares objective function and its gradient. The weighted least squares and the unweighted least-pth cases are formed later as extensions of this case and then

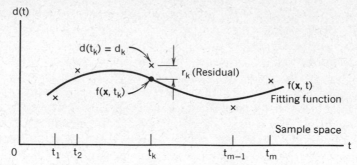

Figure 4.4.1. A curve-fitting example where a function $d(t)$ known by m given data pairs is to be approximated by a nonlinear function $f(x, t)$ that is the sum of given nonlinear basis functions. The problem is to minimize the errors or residuals, r_k, at each point in the sample space.

compared with each other. This definition of the nonlinear least-squares objective function is a vital step in applying optimization to many practical problems, and the reader is encouraged to recognize the problem that it solves.

Recall the previous mention of least-squares objective functions, especially the graphical description in Figure 3.1.3 which is repeated in Figure 4.4.1. Most recently, the mathematical model in (3.1.52) was linear in the coefficients x_j, even though the functions f_j were nonlinear in the independent variable t. For example, $f(x, t) = x_1\cos(2t) + x_2\exp(-3t) + x_3(t^2 - 1)^{1/2}$ is linear in $\mathbf{x}$. This is in contrast to the nonlinear mathematical model described in (1.2.5), where some elements of $\mathbf{x}$ are also involved in the basis functions; that is the present case. Figure 4.4.1 shows the m data pairs at the "×" points in the *sample space* t. The relevant *nonlinear mathematical model* is:

$$f(\mathbf{x}, t) = x_1 f_1(\mathbf{x}, t) + \cdots + x_j f_j(\mathbf{x}, t) + \cdots x_n f_n(\mathbf{x}, t). \quad (4.4.1)$$

The nonlinear mathematical model in (4.4.1) "fits" the data within some residual error r_k at each kth sample point as shown in Figure 4.4.1. Writing the equations for each sample point as if each residual were equal to zero (for $m = 5$ and $n = 3$ without loss of generality):

$$x_1 f_1(\mathbf{x}, t_1) + x_2 f_2(\mathbf{x}, t_1) + x_3 f_3(\mathbf{x}, t_1) = d_1,$$

$$x_1 f_1(\mathbf{x}, t_2) + x_2 f_2(\mathbf{x}, t_2) + x_3 f_3(\mathbf{x}, t_2) = d_2,$$

$$x_1 f_1(\mathbf{x}, t_3) + x_2 f_2(\mathbf{x}, t_3) + x_3 f_3(\mathbf{x}, t_3) = d_3, \quad (4.4.2)$$

$$x_1 f_1(\mathbf{x}, t_4) + x_2 f_2(\mathbf{x}, t_4) + x_3 f_3(\mathbf{x}, t_4) = d_4,$$

$$x_1 f_1(\mathbf{x}, t_5) + x_2 f_2(\mathbf{x}, t_5) + x_3 f_3(\mathbf{x}, t_5) = d_5.$$

These equations may be written as $\mathbf{Ax} = \mathbf{d}$; they are overdetermined and therefore inconsistent. Even worse, they are nonlinear in the variables **x**, since $\mathbf{A} = [f_{kj}] = [f_j(\mathbf{x}, t_k)]$.

The classical approach to fitting the nonlinear function $f(\mathbf{x}, t)$ to the given data vector $\mathbf{d} = (d_1 \ d_2 \ \cdots \ d_k \ \cdots \ d_m)^T$ for m data pairs (t_k, d_k) shown in Figure 4.4.1 is to minimize the two-norm of the *residual* vector

$$\mathbf{r} = \mathbf{A}\mathbf{x} - \mathbf{d}. \tag{4.4.3}$$

Thus the function to minimize is

$$F(\mathbf{x}, t) = \tfrac{1}{2}\mathbf{r}^T\mathbf{r} = \frac{1}{2} \sum_{k=1}^{m} r_k^2. \tag{4.4.4}$$

The factor of $\frac{1}{2}$ in (4.4.4) is for notational simplicity in subsequent expressions. Comparing (4.4.2) to the right-hand side of (4.4.3), it is convenient to define

$$\mathbf{A} = \begin{bmatrix} \mathbf{f}_1 \ \mathbf{f}_2 \ \cdots \ \mathbf{f}_j \ \cdots \ \mathbf{f}_n \end{bmatrix}, \tag{4.4.5}$$

where the typical column of nonlinear *basis functions* is the vector

$$\mathbf{f}_j = \begin{pmatrix} f_{1j} \ f_{2j} \ \cdots \ f_{kj} \ \cdots \ f_{mj} \end{pmatrix}^T. \tag{4.4.6}$$

Then the kth row of (4.4.3) for that residual or sampled error is

$$r_k = x_1 f_{k1} + \cdots + x_j f_{kj} + \cdots + x_n f_{mn} - d_k, \qquad k = 1 \text{ to } m. \tag{4.4.7}$$

Consider the first partial derivative of F as defined by the summation in (4.4.4). The gradient of the least-squares objective function F with respect to x_j is

$$\nabla_j F = \sum_{k=1}^{m} r_k(\nabla_j r_k), \qquad j = 1 \text{ to } n,$$

or $\hspace{10cm}$ (4.4.8)

$$\nabla F = \sum_{k=1}^{m} r_k(\nabla r_k).$$

In turn, the first partial derivative of the kth residual with respect to x_j may be obtained by differentiating (4.4.7):

$$\nabla_j r_k = f_{kj} + x_1(\nabla_j f_{k1}) + \cdots + x_j(\nabla_j f_{kj}) + \cdots + x_n(\nabla_j f_{mn}),$$

$$= f_{kj} + \sum_{q=1}^{n} x_q(\nabla_j f_{kq}), \qquad k = 1 \text{ to } m, \ j = 1 \text{ to } n. \tag{4.4.9}$$

Note that the first term in (4.4.9) results from the differentiation rule for a product, $d(uv) = v(du) + u(dv)$, as applied to the term $x_j f_{kj}$ in (4.4.7). It is again emphasized that *design matrix* $\mathbf{A}$ in the LLS problem was not a function of $\mathbf{x}$. See (3.1.53); there $\mathbf{A} = [a_{kj}] = [f_{kj}]$ in comparison to (4.4.2). Matrix $\mathbf{A}$ in the NLLS problem is a function of both $\mathbf{x}$ and sample parameter t.

Example 4.4.1. Consider the previous illustration of nonlinear curve fitting from Section 1.2.2, namely, the function in (1.2.5) which is repeated:

$$f(\mathbf{x}, t) = x_1 + x_2 \exp(-x_4 t) + x_3 \exp(-x_5 t). \tag{4.4.10}$$

The kth residual is

$$r_k = f(\mathbf{x}, t_k) - d_k, \tag{4.4.11}$$

and the gradient of the kth residual is composed of the following elements:

$$\nabla_1 r_k = 1, \qquad \nabla_2 r_k = \exp(-x_4 t_k), \qquad \nabla_3 r_k = \exp(-x_5 t_k),$$
$$\nabla_4 r_k = -t_k x_2 \exp(-x_4 t_k), \qquad \nabla_5 r_k = -t_k x_3 \exp(-x_5 t_k). \tag{4.4.12}$$

Example 4.4.2. The Rosenbrock function introduced in Section 4.3.2 can be recast as a least-squares problem by rewriting (4.3.1) in terms of residuals:

$$r_1 = 10(x_2 - x_1^2),$$
$$r_2 = (1 - x_1), \tag{4.4.13}$$

so that

$$F(\mathbf{x}) = \tfrac{1}{2}\mathbf{r}^T\mathbf{r} = \frac{1}{2} \sum_{k=1}^{2} r_k^2.$$

Note that a factor of $\frac{1}{2}$ has been introduced, according to (4.4.4). In this case the residuals are not a function of the sample space parameter. Even if they had been, there are no identifiable basis functions as defined by the structure in (4.4.7). Thus, starting with the given residuals, the gradient vectors of the respective residuals are:

$$\nabla r_1 = \begin{bmatrix} -20x_1 \\ 10 \end{bmatrix}, \qquad \nabla r_2 = \begin{bmatrix} -1 \\ 0 \end{bmatrix}. \tag{4.4.14}$$

Example 4.4.1 is a classical function-fitting problem. Example 4.4.2 often arises in an attempt to find a solution for several nonlinear functions implied by the vector function $\mathbf{r}(\mathbf{x}) = \mathbf{0}$. In many practical problems only a set of residual functions are known that are nonlinear functions of both $\mathbf{x}$ and t. Furthermore, the residual functions and their gradients may not be known explicitly, but by the results of an algorithmic process. All of these cases are candidates for the following Gauss–Newton formulation.

To make those developments easier to express, define the *Jacobian matrix*:

$$\mathbf{J}_{mn} = \left[\nabla_j r_k\right] \quad \text{for row } k = 1 \text{ to } m \text{ and column } j = 1 \text{ to } n,$$

$$= \left(\nabla r_1 \ \nabla r_2 \ \cdots \ \nabla r_m\right)^T. \tag{4.4.15}$$

Consider the following example of the Jacobian for $m = 5$ samples and $n = 3$ variables:

$$\mathbf{J} = \begin{bmatrix} \nabla_1 r_1 & \nabla_2 r_1 & \nabla_3 r_1 \\ \nabla_1 r_2 & \nabla_2 r_2 & \nabla_3 r_2 \\ \nabla_1 r_3 & \nabla_2 r_3 & \nabla_3 r_3 \\ \nabla_1 r_4 & \nabla_2 r_4 & \nabla_3 r_4 \\ \nabla_1 r_5 & \nabla_2 r_5 & \nabla_3 r_5 \end{bmatrix}. \tag{4.4.16}$$

This is the same Jacobian matrix described in (3.2.51), except that it is not square in this NLLS case, since there are more equations than variables.

The Jacobian provides immediate simplification, since it is easy to verify from (4.4.8) that the gradient of the least-squares objective function in (4.4.4) is

$$\nabla F = \mathbf{g} = \mathbf{J}^T \mathbf{r}. \tag{4.4.17}$$

4.4.2. Positive-Definite Hessian Approximation. The Hessian matrix $\mathbf{H}$ is composed of all second partial derivatives of $F(\mathbf{x}, t)$; see (3.2.46). Using the vector del operator ∇ from (3.1.40) and the fact that the order of partial differentiation is arbitrary, a valid expression for the Hessian is

$$\mathbf{H} = \nabla(\nabla F)^T = \nabla^2 F. \tag{4.4.18}$$

The notation $\nabla^2 F$ is a symbol for a matrix of second partial derivatives of F, the important defining operations being the middle part of (4.4.18). Thus, (4.4.18) and (4.4.8) yield

$$\mathbf{H} = \nabla\left[\sum_{k=1}^{m} r_k(\nabla r_k)\right]^T$$

$$= \sum_{k=1}^{m} \left[(\nabla r_k)(\nabla r_k)^T + r_k(\nabla^2 r_k)\right]. \tag{4.4.19}$$

This result can be stated in equivalent matrix notation:

$$\mathbf{H} = \mathbf{J}^T \mathbf{J} + \mathbf{M}, \tag{4.4.20}$$

where the $n \times n$ matrix $\mathbf{M}$ containing the residuals and their second derivatives is

$$\mathbf{M} = \sum_{k=1}^{m} r_k \nabla^2 r_k. \tag{4.4.21}$$

It is usually argued that (4.4.20) can be approximated by only its first term, the positive-definite matrix $\mathbf{J}^T\mathbf{J}$, since the second term, $\mathbf{M}$, is nearly zero if the residuals, r_k, are nearly zero. Of course, the residuals may approach zero near a solution, but whether that happens or not depends on particular cases.

Example 4.4.3.   Example 4.4.2 for the Rosenbrock function is continued by forming the gradient and Hessian functions. Using (4.4.17), the gradient is

$$\mathbf{g} = \mathbf{J}^T\mathbf{r} = \begin{bmatrix} -20x_1 & -1 \\ 10 & 0 \end{bmatrix} \begin{bmatrix} 10(x_2 - x_1^2) \\ 1 - x_1 \end{bmatrix}$$

$$= \begin{bmatrix} -200x_1(x_2 - x_1^2) - (1 - x_1) \\ 100(x_2 - x_1^2) \end{bmatrix}. \tag{4.4.22}$$

The two components of the Hessian from (4.4.20) are

$$\mathbf{J}^T\mathbf{J} = \begin{bmatrix} 400x_1^2 + 1 & -200x_1 \\ -200x_1 & 100 \end{bmatrix} \tag{4.4.23}$$

and

$$\mathbf{M} = 10(x_2 - x_1^2)\begin{bmatrix} -20 & 0 \\ 0 & 0 \end{bmatrix}. \tag{4.4.24}$$

It is easily confirmed that $\mathbf{H} = \mathbf{J}^T\mathbf{J} + \mathbf{M}$ and that $\mathbf{M} \to \mathbf{0}$ as the solution $\mathbf{x}' = (1\ \ 1)^T$ is approached.

It is now possible to compare these results for solving the NLLS problem with those obtained for the LLS problem. It will also become apparent that the Hessian expression in (4.2.20) bears an interesting resemblance to the Levenberg–Marquardt formulation. Begin by recalling that the Newton step in the search for a solution $\mathbf{x}'$ that minimizes an objective function $F(\mathbf{x})$ has always been

$$d\mathbf{x} = \mathbf{x} - \mathbf{p} = -\mathbf{H}^{-1}\mathbf{g} \tag{4.4.25}$$

as first defined in (3.2.48), where $\mathbf{p}$ is some point where $\mathbf{H}$ and $\mathbf{g}$ are evaluated. Of course that only works exactly when $F(\mathbf{x})$ is a quadratic function; otherwise there is a sequence of such steps. Earlier in this chapter, methods for reducing that step length and/or interpolating between that step direction and the negative gradient direction were described.

The solution to the *linear* least-squares (LLS) problem was to solve the *normal equations* $\mathbf{A}^T\mathbf{A}\mathbf{x} = \mathbf{A}^T\mathbf{b}$, or according to (3.1.49):

$$\mathbf{x} = (\mathbf{A}^T\mathbf{A})^{-1}\mathbf{A}^T\mathbf{b}. \tag{4.4.26}$$

In the LLS case, $\mathbf{p}$ was not involved because the design matrix $\mathbf{A}$ was comprised of basis functions that did not involve $\mathbf{x}$. According to (4.4.15) and (3.1.37), Jacobian $\mathbf{J} = -\mathbf{A}$. Then equating the gradient expression in (4.4.17) to zero also leads to the normal equations.

The solution to the *nonlinear* least-squares (NLLS) problem is to use (4.4.20) in (4.4.25):

$$\mathbf{dx} = -(\mathbf{J}^T\mathbf{J} + \mathbf{M})^{-1}\mathbf{J}^T\mathbf{r}. \tag{4.4.27}$$

The assumption that the second partial derivatives $\nabla^2 r_k = \mathbf{0}$ in $\mathbf{M}$ according to (4.4.21) is equivalent to assuming that the residuals in (4.4.7) are linear. In that case, (4.4.27) reduces to (4.4.26), so that the NLLS method can be viewed as a sequence of LLS solutions in small neighborhoods. A nice feature is that $\mathbf{J}^T\mathbf{J}$ will always be positive-definite. In relation to Newton's method, Gauss–Newton avoids computing the most expensive part of the Hessian ($\mathbf{M}$) without having to worry about indefinite matrices.

In the Levenberg–Marquardt (LM) method, interpolated directions and reduced step length were both affected by solving (4.2.10), namely,

$$\mathbf{dx} = -(\mathbf{H} + v\mathbf{I})^{-1}\mathbf{g}, \tag{4.4.28}$$

where v is the LM parameter. Comparison of (4.4.28) and (4.4.27) shows that in some sense the second-order term $\mathbf{M}$ has been replaced by the LM term $v\mathbf{I}$. It is shown in Section 4.4.5 that replacing $\mathbf{I}$ by a special diagonal matrix $\mathbf{D}$ can potentially improve the Gauss–Newton algorithm.

4.4.3. Weighted Least Squares and the Least-pth Method.

Each squared residual in the least-squares formulation can receive a unique multiplier or "weight" to emphasize that sample. It is shown that this process amounts to premultiplying both the the design matrix $\mathbf{A}$ and the data vector $\mathbf{d}$ by a diagonal matrix, similar to the matrix scaling Example 3.1.7 in Section 3.1.3. Then the least-pth method is derived by modifying the least-squares derivation. Finally, it is shown that the least-pth method can be viewed as a way to emphasize the larger errors automatically.

Suppose that the unweighted least-squares objective function (4.4.4) is revised to include *weighting coefficients* w_k:

$$\tilde{F} = \frac{1}{2}\sum_{k=1}^{m} w_k r_k^2. \tag{4.4.29}$$

Inspection of (4.4.2) shows that the addition of weights in this way is equivalent to multiplying each row of the nonlinear mathematical model by the square root of the weighting coefficient. It is straightforward to review the preceding derivations for both the LLS and NLLS problems to see that the revised normal equations in the linearized case $(\mathbf{M} = \mathbf{0})$ lead to

$$\mathbf{dx} = -(\mathbf{J}^T \mathbf{W} \mathbf{J})^{-1} \mathbf{J}^T \mathbf{W} \mathbf{r}, \qquad (4.4.30)$$

where the *weighting matrix* $\mathbf{W} = \text{diag}(w_1 \ w_2 \ \cdots \ w_m)$. These w_k are not related to the eigenvalues previously designated by w_i.

Setting the weighting issue aside for now, consider the *least-pth objection function*

$$F = \frac{1}{p} \sum_{k=1}^{m} r_k^p, \qquad p \text{ an even integer.} \qquad (4.4.31)$$

A glance at (4.4.8) shows that the differentiation for the gradient now yields

$$\nabla F = \sum_{k=1}^{m} r_k^{p-1} (\nabla r_k). \qquad (4.4.32)$$

Similarly, the differentiation for the Hessian in (4.4.19) produces

$$\mathbf{H} = \sum_{k=1}^{m} \left[(p-1) r_k^{p-2} (\nabla r_k)(\nabla r_k)^T + r_k^{p-1} (\nabla^2 r_k) \right]. \qquad (4.4.33)$$

Now it can be expected that the second term in (4.4.33) will decay more rapidly for small residuals and higher values of p.

The Gauss–Newton step for the least-pth formulation thus becomes

$$\mathbf{dx} = -(p-1)^{-1} (\mathbf{J}^T \mathbf{B} \mathbf{J})^{-1} \mathbf{J}^T \tilde{\mathbf{r}}, \qquad (4.4.34)$$

where

$$\tilde{\mathbf{r}} = \left(r_1^{p-1} \ r_2^{p-1} \ \cdots \ r_m^{p-1} \right)^T, \qquad (4.4.35)$$

and the diagonal matrix $\mathbf{B}$ is

$$\mathbf{B} = \text{diag}\left(r_1^{p-2} \ r_2^{p-2} \ \cdots \ r_m^{p-2} \right). \qquad (4.4.36)$$

As Breen (1973:688) noted, (4.4.34) may be considered the normal equations for the overdetermined system of linear equations

$$(p-1)\mathbf{R} \mathbf{J} \, \mathbf{dx} = \text{diag}\left(r_1^{p/2} \ r_2^{p/2} \ \cdots \ r_m^{p/2} \right). \qquad (4.4.37)$$

where

$$\mathbf{R} = \text{diag}\left(r_1^{p/2-1} \ r_2^{p/2-1} \ \cdots \ r_m^{p/2-1} \right). \qquad (4.4.38)$$

Comparison of (4.4.30) for weighted least squares and (4.4.34) for unweighted least-pth suggests a strong similarity so that the least-pth method tends automatically to emphasize the larger residuals. However, for values of p in excess of 10, very large and very small numbers may cause over/underflow problems in computations. This aspect of the least-pth method is discussed in Section 4.5 for program LEASTP.

4.4.4.   Numerical Integration As a Sampling Strategy.   Any user of optimization routines for curve- or model-fitting purposes such as illustrated in Figure
4.4.1 is faced with selecting a number of sample points and perhaps weights at
those points. A generally recommended rule of thumb is that at least $2n$
samples be employed, where n is the number of variables. As valid as that
advice may be, it is vitally important to understand a few key points concerning numerical integration or *quadrature*. Referring to Figure 4.4.1, the ideal
measure of error between model function $f(\mathbf{x}, t)$ and data function $d(t)$ is the
continuous residual function

$$r(\mathbf{x}, t) = f(\mathbf{x}, t) - d(t). \tag{4.4.39}$$

The scalar value best representing that error is the classical *integral p-norm*:

$$\|r(\mathbf{x}, t)\|_p = \left[\int_I |r(\mathbf{x}, t)|^p \, dI \right]^{1/p}, \tag{4.4.40}$$

where the approximation is over the closed continuum I in one or many
variables. The discrete representation of (4.4.40) has been considered previously in (2.1.39), the p norm of an n vector. The exponent $1/p$ is sometimes
omitted in mathematical analysis, since the minimum of the norm in (4.4.40) is
unaffected without it. Also, exponent p is taken to be even so that absolute
values are not required in the integrand. A further simplification is to take
$p = 2$, the least-squares case.

Example 4.4.4.   As explained by Forsythe (1970), consider the integral error
function

$$T(\mathbf{x}) = \int_0^1 [f(\mathbf{x}, t) - q(t)]^2 \, dt, \tag{4.4.41}$$

where

$$f(\mathbf{x}, t) = x_1 + x_2 t + \cdots + x_n t^{n-1}, \tag{4.4.42}$$

and $q(t)$ is a given continuous function. The minimum of $T(\mathbf{x})$ with respect to
all n of the x_j may be found simply by equating to zero each of the n first
derivatives of T with respect to x_j. (Recall that Leibnitz' rule states that the
order of differentiation and integration may be interchanged.) That leads to a
system of *normal equations* $\mathbf{Ax} = \mathbf{b}$, $\mathbf{A} = [a_{ij}]$, where

$$\begin{aligned} a_{ij} &= \int_0^1 [t^{i-1} t^{j-1}] \, dt, \\ &= (i + j - 1)^{-1}, \end{aligned} \qquad \text{for } i, j = 1, 2, \ldots, n, \tag{4.4.43}$$

and

$$b_i = \int_0^1 [t^{i-1} q(t)] \, dt, \qquad \text{for } i = 1, 2, \ldots, n. \tag{4.4.44}$$

Matrix **A** in (4.4.43) is the notorious *Hilbert matrix*:

$$\mathbf{A} = \begin{bmatrix} \dfrac{1}{1} & \dfrac{1}{2} & \dfrac{1}{3} & \cdots & \dfrac{1}{n} \\[2ex] \dfrac{1}{2} & \dfrac{1}{3} & \dfrac{1}{4} & \cdots & \dfrac{1}{n+1} \\[2ex] & & \cdots & & \\[1ex] \dfrac{1}{n} & \dfrac{1}{n+1} & \dfrac{1}{n+2} & \cdots & \dfrac{1}{2n-1} \end{bmatrix}. \qquad (4.4.45)$$

The use of this result appears simple: Given a function $q(t)$, use some method for numerically integrating (4.4.44) and then solve $\mathbf{x} = \mathbf{A}^{-1}\mathbf{b}$. However, the Hilbert matrix is notorious because it is illconditioned. Program C2-1, MATRIX, can be used to invert the 6×6 Hilbert matrix by entering nine significant digits for the matrix elements. Also, the exact inverse of the Hilbert matrix can be derived; see Knuth (1968:37). The fifth rows of the exact and computed Hilbert inverse are shown by the respective lines in Table 4.4.1. Of course, there is roundoff error in entering the rational values in the Hilbert matrix. That adds to the illconditioning to produce the differences in the exact and computed matrix inverse values. Values as small as 36 are also in the matrix inverse, so there is a huge range of element values represented. Since the fifth row of **A** multiplies the respective elements of **b** to yield x_5, a change in b_5 of only 10^{-6} would change x_5 by 4.41!

There is a clear message in this example: avoid forming illconditioned problems. There are alternatives; for example, the approximating function could be a linear sum of Chebyshev functions of the first kind instead of a linear sum of exponentials. The interested reader is referred to Cuthbert (1983:20).

Numerical integration formulas estimate the value of definite integrals as a linear sum of function values obtained at certain samples:

$$T = \int_0^1 f(t)\, dt \doteq \sum_{k=1}^{m} w_k f(t_k) = w_1 f_1 + w_2 f_2 + \cdots + w_m f_m. \quad (4.4.46)$$

The w_k coefficients are called *weights* in the sense used in the preceding section. The Newton–Cotes integration formulas employ evenly spaced samples. The *trapezoidal rule* has $m = 2$, $w_1 = w_2 = \frac{1}{2}$, $t_1 = 0$, and $t_2 = 1$; it is

Table 4.4.1.   Fifth Row of the Exact and Computed Hilbert Matrix Inverse

7560	−220500	1512000	−3969000	4410000	−1746360
7489.99	−218591.15	1499500.29	−3937278.02	4375654.71	−1733035.70

exact only for functions $f(t)$ that are linear on the interval $t = 0$ to 1. Similarly, *Simpson's rule* has $m = 3$, $w_1 = w_3 = \frac{1}{6}$, $w_2 = \frac{2}{3}$, $t_1 = 0$, $t_2 = \frac{1}{2}$, and $t_3 = 1$; it is exact only for functions $f(t)$ that are quadratic on the interval $t = 0$ to 1.

As seen from the above rules, a polynomial function of degree $m - 1$ or less can be integrated exactly by m evenly spaced sample points, t_k. It turns out that polynomial functions of degree $2m - 1$ or less can be integrated exactly by only m *unevenly spaced sample points*, using certain unique weighting coefficients. Since this is the best that can be obtained, those interested in curve- and model-fitting applications should be able to determine these unique sample points and their respective weights. Consider the *Gaussian integration* formulation of a squared residual function on the interval $t = -1$ to $+1$:

$$F(t) = \int_{-1}^{+1} r^2(t)\, dt \doteq w_1 r^2(t_1) + w_2 r^2(t_2) + \cdots + w_m r^2(t_m). \quad (4.4.47)$$

By making (4.4.47) exact for the $2m - 1$ function values $r^2(t) = 1$, $t, t^2, t^3, \ldots, t^{2m-1}$, $2m$ equations are obtained. Each equation has m sample points:

$$w_1 t_1^{k-1} + w_2 t_2^{k-1} + \cdots + w_m t_m^{k-1} = \int_{-1}^{+1} t^{k-1}\, dt, \qquad k = 1 \text{ to } 2\,m. \quad (4.4.48)$$

The definite integral on the right-hand side of (4.4.48) has the value zero if k is even and the value $2/k$ if k is odd. It is assumed that equal importance is assigned the symmetric intervals about zero, namely, $t = [-1, 0]$ and $t = [0, +1]$. Therefore, the samples and weights are assigned as illustrated in Figure 4.4.2; note that the sample $t = 0$ is used when there are an odd number of samples. Because of the assumed symmetry, only the odd-numbered equations are required. Two sets of m nonlinear equations that define the Gaussian integration method according to (4.4.48) are shown in Table 4.4.2 for $m = 2$ and $m = 3$. These equations are written in terms of variable x_j for the corresponding w_i and t_i values shown in Figure 4.4.2.

Each set of nonlinear equations in Table 4.4.2 can be solved by treating the defined residuals as those in a least-squares problem according to (4.4.4), at least in principle. The required first partial derivatives are also shown in Table 4.4.2. This is accomplished in Section 4.5.2, using program LEASTP. However, in practice the solutions are usually obtained by an entirely different method. Ralston (1965:87) described an analytic approach that shows that the m sample points for Gaussian integration are the roots of the Legendre polynomial of degree m. Wilf (1962:69) gives an elementary expression for the corresponding weights in terms of the first derivative of the same polynomial. However these Gaussian integration parameters are calculated, it need only be accomplished once. Table 4.4.3 shows these values that satisfy the sets of nonlinear equations in (4.4.48) for two to six samples. Tables are available for

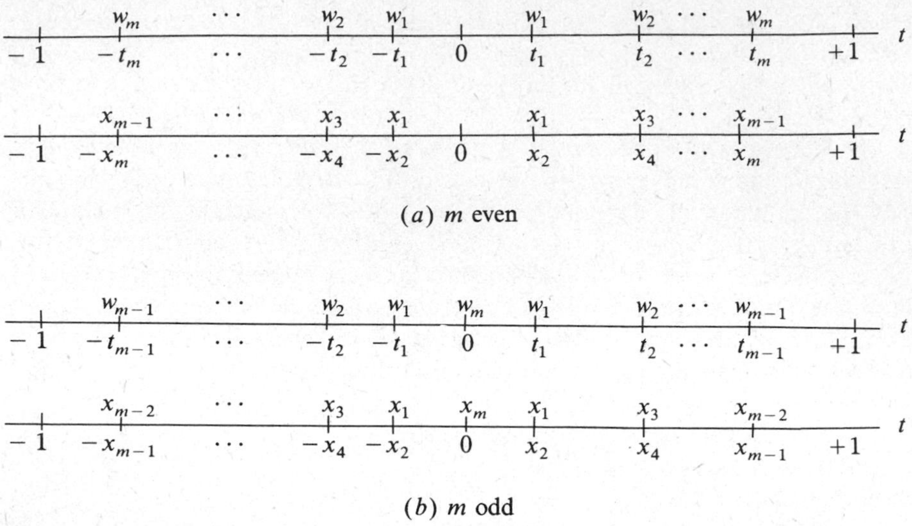

$$
\begin{array}{ccccccccccc}
 & w_m & \cdots & w_2 & w_1 & & w_1 & w_2 & \cdots & w_m & \\
-1 & -t_m & \cdots & -t_2 & -t_1 & 0 & t_1 & t_2 & \cdots & t_m & +1
\end{array} \quad t
$$

$$
\begin{array}{ccccccccccc}
 & x_{m-1} & \cdots & x_3 & x_1 & & x_1 & x_3 & \cdots & x_{m-1} & \\
-1 & -x_m & \cdots & -x_4 & -x_2 & 0 & x_2 & x_4 & \cdots & x_m & +1
\end{array} \quad t
$$

(a) m even

$$
\begin{array}{ccccccccccc}
 & w_{m-1} & \cdots & w_2 & w_1 & w_m & w_1 & w_2 & \cdots & w_{m-1} & \\
-1 & -t_{m-1} & \cdots & -t_2 & -t_1 & 0 & t_1 & t_2 & \cdots & t_{m-1} & +1
\end{array} \quad t
$$

$$
\begin{array}{ccccccccccc}
 & x_{m-2} & \cdots & x_3 & x_1 & x_m & x_1 & x_3 & \cdots & x_{m-2} & \\
-1 & -x_{m-1} & \cdots & -x_4 & -x_2 & 0 & x_2 & x_4 & \cdots & x_{m-1} & +1
\end{array} \quad t
$$

(b) m odd

Figure 4.4.2. Unevenly spaced, symmetric sample points and weights for Gaussian integration.

Table 4.4.2. The Nonlinear Equations That Define Gaussian Integration for Two and Three Sample Points

k	Equations	Derivatives
1	$r_1 = 2x_1 - 2/k = 0$	$\nabla_1 r_1 = 2,\ \nabla_2 r_1 = 0$
3	$r_2 = 2x_1 x_2^2 - 2/k = 0$	$\nabla_1 r_2 = 2x_2^2,\ \nabla_2 r_2 = 4x_1 x_2$
1	$r_1 = 2x_1 + x_3 - 2/k = 0$	$\nabla_1 r_1 = 2,\ \nabla_2 r_1 = 0,\ \nabla_3 r_1 = 1$
3	$r_2 = 2x_1 x_2^2 - 2/k = 0$	$\nabla_1 r_2 = 2x_2^2,\ \nabla_2 r_2 = 4x_1 x_2,\ \nabla_3 r_2 = 0$
5	$r_3 = 2x_1 x_2^4 - 2/k = 0$	$\nabla_1 r_3 = 2x_2^4,\ \nabla_2 r_3 = 8x_1 x_2^3,\ \nabla_3 r_3 = 0$

Table 4.4.3. Gaussian Integration Sample Points and Weights for m = 2 to 6

m	Sample $\pm t_i$	Weight w_i	m	Sample $\pm t_i$	Weight w_i
2	$(3)^{-1/2}$	1	5	0.906179846	0.236926885
				0.538469310	0.478628670
3	$(0.6)^{1/2}$	5/9		0	0.568888889
	0	8/9			
			6	0.932469514	0.171324492
4	0.861136312	0.347854845		0.661209386	0.360761573
	0.339981044	0.652145155		0.238619186	0.467913935

up to 24 sample points with as many as 21 significant figures in Abramowitz (1972:916). Notice that no samples are placed at the ends of the sample domain, that is, at $t = -1$ or $t = -1$.

To employ Gaussian integration over an interval $x = [a, b]$ instead of $t = [-1, +1]$, a linear change of variables is required:

$$x = \frac{t(b - a) + (b + a)}{2}.$$ (4.4.49)

Therefore, (4.4.47) can be expressed in terms of independent variable x as:

$$T = \int_a^b f(x)\, dx = \frac{b - a}{2}\left[w_1 f(x_1) + w_2 f(x_2) + \cdots + w_m f(x_m)\right].$$ (4.4.50)

Example 4.4.5.   Gaussian integration (quadrature) is employed to evaluate the definite integral

$$T = \int_1^2 \left(\frac{1}{x}\right) dx.$$

An arbitrary decision is to use three sample points, implying that the Gaussian integration is exact for a fifth-degree polynomial that approximates the function $1/x$. Using the data for $m = 3$ from Table 4.4.3 in (4.4.49), the normalized and translated samples and weights are:

t_i	x_i	w_i
$-(0.6)^{1/2}$	1.112701666	$\frac{5}{9}$
$+(0.6)^{1/2}$	1.887298335	$\frac{5}{9}$
0	1.5	$\frac{8}{9}$

Then the estimated value of the definite integral is found from (4.4.50):

$$T = \tfrac{1}{2}\left[\tfrac{5}{9}(1.112701666)^{-1} + \tfrac{8}{9}(1.5)^{-1} + \tfrac{5}{9}(1.887298335)^{-1}\right]$$

$$= 0.693121693.$$

It is well known by integral calculus that the exact answer to this problem is $\ln(2) - \ln(1) = 0.693147181$ to nine significant figures. Thus, only three sample points obtained agreement for four significant figures. In general, Gaussian integration obtains a given accuracy with about half as many unevenly spaced, weighted sample points than by the more elementary Newton–Cotes formulas. Readers interested in a more detailed derivation of Gauss quadrature are referred to McCalla (1967:280).

The main points of this discussion about weights and sampling are now summarized. Optimization that seeks to fit a set of data points with the results of some nonlinear model function of many variables requires that the number and location of sample points be selected whenever possible. The objective is accurately to approximate the integral of the residual function over the entire sample domain so that it can be minimized. In fact, the sampling scheme is in itself a modeling process.

A rule of thumb is that at least twice as many sample points as there are variables should be employed; otherwise, the approximating model function may misbehave between sample points. If that occurs, the sampled error may be very low but the actual approximation overall may be terrible. Excessive sampling can lead to an illconditioned system of equations and extended computing time.

There is much to be gained by systematically spacing the samples unevenly according to Gaussian constants that may easily be stored in the computer. It is not optimal to place samples at the ends of the sample domain, and it is about half as effective to space sample points evenly as it is to space them according to Gaussian integration rules. However, the model may be intrinsically illconditioned so that more sophisticated sampling and weight choices won't help.

Adaptive integration methods (Forsythe 1977:92) use one or two basic rules (such as the trapezoidal rule) to determine dynamically the subinterval samples so that some specified accuracy can be obtained in the result. That method is not recommended for optimization because it tends to confuse gradient optimization algorithms, where smoothness is more important than accuracy when far from the minimum. The interested reader is referred to Lyness (1976).

A new and interesting method has been suggested by Davidon (1976) in which the variables are adjusted after each sample instead of after all samples. This method fluctuates about the minimum without ultimately converging to it. However, solutions of linear systems are not required as with Newton methods, and linear bounds on variables may be included even when the Hessian is singular.

Section 4.5.3 will provide more information about the properties of least-pth objective functions and how equivalent results may be obtained by other means, especially as $p \to \infty$, which is equivalent to the minimax case.

4.4.5. Controlling the Levenberg–Marquardt Parameter.

This section extends considerations for the Levenberg–Marquardt (LM) parameter for the Gauss–Newton method, where an approximate positive-definite Hessian matrix is available. Certain of these considerations have been implemented in program LEASTP, which are described in Section 4.5.1. By using a diagonal matrix instead of the unit matrix in conjunction with the LM parameter, the basis of a scaling method for the variables is available as a valuable feature.

Fletcher (1971a) discussed a number of considerations for a Gauss–Newton computer program based on ideas from Levenberg (1944) and Marquardt (1963). Equation (4.2.6) gave the set of linear equations to be solved in determining a Newton step $\mathbf{dx}$ to be taken from a point where the Hessian matrix $\mathbf{H}$ and gradient vector $\mathbf{g}$ had been evaluated; that is repeated here for convenience:

$$(\mathbf{H} + v\mathbf{I})\,\mathbf{dx} = -\mathbf{g}. \tag{4.4.51}$$

In the Gauss–Newton case, $\mathbf{H} = \mathbf{J}^T\mathbf{J}$ is a positive-definite matrix (except for roundoff error) that approximates the true Hessian according to (4.4.20), when certain second-derivative terms are omitted. As described in Section 4.2.2, the LM parameter v interpolates between the Newton step ($v = 0$) and an infinitesimal steepest descent step ($v \to \infty$).

Levenberg suggested that $\mathbf{dx}$ should be treated as a search direction and that v should be used as a search parameter to estimate a minimum along that trajectory; see Figure 4.2.1. To the contrary, Fletcher stated that it is more efficient to obtain a reasonable decrease in function value on the trajectory and then begin a new iteration with more current values of $\mathbf{H}$ and $\mathbf{g}$.

Marquardt was concerned that Levenberg's method overemphasized the steepest-descent direction; Fletcher agreed. Let $F' = F(\mathbf{x} + \mathbf{dx})$, the function value at a trial step, and $F = F(\mathbf{x})$ be the function value at the current turning or iteration point. Marquardt suggested increasing v by a factor of 10 after unsuccessful ($F' > F$) steps and decreasing v by the same factor after successful ($F' < F$) steps. It was noted in Section 4.2.2 that adjustments in v needed to be made in some geometric progression to be significant. Fletcher concedes that Marquardt's parameter adjustment scheme is relatively effective in practice but suggests the following three desirable actions discussed here: (1) select a reasonable choice for the initial value of v, (2) decide if the factor 10 is a reasonable adjustment under all conditions, and (3) set $v = 0$ near a solution, since convergence is then quadratic instead of superlinear.

Actions (1) and (3) are related in that a value of v greater than zero must be chosen, preferably not arbitrarily. Recall that near the solution point, the function appears quadratic, $\mathbf{H} = \mathbf{J}^T\mathbf{J}$ approaches the true Hessian *if* the residuals are all negligibly small, and full steps with $v = 0$ are appropriate. Therefore, the question arises as to what value of $v > 0$ would cause a significant decrease in step size relative to $v = 0$, say, a halving of the step length. The answer is available from (4.2.13), where that sum of terms is dominated by the term containing the smallest eigenvalue, w_n, when $v \doteq 0$. Referring to (2.1.48), the magnitude of the largest eigenvalue of $\mathbf{H}$ is equal to the spectral norm $\|\mathbf{H}\|_2$, and that should have the same order of magnitude as the other matrix norms. But the smallest eigenvalue of $\mathbf{H}$ is equal to the largest eigenvalue of $\mathbf{H}^{-1}$, so it follows that if v is not equal to zero, then it must be about as large as some norm $\|\mathbf{H}^{-1}\|$. Fletcher noted that both the $\|\cdot\|_\infty$ and

$\| \cdot \|_F$ norms overestimate the $\| \cdot \|_2$ norm and are convenient to compute. However, it requires on the order of $n^3/3$ operations to compute $\mathbf{H}^{-1}$, so that this approach is feasible only for initially choosing the starting value of v and for occasionally computing a value of v below which v is simply set equal to zero (so that quadratic convergence is possible).

The issue of increasing v depends on whether v is small or large. Let this factor be q, so that the new value of v would be qv. If v is small, then Fletcher argues that a factor of $q = 2$ is adequate except during early iterations far from a minimum where v might be far too small and a factor $q = 10$ would be much more appropriate (see Figure 4.2.2). For large values of v, (4.2.13) and (4.4.51) both show that increasing v to qv decreases $\mathbf{dx}$ to $\mathbf{dx}/q$. Fletcher (1971a:4) makes this decision for $2 \le q \le 10$ based on the step length to a minimum in an arbitrary direction $\mathbf{s}$ on a quadratic surface. That has already been derived with the result as t^* in (3.2.19) and is repeated for convenience:

$$t^* = -\frac{\mathbf{s}^T\mathbf{g}}{\mathbf{s}^T\mathbf{Hs}}. \tag{4.4.52}$$

In the present application, suppose that a sum of squares $F = F(\mathbf{x})$ and a gradient $\mathbf{g}(\mathbf{x})$ have been obtained at a turning point. Further suppose that a trial step $\mathbf{dx}$ has been computed with some value v and that step taken with the unsuccessful result that $F' = F(\mathbf{x} + \mathbf{dx}) > F$. Rather than compute the quadratic form in the denominator, Fletcher noted that t^* can be computed using $\mathbf{s} = \mathbf{dx}$, F, and F', all of which have been computed. To obtain that expression, note that a three-term Taylor series approximation of F about $\mathbf{x} + \mathbf{dx}$ is

$$F' = F + \mathbf{dx}^T\mathbf{g} + \tfrac{1}{2}\mathbf{dx}^T\mathbf{H}\,\mathbf{dx}. \tag{4.4.53}$$

Therefore, the denominator of (4.4.52) may be replaced by the right-hand term of (4.4.53) with the result that

$$t^* = \frac{1}{2 - (F' - F)/(\mathbf{dx}^T\mathbf{g})}, \tag{4.4.54}$$

where F is defined by (4.4.4) and $\mathbf{g} = \nabla F$ by (4.4.17). Fletcher computes the factor $q = 1/t^*$; if it is greater than 10 it is set equal to 10, and if less than 2 it is set equal to 2.

Instead of deciding whether a step is successful based on a simple increase or decrease in function value as employed in the preceding analyses, Fletcher used the criterion similar to that described by (4.2.2), in this case the ratio

$$r = \frac{F' - F}{\mathbf{dx}^T\mathbf{g} + \tfrac{1}{2}\,\mathbf{dx}^T\mathbf{H}\,\mathbf{dx}}. \tag{4.4.55}$$

The denominator of (4.4.55) is clearly the *ideal* change in function value according to (4.4.53), whereas the numerator of (4.4.55) is the *actual* change obtained by the step **dx**. Near convergence, $r = 1$ is expected. Clearly, $r \gg 1$ is good news and $r \ll 1$ is bad news. Fletcher left the value of v unchanged if $0.25 \le r \le 0.75$. If $r > 0.75$, then he reduced v to $v/2$, and if the new v was less than $1/\|\mathbf{H}^{-1}\|_2$, then it was equated to zero for quadratic convergence. If $r < 0.25$, then v was increased to qv, $2 \le q \le 10$ as described earlier.

There is still the problem of solving the augmented normal equations for step **dx**:

$$(\mathbf{J}^T\mathbf{J} + v\mathbf{I})\,\mathbf{dx} = -\mathbf{g}. \tag{4.4.56}$$

These equations are likely to be very illconditioned. In fact, since $\mathbf{g} = \mathbf{J}^T\mathbf{r}$, where $\mathbf{J}$ is the Jacobian and $\mathbf{r}$ the residuals, it can be seen that at a minimum where the gradient $\mathbf{g} = \mathbf{0}$, if $\mathbf{r} \ne \mathbf{0}$ then $\mathbf{J}$ must be singular! Therefore, as a minimum is approached, the steps theoretically become more and more accurate but the Jacobian matrix deteriorates progressively. Since the condition number of $\mathbf{J}^T\mathbf{J}$ is equal to the square of the condition number of Jacobian $\mathbf{J}$ alone, it is highly desirable to obtain a solution for **dx** in (4.4.56) that uses only $\mathbf{J}$. One way to do this is to employ orthogonal decomposition as described by (3.1.47).

An even better approach for computing the Newton step **dx** in (4.4.56) is to use singular value decomposition (SVD). As described in Section 3.1.3, it is always possible to decompose the Jacobian matrix as

$$\mathbf{J} = \mathbf{U}\mathbf{S}\mathbf{V}^T, \tag{4.4.57}$$

where $\mathbf{U}$ and $\mathbf{V}$ are orthogonal matrices and $\mathbf{S}$ contains the singular values of $\mathbf{J}$ on a diagonal. Since $\mathbf{U}^T\mathbf{U} = \mathbf{I}$ and $\mathbf{V}^T\mathbf{V} = \mathbf{I}$, $\mathbf{J}^T\mathbf{J} = \mathbf{V}\mathbf{S}^T\mathbf{S}\mathbf{V}^T$, so that

$$\mathbf{J}^T\mathbf{J} + v\mathbf{I} = \mathbf{V}(\mathbf{S}^T\mathbf{S} + v\mathbf{I})\mathbf{V}^T = \mathbf{V}\mathbf{W}\mathbf{V}^T. \tag{4.4.58}$$

Matrix $\mathbf{W}$ defined by (4.4.58) is clearly diagonal. Therefore, step **dx** is

$$\mathbf{dx} = -\mathbf{V}\mathbf{W}^{-1}\mathbf{V}^T\mathbf{g}. \tag{4.4.59}$$

It is easy to obtain $\mathbf{W}^{-1}$ and to update $\mathbf{W}$ using different values of LM parameter v. As Forsythe (1977) noted, "If $\mathbf{J}$ has full rank, then the solution $\mathbf{x}'$ is unique and can be reliably computed by several different algorithms, some of which are faster than the SVD. But the SVD also handles the rank-deficient case and, except for some very large problems, is not much more costly than the other reliable methods. (It is less costly than a fast algorithm which may give the wrong answer.)"

There is one additional consideration for the LM parameter method that involves implicit scaling of the variables in **x**. Suppose that there is diagonal matrix $\mathbf{D}_{nn} = [d_{jj}]$ having all positive elements. Then a linear transformation of the variable space is

$$\tilde{\mathbf{x}} = \mathbf{D}\mathbf{x} \quad \text{or} \quad \mathbf{x} = \mathbf{D}^{-1}\tilde{\mathbf{x}}. \tag{4.4.60}$$

Recall that the Jacobian matrix contains all first partial derivatives of the residuals evaluated at each sample point, t_k:

$$\mathbf{J} = [\nabla_j r_k], \quad \text{row } k = 1 \text{ to } m, \text{ column } j = 1 \text{ to } n. \tag{4.4.61}$$

By the chain rule,

$$\frac{\partial r_k}{\partial \tilde{x}_j} = \frac{\partial r_k}{\partial x_j} \frac{\partial x_j}{\partial \tilde{x}_j}, \tag{4.4.62}$$

where

$$\frac{\partial x_j}{\partial \tilde{x}_j} = d_{jj}^{-1}. \tag{4.4.63}$$

Therefore, the Jacobian matrix of partial derivatives in the new variable space $\tilde{\mathbf{x}}$ is

$$\tilde{\mathbf{J}} = \mathbf{J}\mathbf{D}^{-1}. \tag{4.4.64}$$

In the new variable space, consider the Gauss–Newton step $d\tilde{\mathbf{x}}$ defined by

$$[\tilde{\mathbf{J}}^T\tilde{\mathbf{J}} + v\mathbf{I}]\, d\tilde{\mathbf{x}} = -\tilde{\mathbf{J}}^T\mathbf{r}. \tag{4.4.65}$$

Substituting from (4.4.64) and (4.4.60) into (4.4.65):

$$\left[(\mathbf{J}\mathbf{D}^{-1})^T(\mathbf{J}\mathbf{D}^{-1}) + v\mathbf{I}\right]\mathbf{D}\, d\mathbf{x} = -(\mathbf{J}\mathbf{D}^{-1})^T\mathbf{r}, \tag{4.4.66}$$

which reduces to

$$[\mathbf{J}^T\mathbf{J} + v\mathbf{D}^2]\, d\mathbf{x} = -\mathbf{J}^T\mathbf{r}. \tag{4.4.67}$$

The Gauss–Newton step differs from the original definition only in the substitution of a diagonal matrix $\mathbf{D}^2$ for the unit matrix $\mathbf{I}$. However, the preceding development shows that (4.4.67) represents an *implicit scaling of variables*. This development is very important in connection with the trust neighborhood interpretation of the Levenberg method developed in Section 4.2.2, because it deals with a prediction $d\mathbf{x}$ on an ideal quadratic function model such that $\|d\mathbf{x}\|_2 \leq R$, R a constant. This used the two-norm or vector length, which depends on the scales chosen in the $\mathbf{x}$ space.

The use of (4.4.67) instead of (4.4.56) with $\mathbf{g} = \mathbf{J}^T\mathbf{r}$, has proved beneficial in practice. Many algorithms employ Marquardt's suggestion that a good choice of $\mathbf{D}^2$ is to equate it to the diagonal values of $\mathbf{J}^T\mathbf{J}$ at the initial choice of variables, $\mathbf{x}^{(0)}$; see Fletcher (1971a:7). For the linear transformation of variables in (4.4.60), that choice means that the elements of $\mathbf{D} = [d_{jj}]$ are the root mean square values of the first derivatives of the residuals:

$$d_{jj} = \left[\left(\nabla_j r_1 \right)^2 + \left(\nabla_j r_2 \right)^2 + \cdots + \left(\nabla_j r_m \right)^2 \right]^{1/2}. \qquad (4.4.68)$$

This can be verified from (4.4.19) and (4.4.20).

Within a constant, (4.4.68) is the *root mean square* of the derivatives $\nabla_j r_k$, taken over the sample points $k = 1$ to m. Marquardt (1963:437) has noted the statistical significance of this choice; in fact, the LLS problem often has been described as a statistical tool. In the present context the use of an average derivative or slope in each coordinate direction according to (4.4.68) as a scaling factor makes sense. Consider the linear transformation $\tilde{\mathbf{x}} = \mathbf{Dx}$ in (4.4.60): If the function is changing very rapidly in the x_j direction, the $|\nabla_j F|$ is large and the level curves are bunched together. Then increasing the number of units in $\tilde{x}_j$ is the appropriate scaling action necessary to increase the spacing between level curves or contours.

4.5. Program LEASTP

Program C4-5, LEASTP, provides an illustration of both the Gauss–Newton approximation to the Hessian and the implementation of a particular Levenberg–Marquardt policy. The menu choices are similar to those in program NEWTON. However, subroutines 5000 and 7000 supply only the necessary ingredients of the objective function and its gradient, respectively, because of the different Gauss–Newton formulation. Program LEASTP includes a simple implementation of the least-pth objective function. Section 4.5.2 illustrates some of its limitations, and Section 4.5.3 contains brief discussions of more sophisticated least-pth techniques and their relationships to the minimax problem.

4.5.1. The Algorithm and Its Implementation.
The listing for program C4-5, LEASTP, is contained in Appendix C. The major variable names are shown in the remarks contained in lines 130 to 180; there is also a complete list of all variable names used at the end of the program listing. As in the preceding program NEWTON, the programming variables are contained in the single-subscripted array $X()$, and the gradient of objective function F with respect to the variables is contained in $G()$. In this Gauss–Newton implementation, it is necessary to store the residuals in the array $R()$ in order to compute both F and the gradient $\mathbf{g}$. Similarly, the modified Gauss–Newton approximation to the Hessian matrix is stored in vector form in array $H()$, dimensioned $n(n + 1)/2$, where n is the number of problem variables.

Variables in program LEASTP are dimensioned in line 320; problems may involve as many as $n = 20$ variables and $m = 40$ sample points (see Figure 4.4.1). The user can change these dimensions, of course. There are several large arrays required in the Gauss–Newton method, indicative of the usual tradeoff between computation and memory in the choice of algorithms. Array $A(,)$ contains the $m \times n$ Jacobian matrix of first partial derivatives, and array $S(,)$ contains the m sample points and their respective data values.

A list of the major subroutines and their line numbers in program LEASTP is given in Table 4.5.1. Program LEASTP is similar in structure and flow to program NEWTON that was described earlier in this chapter. However, no explicit trust neighborhood radius must be entered into LEASTP, since the Levenberg parameter v controls both the step direction and its length. Also, control parameters do not include the finite difference factor, since the second partial derivatives are approximated using the structure inherent in the Gauss-Newton least-pth objective function.

Because LEASTP is a curve-fitting program, the user must supply the sample values and their respective data values. This must be furnished by the user in program lines 400 to 600; an illustration is included in the program C4-5 listing, lines 400 to 430. These data are read by lines 340 to 370 whenever the program is run and may be reviewed at any time using menu command 6. Of course, the program can be modified to read data from disk files as in program C2-1, MATRIX. However, lines 400 to 600 can easily be created by an editor such as IBM's EDLIN, saved on disk, and then MERGE'd into LEASTP before it is run. In fact, both subroutines 5000 and 7000 also must be merged into LEASTP, so it makes sense to combine both subroutines *and* the data statements all into the same file to be merged. Of course, all three sets of information are unique to the particular problem at hand.

Table 4.5.1. Major Subroutines in Optimizer Program C4-5, LEASTP

Name	Lines
Enter Number and Value of Variables	1200–1270
Enter/Revise Control Parameters	1280–1360
Main Optimization Algorithm	1400–2490
Display Function, Gradient, and Variables	2500–2580
Calculate and Store Normal matrix in $H()$	2590–2730
Calculate Gradient and Its Length	2740–2850
Add Levenberg Parameter to Normal Matrix	2860–2970
Display Sample Data from Lines 400–600	2980–3080
$\mathbf{LDL}^T$ Factorization of Hessian in situ	3090–3360
Solution for Step $\mathbf{dx}$ in $(\mathbf{J}^T\mathbf{J} + v\mathbf{D}^2)\,\mathbf{dx} = -\mathbf{g}$	3370–3600
Residuals $\mathbf{r}$ for F and $\mathbf{g}$ (user supplied)	5000–6999
Jacobian Matrix $\mathbf{J} = [\nabla_j r_k]$ (user supplied)	7000–8999

The flow chart for program C4-5, LEASTP, is shown in Figure 4.5.1. It starts in the same way as the preceding major programs, namely, with some notes and then a command menu. Command 5 has been reserved as a "spare" for later use as suggested by Problem 4.14. Command 1 must be used to input the number and values of problem variables before the *first* optimization cycle is started, using command 3. Command 2 to revise the program parameters is optional, since default values have been set by the program. (Use command 2 to review these default values.) Several initial tasks are performed before iterations can begin.

Upon selection of command 3, the user is asked to select a value for p, the exponent for the residuals in the least-pth objective function. This allows the user to optimize with $p = 2$, then use command 3 again to optimize for $p > 2$, starting from the variable values obtained from the preceding minimization. Next, subroutine 5000 is called to obtain the residuals and to check the *number* of residuals the user has programmed there, set by the BASIC variable named "M". This should correspond to the same number of samples indicated by the user as "$M7$" in the DATA statement on line 350. The program pauses to let the user ascertain that $M = M7$ (i.e., that this value is consistent with the problem at hand). LEASTP then computes the exact gradient $\mathbf{g} = \nabla F$ using the residuals from subroutine 5000 and the Jacobian from subroutine 7000; see (4.4.17).

The root mean square values of the gradient are then placed in diagonal matrix $\mathbf{D}^2$, appearing in the program as BASIC variable $D(\ )$. These are not changed afterward. Then the program obtains the approximate gradient by finite differences and displays these alongside the exact values. This is an important feature, because it at least checks the programming the user furnished in subroutine 7000 for the Jacobian. To the extent that either set of values make any sense, it also provides a check on the programming in subroutine 5000 for the residuals.

Reentry point 1890 in Figure 4.5.1 marks the beginning of the iteration loop for the sequential search directions in E^n. The first order of business on any iteration is to adjust the value of LM parameter v. It is simply started at $v = 0.001$ in line 290, as opposed to $\|\mathbf{H}^{-1}\|_2$ as proposed by Fletcher and described in Section 4.4.5. The policy implemented in program LEASTP for increasing or decreasing v is programmed in lines 1900 to 1930 and 2020 and listed in Table 4.5.2.

This policy for the Levenberg–Marquardt parameter v is considerably more elementary than Fletcher's policy described in Section 4.4.5. However, this policy works reasonably well and its implementation does not lengthen the program code to the point of obscuring more essential concepts. One price paid for this policy is that the user will often see v oscillate by a factor of 10 up and down for several iterations. Fletcher's policy stops that, but so does a policy of decreasing v by say 2.5 while increasing v by 10. There is some question as to whether the total number of function evaluations is seriously

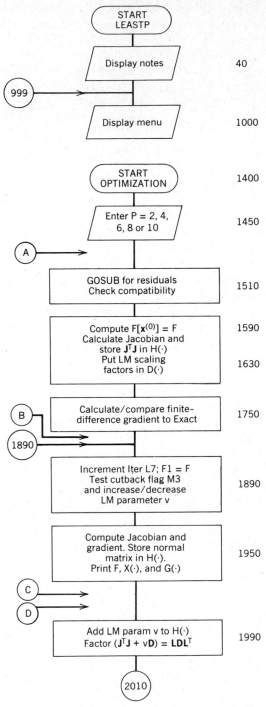

Figure 4.5.1. Flow chart of optimizer program C4-5, LEASTP. *Note*: Ⓐ – Ⓕ apply to variable bounds. See Table 6.2.3.

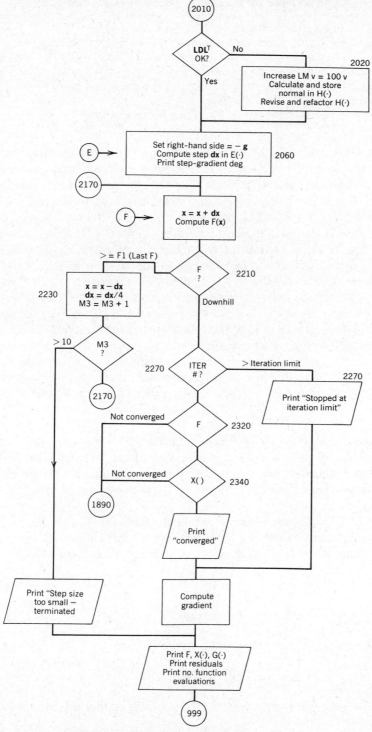

Figure 4.5.1. *Continued*

Table 4.5.2.   LEASTP Policy for Increasing / Decreasing LM Parameter v

At the beginning of each iteration:
1. If there were no step length cutbacks in the preceding iteration, then $v = v/10$.
2. If $v < 10^{-20}$ then $v = 10^{-20}$, since $v = 0$ is not allowed.
3. If there were cutbacks in the preceding iteration, then $v = 10v$.
4. Anytime the approximate Hessian is not positive-definite or its determinant $< 10^{-6}$, then $v = 100v$.

affected by the brief number of oscillations in v that may be observed. It is assumed that generally it is much more expensive to recompute the function and its gradient than it is to perform the $\mathbf{LDL}^T$ factorization and solution. In any event, downhill progress is obtained with or without oscillation.

The flow chart for LEASTP in Figure 4.5.1 shows the modification of the approximate Hessian $(\mathbf{J}^T\mathbf{J} + v\mathbf{D}^2)$ according to (4.4.67). The implicit-scaling, diagonal matrix $\mathbf{D}^2$ is fixed at the root mean square values of the gradient as previously mentioned. If the factorization of the Hessian in subroutine 3110 returns with BASIC variable $N5 = 1$, then v is increased to $100v$, and the Hessian is recreated and refactored before proceeding; see case 4 in Table 4.5.2. The remainder of the flow chart in Figure 4.5.1 is essentially the same as Figure 4.3.1 for optimizer NEWTON, that is, the steps, cutback, and convergence tests are similar. Program LEASTP does require a few more steps to assemble the objective function and the gradient, since the user is only burdened with providing the residuals and their derivatives in subroutines 5000 and 7000, respectively. Typical user subroutines are described by the examples.

4.5.2.   Some Examples Using Program LEASTP.
Some previous examples are reconsidered in order to highlight differences between the least-pth Gauss–Newton method and previous optimization techniques. A new example will illustrate some of the pitfalls awaiting the careless user of $p > 2$.

Example 4.5.1.   The first example recasts the Rosenbrock function with two variables as previously defined by (4.3.1) into a form employing two explicit residuals. This is the solution of two nonlinear equations from (4.4.13):

$$r_1 = 10(x_2 - x_1^2), \tag{4.5.1}$$

$$r_2 = 1 - x_1, \tag{4.5.2}$$

so that the objective function is to be minimized

$$F(\mathbf{x}) = \tfrac{1}{2}\left[r_1^2 + r_2^2\right]. \tag{4.5.3}$$

Notice that a factor of $\tfrac{1}{2}$ has been added to (4.5.3) so that it is now consistent with the defined least-squares objective function (4.4.4). The user will have to program subroutine 5000 for the residuals defined by (4.5.1) and (4.5.2). Also, the Jacobian matrix containing the first partial derivatives of the residuals must be programmed in subroutine 7000 by the user. *Great care is required to ensure that the row-column convention in* (4.4.15) *is properly related to the BASIC array*: $A(K, J) = [\nabla_j r_k]$. The expressions for these are:

$$\nabla_1 r_1 = -20x_1, \qquad \nabla_2 r_1 = 10,$$
$$\nabla_1 r_2 = -1, \qquad \nabla_2 r_2 = 0.$$

(4.5.4)

Program LEASTP uses subroutine 5000 to calculate $F(x)$ according to (4.5.3); it also computes the gradient using subroutine 7000:

$$
\mathbf{g} = \mathbf{J}^T \mathbf{r} = \begin{bmatrix} -20x_1 & 10 \\ -1 & 0 \end{bmatrix}^T \begin{bmatrix} 10(x_2 - x_1^2) \\ 1 - x_1 \end{bmatrix}
$$
$$
= \begin{bmatrix} -200x_1(x_2 - x_1^2) - (1 - x_1) \\ 100(x_2 - x_1^2) \end{bmatrix}.
$$

(4.5.5)

Subroutines for this form of the Rosenbrock problem are contained in Appendix C, program C4-6, ROSENPTH. In this case no DATA statements for lines 400 to 600 in LEASTP are required. It is important to note that the number of residuals must be set in subroutine 5000, in this case $M = 2$ on line 5020.

Variable $L1 = 0$ is included in subroutine 5000 to indicate to LEASTP that variables in array $X()$ are mathematically feasible for the problem, for example, negative variable values are not involved in root or logarithm calculations. This is checked at line 1560 of LEASTP only on the first call; it may be a good idea to check $L1$ at every call of both subroutines 5000 and 7000. In any event, $L1 = 1$ must be set dynamically in subroutine 5000 if the calculation cannot accept some value of $X()$ passed to it. Any such programming details in both LEASTP and the subroutines 5000 and 7000 are left to the user as required. In the case of ROSENPTH, $L1 = 0$ is set since there can be no problems of this sort. The concept of a flag variable in optimization subroutines to avoid misuse of variables assigned by the optimizer was suggested by Fletcher (1971a).

Note in subroutine 7000 in ROSENPTH that the Jacobian matrix is in $A(,)$. The subscripts for each element value are defined by (4.4.16) and must be correct. In this case, the matrix is composed as positioned in (4.5.4). The main reason for checking the gradient computed by using the Jacobian versus the gradient computed by using finite-differenced residuals is to be certain that the user correctly programmed subroutine 7000. It is easy to make errors in subscripts, and the optimizer certainly cannot perform with incorrect gradient information.

The essential output from LEASTP for the Rosenbrock problem is shown in Table 4.5.3. The standard start from $\mathbf{x}^{(0)} = (-1.2\ 1)^T$ was used with $p = 2$. The relative differences between the exact (SUB7000) and finite-differenced gradient values is typical. These data indicate that (1) subroutine 7000 was programmed correctly and (2) the magnitudes of the partial derivatives are reasonable. The latter fact suggests that subroutine 5000 was programmed correctly and that the problem is reasonably well scaled (units of each x_j are reasonably related to each other). Iterations 3 through 22 are not shown in Table 4.5.3. The behavior of the Levenberg parameter v was typical; it steadily increased from its starting value of $v = 0.001$ to $v = 10$, then oscillated between $v = 10$ and $v = 1$, finally decreasing to a relatively small number at convergence. In this case, the root-mean-square values assigned to scaling matrix $\mathbf{D}^2$ in (4.4.67) were 0.9853 and 0.1708 for x_1 and x_2, respectively; these are easy to obtain after terminating execution by entering keyboard characters $\langle ?D(1);\ D(2)\ \text{Rtn}\rangle$. The gradient value shown for iteration 1 in Table 4.5.3 confirms the proper trend for correcting the scale for this problem and the particular starting point.

The performance obtained on the Rosenbrock problem from the standard starting point by LEASTP is comparable to that obtained by NEWTON, but in this case the Hessian matrix of second derivatives was only approximate. On the other hand, NEWTON used finite differences of the gradient to compute the Hessian, requiring n extra function evaluations. For a large number of variables (say 25 to 50) LEASTP is much more efficient without any substantial penalty. Other data from LEASTP for other starting points on the Rosenbrock function show it to be superior to NEWTON; see Table 5.3.3 in Chapter Five.

It is useful in the Rosenbrock case to compare expressions for the exact and approximate second derivatives. The exact Hessian matrix $\mathbf{H}$ from (3.2.46) is

$$\mathbf{H} = \begin{bmatrix} \nabla_1 g_1 & \nabla_2 g_1 \\ \nabla_1 g_2 & \nabla_2 g_2 \end{bmatrix} = \begin{bmatrix} 600x_1^2 - 200x_2 + 1 & -200x_1 \\ -200x_1 & 100 \end{bmatrix}, \quad (4.5.6)$$

where the elements of the gradient are obtained from (4.5.5). Another exact expression for the Hessian matrix for least-pth objective functions is $\mathbf{H} = \mathbf{J}^T\mathbf{J} + \mathbf{M}$, using $\mathbf{J}$ from (4.4.16) and $\mathbf{M}$ from (4.4.21). In the Rosenbrock case

$$\mathbf{J} = \begin{bmatrix} 400x_1^2 + 1 & -200x_1 \\ -200x_1 & 100 \end{bmatrix},$$

and $\hspace{9cm}$ (4.5.7)

$$\mathbf{M} = 10(x_2 - x_1^2)\begin{bmatrix} -20 & 0 \\ 0 & 0 \end{bmatrix} + (1 - x_1)[0].$$

It can be seen that $\mathbf{J}^T\mathbf{J}$ in (4.5.7) approaches $\mathbf{H}$ in (4.5.6) as $\mathbf{x} \to \mathbf{x}' = (1\ 1)^T$, that is, $\mathbf{M} \to \mathbf{0}$, and that the sum $\mathbf{J}^T\mathbf{J} + \mathbf{M}$ in (4.5.7) is equal to $\mathbf{H}$ in (4.5.6).

Table 4.5.3. Output from Program LEASTP for the Standard Rosenbrock Problem

```
NUMBER OF VARIABLES = ? 2
ENTER STARTING VARIABLES X(I):
   X( 1 )=? -1.2
   X( 2 )=? 1
PRESS <RETURN> KEY TO CONTINUE -- READY? 3
EXPONENT P (2,4,6,8, OR 10) =? 2
USER SET NUMBER OF SAMPLES M = 2  IN SUBROUTINE 5000
         IS THIS CONSISTENT WITH THIS PROBLEM (Y/N)? Y
GRADIENT    VIA SUB7000          VIA DIFFERENCES
            -107.80000234           -107.77513186
             -44.00000095            -43.99299622
PRESS <RETURN> KEY TO CONTINUE -- READY?
AT START OF ITERATION NUMBER 1
   FUNCTION VALUE = 12.1
  I        X(I)              G(I)
  1     -1.20000000    -107.80000234
  2      1.00000000     -44.00000095
                            LM PARAM V= 1.0D-04
                STEP-TO-GRADIENT DEGREES= 87.7591
  ###### CUT BACK STEP SIZE BY FACTOR OF 4 ######
  ###### CUT BACK STEP SIZE BY FACTOR OF 4 ######
AT START OF ITERATION NUMBER 2
   FUNCTION VALUE = 11.43226
  I        X(I)              G(I)
  1     -1.06252594     -93.73706397
  2      0.69756230     -43.13990593



                        •



                        •



                        •



AT START OF ITERATION NUMBER 23
   FUNCTION VALUE = 4.501056E-12
  I        X(I)              G(I)
  1      0.99999890      0.00005471
  2      0.99999752     -0.00002791
                            LM PARAM V= 1.0D-04
                STEP-TO-GRADIENT DEGREES= 86.9059
CONVERGED; SOLUTION IS:
AT START OF ITERATION NUMBER 24
   FUNCTION VALUE = 1.882328E-20
  I        X(I)              G(I)
  1      1.00000000      0.00000000
  2      1.00000000     -0.00000000
PRESS <RETURN> KEY TO CONTINUE -- READY?
RESIDUALS ARE:
  1           -1.641228819515561D-11
  2            1.933318505287218D-10
TOTAL NUMBER OF FUNCTION EVALUATIONS = 39
EXPONENT P = 2
PRESS <RETURN> KEY TO CONTINUE -- READY?
```

Table 4.5.4. Output from Program LEASTP for the GAUSS Problem with $m = 4$

```
AT START OF ITERATION NUMBER 1
    FUNCTION VALUE = 8.407742E-02
  I              X(I)                 G(I)
  1          0.50000000           -0.12566035
  2          0.50000000           -0.35187874
  3          0.50000000           -0.12566035
  4          0.50000000           -0.35187874
                             LM PARAM V= 1.0D-04


  2          0.63855212           -0.00000207
  3          0.49180462           -0.00000003
  4          0.63855212           -0.00000207
                             LM PARAM V= 1.0D-06
HESSIAN NOT PD OR TOO SMALL DETERMINANT = 6.8349D-12
                             LM PARAM V= 1.0D-04
                STEP-TO-GRADIENT DEGREES= 17.1973
CONVERGED; SOLUTION IS:
AT START OF ITERATION NUMBER 4
    FUNCTION VALUE = 2.397782E-02
  I              X(I)                 G(I)
  1          0.49180450           -0.00000014
  2          0.63855252           -0.00000056
  3          0.49180450           -0.00000014
  4          0.63855252           -0.00000056
PRESS <RETURN> KEY TO CONTINUE -- READY?
RESIDUALS ARE:
  1              -3.2782003710127350-02
  2               .1354650063976287
  3              -7.2931389240677460-02
  4              -.1523522776882509
TOTAL NUMBER OF FUNCTION EVALUATIONS = 4
EXPONENT P = 2
```

Example 4.5.2. This example solves the sets of nonlinear equations such as those in Table 4.4.2 that define Gauss integration. Since the number of sample points and the number of variables are equal, insight into the structure of the general case can be obtained by also writing out the equations for $m = 4$ and $m = 5$. The pertinent BASIC subroutines 5000 and 7000 are contained in Appendix C in program GAUSS (C4-7). Merging that into LEASTP enables input of the number of variables (and samples) and some starting guess for the variables. Recall from Figure 4.4.2 that the variables $x_1, x_2, x_3, x_4, x_5, \ldots$ correspond to weights and samples $w_1, t_1, w_2, t_2, w_3, \ldots$, respectively. By running the program and starting with the reasonable choices $x_j = 0.5$, $j = 1$ to n, the answers for $m = 2$ and $m = 3$ in Table 4.4.3 are easily obtained.

The essential results for trying that starting point for $m = 4$ are shown in Table 4.5.4. The necessary condition for a minimum seems satisfied, namely, the gradient is nearly equal to zero, but clearly the residuals and their squared sum are not anywhere near equal to zero. The data represent a local minimum of the sum of squares but not a solution to the system of nonlinear equations. It is easily verified that starting with the first significant digits of the known solution in Table 4.4.3, namely, $\mathbf{x}^{(0)} = (0.3 \quad 0.8 \quad 0.6 \quad 0.3)^T$, will converge to the known solution with zero residuals. The same approach will confirm solutions for the $m = 5$ and $m = 6$ sets of nonlinear equations as well.

Having to start the optimization that close to a valid solution is a severe limitation. This requirement is why the author calls optimization (nonlinear programming) *computer-aided redesign*! Users of optimization can expect frequent failures when using random guesses for starting points. Not only must the variables in the problem be well scaled (reasonably related units of measure), but the starting variable vector must be within a *unimodal neighborhood* of a useful minimum. Sometimes that neighborhood is quite small. This is an important limitation and one that may be met by the kind of problem estimation procedures that are in the tool kits of good engineers. The results from those educated starting points are often astounding and unattainable by any other means. But problems must be well scaled, formulated to be well conditioned, and started near a potentially useful solution. Optimization may appear to be art, but its real power is gained through the insight of an informed artist.

Regarding illconditioning, the formulation of the Gauss integration solution as in Table 4.4.2 has the same defect as the integral least squares in Example 4.4.4 and for the same reason. The power series representation is intrinsically illconditioned; Program LEASTP with GAUSS cannot solve this particular problem with large numbers of sample points. As mentioned in Section 4.4.4, there are at least two better means for finding those solutions.

Example 4.5.3.   Sargeson's least-squares problem introduced in Chapter One with fitting function (1.2.5) and results in Table 1.2.1 is solved with program LEASTP. In this chapter, Example 4.4.1 in Section 4.4.1 provided expressions for the nonlinear model, residuals, and their first derivatives; see (4.4.10) to (4.4.12). These have been coded in Appendix C, program C4-8 SARGESON.

The 33 independent sample points are on the domain $0 \leq t \leq 320$ at intervals of 10 as entered by the DATA statements in lines 420 to 450. The corresponding dependent data are entered in lines 460 to 480. Subroutine 5000 programs the residuals in (4.4.10) to (4.4.11), and subroutine 7000 programs their partial derivatives according to (4.4.12).

It often happens that certain quantities computed for the residuals are also required for computing their derivatives. Also, the derivatives are often computed several times more than the residuals during the course of an optimization. Therefore, it is important to save the residuals rather than to recalculate them in subroutine 7000. In this case two sets of exponentials also are saved in program variables $Y4(\,)$ and $Y5(\,)$ on lines 5040 and 5050, respectively. Then they may be used in subroutine 7000 as required, since the current point in $\mathbf{x}$ space will have been used in subroutine 5000 first.

Another important fact concerns IBM BASIC release 2.0. If the user executes $\langle \text{BASICA} \ /\text{D} \rangle$, then releases 2.0 and subsequent compute the ATN, COS, EXP, LOG, SIN, SQR, and TAN functions in double precision. This was not possible in earlier releases. In this example the EXP function is crucially involved, so the following data were obtained with the /D option actuated.

This five-variable problem was started with and converged to the values shown in Table 1.2.1. The diagonal elements in scaling matrix $\mathbf{D}^2$ were 0.000612, 0.000102, 0.0000562, 0.998, and 0.0579, indicating that the scales of variables x_4 and x_5 were grossly out of balance with the other variables. This is not unexpected, since both x_4 and x_5 appear in the exponents of the nonlinear model function in (4.4.10).

The before/after curve fit was shown in Figure 1.2.1. Sargeson's input and performance data have been given in Lootsma (1972:185) in an article by Osborne. Osborne's algorithm required 27 iterations. Using LEASTP with the stopping criterion set equal to 10^{-6} and $p = 2$, convergence to a zero gradient valid to eight decimal places was obtained in 11 iterations requiring 16 function evaluations. This program was run three ways:

1. IBM Interpreted BASICA: 290 seconds.
2. IBM Compiled BASIC: 42 seconds.
3. IBM/Microway 8087 BASIC: 22 seconds.

The numerical results were nearly identical among the three methods.

The residual error at the sample points cannot be seen in Figure 1.2.1 because of its scale, so it has been plotted in Figure 4.5.2. The uneven error over the sample space suggests that higher values of exponent p should automatically emphasize and therefore suppress the higher error peaks.

The reader can try this example with $p = 4$, starting with the $\mathbf{x'}$ value obtained from $p = 2$ optimization (in Table 1.2.1). After several futile iterations, program LEASTP announces convergence with essentially no change. The difficulty is that many functions are simply not amenable to an equal-ripple error approximation, and apparently this is one of them. The next example illustrates the careful selection and construction of a problem that is amenable to an equal-ripple error approximation.

Example 4.5.4.   Consider the even polynomial

$$F(y) = -1 + x_1 y^2 + x_2 y^4 + x_3 y^6. \qquad (4.5.8)$$

When the coefficients are $\mathbf{x} = (18 \ \ -48 \ \ 32)^T$, this is the *Chebyshev function of the first kind* of degree 6. This class of Chebyshev functions is noted for its equal-ripple approximation of zero over the domain $-1 \le y \le +1$ with an error of unity magnitude. The sixth-degree Chebyshev function is shown by the solid line in Figure 4.5.3. Program C4-9, CHEBY, in Appendix C, has been written to be merged with program LEASTP to optimize (4.5.8), starting from an arbitrary $\mathbf{x}$ vector. The eleven evenly spaced sample points and the dependent data "targets" are determined by lines 420 and 430, respectively.

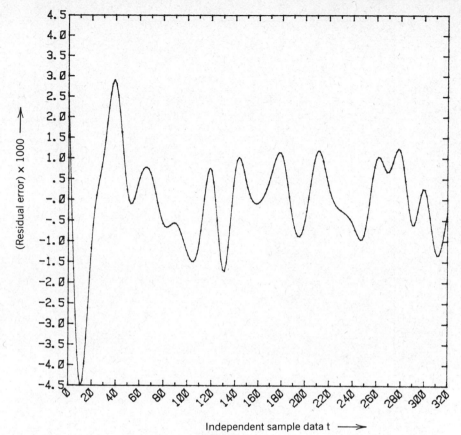

Figure 4.5.2.   Residual error at the optimal least-squares solution of Sargeson's fitting problem.

Subroutine 5000 computes the residual errors between the approximation in (4.5.8) and the target data values. Notice that the target at $y = 0$ is necessarily -1, since (4.5.8) can assume no other value at that sample. Subroutine 7000 computes the first partial derivatives of the residuals with respect to the three variables.

Program LEASTP stopping criterion was set to 10^{-6} and run in succession for $p = 2, 4, 6, 10$, and 20, starting from $\mathbf{x} = (0\ 0\ 0)^T$ first and then from each preceding minimization $\mathbf{x}$ vector. The curves of (4.5.8) for the $\mathbf{x}$ vector of coefficients obtained after minimization with $p = 2$ and $p = 10$ are shown in Figure 4.5.3. Clearly, there is a trend for large values of p to approach the Chebyshev case.

Consider the graph of each of the three variables versus $1/p$ shown in Figure 4.5.4. The data points obtained with LEASTP are shown by the circles; the $1/p = 0$ ($p \to \infty$) data are the Chebyshev coefficients previously mentioned. Figure 4.5.4 suggests that these variables are smooth functions of $1/p$

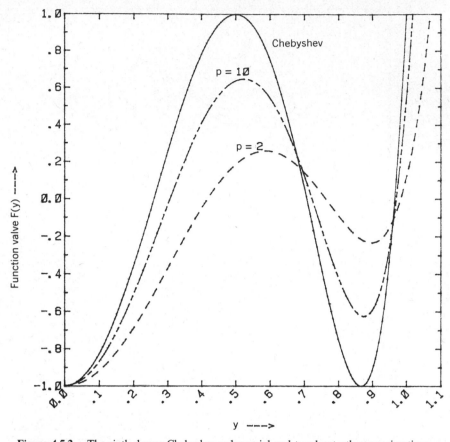

Figure 4.5.3.   The sixth-degree Chebyshev polynomial and two least-pth approximations.

and might be extrapolated to $1/p = 0$. This and other possibilities are discussed next.

4.5.3.   Approaches to Minimax Optimization.   The *minimax objective function* is

$$\min_{\mathbf{x}} \max_{k} |r_k(\mathbf{x})|, \qquad k = 1 \text{ to } m. \tag{4.5.9}$$

As defined by (2.1.39) in Section 2.1.5, the p norm of any vector $\mathbf{r}$ in E^m is

$$\|\mathbf{r}\|_p = \left[\sum_{k=1}^{m} |r_k|^p \right]^{1/p} \tag{4.5.10}$$

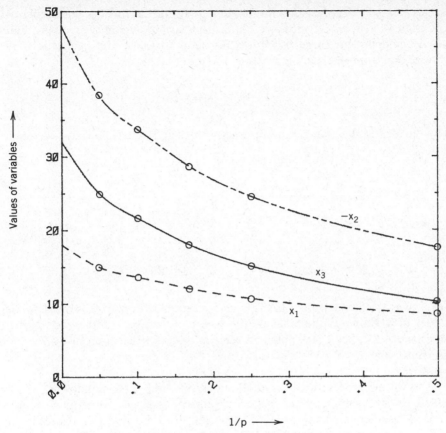

Figure 4.5.4. The three variables in Example 4.5.4 versus $1/p$.

For $p \to \infty$, the infinity norm is

$$\|\mathbf{r}\|_\infty = \max_k |r_k|, \qquad k = 1 \text{ to } m. \tag{4.5.11}$$

It is not possible to use gradient optimizers such as NEWTON and LEASTP to minimize only the maximum residual at each iteration, because the sample point for that residual will change between iterations, causing discontinuities in the gradient. Several alternative methods that will achieve an equivalent effect are discussed.

Bandler (1974b) and others have suggested extrapolation of results obtained with increasing values of exponent p in order to predict the minimax result for $p \to \infty$. A practical method for doing that is to extrapolate in the new variable $q = 1/p$, as $q \to 0$.

Example 4.5.5. The usual extrapolation procedure is to fit a polynomial of low degree to successive subsets of data. Each such polynomial predicts a limit as the independent variable approaches zero. Consider the quadratic polynomial that approximately represents x_3 from (4.5.8):

$$x_3(q) = y_1 + y_2 q + y_3 q^2. \tag{4.5.12}$$

This quadratic function can be fit exactly to each of the following two sets of data:

		Set 1				Set 2	
k	1	2	3	k	1	2	3
q	0.5	0.25	0.167	q	0.25	0.167	0.100
d	10.209	15.087	18.125	d	15.087	18.125	21.667

Set 1 was obtained from LEASTP optimization with $p = 2$, 4, and 6; the values of d shown in the table are the values of variable x_3; these are plotted in Figure 4.5.4. Set 2 was obtained after another optimization using $p = 10$; the data for $p = 2$ were discarded and the set for $p = 4$, 6, and 10 was employed.

The fitting process involves three equations and three unknowns, as opposed to the overdetermined system of equations previously applied to this kind of problem in Example 3.1.5, especially (3.1.55) and (3.1.56). Still, the *Vandermonde matrix* $\mathbf{A}$ is created for the linear system of equations $\mathbf{Ay} = \mathbf{d}$, where $\mathbf{A}_1$ and $\mathbf{A}_2$ are created from sets 1 and 2, respectively:

$$\mathbf{A}_1 = \begin{bmatrix} 1 & 0.5000 & 0.250000 \\ 1 & 0.2500 & 0.062500 \\ 1 & 0.1667 & 0.027789 \end{bmatrix}, \quad \mathbf{A}_2 = \begin{bmatrix} 1 & 0.2500 & 0.062500 \\ 1 & 0.1667 & 0.027789 \\ 1 & 0.1000 & 0.010000 \end{bmatrix}.$$

$$\tag{4.5.13}$$

The $\mathbf{d}$ vectors contain the elements shown in the two data sets, so that the solutions for the coefficients in (4.5.12) may be obtained as $\mathbf{y} = \mathbf{A}^{-1}\mathbf{d}$, using program MATRIX. It is seen in Figure 4.5.4 that the only coefficient in (4.5.12) of interest is y_1, since that equals the predicted value of x_3 when $q = 0$ and $p \to \infty$. The values are $y_1 = 26.325$ and $y_1 = 28.826$ for data sets 1 and 2, respectively. In other words, after minimizations using $p = 2$, 4, and 6, it can be predicted that $x_3 = 26.325$ for $p \to \infty$, and that value could be used to start the minimization using $p = 10$. After minimizing with $p = 10$, it is predicted that $x_3 = 28.826$ for $p \to \infty$.

One difficulty with extrapolation of data obtained by an iterative procedure with somewhat unpredictable stopping criteria is that the data may be in error. For example, the $1/p = 0.1$ data in Figure 4.5.4 appear to be somewhat high. Data averaging might be considered in the form of a least-squares determination of the fitting polynomial as in Example 3.1.5, but that still might not avoid risks inherent in extrapolation.

In practice, matrix computations are avoided in the exactly determined polynomial fitting process by using "repeated linear interpolation." That and the concept of extrapolation to zero ("Richardson extrapolation") are the essential components of the highly effective Romberg integration method. Readers interested in learning more about techniques for extrapolation should consult McCalla (1967) for the topics just mentioned.

Another problem connected with large values of exponent p is that large residuals ($r_k \gg 1$) tend to cause numerical overflow in the computer while small residuals tend to underflow (become zero). Bandler (1975) proposed to alter slightly the least-pth objective function in (4.4.31); he proposed to identify the largest residual among the sample points at the start of each iteration:

$$R(\mathbf{x}) = \|\mathbf{r}\|_\infty = \max|r_k(\mathbf{x})|, \qquad k = 1 \text{ to } m. \tag{4.5.14}$$

Then, for strictly positive $R(\mathbf{x})$, Bandler's definition of the least-pth objective function is

$$F(\mathbf{x}, t) = R(\mathbf{x}) \left[\sum_{k=1}^{m} \left| \frac{r_k(\mathbf{x})}{R(\mathbf{x})} \right|^p \right]^{1/p}, \tag{4.5.15}$$

so that at least one of the numbers raised to the pth power is unity. Of course, the value of $F(\mathbf{x})$ and its gradient are not altered by this normalization, but Bandler reports that values of p as high as 1000 are feasible when using double-precision computation.

Another way to compute minimax solutions is by solving a sequence of weighted least-squares problems with an objective function previously defined by (4.4.29). According to Gill (1981:98), the procedure is to minimize

$$F(\mathbf{x}) = \sum_{k=1}^{m} w_k^{(j)} (r_k)^2, \qquad \text{for } j = 1, 2, 3, \dots. \tag{4.5.16}$$

Let $\tilde{\mathbf{x}}^{(j)}$ represent the solution of the jth minimization of (4.5.16). The starting weights are $w_k^{(1)} = 1/m$, $k = 1$ to m. Then the subsequent sets of weights in the sequence of minimizations must be

$$w_k^{(j+1)} = \frac{w_k^{(j)} r_k^2(\tilde{\mathbf{x}}^{(j)})}{S}, \tag{4.5.17}$$

where the quantity S is chosen to make the sum of the weights during a minimization equal to unity:

$$S = \sum_{k=1}^{m} w_k^{(j)} r_k^2(\tilde{\mathbf{x}}^{(j)}). \tag{4.5.18}$$

It has been reported that it is not necessary to obtain each $\tilde{\mathbf{x}}^{(j)}$ to high accuracy, so that only a few iterations are required for each minimization. That is in contrast to the extrapolation method where accurate minimization is extremely important. An even more general method along these lines has been described by Charalambous (1978).

As previously noted, the success of any of the methods in this section depends on whether the mathematical model allows a close fit to the data. Generally, if that is true, then the least-squares solution is close to the least-pth solution in the two-norm sense, and only a small amount of additional computational is required to go from one solution to the other.

Finally, it is noted that an obvious way to solve the minimax problem (4.5.9) is to convert it into an equivalent constrained problem that has an added variable, x_{n+1}:

Minimize x_{n+1} such that

$$x_{n+1} - r_k^2(\mathbf{x}) \geq 0, \quad \text{for } k = 1 \text{ to } m, \tag{4.5.19}$$

While (4.5.19) deals with residuals without regard to sign so that $r_k^2(\mathbf{x}) \leq x_{n+1}$, the minimization could be constrained without squaring the residuals so that upper and lower bounds are enforced. The interested reader is referred to Bandler (1970a). A more generalized Kuhn–Tucker analysis can be applied to this inequality case.

Conversely, the constrained minimization problem

Minimize $F(\mathbf{x})$ such that

$$c_k \geq 0, \quad k = 1 \text{ to } m, \tag{4.5.20}$$

can be shown to be equivalent to the minimax problem

$$\min_{\mathbf{x}} \max_{k} \left[F(\mathbf{x}), F(\mathbf{x}) - q_k c_k(\mathbf{x}) \right], \quad q_k \geq 0. \tag{4.5.21}$$

It is necessary to guess the values of the positive coefficients, q_k; if they are too small some of the constraints are violated, and if too large the problem is badly conditioned. The interested reader is referred to Brayton (1980:322) and Bandler (1974a). Further discussion of constrained optimization is deferred until the last part of Chapter Five.

Problems

4.1. Consider a system of nonlinear equations $\mathbf{r}(\mathbf{x}) = \mathbf{0}$ for $\mathbf{x}$ in E^n and $\mathbf{r}$ in E^m. A well-determined system of equations $(m = n)$ is

$$r_1 = 3x_1 + 2x_2^2 + x_3 + 4x_4^3,$$

$$r_2 = 2x_1x_4 + 2x_2^2x_3 + 7x_3 + x_4^2,$$

$$r_3 = 5x_1 + 7x_1x_2 + x_3^5 + 6x_4,$$

$$r_4 = 3x_1^3 + 2x_2 + 2x_3^3x_4^2 + 2x_4^4.$$

(a) Compute the *Newton–Raphson step* $\mathbf{dx}$ from $\mathbf{x} = (1\ 1\ 1\ 1)^T$ using $\mathbf{J\,dx} = -\mathbf{r}$, where $\mathbf{J}$ is the Jacobian matrix.

(b) Is $\mathbf{J}$ positive-definite at $\mathbf{x} = (1\ 1\ 1\ 1)^T$? How do you know?

(c) Notice that the *Gauss–Newton formula* $\mathbf{J^TJ\,dx} = -\mathbf{J^Tr}$ can be obtained from the Newton–Raphson formula in the special case that $m = n$ by premultiplying both sides by $\mathbf{J^T}$. Is $\mathbf{J^TJ}$ positive-definite for the system of equations given above? How do you know? Is the Gauss–Newton step $\mathbf{dx}$ the same?

(d) Add an additional equation to the system given above:

$$r_5 = x_1 + 15x_2x_3 - 3x_4^5.$$

Compute the new Gauss–Newton step $\mathbf{dx}$ from $\mathbf{x} = (1\ 1\ 1\ 1)^T$. Is the new $\mathbf{J^TJ}$ positive-definite? How do you know?

4.2. From Powell (1967:145), compute the Newton–Raphson step from $\mathbf{x} = \mathbf{0}$ for

$$F(\mathbf{x}) = x_1^4 + x_1x_2 + (1 + x_2)^2.$$

(a) What is the slope in the $\mathbf{dx}$ direction at $\mathbf{x} = \mathbf{0}$? Is it possible to conduct a normal line search from that point? Why?

(b) Find a *negative curvature descent direction* $\mathbf{s}$ at $\mathbf{x} = \mathbf{0}$ such that $\mathbf{s^Tg} < 0$ when $\mathbf{s^THs} < 0$.

(c) Does this function have a minimum? If so, where is it?

(d) At $\mathbf{x} = \mathbf{0}$, what is the range of values of v for which $\mathbf{H} + v\mathbf{I}$ is positive-definite?

(e) Verify that the range of values of v for which step $\mathbf{dx} = -(\mathbf{H} + v\mathbf{I})^{-1}\mathbf{g}$ reduces $F(\mathbf{x})$ from $\mathbf{x} = \mathbf{0}$ is $v \geq 0.9$, and that the optimal reduction in $F(v)$ is about $v \doteq 1.2$.

4.3. Use the exact gradient expressions for the Rosenbrock functions in (4.3.2) to (4.3.3) to estimate the Hessian matrix according to (4.1.3); use $dx_j = 0.0001 |x_j|$ and $\mathbf{x} = (-1.2\ 1)^T$. Compare the results with the exact Hessian matrix [twice (4.5.6)]. Carry at least six decimal places in your calculations.

4.4. For $\mathbf{A} = \mathbf{H} + v\mathbf{I}$, where

$$\mathbf{H} = \begin{bmatrix} 3 & 2 & 1 & 9 \\ 2 & 4 & 7 & 2 \\ 1 & 7 & 5 & 6 \\ 9 & 2 & 6 & 8 \end{bmatrix},$$

is $\mathbf{A}$ positive-definite for $v = 0$? If not, what is the minimum value of v that will make $\mathbf{A}$ positive-definite?

4.5. For the function

$$F(\mathbf{x}) = x_1^3 + 2x_2^2,$$

find the point on a unit circle centered at the origin (i.e., the two-norm) that minimizes a *linear model* of the function at the point $\mathbf{x} = (1\ 1)^T$.

4.6. For the same linear model of the function in Problem 4.5, find the minimum on the ellipsoidal norm $(\mathbf{x}^T\mathbf{H}\mathbf{x} = 1)$ about the point $\mathbf{x} = (1\ 1)^T$.

4.7. Apply the modified Gauss–Newton step formula $\mathbf{dx} = -(\mathbf{H} + v\mathbf{I})^{-1}\mathbf{g}$ to the function

$$F(\mathbf{x}) = x_1^3 + 2x_2^2.$$

(*a*) What is $\mathbf{dx}$ when $v = 0$?

(*b*) What is $\mathbf{dx}$ when $v = 4$?

(*c*) Find the eigensolution for $\mathbf{H}$ and verify both (4.1.13) and (4.2.13) for this problem.

(*d*) Approximately what is the smallest positive value of v that would halve the step size when $v = 0$?

4.8. Use program NEWTON to find minima of the following functions:

(*a*) $F(\mathbf{x}) = x_1^2 x_2^2 - 4x_1^2 x_2 + 4x_1^2 + 2x_1 x_2^2 + x_2^2 - 8x_1 x_2 + 8x_1 - 4x_2$.

(*b*) $F(\mathbf{x}) = x_1^2 + 4x_2^2 - 4x_1 - 8x_2$.

(*c*) $F(\mathbf{x}) = x_1^2 + 2x_2^2 + 4x_1 + 3x_2$.

4.9. Use program LEASTP to find minima of the following functions:

(*a*) $F(\mathbf{x}) = (x_1 + x_2)^2 + [2(x_1^2 + x_2^2 - 1) - \frac{1}{3}]^2$.

(*b*) $F(\mathbf{x}) = (x_1 - x_2^3 + 5x_2^2 - 2x_2 - 13)^2 + (x_1 + x_2^3 + x_2^2 - 14x_2 - 29)^2$.

(c) $F(\mathbf{x}) = \dfrac{r_1^2 + r_2^2}{2}$, where $r_1 = x_1^4 - x_1^2 x_2 - 52x_1 + 11x_2 + 23$, and $r_2 = 51x_1 - x_1 x_2^2 - 94x_2 + x_2^3 + 325$.

(d) Find a minimum for $F(\mathbf{x})$ in part (c), starting from $\mathbf{x} = (-5 \ -5)^T$. Does the minimum obtained indicate a solution for the nonlinear system of equations $r_1 = 0$ and $r_2 = 0$?

4.10. *Chebyshev (equal weights) m-point quadrature* is exact for functions $f(t)$ of degree m or less. The formula is

$$\int_0^1 f(x)\, dx = \frac{1}{m} \sum_{k=1}^m f(x_k).$$

The set of sample points $\mathbf{x} = (x_1 \ x_2 \ \cdots \ x_k \ \cdots \ x_m)^T$ are in the open domain $0 < x_k < 1$. For an arbitrary $\mathbf{x}$, define the residual error as the difference between the integral and its approximation:

$$r_k(\mathbf{x}) = \int_0^1 T_k^*(x)\, dx - \frac{1}{m} \sum_{j=1}^m T_k^*(x_j),$$

where $T_k^*(x)$ is a polynomial in x. Let there be two samples, $m = 2$, and use the polynomials

$$T_1^*(x) = 2x - 1,$$

$$T_2^*(x) = 2(2x - 1)^2 - 1.$$

Show that the residuals to be minimized to find samples x_1 and x_2 are:

$$r_1 = x_1 + x_2 - 1,$$

$$r_2 = (2x_1 - 1)^2 + (2x_2 - 1)^2 - \tfrac{2}{3}.$$

4.11. The Chebyshev quadrature described in Problem 4.10 is the basis for the well-known *Chebyquad* optimization test function by Fletcher (1965:36). The kth degree polynomials $T_k^*(x)$ are the *shifted Chebyshev polynomials of the first kind*. They may be generated from the *Chebyshev polynomials of the first kind*, $T_k(x)$, defined by the recursion:

$$T_k(x) = 2x T_{k-1}(x) - T_{k-2}(x),$$

starting from

$$T_1(x) = x \quad \text{and} \quad T_2(x) = 2x^2 - 1.$$

The shifted Chebyshev polynomials $T_k^*(x) = T_k(2x - 1)$, that is, the argument x is simply replaced by $2x - 1$.

(*a*) Verify the sixth-degree Chebyshev polynomial is (4.5.8).

(*b*) Verify the two shifted Chebyshev polynomials given in Problem 4.10.

(*c*) Starting from $\mathbf{x}^{(0)} = (\frac{1}{3} \ \frac{2}{3})^T$, minimize the two residuals given in Problem 4.10 using program LEASTP to find the two sample points, $\mathbf{x} = (0.21132486 \ 0.78867514)^T$.

(*d*) Obtain the residuals associated with three Chebyshev samples ($m = 3$).

(*d*) Write a general program to find samples for $m = 2$ through 9. Note that there is no exact solution ($r_k = 0$) for $m = 8$; since the residuals are not zero, the Jacobian is singular at the minimum. Elements of the standard starting vector of unknowns is $x_j = j/(n + 1)$.

4.12. Use program LEASTP with $p = 2$ to fit the nonlinear model

$$F(\mathbf{x}, t) = x_1\exp(-x_2 t)$$

to the following data pairs:

k	1	2	3	4
t_k	-1	0	1	2
d_k	2.7	1.0	0.4	0.1

Start with $\mathbf{x}^{(0)} = (1 \ 1)^T$.

4.13. Use program LEASTP with $p = 2$ to solve a fitting problem by Walsh (1975):

$$f(\mathbf{x}, t) = \left(1 - \frac{x_1 t}{x_2}\right)^{[1/(x_1 c) - 1]},$$

where $c = 96.05$. The data set is:

k	1	2	3	4	5	6
t_k	2000	5000	10000	20000	30000	50000
d_k	0.9427	0.8616	0.7384	0.5362	0.3739	0.3096

4.14. Flag variable $L1$ in user subroutine 5000 of program LEASTP can be set to $L1 = 1$ to indicate that something is wrong in the computations for the residuals. Often the problem is an unallowable argument of an intrinsic function, such as the square root of a negative number. However, Fletcher (1971a) has suggested an elementary way such a flag variable might indicate a constraint violation. Use spare command 5 in LEAST to add the ability to branch to a subroutine that tests for violation of one or more constraints (linear or nonlinear). Assuming that the optimizer is started in the *feasible region* (i.e., where $\mathbf{x}^{(0)}$ violates no constraints), then the program LEASTP could pull back from any $\mathbf{dx}$ step that violates a constraint. Add this additional test to the test on line 2210, so that the length of $\mathbf{dx}$ is cut back to $\mathbf{dx}/4$ whenever a constraint is violated. Test this on some simple function of your own.

4.15. Consider the Newton step $\mathbf{dx}$ with the Levenberg–Marquardt parameter v defined by (4.4.51). For purposes of adjusting the length of that step to some assigned value, say R, Hebden (1973:8) shows that the derivative of the step length with respect to v is

$$\nabla_v \|\mathbf{dx}\|_2 = \frac{\mathbf{g}^T(\mathbf{H} + v\mathbf{I})^{-3}\mathbf{g}}{\|\mathbf{dx}\|_2}.$$

Show that by solving the two linear systems of equations (having the same coefficient matrix)

$$(\mathbf{H} + v\mathbf{I})\,\mathbf{dx} = \mathbf{g},$$

$$(\mathbf{H} + v\mathbf{I})\mathbf{q} = \mathbf{dx},$$

the preceding derivative equation is equivalent to

$$\nabla_v \|\mathbf{dx}\| = \frac{\mathbf{q}^T\mathbf{dx}}{\|\mathbf{dx}\|}.$$

4.16. Write a general subroutine 7000 for program LEASTP to calculate the partial derivatives of residuals by finite differences; see (4.1.2). Have this subroutine call subroutine 5000 as required, but first save the nominal values of residuals in an array, say $RR(I)$. Be sure to restore the unperturbed residuals from $RR(I)$ back into $R(I)$ before leaving subroutine 7000. The use of this routine allows the user to supply only subroutine 5000 to generate the residuals in $R(I)$ for particular least-pth optimization problems.

4.17. Run program NEWTON using subroutines 5000 and 7000 in file ROSEN starting with the *four* (4) variables, $\mathbf{x} = (-1.2 \ 1 \ 0 \ 0)^T$. The fact that x_3 and x_4 are not utilized in calculations for the objective function and its gradient simply increases the overhead for the main optimizer program, especially the inefficient finite-difference calls to subroutine 7000. Make a graph of the optimization run time in seconds for two, four, six, and eight variables. Perform the same tests with program LEASTP, merging ROSENPTH for the functions. Plot and compare these results, and comment on the price paid for the finite-differencing technique in NEWTON.

Chapter Five

Quasi-Newton Methods and Constraints

Quasi-Newton methods are the most effective nonlinear optimization methods for general problems. The quasi-Newton family and the most successful technique in that family are described in this chapter. Quasi-Newton methods are named for the fact that successive estimates of the Hessian matrix are updated (revised) so as to retain a key Newton property. Several other important issues are also developed, including various line search methods, lower and upper bounds on variables, linear constraints, and nonlinear constraints. Program QNEWT is introduced to implement the most popular approach to nonlinear optimization, and program BOXMIN may be merged with QNEWT to provide for lower and upper bounds on variables. The multiplier penalty (augmented Lagrangian) method for nonlinear constraints is implemented in program segment MULTPEN to be merged into the QNEWT and BOXMIN combination. Discussions of the underlying theory illustrate both the basis of these methods and the more complicated methods that also address nonlinearly constrained optimization problems.

Readers should be able to appreciate the evolution of these powerful methods as well as to apply programs QNEWT and BOXMIN to a wide variety of practical problems. Like some other methods, only the objective function and its first partial derivatives are generally required for quasi-Newton methods. However, the particular quasi-Newton method and one of several optional line searches implemented in program QNEWT retain their desirable properties when first derivatives are approximated by finite differences, so this method can easily be adapted to work without any explicit derivatives.

5.1. Updating Approximations to the Hessian

Like Newton methods, quasi-Newton methods are based on the assumption that the function being optimized is quadratic. The *conjugate gradient method*

233

described in Section 3.2.1 utilized first partial derivatives and was based on that assumption and two others: (1) the initial line search was in the direction of steepest descent (negative gradient), and (2) exact line searches were accomplished in each direction. Conjugate gradient methods have the property of *quadratic termination*: The minimum of a quadratic surface will be found in n exact line searches, where the function involves n variables.

Newton's method (Chapter Four) utilized the vector of first partial derivatives and the Hessian matrix of second partial derivatives to determine the exact single step to the minimum from any point on a quadratic surface. One way to estimate the Hessian matrix is to perturb each variable in each of the n coordinate directions and then to utilize the n different perturbed gradient vectors. The Gauss–Newton method was a variation for objective functions with a specific structure that utilized first partial derivatives efficiently to provide a positive-definite estimate of the Hessian matrix that converged to exact values for ideal problems.

Quasi-Newton methods do not require a steepest-descent start or exact line searches, although these may be employed. They start with some positive-definite estimate of the Hessian matrix, $\mathbf{B}^{(0)}$ and employ rank 1 or rank 2 updates to successive estimates $\mathbf{B}^{(k)}$ following each line search in a sequence of search directions. On quadratic surfaces, quasi-Newton methods that employ exact line searches also possess the *quadratic termination property*, because the estimate of the Hessian matrix becomes exact after n line searches, enabling a Newton step if required. On quadratic surfaces, quasi-Newton methods that employ exact line searches *and* are started in the steepest-descent direction $[\mathbf{B}^{(0)} = \mathbf{I}]$ are equivalent to the conjugate gradient method, Table 3.2.3 in Section 3.2.1.

Members of the quasi-Newton family of learning or adaptive methods are also called variable metric methods, modification methods, and update methods. They are reasonably efficient and very robust (hardy) optimizers of nonquadratic functions, requiring about $5n$ to $50n$ iterations to converge in typical cases, depending on the proximity of the starting point to a minimum. Their origins and important properties are developed.

5.1.1. General Secant Methods.

Quasi-Newton optimization algorithms are an extension of the Newton–Raphson and secant search methods commonly encountered in problems of a single variable. The *Newton–Raphson algorithm* in one variable was previously encountered in Example 1.3.1 in Section 1.3.1. The iteration formula to search for a root of function $g(x)$ is

$$x^{(k+1)} - x^{(k)} = \frac{-g^{(k)}}{g'^{(k)}}, \qquad (5.1.1)$$

as illustrated in Figure 5.1.1*a*. The formula employs the function value $g^{(k)}$ at $x = x^{(k)}$ and predicts a zero of $g(x)$ at $x = x^{(k+1)}$ based on linear extrapolation of the slope $g' = dg/dx$ evaluated at $x^{(k)}$. The process is repeated for $k = 0, 1, 2, \ldots$, and it is known to converge at a quadratic rate.

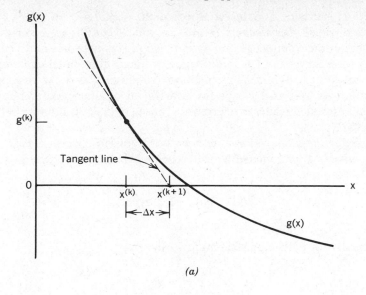

(a)

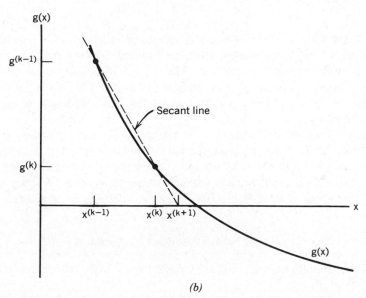

(b)

Figure 5.1.1. Second-order zero-finding algorithms for one variable. (*a*) Newton–Raphson. (*b*) Secant method.

Safeguards on step length Δx are necessary, since the slope may approach zero, causing long, diverging steps. A commonly employed variation on the Newton–Raphson method employs a small "damping factor" t multiplying the right-hand side of (5.1.1) to shorten the step length. It is more interesting, however, if both sides of (5.1.1) are then divided by t, so that t appears in the denominator of the left-hand side of (5.1.1). When t is allowed to approach zero, (5.1.1) becomes a differential equation: $dx/dt = -g(x)/g'(x)$. This approach is called *Davidenko's method*, a widely convergent algorithm that will approach the solution $g(x) = 0$ at $t \to \infty$. This approach may be generalized to form a system of n initial-value ordinary differential equations that can be integrated. Davidenko's method has been the basis of nonlinear optimization as well as a means for solution of nonlinear differential equations. The interested reader is referred to Talisa (1985:967), Branin (1972), and Zirilli (1982).

The *secant algorithm* follows the Newton–Raphson concept closely, except the derivative is approximated by differencing. The secant recursion is

$$x^{(k+1)} - x^{(k)} = \frac{g^{(k+1)} - g^{(k)}}{B^{(k)}}, \tag{5.1.2}$$

where the divisor is the approximation to the slope:

$$B^{(k)} = \frac{g^{(k)} - g^{(k-1)}}{x^{(k)} - x^{(k-1)}}. \tag{5.1.3}$$

In (5.1.2), the term $g^{(k+1)} = 0$ is always assumed in search of a zero of $g(x)$; however, this *divided-difference* form is retained for comparison with the following quasi-Newton expressions. The secant algorithm must be started with two points, $x^{(-1)}$ and $x^{(0)}$, as apparent in Figure 5.1.1b; then it proceeds with $k = 0, 1, 2, \ldots$, in the same fashion as the Newton–Raphson algorithm for one variable.

For more than one variable, $n > 1$, the *quasi-Newton algorithm* or generalized secant search seeks a zero of the vector function $\mathbf{g(x)}$ and consists of the steps shown in Table 5.1.1. Compared to the Newton method, the quasi-Newton method does not require second partial derivatives of $F(\mathbf{x})$, and the estimated Hessian matrix $\mathbf{B}$ is always positive definite so that each search

---

Table 5.1.1. Steps in the Quasi-Newton Algorithm for Nonlinear Functions

1. $\mathbf{x}^{(0)}$ is an arbitrary starting point and $\nabla F[\mathbf{x}^{(0)}] = \mathbf{g}^{(0)}$ is the related gradient vector. For $k = 0$:
2. Solve the system of linear equations (Newton step) $\mathbf{B}^{(k)}\mathbf{s}^{(k)} = -\mathbf{g}^{(k)}$ for search direction $\mathbf{s}^{(k)}$, where $\mathbf{B}^{(k)}$ is a special positive-definite estimate of the Hessian matrix,
3. Perform a line search $\mathbf{x}^{(k+1)} = \mathbf{x}^{(k)} + t_k \mathbf{s}^{(k)}$ by varying t_k so that objective function $F[(\mathbf{x}^{(k+1)}] < F[\mathbf{x}^{(k)}]$,
4. Terminate if converged; or else
5. Perform one or two rank 1 updates to $\mathbf{B}^{(k)}$, giving a positive-definite estimate $\mathbf{B}^{(k+1)}$,
6. Increment $k = k + 1$ and go to step 2.

direction is a descent direction. The behavior of quasi-Newton search methods on general nonlinear functions far from a minimum is robust, which is to say that the algorithm adapts to nonquadratic surfaces to produce a controlled descent toward the minimum. The main feature of the method is in the nature of the family of formulas that update the sequential estimates of the Hessian matrix.

The general secant method for one variable is now related to the same process in many variables. Repeating (3.2.7) to (3.2.9), a quadratic model is assumed for the function:

$$F(\mathbf{x}) = c + \mathbf{b}^T\mathbf{x} + \tfrac{1}{2}\mathbf{x}^T\mathbf{H}\mathbf{x}, \tag{5.1.4}$$

with linear gradient equations

$$\nabla F(\mathbf{x}) = \mathbf{g} = \mathbf{b} + \mathbf{H}\mathbf{x}. \tag{5.1.5}$$

The Hessian matrix $\mathbf{H}$ is a constant for quadratic functions, so consideration of two points $\mathbf{x}^{(k)}$ and $\mathbf{x}^{(k+1)}$, in (5.1.5) yields the linear mapping between changes of gradient and corresponding changes in position:

$$\mathbf{H}[\mathbf{x}^{(k+1)} - \mathbf{x}^{(k)}] = \mathbf{g}^{(k+1)} - \mathbf{g}^{(k)}. \tag{5.1.6}$$

For convenience and historical reasons, (5.1.6) can be expressed as

$$\mathbf{H}\mathbf{d}^{(k)} = \mathbf{y}^{(k)}, \tag{5.1.7}$$

where the following definitions have been employed:

$$\mathbf{d}^{(k)} = \mathbf{x}^{(k+1)} - \mathbf{x}^{(k)}, \tag{5.1.8}$$

$$\mathbf{y}^{(k)} = \mathbf{g}^{(k+1)} - \mathbf{g}^{(k)}. \tag{5.1.9}$$

Now consider a nonquadratic function and define matrix $\mathbf{B}^{(k)}$ to be the estimate of the Hessian matrix $\mathbf{H}$ at $\mathbf{x} = \mathbf{x}^{(k)}$, where k is the iteration number. Since $\mathbf{d}^{(k)}$ and $\mathbf{y}^{(k)}$ can be computed only *after* the line search (step 3 in Table 5.1.1), the estimated Hessian matrix $\mathbf{B}^{(k)}$ does not represent this mapping correctly in the sense of (5.1.7). To force that, the *quasi-Newton condition* is assumed:

$$\mathbf{B}^{(k+1)}\mathbf{d}^{(k)} = \mathbf{y}^{(k)}. \tag{5.1.10}$$

The quasi-Newton condition merits emphasis: it maintains the linear mapping between corresponding changes in gradient and position that underlie Newton's method.

Notice that quasi-Newton condition (5.1.10) corresponds exactly to the divided-difference relationship, (5.1.2), in the secant algorithm for the kth iteration. In fact, (5.1.10) is simply the finite-difference approximation to $\mathbf{B}^{(k+1)}\mathbf{d}^{(k)}$ computed in the $\mathbf{d}^{(k)}$ direction; see Gill (1974b:68). The name "quasi-Newton" has been assigned because $\mathbf{B}^{(k+1)}$ in (5.1.10) is analogous to the approximate derivative in (5.1.3) and the exact derivative in (5.1.1). The reader interested in a more sophisticated mathematical treatment of the relationships between secant methods and quasi-Newton methods is referred to Rheinboldt (1974:25–54). It is next shown that the quasi-Newton condition may be enforced in several ways, leading to families of formulas for updating the sequential estimates $\mathbf{B}^{(k)}$ for the Hessian matrix $\mathbf{H}$.

5.1.2. Families of Quasi-Newton Matrix Updates.
A general definition of a matrix update is

$$\mathbf{B}^{(k+1)} = \mathbf{B}^{(k)} + \Delta\mathbf{B}^{(k)} \quad \text{or} \quad \mathbf{B}^* = \mathbf{B} + \Delta\mathbf{B}. \quad (5.1.11)$$

The right-hand representation for $\mathbf{B}^*$ is used to avoid excessive notation, since the concern here is only with changes that occur during each iteration. ($\mathbf{B}^*$ does not denote complex conjugation in this chapter). The Sherman–Morrison–Woodbury formula (2.1.36) applies to this case and can provide an explicit expression for the update term, $\Delta\mathbf{B}$, showing that its rank may be as great as n, the number of variables. However, in practice the updates are usually only of rank 1 or 2.

A rank 1 update simply adds a scaled outer product as defined by (2.1.22):

$$\mathbf{B}^* = \mathbf{B} + q\mathbf{z}\mathbf{z}^T. \quad (5.1.12)$$

If the quasi-Newton condition in (5.1.10) is to be satisfied, then

$$\mathbf{B}\mathbf{d} + q\mathbf{z}\mathbf{z}^T\mathbf{d} = \mathbf{y}. \quad (5.1.13)$$

Note that $\mathbf{z}^T\mathbf{d}$ in the second term is a scalar, so that $\mathbf{z}$ must be proportional to $\mathbf{y} - \mathbf{B}\mathbf{d}$. A simple choice for $\mathbf{z}$ and a consequent value for q are:

$$\mathbf{z} = \mathbf{y} - \mathbf{B}\mathbf{d}, \quad (5.1.14)$$

$$q = \frac{1}{\mathbf{z}^T\mathbf{d}}. \quad (5.1.15)$$

Therefore, the unique *symmetric rank* 1 *update formula* (*SR*1) that satisfies the quasi-Newton condition in (5.1.10) is

$$\mathbf{B}^* = \mathbf{B} + \frac{(\mathbf{y} - \mathbf{B}\mathbf{d})(\mathbf{y} - \mathbf{B}\mathbf{d})^T}{(\mathbf{y} - \mathbf{B}\mathbf{d})^T\mathbf{d}}. \quad (5.1.16)$$

Update formula (5.1.16) has been attributed to several investigators, especially Broyden (1965). It has two major deficiencies. It does not maintain positive definiteness, even when employed on a quadratic function; therefore, it may not result in a downhill search direction. Also, the denominator in (5.1.16) may approach zero, requiring various safeguards in the algorithm. However, Fletcher (1980:41) showed that this rank 1 method terminates on a quadratic surface in at most $n + 1$ searches with $\mathbf{B}^{(n+1)} = \mathbf{H}$ *without exact line searches*. The fact that the SR1 formula in (5.1.16) does not force the approximation $\mathbf{B}^*$ to be positive-definite has been viewed as an advantage in certain optimization programs. See Brayton (1977).

Rank 2 updates can be written as the sum of two rank 1 updates:

$$\mathbf{B}^* = \mathbf{B} + q_1 \mathbf{z}_1 \mathbf{z}_1^T + q_2 \mathbf{z}_2 \mathbf{z}_2^T. \tag{5.1.17}$$

The quasi-Newton condition in (5.1.10) must be satisfied in this case also, leading to

$$\mathbf{B}\mathbf{d} + q_1 \mathbf{z}_1 \mathbf{z}_1^T \mathbf{d} + q_2 \mathbf{z}_2 \mathbf{z}_2^T \mathbf{d} = \mathbf{y}. \tag{5.1.18}$$

Now vectors $\mathbf{z}_1$ and $\mathbf{z}_2$ are not unique, but useful choices for these two free parameters are

$$\mathbf{z}_1 = \mathbf{y}, \tag{5.1.19}$$

$$\mathbf{z}_2 = \mathbf{B}\mathbf{d}. \tag{5.1.20}$$

Use of these two choices in (5.1.18) requires the scalar equalities $q_1 \mathbf{z}_1^T \mathbf{d} = 1$ and $q_2 \mathbf{z}_2^T \mathbf{d} = -1$.

The resulting rank 2 update formula is

$$\mathbf{B}^*_{\text{BFGS}} = \mathbf{B} + \frac{\mathbf{y}\mathbf{y}^T}{\mathbf{y}^T\mathbf{d}} - \frac{(\mathbf{B}\mathbf{d})(\mathbf{B}\mathbf{d})^T}{(\mathbf{B}\mathbf{d})^T\mathbf{d}}. \tag{5.1.21}$$

The name *BFGS formula* is commonly used to indicate that it was independently discovered by Broyden, Fletcher, Goldfarb, and Shanno, all in 1970. In computations, (5.1.21) is more conveniently arranged as

$$\mathbf{B}^*_{\text{BFGS}} = \mathbf{B} + \frac{\mathbf{y}\mathbf{y}^T}{t\mathbf{y}^T\mathbf{s}} + \frac{\mathbf{g}\mathbf{g}^T}{\mathbf{g}^T\mathbf{s}}, \tag{5.1.22}$$

since $\mathbf{d} = t\mathbf{s}$ and $\mathbf{B}\mathbf{s} = -\mathbf{g}$ according to steps 2 and 3 in Table 5.1.1 and (5.1.8). Notice that no matrix-vector products are required for the BFGS formula in (5.1.22), as opposed to the formula in (5.1.16) and others to be considered. Therefore, the condition of the old approximate Hessian matrix $\mathbf{B}$ does not affect either of the two rank 1 updates in (5.1.22). The BFGS formula is the one used in program QNEWT later in this chapter.

So far, *direct update formulas* have been considered, that is, those where matrix $\mathbf{B}$ is an approximation of the exact Hessian matrix $\mathbf{H}$. The *Broyden family* of direct updates that satisfy (5.1.10) may be expressed by

$$\mathbf{B}^* = \mathbf{B}^*_{\mathrm{BFGS}} + r\mathbf{w}\mathbf{w}^T,$$

$$\mathbf{w} = (\mathbf{d}^T\mathbf{B}\mathbf{d})^{1/2}\left[\frac{\mathbf{y}}{\mathbf{d}^T\mathbf{y}} - \frac{\mathbf{B}\mathbf{d}}{\mathbf{d}^T\mathbf{B}\mathbf{d}}\right]. \tag{5.1.23}$$

Generally, the mathematical properties to be discussed for any member of this family apply to the entire family. There are also several unique values of r that make $\mathbf{B}^*$ singular. It can be shown that the SR1 formula in (5.1.16) is in the Broyden family and that its value of r in (5.1.23) is not in the closed range [0, 1], assuming that $\mathbf{d}^T\mathbf{y} > 0$ and $\mathbf{B}^*$ is positive-definite.

There is a corresponding Broyden family of *inverse update formulas* based on a matrix, say $\mathbf{R}$, that approximates the inverse of the exact Hessian matrix, $\mathbf{H}^{-1}$. Observe that the *quasi-Newton condition* in (5.1.10) might just as well have been expressed in the form

$$\mathbf{d}^{(k)} = \mathbf{R}^{(k+1)}\mathbf{y}^{(k)}. \tag{5.1.24}$$

The preceding development leading to (5.1.21) would simply have interchanged $\mathbf{R}$ for $\mathbf{B}$, $\mathbf{y}$ for $\mathbf{d}$, and $\mathbf{d}$ for $\mathbf{y}$. The result is the famous and first quasi-Newton formula introduced by Davidon (1959) and simplified by Fletcher and Powell (1963):

$$\mathbf{R}^*_{\mathrm{DFP}} = \mathbf{R} + \frac{\mathbf{d}\mathbf{d}^T}{\mathbf{d}^T\mathbf{y}} - \frac{\mathbf{R}\mathbf{y}\mathbf{y}^T\mathbf{R}}{\mathbf{y}^T\mathbf{R}\mathbf{y}}. \tag{5.1.25}$$

Formulas related by interchanges of variables as appropriate for (5.1.24) are said to be *duals*. The SR1 formula in (5.1.16) is self dual. Furthermore, if the inverse of the dual is obtained by the Sherman–Morrison–Woodbury formula (2.1.36), the result is called the *complementary formula*. See Problem 5.4. The direct approximation of the Hessian $\mathbf{B}_{\mathrm{DFP}}$ that corresponds to the indirect DFP formula in (5.1.25) is described for $r = 1$ in the Broyden family expression in (5.1.23).

To summarize, the Broyden family of rank 2 formulas update either the estimated Hessian or the estimated inverse Hessian. The DFP inverse formula (5.1.25) was discovered first; many years later it was concluded that it did not perform as well with inexact line searches as either the inverse or direct BFGS formula (5.1.22). The BFGS and DFP formulas anchor the ends of the Broyden family with $r = 0$ and $r = 1$ in (5.1.23), respectively; clearly, the difference between any two formulas in the family is rank 1. Because the DFP update formula (5.1.25) was discovered first, the BFGS update formula (5.1.21) was and sometimes still is called the *complementary DFP formula* (CDFP).

After many years of improved results claimed for a variety of new formulas in the Broyden family, Dixon (1972a) published an incredible theorem that applied to the behavior of all family formulas on any continuous *nonlinear* function: Assuming *exact* line searches, the search directions for all Broyden formulas vary only in length, not direction. In theory, there should be no difference in the performance of any of the Broyden formulas! Interested readers are referred to Fletcher (1980:50–53).

In practice, the direct BFGS update (5.1.22) has consistently performed much better than other members of the Broyden family, especially when inexact line searches are employed. So far, there has been no convincing explanation of why this is the case. The BFGS direct update requires no matrix-vector products. However, any inverse update only requires about n^2 operations to solve for the search direction, step 2 in Table 5.1.1, since the inverse is already on hand. But the ability to perform two rank 1 updates to an $\mathbf{LDL}^T$ matrix factorization (see Section 3.1.1) using about $3n^2$ operations eliminates any serious differences between direct and inverse update formulas. Also, the $\mathbf{LDL}^T$ method enables a guarantee of positive-definite estimates; even though that is theoretically accounted for, roundoff errors and inaccurate line searches can produce violations.

There are two other quasi-Newton families that deserve mention. Huang (1970) has suggested a modification to the quasi-Newton condition in (5.1.10):

$$p^{(k)}\mathbf{B}^{(k+1)}\mathbf{d}^{(k)} = \mathbf{y}^{(k)}. \tag{5.1.26}$$

Parameter $p^{(k)}$ is a scalar and $p = 1$ for all k is the Broyden family, which is by far the most important subset. The *Huang family* thus has three parameters, and the direct or inverse estimates of the Hessian matrix may be unsymmetric.

Fletcher (1980:50) has shown that search directions that belong to the Broyden family possess quadratic termination and two other special properties when employed on quadratic functions *using exact line searches*:

$$\mathbf{B}^{(k+1)}\mathbf{d}^{(j)} = \mathbf{y}^{(j)}, \qquad j = 1, 2, \ldots, k, \ k = 1 \text{ to } n, \tag{5.1.27}$$

$$\mathbf{s}^{(k)T}\mathbf{H}\mathbf{s}^{(j)} = 0, \qquad j = 1, 2, \ldots, k - 1, \ k = 1 \text{ to } n. \tag{5.1.28}$$

The *hereditary property* is expressed by (5.1.27), which is an extension of the quasi-Newton condition (5.1.10) to span all the iterations. The *conjugacy condition* in (5.1.28) also applies to the Broyden family when exact line searches are used on quadratic functions. Thus the quasi-Newton methods are conjugate gradient methods when exact line searches are used on quadratic surfaces. Recall that the latter methods are equivalent to exact line searches in eigenvector directions, that is, on the canonical form (3.2.5).

One reason for mentioning the hereditary property is that Davidon (1975) has defined a family of formulas that preserve the hereditary property when *inexact* line searches are used. The SR1 formula in (5.1.16) is a member of that family, but other more important members do not degenerate in that way and do maintain positive definiteness of the updated matrices that estimate the Hessian. Davidon's new method is substantially more complicated and will not be described here. In spite of the promising formulation of his approach, numerical comparisons have not indicated that Davidon's new concepts provide better performance than the BFGS formulas.

5.1.3. Invariance of Newton-Like Methods to Linear Scaling.

The purpose of this section is to show that search methods that are derived from the Newton step $\mathbf{d} = -\mathbf{H}^{-1}\mathbf{g}$ possess an important independence or *invariance property* with respect to a general linear transformation of variables. A more comprehensive alteration of scale for each variable is contemplated than was accomplished by the diagonal matrix employed in the linear transformation of (4.4.60) in connection with the Levenberg modification of the Gauss–Newton method. The significance of the invariance property is that a quasi-Newton method becomes invariant to the scales chosen for the variables as the search progresses, which implies a robustness that algorithms like steepest descent do not have. There are differing viewpoints on this matter, and the reader is referred to Dennis (1983:155, 203), Fletcher (1980:45), and Fletcher (1969:342).

The concept of *scaling* of variables arises in the choice of units for the variables. Optimization of components in an electrical network may require capacitor values in the range 10^{-10} to 10^{-14} farads and inductor values in the range 10^{-7} to 10^{-11} henrys. A natural choice of units might be picofarads (10^{-12} farad) and nanohenrys (10^{-9} henry), respectively, so that both scales would be in the range 0.01 to 100. However, the choice of millihenrys (10^{-3} henry) as the unit of inductance would cause the scales to differ by 10^6, certainly a gross distortion of the function surface and a great impediment to algorithms on finite-word-length computers that must take discrete steps in the variable space.

Bad scaling impacts optimization algorithms that depend on the two-norm $\|\mathbf{x}\|_2$ for measuring Euclidean distance:

$$\|\mathbf{x}\|_2^2 = \mathbf{x}^T\mathbf{x} = x_1^2 + x_2^2 + \cdots + x_n^2. \tag{5.1.29}$$

If $x_1 = 10^{-6}x_2$, then x_1 certainly would not effect the Euclidean length of $\mathbf{x}$. The most general linear transformation from one variable space to another is accomplished by an unrestricted real matrix $\mathbf{T}$:

$$\tilde{\mathbf{x}} = \mathbf{T}\mathbf{x}. \tag{5.1.30}$$

If the Euclidean length of a vector is to remain constant in variables spaces $\mathbf{x}$ and $\tilde{\mathbf{x}}$, then

$$\tilde{\mathbf{x}}^T\tilde{\mathbf{x}} = (\mathbf{T}\mathbf{x})^T(\mathbf{T}\mathbf{x}) = \mathbf{x}^T\mathbf{T}^T\mathbf{T}\mathbf{x}, \tag{5.1.31}$$

which implies that $\mathbf{T}^T\mathbf{T} = \mathbf{I}$, that is, $\mathbf{T}$ must be an orthonormal matrix as described in Section 2.2.2. Notice that a simple diagonal matrix $\mathbf{T} = $ diag(t_{11} t_{22} $\ldots$, t_{nn}) cannot satisfy (5.1.31).

Of greater consequence is the effect of linear transformations on the quadratic function model that is the basis for all effective search algorithms:

$$F(\mathbf{x}) = c + \mathbf{b}^T\mathbf{x} + \tfrac{1}{2}\mathbf{x}^T\mathbf{H}\mathbf{x}. \tag{5.1.32}$$

Hessian matrix $\mathbf{H}$ is symmetric and positive definite. To study this problem, recall that the similarity transformation $\mathbf{P}$ in (2.2.46) was defined by

$$\mathbf{P}^{-1}\mathbf{H}\mathbf{P} = \mathbf{W}, \tag{5.1.33}$$

and that matrix $\mathbf{W}$ is diagonal if $\mathbf{H}$ is symmetric. Recall that a special similarity transformation in Section 3.2.1 employed the orthonormal matrix

$$\mathbf{V} = (\mathbf{v}_1\ \mathbf{v}_2\ \cdots\ \mathbf{v}_n) \tag{5.1.34}$$

having columns that are the normalized eigenvectors, $\mathbf{v}_j$, of matrix $\mathbf{H}$. It was shown in (3.2.2) that a similarity transformation is

$$\mathbf{V}^T\mathbf{H}\mathbf{V} = \mathbf{W}, \tag{5.1.35}$$

where $\mathbf{W}$ is diagonal with elements that are the eigenvalues of $\mathbf{H}$.

The most important consequence is that the linear transformation

$$\tilde{\mathbf{x}} = \mathbf{V}^T\mathbf{x} \tag{5.1.36}$$

transforms $F(\mathbf{x})$ in (5.1.32) to the equivalent function

$$F(\tilde{\mathbf{x}}) = c + (\mathbf{V}^T\mathbf{b})^T\tilde{\mathbf{x}} + \tfrac{1}{2}\tilde{\mathbf{x}}^T\mathbf{W}\tilde{\mathbf{x}}. \tag{5.1.37}$$

The function (5.1.37) in $\tilde{\mathbf{x}}$ space is in *canonical form* since the axes of its elliptical contours are aligned with the coordinate axes. Therefore, only n exact linear searches are required from any starting point in order to locate the minimum in the $\tilde{\mathbf{x}}$ space. The only other improvement would be if matrix $\mathbf{W} = \mathbf{I}$ in (5.1.37) so that the elliptical contours are in fact circles. Then only one exact line search in the steepest descent (negative gradient) direction would locate the minimum.

To aid in further analysis of the impact of the linear transformation $\tilde{\mathbf{x}} = \mathbf{T}\mathbf{x}$ on a quadratic model, it is necessary to relate the gradient vectors and Hessian matrices in the $\mathbf{x}$ and $\tilde{\mathbf{x}}$ spaces. Consider matrix $\mathbf{T}_{n,\,n} = [t_{ij}]$ and the linear transformation of (5.1.30). By writing a few of those equations, it can be seen that the kth equation is

$$\tilde{x}_k = \sum_{j=1}^{n} t_{kj}x_j. \tag{5.1.38}$$

Then the first partial derivative of some $\tilde{x}_k$ with respect to any x_j is

$$\frac{\partial \tilde{x}_k}{\partial x_j} = t_{kj}. \tag{5.1.39}$$

Function F in (5.1.32) is a function of $\mathbf{x}$ and is also a function of $\tilde{\mathbf{x}}$ according to (5.1.30). Therefore, the chain rule yields

$$\frac{\partial F}{\partial x_j} = \sum_{k=1}^{n} \frac{\partial \tilde{x}_k}{\partial x_j} \frac{\partial F}{\partial \tilde{x}_k} = \sum_{k=1}^{n} t_{kj} \frac{\partial F}{\partial \tilde{x}_k}, \tag{5.1.40}$$

where (5.1.39) has been substituted. Since the left-hand side of (5.1.40) describes the elements of $\nabla F(\mathbf{x})$ and the right-hand side describes the elements of $\nabla F(\tilde{\mathbf{x}})$, subscripts are used with ∇ to indicate these spaces:

$$\nabla_{\mathbf{x}} F = \mathbf{T}^T \nabla_{\tilde{\mathbf{x}}} F. \tag{5.1.41}$$

An expression for the Hessian from (4.4.18) is $\mathbf{H} = \nabla(\nabla F)^T$; this yields

$$\mathbf{H}_{\mathbf{x}} = \nabla_{\mathbf{x}}(\nabla_{\mathbf{x}} F)^T = \mathbf{T}^T \nabla_{\tilde{\mathbf{x}}} (\mathbf{T}^T \nabla_{\tilde{\mathbf{x}}} F)^T$$

$$= \mathbf{T}^T (\nabla_{\tilde{\mathbf{x}}} \nabla_{\tilde{\mathbf{x}}}^T F) \mathbf{T}. \tag{5.1.42}$$

Therefore, the relationship between Hessian matrices in $\mathbf{x}$ and $\tilde{\mathbf{x}}$ is

$$\mathbf{H}_{\mathbf{x}} = \mathbf{T}^T \mathbf{H}_{\tilde{\mathbf{x}}} \mathbf{T}. \tag{5.1.43}$$

It is now possible to show that many forms of Newton's method with line searches on a quadratic model are invariant with respect to a linear transformation of variables. Recall from (5.1.8) that $\mathbf{x}^*$ and $\mathbf{x}$ represent the turning points $\mathbf{x}^{(k+1)}$ and $\mathbf{x}^{(k)}$ in line search k. Thus the Newton step in the $\mathbf{x}$ space is

$$\mathbf{x}^* = x - t\mathbf{H}_{\mathbf{x}}^{-1} \mathbf{g}_{\mathbf{x}}, \tag{5.1.44}$$

where Hessian $\mathbf{H}$ is both symmetric and positive definite. Line search metric parameter t assumes a value that obtains at least a function decrease if not an approximate minimum in the Newton direction. Line searches are more fully described in Section 5.2.

Consider a Newton step in the $\tilde{\mathbf{x}}$ space:

$$\tilde{\mathbf{x}}^* = \tilde{x} - t\mathbf{H}_{\tilde{\mathbf{x}}}^{-1} \mathbf{g}_{\tilde{\mathbf{x}}}. \tag{5.1.45}$$

Substituting (5.1.30), (5.1.43), and (5.1.42) into (5.1.45),

$$\tilde{\mathbf{x}}^* = \mathbf{T}\mathbf{x} - t(\mathbf{T}\mathbf{H}_{\mathbf{x}}^{-1}\mathbf{T}^T)(\mathbf{T}^{-T}\mathbf{g}_{\mathbf{x}}) = \mathbf{T}(\mathbf{x} - t\mathbf{H}_{\mathbf{x}}^{-1}\mathbf{g}_{\mathbf{x}}). \tag{5.1.46}$$

It is seen from the left and rightmost expressions in (5.1.46) and (5.1.44) that

$$\tilde{\mathbf{x}}^* = \mathbf{T}\mathbf{x}^*, \qquad \text{that is,} \qquad \tilde{\mathbf{x}}^{(k+1)} = \mathbf{T}\mathbf{x}^{(k+1)}. \tag{5.1.47}$$

The important conclusion is that the corresponding line searches in either space $\mathbf{x}$ or $\tilde{\mathbf{x}}$ arrive at the same corresponding turning points so that Newton line searches on a quadratic model are *invariant* with respect to the general linear transformation $\tilde{\mathbf{x}} = \mathbf{T}\mathbf{x}$. It is left to problem 5.5 to show in a similar way that derivatives (slopes) in corresponding directions $\mathbf{s}_\mathbf{x}$ and $\mathbf{s}_{\tilde{\mathbf{x}}}$ are equal:

$$\mathbf{g}_\mathbf{x}^T \mathbf{s}_\mathbf{x} = \mathbf{g}_{\tilde{\mathbf{x}}}^T \mathbf{s}_{\tilde{\mathbf{x}}}. \tag{5.1.48}$$

The fact that the corresponding function, gradient, and directional derivative values in the two spaces are equal means that line search algorithms that employ those values will obtain identical results in either $\mathbf{x}$ or $\tilde{\mathbf{x}}$ space.

The significance of the invariance property of Newton methods is that those algorithms should perform just as well with or without scaling of the variable space (within the numerical capability of the computer). The Hessian matrix may be quite illconditioned, but the implication of the invariance property is that it is not necessary to introduce scaling so that $\mathbf{H}_{\tilde{\mathbf{x}}} = \mathbf{I}$.

It is left to problem 5.5 to show that the steepest descent method is not invariant, and it is known to perform badly when the Hessian is illconditioned. The Gauss–Newton step does not have the invariance property because $\mathbf{J}^T\mathbf{J}$ only approximates the Hessian except when all residuals are zero. The Levenberg step $\mathbf{d} = -(\mathbf{J}^T\mathbf{J} + v\mathbf{I})^{-1}\mathbf{g}$ and the modification of replacing $\mathbf{I}$ by diagonal scaling matrix $\mathbf{D}^2$ do not have the invariant property. Perhaps it is intuitively reasonable that non-Newton methods do not have the invariance property: A perfect Newton step always goes to the global (only) minimum on a quadratic surface. A single line search in any other direction will find only a minimum in that direction.

Quasi-Newton methods that employ updated estimates to the Hessian are often started with the estimate $\mathbf{B}^{(0)} = \mathbf{I}$. In cases such as this the initial searches are not invariant, but tend to become so as the estimate $\mathbf{B}^{(k)} \to \mathbf{H}$ as $k \to n$. For that matter, the general nonlinear function only tends to the quadratic model in a neighborhood of a minimum, so this particular limitation of quasi-Newton searches should be of only limited concern. Furthermore, it is straightforward to show that the BFGS updating formula in (5.1.21) preserves the linear transformation $\tilde{\mathbf{x}} = \mathbf{T}\mathbf{x}$, that is, $\mathbf{B}_\mathbf{x}^* = \mathbf{B}_{\tilde{\mathbf{x}}}^*$; see problem 5.5. In fact, a similar result may be obtained for any updating formula from the Broyden family that is a sum of rank 1 corrections (5.1.11) that utilize vectors $\mathbf{y}$ and $\mathbf{Bd}$.

There is still motivation for using the analysis of this section for scaling when the invariance property does not apply to the search algorithm. The equivalence for Hessian matrices in $\tilde{\mathbf{x}}$ and $\mathbf{x}$ given by (5.1.43) shows that perfect scaling (condition) in the $\tilde{\mathbf{x}}$ space, $\mathbf{H}_{\tilde{\mathbf{x}}} = \mathbf{I}$, requires that

$$\mathbf{T}^T\mathbf{T} = \mathbf{H}_\mathbf{x}, \qquad \text{when } \mathbf{H}_{\tilde{\mathbf{x}}} = \mathbf{I}. \tag{5.1.49}$$

Thus transformation matrix $\mathbf{T}$ might assume an upper triangular form $\mathbf{U}$, so that $\mathbf{U}^T\mathbf{U} = \mathbf{H}_x$, which is the Cholesky factorization of a positive-definite $\mathbf{H}$ matrix described in Section 3.1.1. It is left to problem 5.5 to show that the scaling transformation $\tilde{\mathbf{x}} = \mathbf{U}\mathbf{x}$ simply converts steepest descent into Newton's method. Notice that this corresponds exactly to the result from the Lagrangian function in (4.1.10) that obtained the minimum on a linear model constrained by an assigned value of the *elliptic norm*

$$\|\mathbf{d}\|_{\mathbf{H}} = \mathbf{d}^T\mathbf{H}\mathbf{d}. \tag{5.1.50}$$

The same conclusion is obtained either way: a Newton step in $\mathbf{x}$ is equivalent to a unit steepest descent step in the transformed space $\tilde{\mathbf{x}} = \mathbf{U}\mathbf{x}$. It is easily shown that $\|\mathbf{x}\|_{\mathbf{H}_x} = \|\tilde{\mathbf{x}}\|_{\mathbf{H}_{\tilde{x}}}$; see problem 5.5. The elliptic norm in quasi-Newton methods is $\|\mathbf{x}\|_{\mathbf{B}}$, where $\mathbf{B}$ is the current updated estimate of the Hessian matrix. This is why quasi-Newton methods are often called *variable metric methods*.

For simple scaling of variables using a diagonal matrix, $\tilde{\mathbf{x}} = \mathbf{D}\mathbf{x}$, (5.1.49) indicates that $\mathbf{D}^2$ should approximate the main diagonal of Hessian matrix $\mathbf{H}$ in some way. This has been considered for the Gauss–Newton–Levenberg case in (4.4.68), where the elements d_{ii} were equated to the root mean square value of the first derivatives of the corresponding residuals. Regardless of how a simple scaling matrix $\mathbf{D}$ is obtained, both the steepest descent and quasi-Newton methods benefit from starting with an initial estimate $\mathbf{B}^{(0)} = \mathbf{D}^2$ instead of $\mathbf{I}$.

Usually, a linear translation of variables should be added to any diagonal scaling employed in order to preserve relative precision. The conventional rule of thumb is that values of variables should be transformed so that they have a magnitude of approximately unity. Therefore, the elements of diagonal matrix $\mathbf{D}$ in $\tilde{\mathbf{x}} = \mathbf{D}\mathbf{x}$ should be set to the inverse of the related nominal variable value. It is easy to show that this is insufficient by itself. Consider a variable x with a small expected range of values, say, $900.1234 \le x \le 900.4321$. Simple transformation (normalization) that divides by the lower value produces a new range $1.000000 \le \tilde{x} \le 1.000343$ on a computer carrying seven significant figures. In this case only the three least significant figures are available to indicate any changes in the variable.

Linear transformation *and* translation of variables are easily accomplished by

$$\tilde{\mathbf{x}} = \mathbf{D}\mathbf{x} + \mathbf{c}, \tag{5.1.51}$$

which retains all the properties of transformation (5.1.30) with respect to Newton-like methods. Usually lower and upper bounds on the variables can be anticipated for the problem at hand; suppose that these are

$$p_i \le x_i \le q_i, \qquad i = 1 \text{ to } n. \tag{5.1.52}$$

Then the transformation and translation indicated by (5.1.51) is

$$\tilde{x}_i = \frac{2x_i}{q_i - p_i} - \frac{q_i + p_i}{q_i - p_i} \qquad i = 1 \text{ to } n. \tag{5.1.53}$$

It is seen that (5.1.53) maps each variable to the range $-1 \le \tilde{x}_i \le +1$. In general, the ith elements of the respective transformation matrix $\mathbf{D}$ and translation vector c are

$$d_{ii} = 2(q_i - p_i)^{-1}, \qquad c_i = \frac{p_i + q_i}{p_i - q_i}. \tag{5.1.54}$$

The range of the variable previously considered will map to the range -1.000000 to $+1.000000$ from the values 900.1234 to 900.4321, respectively, using $d_{ii} = 6.478782$ and $c_i = -5832.703$ as obtained by computing with seven significant digits. Clearly, the result of translation as well as transformation is a substantial improvement over transformation alone. Of course, the first derivatives and any explicit second derivatives must be rescaled according to (5.1.41) and (5.1.43), respectively. The squared effect of scaling on the Hessian indicated by (5.1.43) shows that even moderate scaling factors can produce drastic changes in the convergence rate of Newton-like optimization algorithms.

Simple nonlinear transformations that change the nature of the problem (from bounded variables to unconstrained) or that equalize partial derivatives may be employed. For example, the transformation $v = x^2$ for optimization in the x space will maintain $v \ge 0$. Similarly, $v = \sin(x)$ maintains $-1 \le v \le +1$. Transformations of variables and derivatives are required at all interfaces between the optimization algorithm and the subroutines that define the objective function and its derivatives. Details and a concrete example of this procedure in a BASIC program have been given by Cuthbert (1983:73). Some graphic illustrations of distortion in the variable space are contained in Wright (1976:73); see problem 1.8. In light of the highly developed and successful treatment of linear constraints by projection as described in Section 5.4.1, nonlinear transformations that impose bounds on variables are not recommended. Nonlinear transformations are easy to incorporate into computer programs but often cause severe illconditioning in the resulting unconstrained minimization that is totally unnecessary.

5.2. Line Searches

Line searches are required because general nonlinear functions are not quadratic, the variable values are not close to a minimum, and the Hessian matrix is unknown. There are only a few concerns in line searches once a suitable search direction has been obtained. The early search algorithms employed conjugate gradient directions so that exact line searches were required to maintain the theoretical hereditary property and practical performance. The more recent BFGS update formula has delivered superior results without exact line searches. In order to compare effects of inaccurate and more nearly perfect line searches when using the BFGS update formula, cutback (backtracking), quadratic interpolation, and cubic interpolation line search strategies are explored.

The cutback strategy is the method employed in both the NEWTON and LEASTP optimization programs previously developed. It simply tests for a decrease in function value for a full Newton step, and the step length is reduced by a factor of 4 whenever that test fails. Its simplicity is attractive when constraints are present in that the step easily can be modified so that limits of constraint feasibility are not violated.

Both the quadratic and cubic interpolation line search techniques assume that a minimum has been bracketed and greater accuracy is desired. The virtue of quadratic interpolation is that only three function values along the line allow a quadratic fit and prediction of a minimum without requiring derivatives. This is attractive because it has been found that the BFGS update formula in a quasi-Newton algorithm is as good an optimizer using finite differencing as any direct-search optimizer that uses no derivatives whatever.

From Davidon's first quasi-Newton optimization algorithm onward, cubic interpolation has been employed when "exact" line searches were desired. It is constructed using two points on the line for which both function values and gradient vectors are available. These line searches also employ the gradient to bracket the minimum before interpolation, so cubic interpolation makes sense when the gradient calculation is not too expensive.

The approach in this section is to discuss these three line search possibilities in the context of the specific optional segments of code in optimizer program QNEWT to follow. Therefore, comparative test results are presented. Respected line search algorithms that employ quadratic and cubic interpolation techniques are based on concepts advocated by Fletcher (1972).

The class of direct sequential line search methods, especially the Golden and Fibonacci search methods, will not be discussed here in spite of their popularity in the late 1960s. They are based on a systematic accumulation of function value information, each additional sample narrowing the range of interest for unimodal (single minimum) line functions. At best, they provide the accuracy of cubic interpolation without requiring derivatives. Since the 1960 era, however, accurate line searches have become far less important. The interested reader is referred to Wilde (1967:230).

5.2.1. *The Cutback Line Search.*

All line searches employed in descent algorithms are based on the negative slope condition existing at the starting (turning) point $\mathbf{x}$:

$$\mathbf{g}^T\mathbf{s} < 0. \tag{5.2.1}$$

Any point x^* along the line in the direction s measured by scalar t is:

$$x^* = x + ts. \qquad (5.2.2)$$

By the definition of a derivative, there is some small value of t such that the function value is reduced, that is, $F(x^*) < F(x)$. However, Newton-like methods are based on steps $d = x^* - x = -ts$ where s satisfies

$$s = -B^{-1}g, \qquad (5.2.3)$$

and B approximates the Hessian matrix. Near a minimum the step metric t equals 1.

The most rudimentary line search is the one used in programs NEWTON and LEASTP in Chapter Four: An initial step metric $t = 1$ is assumed, and if $F(1) > F(0)$, then t is divided by 4 until either the function is reduced from its value at the turning point or this "cutback" has been performed 10 times $(4^{10} = 1.05\text{E}6)$.

The search direction used in this chapter is the quasi-Newton BFGS update formula (5.1.22). Unless it is known that the line search is conducted over a range of t where only positive curvature exists, it cannot be guaranteed that the estimate B^* of the Hessian matrix will remain positive definite. In the present case the LDL^T factorization of B^* and the two rank 1 updates will test and force positive definiteness as described in Section 3.1.1. Since B will thus be positive definite, substitution of (5.2.3) into (5.2.1) shows that the slope at the beginning of the line search will always be negative, that is, downhill in the s direction.

Figure 5.2.1 shows the "cutback" algorithm to be used later in optimizer QNEWT. The four-digit line numbers correspond to the BASIC statements in the computer code. The information available at the beginning of any line search includes the variables in x, the function value $F(x)$ at that point, the gradient vector g at that point, and the search direction vector s.

All update formulas in the Broyden family employ the differences in the x and g vectors from turning point to turning point; see the definitions for d and y in (5.1.8) and (5.1.9), respectively. Therefore, the first operations in the cutback algorithm are to compute the slope and save the turning point and gradient. Step $t = 1$ is set initially, and a trial step is taken at line number 1770 in Figure 5.2.1. The line search is ended if the function decreased, otherwise t is replaced by $t/4$ and a shorter step is again tried from the beginning turning point. The algorithm is terminated if $t < 10^{-6}$ because of ten successive cutbacks.

The BASIC program code that implements the cutback line search is contained in lines 1650 to 1870 of program C5-1, QNEWT. Its behavior is discussed in Section 5.3.2.

5.2.2. Quadratic Interpolation Without Derivatives. Experience has shown that finding only an approximate minimum along a line in variable space and

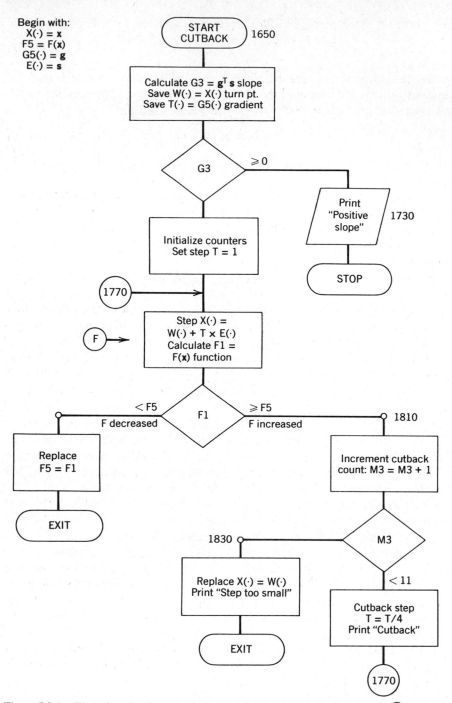

Figure 5.2.1. Flow chart for the cutback line search in optimizer QNEWT. *Note*: Ⓕ applies to BOXMIN. See Table 5.4.2.

250

then updating the estimate of the Hessian matrix (or its inverse) requires fewer total function evaluations than accomplishing quite precise line searches. However, it could be argued that the preceding cutback algorithm may be too crude and ignores some essential characteristics of both the function and the update formula. A compromise is to use various strategies based on only functions values and not derivatives except at the initial turning point. Historically, this kind of line search has employed a quadratic interpolation of a bounded (bracketed) minimum on the line, using three function values. Since derivatives are not required except to determine the required negative slope at the start, optimizers utilizing this line search method are candidates for minimizing functions whose gradient is not known explicitly.

Fletcher (1972) has developed a surprisingly sophisticated line search based solely on the assumption that the function along the line is quadratic and that derivatives along the line are not available except at the start. Thus, it is worthwhile to derive the three main issues: (1) an initial step size, (2) an extrapolation procedure when the first step decreases the function value, and (3) a quadratic interpolation procedure when the first step increases the function value.

From (5.2.2) and (5.2.3), it was noted that ultimately the line search metric t is unity for Newton-like optimizers. Also, the function $F(\mathbf{x})$ in the search direction $\mathbf{s}$ is in fact a function $F(t)$ of the scalar line metric t according to (5.2.2). For analysis the line metric is normalized by defining

$$q = \frac{t}{t^*}, \qquad (5.2.4)$$

where t^* is the length of the first step in the line search (5.2.2). An expression for t^* is developed such that $t = t^*$ is the minimum of a *quadratic* function $F(t)$ in the direction $\mathbf{s}$. Thus $q = 1$ is comparable to $t = t^*$, and the quadratic function to be analyzed is

$$F(t) = f(q) = f_q = a_0 + a_1 q + a_2 q^2. \qquad (5.2.5)$$

Of course, the actual function will not be quadratic, so Figure 5.2.2 illustrates the case when the first step ($q = 1$) produces a reduction in function value but not a minimum. The point in $\mathbf{x}$ space where the current line search starts corresponds to $q = 0$, $q = 1$ represents the first step, and the actual minimum along the line occurs at $q = q_m$.

Previously, (3.2.19) was obtained as an expression for t^* in terms of the gradient vector and the Hessian matrix. Since it is now assumed that these data are not available, an alternative expression for t^* is developed that uses only function values. The first derivative of $f(q)$ in (5.2.5) is

$$f'(q) = a_1 + 2a_2 q. \qquad (5.2.6)$$

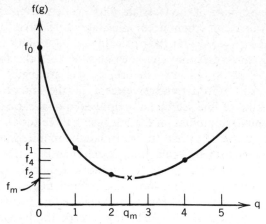

Figure 5.2.2. A general line search function versus normalized line search metric q for the case when the first step produces a decrease in function value.

Similarly, the second derivative of $f(q)$ is

$$f''(q) = 2a_2. \tag{5.2.7}$$

The necessary and sufficient conditions for a minimum are that $f'(q) = 0$ and $f''(q) > 0$; from (5.2.6) and calling that point $q = q_m$,

$$q_m = \frac{-a_1}{2a_2}. \tag{5.2.8}$$

Then the function value at the minimum of a quadratic function is

$$f(q_m) = a_0 - \frac{a_1^2}{4a_2}. \tag{5.2.9}$$

Fletcher observed that the decreases in function values from one iteration to the next were fairly constant when not close to a minimum. Therefore, it is assumed that $F^{(k+1)} - F^{(k)} = F^{(k)} - F^{(k-1)}$, where $F^{(k-1)}$ was the function value at the beginning of the previous iteration and $F^{(k)} = f(0)$ is the function value at the beginning of the current (kth) iteration. Using (5.2.5) through (5.2.9), it may be confirmed that

$$F^{(k+1)} - F^{(k)} = f(q_m) - f(0) = \frac{q_m f'(0)}{2}. \tag{5.2.10}$$

Since q is normalized according to (5.2.4) where $q = 1$ is intended to represent a minimum on a quadratic function, the unnormalized line search metric

$t = t^*$ that estimates the minimum is

$$t^* = \frac{2[F^{(k)} - F^{(k-1)}]}{g^{(k)T}s^{(k)}}.$$ (5.2.11)

The denominator in (5.2.11) is the negative slope $F'(0)$. The initial step size for the line search that employs quadratic interpolation is t^* with the constraint that $t^* \le 1$.

When the function value at the first step is less than that at the turning point ($f_1 < f_0$ in Figure 5.2.2), then Fletcher (1972) takes a second step of the same size. If the related function value increases ($f_2 > f_1$), then the line search is terminated at $q = 1$. That is not the case illustrated in Figure 5.2.2: the three successive points have decreasing function values, and a quadratic function can be found that passes through those three points. Fletcher is willing to double the step size (extrapolate) so that the next sample occurs at $q = 4$ *unless* the minimum predicted by a quadratic function falls in the range $2 \le q < 4$ as shown in Figure 5.2.2.

There is a surprisingly simple test for $2 \le q_m \le 4$. The three consecutively decreasing function values on a quadratic function indicate that the minimum $q_m > 2$. Consider the inequality

$$7f_2 + 5f_0 > 12f_1,$$ (5.2.12)

where the function values f_q correspond to $f(q)$ on a quadratic function defined by (5.2.5) that passes through $q = 0$, $q = 1$, and $q = 2$. For $q = 0$:

$$f_0 = a_0.$$ (5.2.13)

Similarly,

$$f_1 = a_0 + a_1 + a_2,$$ (5.2.14)

and

$$f_2 = a_0 + 2a_1 + 4a_2.$$ (5.2.15)

Substitution of (5.2.13) through (5.2.15) into (5.2.12) yields

$$\frac{-a_1}{2a_2} < 4.$$ (5.2.16)

However, by (5.2.8), the lefthand side of (5.2.16) is equal to the minimum predicted by the quadratic. It is concluded that satisfaction of the inequality in (5.2.12) guarantees that $q_m < 4$, in which case q will not be extrapolated beyond $q = 2$.

Summarizing the situation illustrated in Figure 5.2.2, the first step ($q = 1$) decreased the function value, so the step was repeated and $q = 2$ decreased the function value again. It was then clear that a quadratic would have its minimum to the right of $q = 2$, so the step reference position is moved to $q = 2$. If the quadratic prediction is that $q_m < 4$, then the line search is terminated. Otherwise, $q = 4$ is tested; if $f_4 \geq f_2$, then the line search is also terminated. But if $f_4 < f_2$, then the same series of tests that involved $q = 0, 1$, and 2 is repeated with $q = 0, 2$, and 4. The result of this strategy is that the line search is extrapolated just as long as quadratic predictions indicate that downhill progress is both likely and subsequently confirmed by testing.

The left-hand side of the flow chart for the line search without derivatives in Figure 5.2.3 implements the decisions just described for the case when the first step from the turning point produces a decrease in function value $F1$. The interested reader can review the preceding three paragraphs to follow the programming sequence instead of the analytical development based on quadratic functions.

The right-hand side of Figure 5.2.3 implements a strategy when the first step from the turning point increases the function value $F1$. The step size is halved when the first step from the turning point increases the function value, as illustrated by f_1 in Figure 5.2.4. Then the function value $f_{.5} = f(.5)$ is obtained; if the function value is less than that at the turning point ($f_{.5} < f_0$), the line search is terminated. However, if the $f_{.5} > f_0$, then a quadratic interpolation is considered if the function is convex. Convexity is assured if $f_{.5}$ is less than a linear interpolation between f_0 and f_1; see the dashed line in Figure 5.2.4. If the function passing through $q = 0, 0.5$, and 1 is not convex, the step size is reduced by a factor of 10 and the next function sample is obtained.

The prediction of the minimum of a quadratic function passing through $q = 0, 0.5$, and 1 can be expressed in terms of the three function values, two of which are expressed by (5.2.13) and (5.2.14). It is not difficult to verify that the q coordinate *after q is halved* turns out to be

$$q_m = 1 + \frac{f_0 - f_1}{2(f_0 + f_1 - 2f_{.5})}, \qquad (5.2.17)$$

which is equivalent to (5.2.8). If this value of q is less than 0.1, then the step size is reduced by a factor of 10.

The right-hand side of the flow chart in Figure 5.2.3 indicates that BASIC variable $I3$ counts the number of interpolations. The center of Figure 5.2.3 contains a test so that only three interpolations are allowed because of failures to obtain a reduction in function value. Usually, this condition is obtained only a the end of the final iteration when the optimization procedure has converged. Also, the test of $I3$ on the lefthand side of Figure 5.2.3 prevents an extrapolation once an interpolation has occurred. The BASIC program code that implements the line search without derivatives using quadratic interpolation is C5-2, LINQUAD. Its use is described in Section 5.3.2.

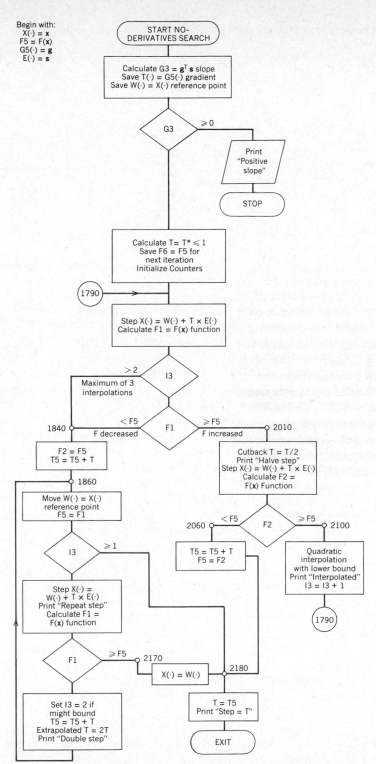

Figure 5.2.3. Flow chart for the optional quadratic interpolation line search in optimizer QNEWT.

255

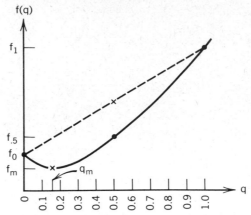

Figure 5.2.4. A general line search function versus normalized line search metric q for the case when the first step produces an increase in function value.

5.2.3. Cubic Interpolation Using Derivatives.

Finding a minimum of the nonlinear function $F(t)$ is equivalent to solving the nonlinear equation $F'(t) = dF/dt = 0$, certainly not a trivial task. Clearly, knowledge of derivative values at various trial solutions is necessary if such solutions are to be qualified. Otherwise, fairly simple examples may be constructed to show that the condition $F^{(k+1)} < F^{(k)}$, $k = 1, 2, \ldots$, does not guarantee convergence to a minimum, only to an *accumulation point* (limit point of a subsequence). See Dennis (1983:118) for examples. Several conventional considerations are described in this section as found in most recent articles and books on this subject, and an algorithm based on Fletcher (1972) is described.

A general nonlinear function $F(t)$ is shown in Figure 5.2.5. Two important goals in any linear search are to avoid estimated solutions near the extremes where $F(t) \doteq F(0)$, that is, when $t \doteq 0$ or at point $t \doteq d$ in Figure 5.2.5. The latter trivial solution may be avoided by requiring that the average rate of decrease from $F(0)$ to $F(t)$ exceed some specified fraction of the initial rate of decrease:

$$F(t) \le F(0) + r\mathbf{g}^T\mathbf{s}, \qquad (5.2.18)$$

where gradient $\mathbf{g} = \mathbf{g}(0)$ and $0 < r < 1$. The dashed lines passing through $F(0)$ with slopes zero and $F'(0)$ in Figure 5.2.5 define the limits of (5.2.18) for $r = 0$ and $r = 1$, respectively. For example, $r = \frac{1}{2}$ in requirement (5.2.18) would result in $t < c$.

The trivial solution where step t is nearly zero can be avoided by requiring that the inequality

$$F'(t) > F'(0) \qquad (5.2.19)$$

is satisfied by some margin, keeping in mind that $F'(0) < 0$. A point where $F'(t) = F(0)$ is shown at $t = a$ in Figure 5.2.5.

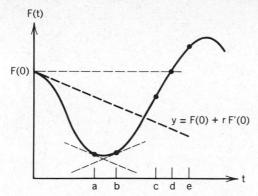

Figure 5.2.5. A general line search function versus line search metric t showing points defined using derivatives.

A single requirement that avoids both problems just described is

$$|F'(t)| \leq -hF'(0), \qquad 0 < h < 1. \tag{5.2.20}$$

Figure 5.2.5 illustrates (5.2.20) with $h = 1$ on the range $a < t < b$, which defines a "bowl" that necessarily contains a minimum of $F(t)$. An equivalent statement of (5.2.20) is the convergence test

$$\frac{\left|\left[\mathbf{g}^{(k+1)}\right]^T\mathbf{s}\right|}{\left|\left[\mathbf{g}^{(k)}\right]^T\mathbf{s}\right|} < h, \qquad 0 < h < 1. \tag{5.2.21}$$

Because (5.2.21) enforces the inequality in (5.2.19), it guarantees that

$$\mathbf{y}^T\mathbf{s} > 0, \tag{5.2.22}$$

where $\mathbf{y}$ is defined by (5.1.9) as the difference between the new and the old gradient values. The significance of (5.2.22) is that it contributes to a positive-definite update in the BFGS search direction formula, where it appears in the second term of (5.1.22). Small values of h such as $h = 0.1$ will force accurate line searches; however, $h = 0.9$ for inexact line searches is often employed with the especially tolerant BFGS update formula. The line search may be terminated for any search metric t in (5.2.2) that satisfies (5.2.22).

The search algorithm using derivatives begins with an estimate of t according to (5.2.11), but limited to $t \leq 1$. The flow chart in Figure 5.2.6 shows that the convergence test in (5.2.21) is performed if $F(t) < F(0)$. If (5.2.21) fails and if slope $F'(t) > 0$ as at point $t = e$ in Figure 5.2.5, then it is assumed that the minimum has been bounded (bracketed) so that interpolation is required. Otherwise, the slope is even steeper (more negative) than at the turning point so that extrapolation is appropriate. Since the model quadratic function has

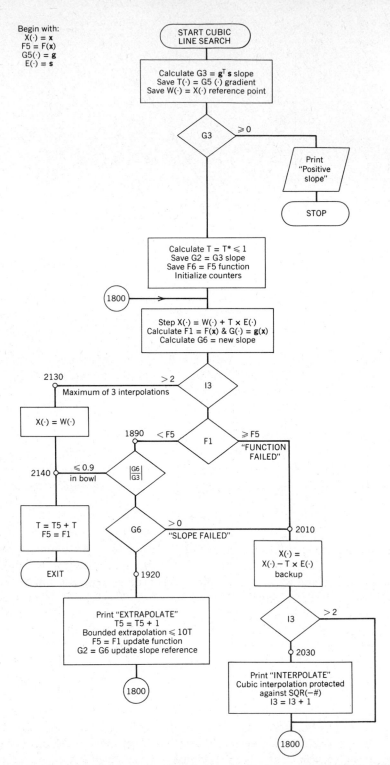

Figure 5.2.6. Flow chart for the optional cubic interpolation line search in optimizer QNEWT.

linear slope, the slope is linearly extrapolated to zero to determine a new step size t, but t is limited to an increase not greater than a factor of 10.

If the minimum has been bounded by encountering a large positive slope or by $F(t) \geq F(0)$, then cubic interpolation is employed as indicated on the right-hand side of Figure 5.2.6. A cubic function

$$y(t) = y_t = b_0 + b_1 t + b_2 t^2 + b_3 t^3 \qquad (5.2.23)$$

can be determined by the four values $y_0 = y(0)$, y_e, y_0', and y_e', where $0 \leq t \leq e$ and e is some positive upper limit for line search metric t. Davidon (1959:10) gave a series of equations for the minimum of a cubic interpolating function. A series of simple substitutions will confirm that the first derivative of (5.2.23) can be written as

$$y' = y_0' - \frac{2t(y_0' + z)}{e} + \frac{t^2(y_0' + y_e' + 2z)}{e^2}, \qquad (5.2.24)$$

where the defined constant z is

$$z = \frac{3(y_0 - y_e)}{e} + y_0' + y_e'. \qquad (5.2.25)$$

Then the minimum occurs for the zero of (5.2.24) in the range $0 < t < e$:

$$t_m = e(1 - c), \qquad (5.2.26)$$

where a second defined constant is

$$c = \frac{y_e' + Q - z}{y_e' - y_0' + 2Q}, \qquad (5.2.27)$$

and a third defined constant is

$$Q = \left(z^2 - y_0' y_e'\right)^{1/2}. \qquad (5.2.28)$$

The peculiar forms given were chosen to minimize the effects of roundoff errors because of subtraction of nearly equal quantities.

The BASIC program code that implements the cubic line search using derivatives is C5-3, LINCUBIC. Its use is described in Section 5.3.2.

5.3. Program QNEWT

Program C5-1, QNEWT, provides an illustration of a quasi-Newton method using the rank 2 BFGS update to the estimated Hessian matrix. As with the preceding optimizers, program QNEWT starts with a short menu to make certain initial choices of parameters and variables. The objective function and its gradient are calculated in subroutines 5000 and 7000, respectively. These must be merged into QNEWT and are interchangeable with those for program NEWTON. Therefore, the subroutines in C4-2, ROSEN, and C4-3, WOODS, may be used with QNEWT without modification. Additional test problems for QNEWT are introduced in Section 5.3.2. Program QNEWT is an important vehicle for adding the ability to constrain solutions in all the important ways.

5.3.1. The Algorithm and Its Implementation.

The listing for program C5-1, QNEWT, is contained in Appendix C. As noted by the remarks in lines 120 through 330, there are certain major programming names associated with the mathematics described in this chapter. The vector of variables **x** is contained in array $X(\ )$. The objective function $F(\mathbf{x})$ is named F when newly computed and $F5$ when saved as a preceding value. The gradient vector of first derivatives, $\nabla F = \mathbf{g}$, is contained in $G(\ )$ when newly computed and saved in vector array $G5(\ )$ for later use in computations after a new gradient has been computed. The updated approximation to the Hessian matrix is stored in vector form in array $H(\ )$. This corresponds to the $\mathbf{LDL}^T$ factorization method described in Section 3.1.1 and used in programs NEWTON and LEASTP. In QNEWT, the estimated Hessian is updated with a rank 1 matrix twice each iteration using a vector stored in $B(\ )$. The last major variable is $E(\ )$, which contains the quasi-Newton search direction vector **s**.

A complete list of variable names employed in QNEWT is appended to the program listing so that the user will not violate previous naming assignments when supplying subroutines 5000 and 7000. In general, function and gradient subroutines 5000 and 7000 are not called from loops or involved in current use of integer variables I, J, K, or L, so that these may be employed by the user. Notice that the BASIC function FNACS( ) is defined on line 370 for computing the inverse cosine.

The dimensions of arrays of program variables and other quantities are set to 20 in line 410, so that up to 20 optimization variables can be accommodated in the merged objective and gradient subroutines. The exception to this dimension is for the vector $H(\ )$ that stores the symmetric Hessian matrix in vector form as previously described in Section 3.1.1. The dimension of $H(\ )$ must be equal to $n(n + 1)/2$; when $n = 20$, the dimension of $H(\ )$ is 210 as shown in line 410. There are two arrays $W(\ )$ and $T(\ )$ used to store working data. Users can reduce the memory required for execution of QNEWT by reducing these dimensions to fit their particular application.

A list of major subroutines and their line numbers in program QNEWT is given in Table 5.3.1. In addition to the preceding structure, program QNEWT has a menu scheme similar to that in program C2-1, MATRIX, and in optimizers C4-1, NEWTON, and C4-5, LEASTP.

Table 5.3.2 shows the information displayed on the screen initially and during menu choices 1 and 2. The "NOTES" are similar to those used

Table 5.3.1. Major Subroutines in Optimizer Program C5-1, QNEWT

Name	Lines
Enter Number & Value of Variables	1200–1260
Enter/Revise Control Parameters	1280–1350
Main Optimization Algorithm (Figure 5.3.1)	1400–2750
Display Function, Gradient, and Variables	2770–2840
Store Unit Matrix in Hessian $H(\;)$	2860–2960
LDL^T Factorization of Hessian in situ in $H(\;)$	2980–3240
Solution for s in $Hs = -g$ (Newton step)	3260–3480
Rank 1 Update of Hessian $H(\;)$	3500–3830
Objective Function $F(x)$ (user supplied)	5000–6999
Gradient Vector $\nabla F = g$ (user supplied)	7000–8999

Table 5.3.2. Screen Displays for Notes and Menu Operation for Optimizer QNEWT

```
********* QNEWT OPTIMIZER *************

NOTES:
1.   USE ONLY UPPER CASE LETTERS
2.   IF 'BREAK' OCCURS, RESTART WITH 'GOTO 999'
3.   USER MUST PROVIDE SUBROUTINE 5000 FOR FUNCTION VALUE
          AND SUBROUTINE 7000 FOR THE GRADIENT VECTOR.
4.   ENTER DEFAULT ANSWERS TO QUESTIONS BY <RETURN>.

PRESS <RETURN> KEY TO CONTINUE -- READY?
************* COMMAND MENU ************
1. ENTER STARTING VARIABLES (AT LEAST ONCE)
2. REVISE CONTROL PARAMETERS (OPTIONAL)
3. START OPTIMIZATION
4. EXIT (RESUME WITH 'GOTO 999')
5. SPARE
************************************
INPUT COMMAND NUMBER:? 1
NUMBER OF VARIABLES = ? 2
ENTER STARTING VARIABLES X(I):
    X( 1 )=? -1.2
    X( 2 )=? 1
PRESS <RETURN> KEY TO CONTINUE -- READY?
************* COMMAND MENU ************
1. ENTER STARTING VARIABLES (AT LEAST ONCE)
2. REVISE CONTROL PARAMETERS (OPTIONAL)
3. START OPTIMIZATION
4. EXIT (RESUME WITH 'GOTO 999')
5. SPARE
************************************
INPUT COMMAND NUMBER:? 2
MAXIMUM # OF ITERATIONS (DEFAULT=50):? 75
STOPPING CRITERION (DEFAULT=.0001):?
PRINT EVERY Ith ITERATION (DEFAULT=1):? 10
PRESS <RETURN> KEY TO CONTINUE -- READY?
```

previously in major programs, especially the recovery method after a ⟨Ctrl⟩⟨Break⟩ or EXIT, when ⟨GOTO 999⟩⟨Rtn⟩ will place the program back into menu selection without resetting any program variables. Menu choice 1 sets the number and value of optimization variables for the corresponding subroutines 5000 and 7000 supplied by the user. Menu choice 2 sets the three program parameters: (1) maximum number of iterations, (2) stopping criterion, and (3) screen printing interval for iteration results. All three parameters have default values as shown in the lower lines of Table 5.3.2; these have been set in line 380 so that menu choice 2 need not be exercised unless changes are desired. Also, after optimization (menu choice 3), menu choices 1 and 2 do not have to be executed. This can be useful when selecting choice 3 again to continue optimization (with a reset iteration count).

The flow chart for program QNEWT in Figure 5.3.1 has a structure similar to the flow chart in Figure 1.3.1 for a generic iterative process. The four-digit numbers in Figure 5.3.1 correspond to the BASIC line numbers in the program C5-1, QNEWT, listed in Appendix C.

The initial estimate for the Hessian matrix is made before iterations and updates are made to improve that estimate. As indicated in Table 5.3.1, program lines 2860 to 2960 simply set the initial Hessian matrix to the unit matrix, which amounts to setting all mixed second partial derivatives to zero and the rest to unity. Convergence is obtained much sooner in many cases if a better initial estimate of the Hessian is provided. If the objective function has the least-pth structure, the Gauss–Newton positive-definite approximation to the Hessian as described in Section 4.4.2 could be furnished as a starting estimate. Other preliminary work includes $\mathbf{LDL}^T$ factorization of that estimate and saving the minimum element of $\mathbf{D}$ for use in forcing positive definiteness of the Hessian during factorization (subroutine 2980).

Reentry balloon 1580 is the starting point for each iteration or step in a search direction. At that time the iteration counter is incremented, and a status report is displayed. The major action in each iteration occurs in the line search portion of the optimizer in line numbers 1650 through 2299. Program C5-1 code contains the "cutback" or backtracking line search method described in Section 5.2.1 according to the flow chart in Figure 5.2.1. By using the operating system command MERGE"LINQUAD, the "cutback" line search algorithm is replaced entirely by that optional line search. Alternatively, the command MERGE"LINCUBIC will replace the "cutback" algorithm. See Section 5.2.2 and 5.2.3 and the flow charts in Figures 5.2.3 and 5.2.6, respectively, for details of these alternate line searches. Comparative test results will be given in Section 5.3.2.

The termination tests contained in QNEWT are the same as those employed in both NEWTON and LEASTP; see (1.3.22) and also lines 2660 to 2690 in program C5-1.

5.3.2. Some Examples Using Program QNEWT.

Rosenbrock's function described in Section 4.3.2 and programmed in subroutines 5000 and 7000 in C4-2 is used to display typical output from optimizer QNEWT. Figure 5.3.2

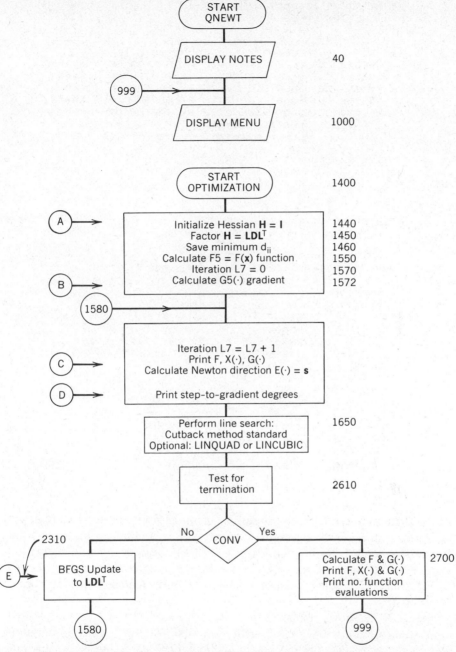

Figure 5.3.1. Flow chart for quasi-Newton optimizer featuring three different line searches and the BFGS update formula. *Note*: Ⓐ – Ⓔ apply to BOXMIN. See Table 5.4.2.

```
AT START OF ITERATION NUMBER 1
   FUNCTION VALUE = 24.2
  I          X(I)              G(I)
  1       -1.20000000    -215.60006561
  2        1.00000000     -88.00000000
     ###### CUT BACK STEP SIZE BY FACTOR OF 4 #####
     ###### CUT BACK STEP SIZE BY FACTOR OF 4 #####
     ###### CUT BACK STEP SIZE BY FACTOR OF 4 #####
     ###### CUT BACK STEP SIZE BY FACTOR OF 4 #####
     ###### CUT BACK STEP SIZE BY FACTOR OF 4 #####
AT START OF ITERATION NUMBER 2
   FUNCTION VALUE = 5.101115
  I          X(I)              G(I)
  1       -0.98945306      38.33806071
  2        1.08593750      21.38402805
                •
                •
                •
AT START OF ITERATION NUMBER 34
   FUNCTION VALUE = 3.845676E-07
  I          X(I)              G(I)
  1        0.99949366       0.01330447
  2        0.99895177      -0.00716059
               STEP-TO-GRADIENT DEGREES= 87.5188
AT START OF ITERATION NUMBER 35
   FUNCTION VALUE = 8.411157E-10
  I          X(I)              G(I)
  1        0.99998304       0.00087378
  2        0.99996373      -0.00047052
               STEP-TO-GRADIENT DEGREES= 87.5124
CONVERGED; SOLUTION IS:
AT START OF ITERATION NUMBER 36
   FUNCTION VALUE = 3.142477E-10
  I          X(I)              G(I)
  1        1.00001773       0.00008649
  2        1.00003543      -0.00000396
TOTAL NUMBER OF FUNCTION EVALUATIONS = 48
PRESS <RETURN> KEY TO CONTINUE -- READY?
```

Figure 5.3.2. Partial data for QNEWT optimization of ROSEN problem using "cutback" line search.

shows the first and the last two iterations using IBM interpreted BASICA with the cutback line search algorithm. Convergence was obtained in 35 iterations from the standard starting point, $x = (-1.2 \ 1)^T$, and 48 function evaluations were required. Because the gradient vector was not required during the line search, only 35 gradient evaluations were required (equal to the number of iterations).

Figure 5.3.3 shows similar output when MERGE"LINQUAD (C5-2) was employed to replace the cutback method in QNEWT, C5-1. Only 34 iterations were required using the line search without derivatives but with quadratic interpolation. However, 73 function evaluations were required, along with the 34 evaluations of the gradient used to begin each of the 34 iterations.

Figure 5.3.4 shows comparable output for the Rosenbrock function when derivatives were used in the line search as well as cubic interpolation (program

```
AT START OF ITERATION NUMBER 1
   FUNCTION VALUE = 24.2
   I          X(I)            G(I)
   1       -1.20000000    -215.60006561
   2        1.00000000     -88.00000000
                   STEP-TO-GRADIENT DEGREES=  0.0000
                                   REPEAT STEP
                                   DOUBLE STEP
                                   REPEAT STEP
                                   DOUBLE STEP
                                   STEP=     0.000357
AT START OF ITERATION NUMBER 2
   FUNCTION VALUE = 9.786829
   I          X(I)            G(I)
   1       -1.12302750    -107.46271109
   2        1.03141734     -45.95468745
                 •
                 •
                 •

AT START OF ITERATION NUMBER 33
   FUNCTION VALUE = 7.034565E-07
   I          X(I)            G(I)
   1        1.00012067       0.03338793
   2        1.00015836      -0.01659993
                   STEP-TO-GRADIENT DEGREES= 84.2245
                                   REPEAT STEP
                                   STEP=     1.000000
AT START OF ITERATION NUMBER 34
   FUNCTION VALUE = 7.350403E-09
   I          X(I)            G(I)
   1        0.99993236       0.00195709
   2        0.99985945      -0.00105351
                   STEP-TO-GRADIENT DEGREES= 87.4670
                                   REPEAT STEP
                                   STEP=     1.000000
CONVERGED; SOLUTION IS:
AT START OF ITERATION NUMBER 35
   FUNCTION VALUE = 1.114715E-11
   I          X(I)            G(I)
   1        1.00000072       0.00012145
   2        1.00000112      -0.00006519
TOTAL NUMBER OF FUNCTION EVALUATIONS = 73
PRESS <RETURN> KEY TO CONTINUE -- READY?
```

Figure 5.3.3. Partial data for QNEWT optimization of ROSEN problem using LINQUAD line search.

segment LINCUBIC merged). In this run, 40 iterations and 51 function and gradient evaluations were required. The reports from within the line search section of program QNEWT indicate the extrapolation and interpolation activity.

No single set of runs gives a valid indication of performance of an algorithm. A series of runs from different starting points is more useful.

```
AT START OF ITERATION NUMBER 1
   FUNCTION VALUE = 24.2
 I        X(I)            G(I)
 1     -1.20000000    -215.60006561
 2      1.00000000     -88.00000000
             ●                   STEP=    0.000089
             ●
             ●
                      STEP-TO-GRADIENT DEGREES= 67.8562
                                   EXTRAPOLATE
                                   FNCN FAILED
                                   INTERPOLATE
                                   STEP=    6.728138
AT START OF ITERATION NUMBER 7
   FUNCTION VALUE = 4.093465
 I        X(I)            G(I)
 1     -0.99934996     -16.38834085
 2      0.96770605      -6.19886045
                   STEP-TO-GRADIENT DEGREES= 86.8944
                                   FNCN FAILED
                                   INTERPOLATE
             ●                     STEP=    0.324191
             ●
             ●

AT START OF ITERATION NUMBER 39
   FUNCTION VALUE = 1.30531E-07
 I        X(I)            G(I)
 1      0.99984479      0.01273622
 2      0.99965697     -0.00652502
                      STEP-TO-GRADIENT DEGREES= 87.1888
                                   STEP=    1.000000
AT START OF ITERATION NUMBER 40
   FUNCTION VALUE = 6.255936E-11
 I        X(I)            G(I)
 1      0.99999491      0.00025903
 2      0.99998922     -0.00012110
                   STEP-TO-GRADIENT DEGREES= 87.5826
                                   FNCN FAILED
                                   INTERPOLATE
                                   FNCN FAILED
                                   INTERPOLATE
                                   FNCN FAILED
                                   INTERPOLATE
                                   STEP=    0.027644
CONVERGED; SOLUTION IS:
AT START OF ITERATION NUMBER 41
   FUNCTION VALUE = 6.255936E-11
 I        X(I)            G(I)
 1      0.99999491      0.00025903
 2      0.99998922     -0.00012110
TOTAL NUMBER OF FUNCTION EVALUATIONS = 51
PRESS <RETURN> KEY TO CONTINUE -- READY?
```

Figure 5.3.4. Partial data for QNEWT optimization of ROSEN problem using LINCUBIC line search.

Table 5.3.3. Performance of QNEWT on the Rosenbrock and Wood Functions Compared to Other Optimizers[b,c]

Starting Point	NEWTON[e]	LEASTP	QNEWT CUTBACK	QNEWT LINQUAD	QNEWT LINCUBIC
		Rosenbrock's Function			
−1.2, 1.0	24/31[a]	23/39	34/49	31/68	39/50
2.0, −2.0	22/28	3/4	47/67	32/76	29/35
−3.635, 5.621	43/48	6/10	34/46	53/125	57/73
6.39, −0.221	26/40	4/4	31/47	76/175	62/77
1.489, −2.547	16/20	4/4	41/56	36/77	21/25
		Wood's Function			
−3, −1, −3, −1	54/79	[d]	53/77	76/186	73/96
1.2, 1, 1.2, 1	24/28	[d]	38/59	36/81	42/53
−3, 1, −3, 1	52/75	[d]	56/81	76/179	84/104
−1.2, 1, −1.2, 1	44/64	[d]	73/108	68/164	70/92

[a] (Number iterations)/(number function evaluations).
[b] Standard termination (0.0001) except 0.000001 for Wood's function using LINQUAD and LINCUBIC.
[c] These data were obtained using IBM compiled BASIC. Data are somewhat different when using interpreted BASICA.
[d] Wood's function is not in the least-pth format.
[e] Number of function evaluation does not include those for finite differences.

Therefore, four more starting points used by several investigators were employed. Also, similar data for four variables was accumulated for the performance of QNEWT on Wood's function also described in Section 4.3.2. The results are tabulated in Table 5.3.3 for the three versions of QNEWT as well as for NEWTON (Table 4.3.4) and LEASTP (Chapter Four). Table 5.3.3 shows the impressive performance of LEASTP when advantage can be taken of the special structure of an objective function. Otherwise, Table 5.3.3 indicates that the three types of line searches performed roughly the same on the two test functions.

In particular, the BFGS search direction does not suffer from the coarse line searches that simply "cut back" the step length until a decrease in function value is obtained. The run illustrated in Figure 5.3.2 only "cut back" 11 times, mostly at the beginning as shown, and only sporadically thereafter. As for all Newton-like methods, the step size that begins each iteration approaches unity as a minimum is approached.

In fairness to the LINCUBIC line search, its accuracy is much more important to algorithms using the DFP search direction. Even though both the LINQUAD and LINCUBIC line search algorithms correspond closely to the ideas of Fletcher (1972), this particular implementation may be detrimental to performance in some unknown way.

5.3.3. Optimization Without Explicit Derivatives.   There are many situations where the user either cannot obtain exact values for derivatives or may not wish to do so. For instance, precision and efficiency may not matter if the problem to be optimized turns out to be senseless. The user may wish to try an initial formulation by providing only an objective function and letting the optimization program generate the gradient by finite differences; see (4.1.2).

There are some important limitations when using finite differences. The objective function must be continuous and smooth (at least the first and second derivatives must be continuous). Finite differences require n more evaluations of the objective function; this can increase the program execution time substantially for large values of n, the number of variables. Also, the finite difference approximations for the elements of the gradient are somewhat inexact so that final convergence to a minimum is noticeably inferior to the case using exact gradients.

Program C5-1, QNEWT, has been constructed to accept gradients approximated by finite differences with minimum degradation in overall performance. The primary consideration is to avoid requiring gradient information within the line search; both the "cutback" line search in QNEWT and the optional program C5-2 "LINQUAD" that can be merged with QNEWT do not require derivatives except at the turning points where the line searches begin. Program C5-3 "LINCUBIC" does require derivatives at every function evaluation and is not recommended when using the finite differencing option. Also, the rank 1 updates to the estimated Hessian matrix of second partial derivatives may be inexact because of errors caused by differencing so that the estimated Hessian may not remain positive definite as required for generation of downhill search directions. However, rank 1 update subroutine 3500 in QNEWT tests for positive definiteness and reports and aborts if violated. This seldom occurs in practice, and the program could easily be modified to refactorize the Hessian when that occurs, with positive definiteness forced by LDL^T factorizing subroutine 2980.

The means for relieving the user of providing explicitly defined first derivatives is simply to replace user-supplied subroutine 7000 with a universal version that performs finite differencing. Program C5-4 "QNEWTGRD", can be merged into QNEWT after the user's subroutine 5000 that defines the objective function. As seen from the brief listing of program C5-4 in Appendix C, it is only necessary to save the nominal value of the objective function, F, before performing a sequence of perturbations of each element of the variable vector x and calling subroutine 5000 for each combination.

The reader is urged to load program QNEWT followed by first merging ROSEN and then QNEWTGRD to run the standard Rosenbrock example from Section 4.3.2 using finite differencing for the gradient. Starting from $x = (-1.2 \ 1)^T$, the optimization terminates in 35 iterations, the same as when using exact first derivatives. However, 120 function evaluations are required as opposed to 48 with exact derivatives. That is not quite an increase of three times, but the optimization with finite differences terminates somewhat prematurely, with an optimum at $\mathbf{x}' = (0.97236820 \ 0.94545264)^T$ as opposed to $\mathbf{x}' = (1 \ 1)^T$, as obtained to five decimal places with exact derivatives and the same standard (0.0001) termination criterion.

Fletcher (1972) describes two quasi-Newton algorithms, VA09A similar to QNEWT and VA10A that does not require derivatives. VA10A differs from VA09A in the line search and the source of derivatives. Particularly, VA10A employs the forward difference approximation of (4.1.2) until near a minimum when it switches to the more accurate central difference approximation that requires both the forward and backward perturbations:

$$\nabla_i F(\mathbf{x}) = \lim_{dx_i \to 0} \frac{F(\mathbf{x} + dx_i \, \mathbf{e}_i) - F(\mathbf{x} - dx_i \, \mathbf{e}_i)}{2 \, dx_i}, \qquad (5.3.1)$$

where dx_i is an increment in the ith element of $\mathbf{x}$ and $\mathbf{e}_i$ is the ith unit vector [e.g., see (2.1.2)]. According to Maron (1982:285), the truncation error in the forward difference approximation is of order $(dx)^1$ while the central difference approximation is in error by order $(dx)^2$. Therefore, the minimum is located quite accurately by VA10A, again at the expense of even more function evaluations. It is not easy for the program to know when to switch differencing formulas; the switch is made in VA10A when the changes in the $\mathbf{x}$ vector are equal to the user's stopping criterion.

Readers should be aware that there are *direct-search algorithms* that also do not require derivatives and usually are not based on the theory of continuous quadratic functions and their gradients. Instead, many direct-search algorithms are based on systematic searches of the variable space $(\mathbf{x})$ using either *patterns* or *simplexes*. Patterns are generated by a sequence of exploratory moves in the coordinate directions with a set of policies based only on changes in objective function values. Simplexes in the direct-search context are n-dimensional polyhedrons whose size and distorted shape are determined by function values obtained at selected vertices. The *Hooke–Jeeves* algorithm is one of the better pattern optimizers, and the interested reader is referred to Himmelblau (1972:142) and Kuester (1973:309). The *Nelder and Mead* algorithm is one of the better simplex optimizers, also described by Himmelblau (1972:148) and Kuester (1973:298).

Direct-search algorithms are appropriate when derivatives cannot be defined, such as when random error is present in the objective function (due to measurement errors or to numerical integration in function definition, for example). Also, it is not difficult to incorporate logic into direct-search algorithms to accommodate both linear and nonlinear inequality constraints; see Section 1.2.1. Since derivatives are not required, the program changes are minimal, but convergence to constrained solutions is often slow and unreliable.

A respected direct-search algorithm by Powell (1964:155 and 1965:303) is based on a special property of quadratic functions. Consider parallel lines that are tangent to any two elliptical level curves of a quadratic function (see Figure 3.2.1): Another line through the two points of tangency always passes through the minimum point, that is, the center of the ellipses. In Figure 3.2.1, that fact is trivially true for a tangent line parallel to either the y_1 or y_2 axis. It is easy to construct other pairs of parallel lines at other inclinations to see that the *parallel tangency* phenomenon is generally true. Powell's direct-search algorithm does assume continuous and smooth functions, but derivative values are not required. Without going into further details, Powell's algorithm can be judged as an improvement on a sequence of one-dimensional line searches. The interested reader is referred to Dixon (1972b:74), Himmelblau (1972:167), and Kuester (1973:331).

Finally, *random-search algorithms* select trial x vectors based on the theory of probability; they are the least elegant and most inefficient methods of optimization. Random searches are especially ineffective in the neighborhood of a solution; however, they are useful when the objective function is discontinuous. Schrack (1972:137) tested three such algorithms that provided no better than linear convergence. However, hybrid combinations of random- and gradient-search algorithms have been employed to obtain successful starting values when only acceptable ranges of variables are known. The interested reader is referred to Himmelblau (1972:177).

The suggestion of using QNEWT with finite differences by QNEWTGRD has been dominated by the side issue of alternative direct-search methods. However, many experienced investigators have lately come to the conclusion that suitable structured gradient optimizers using finite difference approximations are more reliable and efficient than any of the direct methods. It is important that there is a substantial analytical basis for the design and use of gradient optimizers, and that is generally not true for direct-search methods.

5.4.   Constrained Optimization

Program QNEWT is an efficient and effective unconstrained minimizer of nonlinear scalar functions of many variables. In the most practical cases those variables may have lower and/or upper bounds. The variables also may be related among themselves in a more general linear fashion or often in a nonlinear fashion. To restate the general nonlinear programming problem described in Section 1.2.1:

$$\text{Minimize } f(\mathbf{x}) \text{ s.t. } \mathbf{h}(\mathbf{x}) = \mathbf{0} \text{ and } \mathbf{c}(\mathbf{x}) \geq \mathbf{0}, \qquad (5.4.1)$$

where the symbols "s.t." mean "such that." There are q equality constraint functions in vector $\mathbf{h}$ and $m - q$ inequality constraint functions in vector $\mathbf{c}$.

Methods for dealing with both equality and inequality constraints are developed. A projection method is derived for linear constraints and then specialized for simple lower and upper bounds in program segment BOXMIN that can be merged into optimizer QNEWT. Then penalty function methods are reviewed for enforcing all kinds of constraints; some historically important methods are evolved into the more effective multiplier penalty method. The program segment MULTPEN may be merged into the QNEWT-BOXMIN program to implement the powerful multiplier penalty method. Finally, brief descriptions of more recent but fairly complicated nonlinear constraint methods are provided.

After many years of research, it is now clear that almost all methods for dealing with constraints depend on the Lagrangian function in several important ways. Repeating some conclusions from Sections 3.3.2 and 3.3.3: The *classical Lagrangian function* is

$$L(\mathbf{x}, \mathbf{p}) = f(\mathbf{x}) - \mathbf{p}^T \mathbf{h}(\mathbf{x}). \qquad (5.4.2)$$

There are *q equality constraint functions* $h_1(\mathbf{x}), h_2(\mathbf{x}), \ldots, h_q(\mathbf{x})$, and *q* corresponding *Lagrange multipliers* $p_1, p_2, \ldots, p_q$. When $\nabla L(\mathbf{x}, \mathbf{p}) = \mathbf{0}$ with respect to all components of *both* $\mathbf{x}$ and $\mathbf{p}$, then it is known that there is a minimum of $f(\mathbf{x})$ with respect to $\mathbf{x}$ such that $\mathbf{h}(\mathbf{x}) = \mathbf{0}$. Let those values be $\mathbf{x}'$ and $\mathbf{p}'$. Then $L(\mathbf{x}, \mathbf{p}')$ is a minimum with respect to $\mathbf{x}$, and $L(\mathbf{x}', \mathbf{p})$ is a *maximum* with respect to $\mathbf{p}$.

What has just been stated and was shown by Example 3.3.4 in Section 3.3.2 is that the solution of an equality-constrained minimization problem is min-max; that is, it has a saddle point. To see why this is the case, again note that the necessary condition for a minimum of the Lagrangian function in (5.4.2) is the solution in $n + q$ variables of the set of nonlinear equations

$$\nabla L(\mathbf{x}, \mathbf{p}) = \mathbf{0}. \qquad (5.4.3)$$

It is useful in this analysis to consider that $\mathbf{x}$ is a function of $\mathbf{p}$, that is, $\mathbf{x}(\mathbf{p})$, in the sense that for any value selected for $\mathbf{p}$ the solution of (5.4.3) implicitly defines a value of $\mathbf{x}$. Assume for now that $\mathbf{x}(\mathbf{p})$ is the minimizer of $L(\mathbf{x}, \mathbf{p})$ for all possible values of $\mathbf{p}$. Then the inequality

$$L[\mathbf{x}(\mathbf{p}), \mathbf{p}] \leq L[\mathbf{x}', \mathbf{p}] = L[\mathbf{x}', \mathbf{p}'] \qquad (5.4.4)$$

is true by virtue of $\mathbf{h}(\mathbf{x}') = \mathbf{0}$ in (5.4.2). Thus, (5.4.4) is a degenerate form of a saddle point. These are sufficient conditions for (5.4.3) to represent the constrained minimum. These conditions involve the curvature (second derivatives) of both the objective and constraint functions and need to be considered for the analyses in this chapter.

Furthermore, at the constrained optimum point, the gradient vector of $f(\mathbf{x})$ is a linear combination of the gradient vectors of the constraint functions:

$$\nabla f(\mathbf{x}', \mathbf{p}') = \mathbf{g}(\mathbf{x}') = \mathbf{g}' = \mathbf{N}\mathbf{p}', \qquad (5.4.5)$$

where the Lagrange multipliers in $\mathbf{p}'$ may be positive or negative for *equality* constraints. The $n \times q$ matrix $\mathbf{N}$ is similar to the matrix of hyperplane normals in (2.2.88):

$$\mathbf{N} = (\nabla h_1 \ \nabla h_2 \ \cdots \ \nabla h_j \ \cdots \ \nabla h_q). \tag{5.4.6}$$

In this case $\mathbf{N}$ has columns that are the gradient vectors of the corresponding equality constraint functions. When $\mathbf{p}$ is strictly positive, (5.4.5) is the Kuhn–Tucker condition (3.3.22) for inequality constraints: The objective gradient vector $\mathbf{g}$ lies within the cone generated by the gradient vectors of the constraint functions. Also, Farka's lemma verifies that $\nabla f(\mathbf{x}', \mathbf{p}') = \mathbf{0}$ indicates a minimum of $L(\mathbf{x}, \mathbf{p}')$ because the slope or directional derivative is positive in every direction $\mathbf{s}$, that is, $\mathbf{g}^T \mathbf{s} > \mathbf{0}$.

Finally, the sensitivity interpretation of Lagrange multipliers is recalled from (3.3.15). Instead of the equality constraints in the vector $\mathbf{h}(\mathbf{x}) = \mathbf{0}$, consider perturbing the boundary of the ith constraint in $\mathbf{h}$ by the small scalar amount e:

$$h_i(\mathbf{x}) = e. \tag{5.4.7}$$

Then the ith Lagrange multiplier in the vector $\mathbf{p}$ is

$$p_i = \nabla_e f(\mathbf{x}'), \tag{5.4.8}$$

where $\mathbf{x}'$ is the solution vector for the undisplaced constraint ($e = 0$). The significance of (5.4.8) is that the sign of p_i determines the side of the constraint on which the objective function decreases. That is very useful for dealing computationally with *in*equality constraints.

5.4.1. Linear Constraints by Projection.   The linearly constrained problem is

$$\text{Minimize } f(\mathbf{x}) \text{ such that} \tag{5.4.9}$$

$$\mathbf{a}_i^T \mathbf{x} = b_i, \qquad i = 1 \text{ to } q, \tag{5.4.10}$$

$$\mathbf{a}_i^T \mathbf{x} \geq b_i, \qquad i = q + 1 \text{ to } m \leq n. \tag{5.4.11}$$

The objective function $f(\mathbf{x})$ is a nonlinear function, and the linear constraints are defined by sets of coefficients that are the elements of the vectors $\mathbf{a}_i$. Compared with the more general problem (5.4.1), it is seen that the ith equality constraint is $h_i(\mathbf{x}) = \mathbf{a}_i^T \mathbf{x} - b_i$, $1 \leq i \leq q$, and the ith inequality constraint is $c_i(\mathbf{x}) = \mathbf{a}_i^T \mathbf{x} - b_i$, $q + 1 \leq i \leq m < n$, where n is the number of variables. Recalling (2.2.82), it is seen that each equality constraint in (5.4.10) is a *hyperplane* and the coefficient vectors, $\mathbf{a}_i$, are their respective *normal vectors*. Similarly, each inequality constraint in (5.4.11) defines a *half-space* on the side of a hyperplane in the direction of its normal vector.

It is useful to recognize three realities at this time: (1) computations will require an assumed quadratic model for the objective function in the neighborhood of a current value of **x**, (2) it can be assumed that q is known, that is, the *binding constraints* are known (those for which the equality holds), and (3) there remain only $n - q$ degrees of freedom. Therefore, a problem that can be solved is the *quadratic programming problem*:

$$\text{Minimize } Q(\mathbf{x}) = \mathbf{x}^T\mathbf{d} + \tfrac{1}{2}\mathbf{x}^T\mathbf{H}\mathbf{x} \text{ such that} \tag{5.4.12}$$

$$\mathbf{A}^T\mathbf{x} = \mathbf{b}, \tag{5.4.13}$$

where vectors **d** and **x** are in E^n, **H** is the $n \times n$ Hessian matrix, and **b** is in E^q, $q \le n$. Also, the $n \times q$ matrix **A** has rank q and columns composed of vectors $\mathbf{a}_i$:

$$\mathbf{A} = (\mathbf{a}_1 \; \mathbf{a}_2 \; \cdots \; \mathbf{a}_q). \tag{5.4.14}$$

As Fletcher (1981a:80) has pointed out, *direct elimination* is an instructive way to solve (5.4.12) subject to (5.4.13) by using the q constraints to eliminate q of the variables. Suppose that the vectors and matrices are consistently partitioned for that purpose, using the subscript q to denote relationships for the first q dependent variables and subscript s for the remaining $n - q$ independent variables ordered as follows:

$$\mathbf{x} = \begin{bmatrix} \mathbf{x}_q \\ \mathbf{x}_s \end{bmatrix}, \qquad \mathbf{d} = \begin{bmatrix} \mathbf{d}_q \\ \mathbf{d}_s \end{bmatrix}, \tag{5.4.15}$$

$$\mathbf{A} = \begin{bmatrix} \mathbf{A}_q \\ \mathbf{A}_s \end{bmatrix}, \tag{5.4.16}$$

$$\mathbf{H} = \begin{bmatrix} \mathbf{H}_q & \mathbf{C} \\ \mathbf{C}^T & \mathbf{H}_s \end{bmatrix}. \tag{5.4.17}$$

Matrix $\mathbf{A}_q$ is $q \times q$, $\mathbf{A}_s$ is $s \times q$, $\mathbf{H}_q$ is $q \times q$, **C** is $q \times s$, and $\mathbf{H}_s$ is $s \times s$, where $s = n - q$. Then (5.4.13) can be expanded in terms of the partitions of **A** to yield

$$\mathbf{A}_q^T\mathbf{x}_q + \mathbf{A}_s^T\mathbf{x}_s = \mathbf{b}. \tag{5.4.18}$$

Solving for the dependent variables in $\mathbf{x}_q$:

$$\mathbf{x}_q = \mathbf{D} - \mathbf{B}^T\mathbf{x}_s, \tag{5.4.19}$$

where matrices **B** and **D** are defined as

$$\mathbf{B} = \mathbf{A}_s\mathbf{A}_q^{-1}, \qquad \mathbf{D} = \mathbf{A}_q^{-T}\mathbf{b}. \tag{5.4.20}$$

Superscript $-T$ denotes the combined matrix inverse and transpose operations.

The partitioned vectors and matrix in (5.4.15) and (5.4.17) are substituted into (5.4.12) in order to express $Q(\mathbf{x})$ as the equivalent

$$Q(\mathbf{x}_q, \mathbf{x}_s) = \mathbf{x}_q^T \mathbf{d}_q + \mathbf{x}_s^T \mathbf{d}_s + \tfrac{1}{2}\left(\mathbf{x}_q^T \mathbf{H}_q \mathbf{x}_q + \mathbf{x}_q^T \mathbf{C}\mathbf{x}_s + \mathbf{x}_s^T \mathbf{C}^T \mathbf{x}_q + \mathbf{x}_s^T \mathbf{H}_s \mathbf{x}_s\right).$$

$$(5.4.21)$$

Dependent $\mathbf{x}_q$ may be eliminated from (5.4.21) to obtain $Q(\mathbf{x}_s)$ by substituting (5.4.19) everywhere $\mathbf{x}_q$ appears in (5.4.21). After a considerable amount of algebra, the result is:

$$Q(\mathbf{x}_s) = \tfrac{1}{2}\mathbf{x}_s^T\left[\mathbf{H}_s - (\mathbf{BC}) - (\mathbf{BC})^T + (\mathbf{BH}_q)\mathbf{B}^T\right]\mathbf{x}_s$$

$$+\mathbf{x}_s^T\left[(\mathbf{C}^T - \mathbf{BH}_q)\mathbf{D} + (\mathbf{d}_s - \mathbf{Bd}_q)\right] + \mathbf{D}^T(\mathbf{d}_q + \tfrac{1}{2}\mathbf{H}_q\mathbf{D})$$

$$= \tfrac{1}{2}\mathbf{x}_s^T\tilde{\mathbf{H}}\mathbf{x}_s + \mathbf{x}_s^T\tilde{\mathbf{d}} + \tilde{c}. \qquad (5.4.22)$$

By comparisons in (5.4.22):

$$\tilde{\mathbf{H}} = \left[\mathbf{H}_s - (\mathbf{BC}) - (\mathbf{BC})^T + (\mathbf{BH}_q)\mathbf{B}^T\right],$$

$$(5.4.23)$$

$$\tilde{\mathbf{d}} = \left[(\mathbf{C}^T - \mathbf{BH}_q)\mathbf{D} + (\mathbf{d}_s - \mathbf{Bd}_q)\right].$$

A sufficient condition for a minimum of $Q(\mathbf{x}_s)$ is that the matrix $\tilde{\mathbf{H}}$ in (5.4.23) is positive-definite. The necessary condition for a minimum is that $\nabla Q(\mathbf{x}_s) = \mathbf{0}$, where the gradient of $Q(\mathbf{x}_s)$ is $\nabla Q(\mathbf{x}_s) = \tilde{\mathbf{H}}\mathbf{x}_s + \tilde{\mathbf{d}}$ from the definition (3.2.8). This necessary condition is a system of linear equations that can be solved for the minimum, $\mathbf{x}_s'$. Then the remaining variables, $\mathbf{x}_q'$, may be found from (5.4.19).

The Lagrange multiplier vector $\mathbf{p}'$ at the minimum of (5.4.21) may be found from (5.4.5):

$$\mathbf{p}' = \mathbf{A}^+\mathbf{g}', \qquad (5.4.24)$$

where $\mathbf{g}' = \nabla Q(\mathbf{x}')$ and $\mathbf{A}$ is not square. However, those q Lagrange multipliers associated with the q binding constraints may be found using the first partition of $\mathbf{A}$, namely $\mathbf{A}_q$. From (5.4.12), $\mathbf{g} = \mathbf{Hx} + \mathbf{d}$, so that

$$\mathbf{p}_q' = \mathbf{A}_q^{-1}\left(\mathbf{H}_q\mathbf{x}_q' + \mathbf{C}\mathbf{x}_s' + \mathbf{d}_q\right). \qquad (5.4.25)$$

Example 5.4.1. Solve the following problem by eliminating the dependent variables:

Minimize $Q(\mathbf{x}) = \mathbf{x}^T\mathbf{d} + \frac{1}{2}\mathbf{x}^T\mathbf{H}\mathbf{x}$ such that

$$x_1 + 2x_2 + x_3 = 4,$$

$$2x_1 + x_2 - x_3 = 0, \qquad\qquad (5.4.26)$$

$$\mathbf{d} = \begin{bmatrix} 1 \\ 1 \\ \hline -1 \end{bmatrix} = \begin{bmatrix} \mathbf{d}_q \\ \mathbf{d}_s \end{bmatrix}, \qquad \mathbf{H} = \begin{bmatrix} 3 & -1 & \vdots & 0 \\ -1 & 2 & \vdots & -1 \\ \hline 0 & -1 & \vdots & 1 \end{bmatrix} = \begin{bmatrix} \mathbf{H}_q & \mathbf{C} \\ \mathbf{C}^T & \mathbf{H}_s \end{bmatrix}.$$

As in the preceding development, it is assumed that the first two variables are dependent, since there are two constraints. Since the coefficients of the respective constraints form the columns of $\mathbf{A}$,

$$\mathbf{A} = \begin{bmatrix} 1 & 2 \\ 2 & 1 \\ \hline 1 & -1 \end{bmatrix} = \begin{bmatrix} \mathbf{A}_q \\ \mathbf{A}_s \end{bmatrix}, \qquad \mathbf{A}_q^{-1} = \begin{bmatrix} -\frac{1}{3} & \frac{2}{3} \\ \frac{2}{3} & -\frac{1}{3} \end{bmatrix}. \qquad (5.4.27)$$

Using $\mathbf{b} = (4\ 0)^T$ and (5.4.27) in (5.4.20), it is found that

$$\mathbf{B} = (-1\ 1), \qquad \mathbf{D} = \left(-\frac{4}{3}\ \frac{8}{3}\right)^T, \quad \text{and} \quad (\mathbf{B}\mathbf{H}_q) = (-4\ 3). \quad (5.4.28)$$

From (5.4.23), $\tilde{\mathbf{H}} = 10$, which is certainly positive-definite, and $\tilde{\mathbf{d}} = -15$.

Having eliminated the first two (dependent) variables, the remaining task is to minimize the resulting quadratic, $Q(x_3) = \frac{1}{2}x_3^T\tilde{\mathbf{H}}x_3 + x_3^T\tilde{\mathbf{d}} + \tilde{c} = 5x_3^2 - 15x_3 + \tilde{c}$, where $\tilde{c}$ is some constant. The necessary condition is that the gradient of $Q(x_3) = 0$. In the general case the solution is a set of linear equations in the independent variables. In this case there is only one independent variable in the gradient expression, so $10x_3 - 15 = 0$. Therefore, the constrained minimum is at $x_3' = \frac{3}{2}$. Evaluating (5.4.19), the two dependent variables are found to be $x_1 = \frac{1}{6}$ and $x_2 = \frac{7}{6}$. From (5.4.25) the Lagrange multiplier vector for the two constraints is $\mathbf{p}' = (1\ -\frac{1}{3})^T$.

To summarize, the quadratic function in (5.4.26) had three variables and two linear, independent equality constraints. Therefore, there is really only one degree of freedom, so the first two variables were arbitrarily chosen to be dependent. The dependent variables were eliminated from the quadratic objective function so that it could be rewritten in terms of the remaining independent variable, x_3 in this case. Then the gradient of the reduced quadratic function was equated to zero, resulting in a set of linear equations to be solved, but one equation in this case. Finally, the independent variable at the minimum point was used to find the corresponding dependent variable values.

The Lagrange multipliers were calculated according to (5.4.25). However, it is instructive to check that calculation by using (5.4.24):

$$\mathbf{p}' = \mathbf{A}^{+}\mathbf{g}'. \tag{5.4.29}$$

Matrix $\mathbf{A}^{+}$ is the generalized inverse, available from (2.2.97) when $\mathbf{A}$ has full column rank, two in this case. The generalized inverse for the $\mathbf{A}$ matrix in (5.4.27) may be computed using program MATRIX with GENINVP merged (see Example 2.2.11):

$$\mathbf{A}^{+} = \begin{bmatrix} 0 & \frac{1}{3} & \frac{1}{3} \\ \frac{1}{3} & 0 & -\frac{1}{3} \end{bmatrix}. \tag{5.4.30}$$

Since $\mathbf{g}' = \nabla Q(\mathbf{x}') = \mathbf{H}\mathbf{x}' + \mathbf{d}$ and $\tilde{\mathbf{H}} = 10$, $\tilde{\mathbf{d}} = -15$, $\mathbf{x}' = (\frac{1}{6} \ \frac{7}{6} \ \frac{3}{2})^{T}$, then $\mathbf{g}' = (\frac{1}{3} \ \frac{5}{3} \ \frac{4}{3})^{T}$. The Lagrange multipliers previously computed using (5.4.25) are thus verified by substitution into (5.4.29), which involves a generalized inverse.

The solution of the linearly constrained quadratic minimization problem (5.4.12) and (5.4.13) by the method of direct elimination of variables is not necessarily the best method, but it illustrates a very important concept: the number of degrees of freedom (equal to the number of variables) is reduced by the q linearly independent constraints. Therefore, the constrained problem is solved by minimizing an unconstrained quadratic problem in linear *subspace* E^{n-q} of the variable space E^{n}.

A *generalized elimination method* is now described that will introduce projections into the subspace and lay the groundwork for ways to deal with all kinds of constraints. As motivation for this development, note that the partition of $\mathbf{x}$ in (5.4.15) can also be expressed as

$$\mathbf{x} = \sum_{i=1}^{q} x_i \mathbf{e}_i + \sum_{i=q+1}^{n} x_i \mathbf{e}_i, \tag{5.4.31}$$

where $\mathbf{e}_i$ is the ith unit vector in E^{n} similar to (2.1.2). More to the point, (5.4.31) may be expressed as

$$\mathbf{x} = \mathbf{S}\mathbf{x}_q + \mathbf{Z}\mathbf{x}_s, \tag{5.4.32}$$

where $\mathbf{x}$ is in E^{n} and

$$\mathbf{S} = (\mathbf{e}_1 \ \mathbf{e}_2 \ \cdots \ \mathbf{e}_q), \quad \mathbf{x_q} = (x_1 \ x_2 \ \cdots \ x_q)^{T},$$

$$\mathbf{Z} = (\mathbf{e}_{q+1} \ \mathbf{e}_{q+2} \ \cdots \ \mathbf{e}_n), \quad \mathbf{x}_s = (x_{q+1}, x_{q+2}, \ldots, x_n)^{T}. \tag{5.4.33}$$

Matrix $\mathbf{S}$ is $n \times q$ and matrix $\mathbf{Z}$ is $n \times s$ where $s = n - q$.

Recall that the direct elimination method led to an unconstrained optimization method in the subspace reached by the variables in $\mathbf{x}_s$. Let that vector in the subspace be called $\mathbf{y}$ and define a new linear transformation of variables analogous to (5.4.32):

$$\mathbf{x} = \mathbf{S}\mathbf{b} + \mathbf{Z}\mathbf{y}. \tag{5.4.34}$$

Again the matrices are dimensioned $\mathbf{S}_{n,q}$ and $\mathbf{Z}_{n,s}$, where $s = n - q$, and subspace vector $\mathbf{y}$ is in E^s. The generalized elimination method also solves the minimization problem in (5.4.12) subject to the q linear equality constraints in (5.4.13):

$$\mathbf{A}^T\mathbf{x} = \mathbf{b}, \tag{5.4.35}$$

where $\mathbf{b}$ is in E^q and matrix $\mathbf{A}$ is $n \times q$. To define the $\mathbf{S}$ and $\mathbf{Z}$ matrices for the generalized elimination method, substitute (5.4.34) into (5.4.35):

$$\mathbf{A}^T\mathbf{S}\mathbf{b} + \mathbf{A}^T\mathbf{Z}\mathbf{y} = \mathbf{b}, \tag{5.4.36}$$

which reduces to the identity $\mathbf{b} = \mathbf{b}$ if

$$\mathbf{A}^T\mathbf{S} = \mathbf{I}_q, \tag{5.4.37}$$

$$\mathbf{A}^T\mathbf{Z} = \mathbf{0}, \tag{5.4.38}$$

where $\mathbf{I}_q$ is the unit matrix in E^q. Consider the columns of matrices $\mathbf{A}$ and $\mathbf{Z}$:

$$\mathbf{A} = (\mathbf{a}_1 \ \mathbf{a}_2 \ \cdots \ \mathbf{a}_q), \tag{5.4.39}$$

$$\mathbf{Z} = (\mathbf{z}_1 \ \mathbf{z}_2 \ \cdots \ \mathbf{z}_s), \ s = n - q. \tag{5.4.40}$$

The columns of transformation matrix $\mathbf{Z}$ are *basis vectors* that span the subspace containing $\mathbf{y}$; in fact, (5.4.38) defines that subspace to be the *null column space* of constraint matrix $\mathbf{A}$.

The geometric significance of the linear transformation in (5.4.36) can be seen in Figure 5.4.1. In the case illustrated, $\mathbf{A} = (\mathbf{a}_1 \ \mathbf{a}_2)$ and $\mathbf{Z} = (\mathbf{z}_1)$. Then (5.4.38) states that $\mathbf{a}_1^T\mathbf{z}_1 = 0$ and $\mathbf{a}_2^T\mathbf{z}_1 = 0$, which are the conditions for $\mathbf{z}_1$ to be orthogonal to both normal vectors $\mathbf{a}_1$ and $\mathbf{a}_2$. Clearly, $\mathbf{z}_1$ must be parallel to the intersection of the two hyperplanes, that is, the linear manifold known as the null column space of $\mathbf{A}$.

To develop a procedure for finding transformations $\mathbf{S}$ and $\mathbf{Z}$ that appear in (5.4.36), a partitioned matrix product is simply defined:

$$\begin{bmatrix} \mathbf{S}^T \\ \mathbf{Z}^T \end{bmatrix}[\mathbf{A} \ \mathbf{V}] = \begin{bmatrix} \mathbf{S}^T\mathbf{A} & \mathbf{S}^T\mathbf{V} \\ \mathbf{Z}^T\mathbf{A} & \mathbf{Z}^T\mathbf{V} \end{bmatrix} = \mathbf{I}_n. \tag{5.4.41}$$

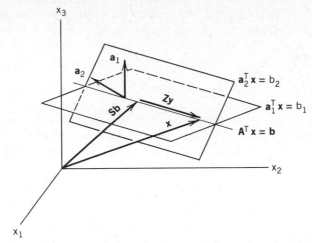

Figure 5.4.1. The null column space of constraint matrix **A** for three variables and two linear constraints.

But in light of (5.4.37) and (5.4.38), (5.4.41) is true if

$$\mathbf{Z}^T\mathbf{V} = \mathbf{I}_s, \qquad \mathbf{S}^T\mathbf{V} = \mathbf{0}. \qquad (5.4.42)$$

Then (5.4.41) states that

$$\begin{bmatrix} \mathbf{S}^T \\ \mathbf{Z}^T \end{bmatrix} = [\mathbf{A} \ \ \mathbf{V}]^{-1}. \qquad (5.4.43)$$

The procedure for determining matrices **S** and **Z** for use in (5.4.36) is to choose any matrix $\mathbf{V}_{n,s}$, $s = n - q$, such that [**A V**] is nonsingular. Then invert [**A V**] and partition between rows q and $q + 1$, thus determining $\mathbf{S}^T$ and $\mathbf{Z}^T$ as indicated in (5.4.43).

The method of direct elimination results from a special choice of **V**:

$$\mathbf{V} = \begin{bmatrix} \mathbf{0} \\ \mathbf{I} \end{bmatrix}. \qquad (5.4.44)$$

Using the partition of **A** in (5.4.16) and the definition of **B** in (5.4.20), it is not difficult to verify this special formulation of (5.4.43):

$$\begin{bmatrix} \mathbf{A}_q & \mathbf{0} \\ \mathbf{A}_s & \mathbf{I} \end{bmatrix}^{-1} = \begin{bmatrix} \mathbf{A}_q^{-1} & \mathbf{0} \\ -\mathbf{B} & \mathbf{I} \end{bmatrix} = \begin{bmatrix} \mathbf{S}^T \\ \mathbf{Z}^T \end{bmatrix}. \qquad (5.4.45)$$

Therefore, the special case for direct elimination requires that

$$\mathbf{S} = \left(\mathbf{A}_q^{-1} \ \ \mathbf{0} \right)^T, \qquad \mathbf{Z} = \left(-\mathbf{B} \ \ \mathbf{I} \right)^T. \qquad (5.4.46)$$

For the simple case where one or more variables are held constant, $x_i = b_i$, $i = 1$ to q, the values of S and Z are especially elementary:

$$A = S = \begin{bmatrix} I_q \\ 0 \end{bmatrix}, \qquad Z = V = \begin{bmatrix} 0 \\ I_s \end{bmatrix}. \tag{5.4.47}$$

In words, (5.4.47) states that A is composed of the first q columns of the $n \times n$ unit matrix, and Z is composed of the remaining $n - q$ columns. The values in (5.4.47) satisfy (5.4.41). The case for constant variable values is more useful later when binding inequality constraints are treated as equality constraints, specifically when dealing with lower and upper bounds on variables. The geometric interpretation of the subspace that represents a problem's degrees of freedom is more easily imagined for fixed variables. For example, equality constraint $x_2 = 0$ is seen in Figure 5.4.1 to produce a subspace (constraint manifold) that is the x_1–x_3 plane.

Rather than select an arbitrary V in (5.4.43) leading to related values for S and Z, there is a way to compute Z so that $S = A^{+T}$ and the vector Sb in (5.4.34) and Figure 5.4.1 is orthogonal to the constraint manifold, which is the subspace representing the problem's degrees of freedom. The subspace basis vectors, z_i, will also be mutually orthogonal. The preceding case where some variables are held constant satisfies these conditions. The computation for Z given $S = A^{+T}$ requires the QR orthogonal decomposition and is well conditioned. Interested readers are referred to Fletcher (1981a:83) and Gill (1974b:61).

Once S is determined by any of the preceding means, the Lagrange multipliers are available; (5.4.5) gives $g' = Ap'$, which can be premultiplied by S^T. Then use of (5.4.37) leads to

$$p' = S^T g'. \tag{5.4.48}$$

Example 5.4.2. Two choices for matrix V are applied to the data previously given in Example 5.4.1. There are three variables and two equality constraints with coefficients appearing as column vectors in A in (5.4.27). First, choose an arbitrary value of $V = (1 \ 2 \ 3)^T$. Then

$$[A \ V] = \begin{bmatrix} 1 & 2 & 1 \\ 2 & 1 & 2 \\ 1 & -1 & 3 \end{bmatrix}, \tag{5.4.49}$$

the inverse matrix is

$$[A \ V]^{-1} = \begin{bmatrix} -\frac{5}{6} & \frac{7}{6} & -\frac{1}{2} \\ \frac{2}{3} & -\frac{1}{3} & 0 \\ \frac{1}{2} & -\frac{1}{2} & \frac{1}{2} \end{bmatrix}, \tag{5.4.50}$$

and (5.4.43) yields

$$
\mathbf{S} = \begin{bmatrix} -\frac{5}{6} & \frac{2}{3} \\ \frac{7}{6} & -\frac{1}{3} \\ -\frac{1}{2} & 0 \end{bmatrix}, \qquad \mathbf{Z} = \begin{bmatrix} \frac{1}{2} \\ -\frac{1}{2} \\ \frac{1}{2} \end{bmatrix}. \tag{5.4.51}
$$

The second case also uses the data from Example 5.4.1 and equations (5.4.46) and (5.4.47) that are applicable to direct elimination. Applying the results for $\mathbf{A}_q^{-1}$ and $\mathbf{B}$ in (5.4.27) and (5.4.28), respectively,

$$
\mathbf{S} = \begin{bmatrix} -\frac{1}{3} & \frac{2}{3} \\ \frac{2}{3} & -\frac{1}{3} \\ 0 & 0 \end{bmatrix}, \qquad \mathbf{Z} = \begin{bmatrix} 1 \\ -1 \\ 1 \end{bmatrix}. \tag{5.4.52}
$$

The transformation of y from the subspace E^1 to $\mathbf{x}$ in E^3 according to (5.4.34) is

$$
\mathbf{x} = \mathbf{Sb} + \mathbf{Z}y = \begin{bmatrix} -\frac{4}{3} \\ \frac{8}{3} \\ 0 \end{bmatrix} + y \begin{bmatrix} 1 \\ -1 \\ 1 \end{bmatrix}. \tag{5.4.53}
$$

Substitution of (5.4.53) into each of the two linear equality constraints in (5.4.26) shows that they are fulfilled regardless of the value of the independent variable y in the subspace.

For both cases in this example it is easy to verify that (5.4.37) and (5.4.38) are true.

The generalized elimination method provides the means for converting a minimization over $\mathbf{x}$ in E^n with q linear constraints to an unconstrained minimization over $\mathbf{y}$ in the subspace E^s, $s = n - q$. Repeating (5.4.34), the linear transformation between variables is

$$
\mathbf{x} = \mathbf{Sb} + \mathbf{Zy}, \tag{5.4.54}
$$

where $\mathbf{S}$ and $\mathbf{Z}$ are chosen according to (5.4.43) so that the constraints are satisfied for all values of $\mathbf{y}$. The quadratic model function assumed valid in a neighborhood of some $\mathbf{x}^{(k)}$ at the start of the kth iteration is repeated from (5.4.12):

$$
Q(\mathbf{x}) = \mathbf{x}^T\mathbf{d} + \tfrac{1}{2}\mathbf{x}^T\mathbf{Hx}, \tag{5.4.55}
$$

where $\mathbf{x} = \mathbf{x}^{(k)}$ and $\mathbf{H}$ is the Hessian matrix evaluated at $\mathbf{x}^{(k)}$. Substituting (5.4.54) into (5.4.55), the quadratic function in $\mathbf{y}$ space is

$$
Q(\mathbf{y}) = \tfrac{1}{2}\mathbf{y}^T(\mathbf{Z}^T\mathbf{HZ})\mathbf{y} + \mathbf{y}^T\mathbf{Z}^T(\mathbf{d} + \mathbf{HSb}) + c, \tag{5.4.56}
$$

where c is some constant.

The quadratic model has always been used to obtain a direction for a line search:

$$\mathbf{x}^{(k+1)} = \mathbf{x}^{(k)} + t\mathbf{s}^{(k)}. \tag{5.4.57}$$

Comparison of (5.4.57) with (5.4.54) shows that $\mathbf{x}^{(k+1)} = \mathbf{x}^{(k)} = \mathbf{Sb}$ when $t = 0$ and $\mathbf{y} = \mathbf{0}$. Furthermore, the gradient of $Q(\mathbf{x})$ in (5.4.55) is $\mathbf{g}(\mathbf{x}) = \mathbf{d} + \mathbf{Hx}$ so that

$$\mathbf{g}^{(k)} = \mathbf{g}[\mathbf{x}^{(k)}] = \mathbf{g}(\mathbf{Sb}) = \mathbf{d} + \mathbf{HSb}. \tag{5.4.58}$$

Substituting (5.4.58) into (5.4.56) yields the desired quadratic function in the $\mathbf{y}$ space:

$$Q(\mathbf{y}) = \tfrac{1}{2}\mathbf{y}^{T}(\mathbf{Z}^{T}\mathbf{HZ})\mathbf{y} + \mathbf{y}^{T}\mathbf{Z}^{T}\mathbf{g} + c, \tag{5.4.59}$$

where both Hessian $\mathbf{H}$ and gradient $\mathbf{g}$ are evaluated at the turning point $\mathbf{x}^{(k)}$, which is coincident with $\mathbf{y} = \mathbf{0}$.

The necessary condition for a minimum over $\mathbf{y}$ in $Q(\mathbf{y})$ in (5.4.59) is that its gradient is equal to zero, therefore:

$$(\mathbf{Z}^{T}\mathbf{HZ})\mathbf{y} = -\mathbf{Z}^{T}\mathbf{g}. \tag{5.4.60}$$

The solution of the linear system of equations in (5.4.60) is just a Newton solution for step $\mathbf{y}$; (5.4.54) and (5.4.57) show that the current Newton step is described by

$$\mathbf{x}^{(k+1)} = \mathbf{x}^{(k)} + \mathbf{Zy}. \tag{5.4.61}$$

Put another way, the search direction at the kth iteration is

$$\mathbf{s}^{(k)} = \mathbf{Zy} = -\mathbf{Z}(\mathbf{Z}^{T}\mathbf{HZ})^{-1}\mathbf{Z}^{T}\mathbf{g}. \tag{5.4.62}$$

Compare (5.4.55) for constrained minimization in the $\mathbf{x}$ space with (5.4.59) for unconstrained minimization in the $\mathbf{y}$ subspace. The *projected or reduced gradient vector* is

$$\nabla_{\mathbf{y}}Q = \mathbf{Z}^{T}\mathbf{g}, \tag{5.4.63}$$

and the *projected or reduced Hessian* is

$$\nabla_{\mathbf{y}}^{2}Q = \mathbf{Z}^{T}\mathbf{HZ}. \tag{5.4.64}$$

Note that (5.4.63) could have been obtained directly from the partial derivative operation on the linear transformation (5.4.54). Previously, the ∇ operation on (5.1.30) produced (5.1.41). In the case of (5.4.63), a similar result is

$$\nabla_y = Z^T \nabla_x, \tag{5.4.65}$$

which confirms (5.4.63).

Again, the necessary condition for minimum $Q(y)$ is that the projected gradient be equal to zero; note that this may be confirmed by premultiplying (5.4.5), $g' = Ap'$, by Z^T. Then (5.4.38) can be substituted to show that $g' = 0$. The additional requirement for a minimum is that the projected Hessian (5.4.64) is positive definite.

The projection concept comes from the original solution of linearly constrained objective functions in a subspace by Rosen (1960). His *gradient projection method* assumed that $H = I$ in the quadratic model, which is equivalent to using a search direction $-Z^T g$ in the y subspace or, by (5.4.61),

$$s^{(k)} = -Z(Z^T Z)^{-1} Z^T g^{(k)}. \tag{5.4.66}$$

However, Gill (1974b:49) shows that

$$\tilde{P} = Z(Z^T Z)^{-1} Z^T = ZZ^+ = I - AA^+. \tag{5.4.67}$$

Projection matrix $\tilde{P}$ as defined by (2.2.89) is $\tilde{P} = ZZ^+$ that projects vectors into the column space of $Z = (z_1 \ z_2 \ \cdots \ z_q)$. Projection matrix P as defined by (2.2.93) projects vectors into the null column space of constraint matrix A, which is also the column space of Z by design. Note that when Z has been constructed by QR orthogonal factorization (or for fixed-variable constraints) that Z is orthogonal. Then $Z^T Z = I$ and (5.4.66) is $s^{(k)} = -ZZ^T g^{(k)}$, which is steepest descent for $y = -Z^T g$ in (5.4.62).

The conclusion available from the preceding conversion from a minimization that has linear equality constraints to an unconstrained minimization with fewer variables is that all the previous minimization methods described in Chapters Four and Five can be used in this way with only minor modifications. For the quasi-Newton methods in this chapter, solution of the Newton equation for y in (5.4.60) at each iteration is actually accomplished using an approximation to the Hessian matrix, namely, B in (5.1.22). Two rank 1 update terms are required at the end of each iteration. There are two vectors required for updating the estimated Hessian matrix (called y and d, but different from those in this section). It can be shown that the operations of projection by Z and rank 1 updates are commutative: updating an estimated Hessian matrix B by any Broyden formula using projected differences $Z^T[g^{(k+1)} - g^{(k)}]$ and $[y^{(k+1)} - y^{(k)}]$ is equivalent to updating $Z^T BZ$ by unprojected differences $[g^{(k+1)} - g^{(k)}]$ and $[x^{(k+1)} - x^{(k)}]$.

When using finite differences for derivatives, there are savings in computation for all linearly constrained minimization methods, since perturbation need be made only in the subspace representing the degrees of freedom. These perturbations are no longer made in the coordinate directions. For quasi-Newton methods, the ith element of the projected gradient would be estimated by

$$\mathbf{Z}^T \mathbf{g} \doteq \frac{f(\mathbf{x}^{(k)} + dx\, \mathbf{z}_i) - f(\mathbf{x}^{(k)})}{dx}, \tag{5.4.68}$$

where $\mathbf{z}_i$ is the ith column of $\mathbf{Z}$ and dx is a small positive number, say 1E–4.

In the finite difference Newton method, (4.1.3), which computes a column of the Hessian matrix, would be replaced by

$$\mathbf{h}_j \doteq \frac{\mathbf{g}(\mathbf{x}^{(k)} + dx\, \mathbf{z}_i) - \mathbf{g}(\mathbf{x}^{(k)})}{dx}. \tag{5.4.69}$$

The greater the number of equality (binding) constraints, the less work there is in estimating derivatives.

Least-pth problems can also be solved by projecting linear equality constraints. Using (4.4.17) and (5.4.65), the reduced gradient for least-squares problems is

$$\nabla_y F = \mathbf{Z}^T \mathbf{J}^T \mathbf{r}, \tag{5.4.70}$$

where $\mathbf{J}$ is the Jacobian matrix and $\mathbf{r}$ is the vector of residuals. Using (4.4.20) and (5.4.64), the reduced approximate Hessian for least squares problems is

$$\nabla_y F = \mathbf{Z}^T \mathbf{J}^T \mathbf{J} \mathbf{Z}. \tag{5.4.71}$$

This section started with minimization of a nonlinear objective function subject to both equality and inequality constraints, (5.4.9) through (5.4.11). The development was restricted to solving only the quadratic function with linear equality constraints, typically for each iteration. This section ends with the addition of the linear inequality constraints.

The most elementary approach is simply to convert the ith linear inequality constraint $c_i(\mathbf{x})$ to an equality constraint $h_i(\mathbf{x})$ by adding an additional *quadratic slack variable*:

$$h_i(\mathbf{x}) = c_i(\mathbf{x}) - x_{n+1}^2 = 0. \tag{5.4.72}$$

The concept is to add one more "slack" variable for each inequality constraint. Because the slack is squared, the added variable(s) will remain positive and the inequality constraint will remain feasible. In addition to increasing the number of variables, this approach introduces a distortion that affects the first-order conditions for convergence according to (5.4.5). Since Fletcher (1981a:8) mentions bad reports for the quadratic slack method in practice, it will not be discussed further.

The most acceptable way to deal with linear inequality constraints is the *active set method*. Define set E as containing the indices $i = 1$ to q of all equality constraints *and* those inequality constraints that are binding:

$$E = \{i \text{ s.t. } h_i(\mathbf{x}) = 0, c_i(\mathbf{x}) \leq 0\}. \qquad (5.4.73)$$

Then the constrained quadratic problem (5.4.12) and (5.4.13) may be solved in the null column space of the active constraints by a line search in the subspace in direction $\mathbf{s}^{(k)}$, (5.4.62):

$$\mathbf{x}^{(k+1)} = \mathbf{x}^{(k)} + t\mathbf{s}^{(k)}, \qquad (5.4.74)$$

where $\mathbf{y} = \mathbf{0}$ corresponds to the point $\mathbf{x}^{(k)}$. Unfortunately, some of the remaining inequality constraints may become active (infeasible), so that the line search must be terminated with $t < 1$ and one or more inequality constraints added to set E (5.4.73). Clearly, only inexact line search algorithms are appropriate for use in the active set method.

Whether the line search terminates with $t = 1$ or $t < 1$, the next iteration is not begun until a test is made on all Lagrange multipliers computed by (5.4.48); if none are negative, then the Kuhn–Tucker conditions (3.3.22) are satisfied and a constrained minimum on the set E has been obtained. Conversely, one or more negative Lagrange multipliers indicate that a decrease in objective function may be achieved without violating its corresponding constraint; see (5.4.8).

The usual practice is to remove the inequality constraint having the most negative Lagrange multiplier from set E before starting the next iteration. When more than one constraint is removed between iterations, there is a tendency for the projected Hessian matrix to become indefinite; however, most $\mathbf{LDL}^T$ factorization algorithms force positive definiteness. Also, there is a possible degenerate condition where the same sets of active constraints will cycle in and out of iterations, but that is usually ignored in most computer programs.

When constraints are deleted from active set E, the new (larger) projected Hessian $\mathbf{Z}^T\mathbf{HZ}$ must have elements added to estimate the related second partial derivatives. Finite differences (5.4.69) may be used, but affected element values of unity and zero are often satisfactory. There have been several exhaustive studies to define the most efficient means for modifying the $\mathbf{LDL}^T$ factorization of an estimated Hessian matrix associated with minimization of a quadratic function subject to linear constraints. Interested readers are referred to Gill (1981) and to Fletcher (1974b) for simple lower and upper bounds on variables ("box" constraints).

A serious problem that may occur in the active set method for linear inequality constraints is *zigzagging*. For quadratic objective function models such as in (5.4.12), the test of Lagrange multipliers after each iteration indicates either that the problem is solved or that a previously inactive

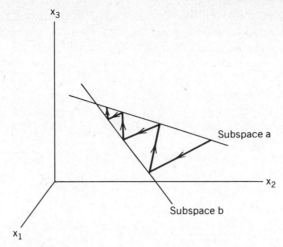

Figure 5.4.2. Zigzagging between two linear manifolds that define subspaces for two active constraints in E^3.

inequality constraint became active during the line search. This is not the case for the general nonlinear objective function in (5.4.9), when the solution is only obtained as the limit of a sequence of iterations *with an unchanged active constraint set E.*

Zigzagging is the situation where the set of active constraints does not settle down during the final iterations, but oscillates between two or more different constraint manifolds that define the subspace. See Figure 5.4.2. Convergence then becomes linear at best. The root of the zigzagging problem is that inexact solutions are obtained during the line searches performed for each iteration. Otherwise, once an inequality constraint index was removed from the active set it would not enter again. Various ad hoc measures to prevent zigzagging have been reviewed by Fletcher in Lootsma (1972:290–291). Perhaps the best of these was proposed by McCormick (1969), who incorporated "bending" along the line search ray (5.4.74). If motion ceases (t bounded) because a previously inactive inequality constraint boundary is encountered, then the bending policy calls for continuing beyond that boundary along the same direction vector **s**, with the appropriate component of **s** set to zero. The different gradient values that exist at the points of encountering constraint boundaries are not utilized.

5.4.2. Program BOXMIN for Lower and Upper Bounds.

Program C5-6, BOXMIN, is designed to be merged into quasi-Newton optimizer program C5-1, QNEWT, which employs the "cutback" line search strategy. BOXMIN adds a command option to set lower and/or upper bounds on variables; otherwise default values of ±10,000 are assumed. More general linear constraints, $\mathbf{Ax} \geq \mathbf{b}$, are not included in BOXMIN, because of the greater programming complexity, less likelihood of need, and the ability to satisfy all kinds of constraints by the multiplier penalty method described in Sections 5.4.3 and 5.4.4. Subroutines 5000 and 7000 usually provided for QNEWT continue to be required to define the problem objective function and its gradient, respectively. This section begins with an illustration of why confinement to a "box" neighborhood is often critical to obtaining a meaningful solution, followed by description of the algorithm in relation to developments in the preceding section. Several examples using BOXMIN merged into QNEWT are provided.

Before describing BOXMIN, an interesting unconstrained optimization problem is described to illustrate the need for providing upper and lower bounds. Consider the "camelback" function described by Branin (1972:521):

$$f(\mathbf{x}) = ax_1^2 + bx_1^4 + cx_1^6 - x_1x_2 + dx_2^2 + ex_2^4. \qquad (5.4.75)$$

Various sets of coefficients a to e produce different numbers of extreme values; in particular, there are six maxima and two minima associated with the coefficients in Table 5.4.1. The rows labeled x_i^0 in Table 5.4.1 are components of the starting variable vector for an optimization to locate an extreme value of f. The surface of $f(\mathbf{x})$ in (5.4.75) with the data in Table 5.4.1 is shown in Figure 5.4.3; the related level curves are shown in Figure 5.4.4. Objective function (5.4.75) and its gradient are programmed in C5-5, CAMEL, to be merged into optimizer QNEWT. It is left to problem 5.11 for the reader to verify the maxima as suggested in Table 5.4.1. Computing the minima indicated in Table 5.4.1 cannot be performed reliably with QNEWT and CAMEL from arbitrary starting points; the reader should verify that fact, trying the starting points, $\mathbf{x}^0$, given in Table 5.4.1. The potential problem can be seen in

Table 5.4.1. Extreme Values for a Set of Coefficients in the Camelback Function

Coefficients: $a = -4$, $b = +2$, $c = -\frac{1}{3}$, $d = +4$, $e = -4$

Maxima:[a]	f	1.0360	0.2155	-2.1040
	x_1	± 0.0898	± 1.7036	± 1.6070
	x_2	± 0.7126	± 0.7960	∓ 0.5686
	x_1^0	± 0.10	± 1.30	± 1.8
	x_2^0	± 0.10	± 0.17	∓ 1.0
Minima:	f	-2.4960		
	x_1	± 1.2300		
	x_2	± 0.1623		
	x_1^0	± 1.5	$\pm 0.5^b$	$\pm 1.0^b$
	x_2^0	± 0.2	$\pm 0.5^b$	$0^{\ b}$

[a] Maxima obtained by reversing signs of a–e and minimizing (5.4.75).
[b] Diverges and overflows unless $0.5 \leq x_1 \leq 1.5$ and $0 \leq x_2 \leq 0.5$.

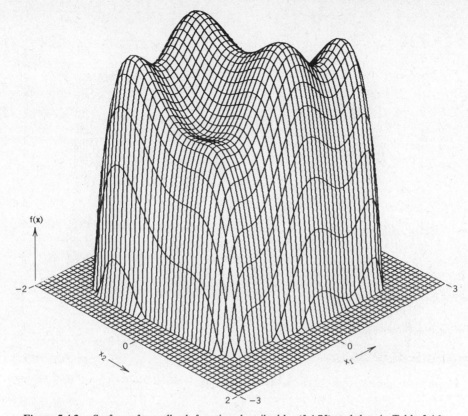

f(x)

−2

3

x_2

0

0

x_1

2 −3

Figure 5.4.3. Surface of camelback function described by (5.4.75) and data in Table 5.4.1.

Figure 5.4.3: there is a good possibility that a line search will fall off the side of the surface and miss the two local minima.

For cases similar to that just cited and because many physical problems allow only limited ranges of variables, lower and upper "box" bounds on variables are very desirable. An examination of Figure 5.4.1 shows that such bounds on variables satisfy the orthogonal factorization conditions for the linear transformation $x = Sb + Zy$ from subspace y to variable space x in a trivial way. In those cases, the columns of Z are some subset of the coordinate axes, as seen in (5.4.47).

Example 5.4.3. Generate a matrix description of the subspace representing the degrees of freedom for box constraints when $n = 5$ and there are three active or binding bounds:

$$x_1 = b_1, \qquad x_2 = b_2, \qquad x_3 = b_3, \qquad (5.4.76)$$

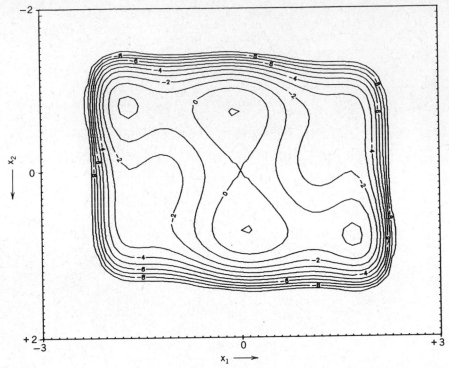

Figure 5.4.4. Level curves for the camelback function described by (5.4.75) and data in Table 5.4.1.

where b_i may represent either a lower or upper bound. For this case, $E = \{1, 2, 3\}$, the subspace is in E^s, $s = n - q = 2$, and (5.4.47) specifies the **A**, **S**, and **Z** matrices:

$$\mathbf{A} = \mathbf{S} = \begin{bmatrix} 1 & 0 & 0 \\ 0 & 1 & 0 \\ 0 & 0 & 1 \\ 0 & 0 & 0 \\ 0 & 0 & 0 \end{bmatrix}, \quad \mathbf{Z} = \begin{bmatrix} 0 & 0 \\ 0 & 0 \\ 0 & 0 \\ 1 & 0 \\ 0 & 1 \end{bmatrix}. \tag{5.4.77}$$

By (5.4.34), the **x** space in E^5 is composed of fixed and variable subspaces:

$$\mathbf{x} = \begin{bmatrix} b_1 \\ b_2 \\ b_3 \\ 0 \\ 0 \end{bmatrix} + \begin{bmatrix} 0 \\ 0 \\ 0 \\ x_4 \\ x_5 \end{bmatrix}. \tag{5.4.78}$$

According to (5.4.63) and (5.4.64), the projected or reduced gradient and Hessian are

$$\nabla_y Q = \begin{bmatrix} g_4 \\ g_5 \end{bmatrix}, \qquad \nabla_y^2 Q = \begin{bmatrix} h_{44} & h_{45} \\ h_{54} & h_{55} \end{bmatrix}. \tag{5.4.79}$$

According to (5.4.62), the search direction in the **x** space is

$$\mathbf{s} = - \begin{bmatrix} 0 & 0 \\ 0 & 0 \\ 0 & 0 \\ 1 & 0 \\ 0 & 1 \end{bmatrix} \begin{bmatrix} h_{44} & h_{45} \\ h_{54} & h_{55} \end{bmatrix}^{-1} \begin{bmatrix} g_4 \\ g_5 \end{bmatrix} = \begin{bmatrix} 0 \\ 0 \\ 0 \\ s_4 \\ s_5 \end{bmatrix}. \tag{5.4.80}$$

The BOXMIN additions for QNEWT have not been implemented with complete generality; to do so would require recording permutations of variables so that the vectors and Hessian matrix illustrated in Example 5.4.3 would have the required structure. Instead, simple modifications to QNEWT have been made that will satisfactorily contain line searches within feasible regions and will behave well under most circumstances. In particular, the entire Hessian in E^n is updated, but with reduced gradient and variable differences (some update elements are zero). Since the full Hessian is positive definite, the reduced Hessian in (5.4.64) will also be positive definite. The interested reader is referred to Fletcher (1974b:166) to see the complications involved in a theoretically correct update of the $\mathbf{LDL}^T$ factorization of the estimated Hessian.

The additions to the QNEWT BASIC code are shown in Table 5.4.2 as keyed to the flow charts in Figures 5.3.1 (QNEWT) and 5.2.1 (cutback line search).

Table 5.4.2.   Additions to QNEWT to Implement BOXMIN for Bounds

Figure	Balloon	Line No.	Purpose
5.3.1	A	1435	Reset/record binding variables
5.3.1	B	1576	Release all bounds
5.3.1	C	1602	Release *all* non-K-T bounds
5.3.1	C	1610	Project gradient into subspace
5.3.1	D	1622	Project search direction into subspace
5.2.1	F	1775	Check/set more bounds in line search
5.3.1	E	2350	Project new gradient into subspace

Table 5.4.3.   Subroutines Included in Code Segment BOXMIN

Name	Lines
Check for Additional Bounds in Subspace	1880–2000
Initialize Flags and Lower/Upper Bounds	2020–2070
See or Reset Lower/Upper Bounds on Variables	2090–2350
Reset and Record Binding Variables	3850–3960

Program segment BOXMIN also adds command 5 to the menu so that lower and/or upper bounds on variables may be reviewed, added, or modified. As seen in the BOXMIN code, program C5-6 in Appendix C, line 415 specifies arrays $L4(20)$, $L5(20)$, and $P5(20, 2)$ for up to 20 lower and upper bound active flags and the actual bound values, respectively. Array $L5(\ )$ contains 0's, $+1$'s, and -1's to indicate that each variable has not been set to a bound, is set to its lower bound, or is set to its upper bound, respectively. The subroutines required to perform these activities are shown in Table 5.4.3.

It is recommended that the reader run QNEWT with BOXMIN and CAMEL merged, starting with the $\mathbf{x}^0$ values subject to footnote b in Table 5.4.1. In addition, the following example from Himmelblau (1972:416) is provided.

Example 5.4.4.   Paviani posed the problem to minimize

$$f(\mathbf{x}) = \sum_{i=1}^{10} \left\{ [\ln(x_i - 2)]^2 + [\ln(10 - x_i)]^2 \right\} - \left\{ \prod_{i=1}^{10} x_i \right\}^{0.2} \quad (5.4.81)$$

such that

$$2.001 \le x_i \le 9.999, \qquad i = 1 \text{ to } 10,$$

starting from

$$x_i^{(0)} = 9, \qquad i = 1 \text{ to } 10.$$

Note that (5.4.81) is a situation where all evaluations of the objective function *must* be feasible, since the computer cannot take the logarithm of a negative number. Program C5-7, PAV17, is listed in Appendix C; it contains subroutine 5000 for (5.4.81) and subroutine 7000 for the gradient. Alternatively, lines 7000– may be deleted after PAV17 is merged and then QNEWTGRD can be merged so that derivatives are obtained by finite differences. Either way, the solution is obtained in six iterations: $x_i' = 9.350267$ using exact derivatives and $x_i' = 9.35516$ using approximate derivatives, $i = 1$ to 10.

Program segment BOXMIN is included in all subsequent developments in this chapter, since lower and upper bounds are frequently required.

5.4.3. Nonlinear Constraints by Penalty Functions. The general nonlinear programming problem from (5.4.1) is modified for convenience:

$$\text{Minimize } f(\mathbf{x}) \text{ s.t. } \mathbf{c}(\mathbf{x}) \geq \mathbf{0}. \tag{5.4.82}$$

The symbols "s.t." mean "such that". There are q *equality* constraints in the first part of vector $\mathbf{c}$ and $m - q$ *inequality* constraints in last part of vector $\mathbf{c}$, for a total of m constraints. It makes no sense to have more than n equality constraints when there are n variables, since in theory the variables could be determined so that optimization is not possible. See problem 5.23. The presentation in this section will depend on a simple one-dimensional example followed by a more comprehensive mathematical description. A powerful program that solves problems having the structure of (5.4.82) is described in Section 5.4.4.

The earliest *penalty function* was suggested by Courant (1943) for equality constraints:

$$F(\mathbf{x}, s) = f(\mathbf{x}) + s\left\{ [\mathbf{c}(\mathbf{x})]^T [\mathbf{c}(\mathbf{x})] \right\}$$
$$= f(\mathbf{x}) + s\sum_{i=1}^{q} [c_i(\mathbf{x})]^2. \tag{5.4.83}$$

The idea is simple enough: if $F(\mathbf{x}, s)$ is minimized as values of $s \to \infty$, then $c_i \to 0$, $i = 1$ to q, and $f(\mathbf{x})$ is a constrained minimum, assuming that the constraints can be satisfied. The penalty function name comes from the fact that any $c_i \neq 0$ penalizes the minimization of F. In practice, a nominal value of s is chosen, then F is minimized with respect to $\mathbf{x}$, a larger value of s is chosen, F is again minimized, and so on. The sequential nature of the process has been given the name Sequential Unconstrained Minimization Technique (SUMT), which was made popular by Fiacco and McCormick (1968). They have shown that if a minimum $F(\mathbf{x}, s)$ occurs at some value $\mathbf{x} = \mathbf{x}'$ given a value for s, then $\mathbf{x}'$ is a continuous and smooth function of s; that is, $\mathbf{x}'(s)$ is differentiable and may be extrapolated to the limit $s \to \infty$.

Example 5.4.5. Consider the optimization problem in one variable subject to one linear constraint:

$$\text{Minimize } x^{1/2} \text{ s.t. } x = 1. \tag{5.4.84}$$

Of course, the answer is $x = 1$, but it is instructive to form the penalty function according to (5.4.83):

$$F(\mathbf{x}, s) = x^{1/2} + s(x - 1)^2. \tag{5.4.85}$$

Figure 5.4.5 shows the penalty function $F(\mathbf{x}, s)$ in (5.4.85) for several values of s; when $s = 0$, the graph is just the unconstrained objective function $f(\mathbf{x})$. Note that large values of the penalty multiplier s steepen the descent surface

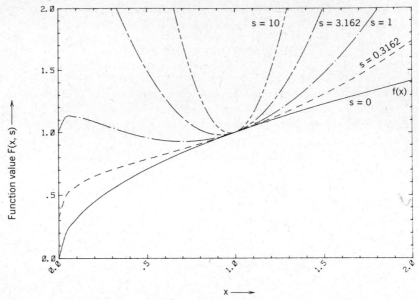

Figure 5.4.5.   The penalty function $F(x, s)$ for values of penalty multipliers.

and move the minima closer to $x = 1$, but there are smaller values of s for which the minimum penalty function value is unrelated to the known solution at $x = 1$.

The first and second derivatives are continuous functions of x:

$$\nabla F = \tfrac{1}{2}x^{-1/2} + 2s(x - 1), \tag{5.4.86}$$

$$\nabla^2 F = -0.25x^{-3/2} + 2s. \tag{5.4.87}$$

It is not too obvious in Figure 5.4.5, but the minimum points on loci for increasing s approach $x = 1.0$ from the left. Similarly, it is important to observe from (5.4.86) that a zero value for the gradient of the penalty function does not produce the correct constrained minimum $x = 1$ unless $s \to \infty$. In that limit the second derivative in (5.4.87) is infinite. Each sequential minimization for increasingly larger values of multiplier s becomes progressively more illconditioned. In practice, the sequence of minimizations with fixed, larger values of s is terminated when the equality constraint, $c(\mathbf{x})$ is sufficiently small.

Inequality constraints, $\mathbf{c}(\mathbf{x}) \geq \mathbf{0}$, can also be treated by *exterior penalty functions* that result from a simple modification of (5.4.83):

$$F(\mathbf{x}, s) = f(\mathbf{x}) + s \sum_{i=q+1}^{m} \left\{ \min[c_i(\mathbf{x}), 0] \right\}^2. \tag{5.4.88}$$

The penalty function $F(\mathbf{x}, s)$ in the equality constraint example just given was shown in Figure 5.4.5, and it would apply to a similar problem, minimize $x^{1/2}$ such that $x \geq 1$, except that the curves to the right of $x = 1$ in Figure 5.4.5 all join the $s = 0$ or $f(\mathbf{x})$ curve.

The first derivative in (5.4.86) for Example 5.4.5 would remain continuous using (5.4.88), but it is seen from (5.4.87) that there is a finite jump in the value of the second derivative as x increases through the value $x = 1$. Unfortunately, this is a defect in the inequality penalty function (5.4.88), since the second derivative is discontinuous exactly at the solution ($s \to \infty$). Of course, the inequality and equality penalty functions both have the defect that the Hessian matrix of second derivatives becomes increasingly illconditioned as the solution is approached, that is, as $s \to \infty$.

Also note that for a finite value of s when (5.4.86) is equal to zero (a minimum), the constraint $c(x) = x - 1 < 0$. This is typical of the inequality constraint penalty function in (5.4.88). Convergence is always approached from the infeasible side of the constraint boundary, thus the name exterior penalty function.

These remarks concerning early formulations of the penalty function apply to $n > 1$ as well. Illconditioning as the constrained minimum is approached is in the form of multidimensional ellipsoids with widely separated eigenvalues. Contrary to the exterior method described where the approach is from the infeasible side of the constraint boundaries, there is a *barrier function* method that remains feasible during its approach to the constraint boundaries. Barrier methods will not be described here (see problem 5.16), since they suffer from the same if not worse illconditioning and other computational difficulties and do not lead to the more successful multiplier penalty method to be introduced next. The interested reader is referred to Fletcher (1981a:126) and Gill (1981:207) for excellent overviews of all these penalty methods with instructive contour graphs of typical two-variable problems in those categories.

Considering the difficulties with illconditioning and discontinuous second derivatives associated with the exterior penalty functions in (5.4.83) and (5.4.88), it is fortunate that Powell (1969) has described a remedy for the basic method. His method is described by extending Example 5.4.5 for the one-variable case. Then a more general mathematical analysis is presented, followed by program segment MULTPEN to be merged into QNEWT as described in Section 5.4.4.

The *multiplier penalty function* for equality constraints is a modification of (5.4.83):

$$F(\mathbf{x}, \mathbf{S}, \mathbf{u}) = f(\mathbf{x}) + \frac{1}{2} \sum_{i=1}^{q} s_i [c_i(\mathbf{x}) - u_i]^2$$

$$= f(\mathbf{x}) + \tfrac{1}{2}[\mathbf{c}(\mathbf{x}) - \mathbf{u}]^T \mathbf{S}[\mathbf{c}(\mathbf{x}) - \mathbf{u}].$$

(5.4.89)

This penalty function has q separate multipliers, s_i, for each of the q equality constraints; these are collected as the elements in the *diagonal* matrix $S = \text{diag}[s_i]$ to provide more compact notation in subsequent analysis. Vector u is a set of q offset values that shift the origin in the constraint or c space; $u = 0$ makes (5.4.89) essentially the same exterior penalty function as (5.4.83), except that the latter assumes that the constraints are all equally scaled (so that only one multiplier is appropriate).

There are two sets of parameters in (5.4.89): S and u. Suppose that $x = x'(S, u)$ is the value of x that minimizes $F(x, S, u)$, for arbitrary values of S and u. Powell (1969:284) showed that the *particular* values S' and u' that make $c[x'(S', u')] = 0$ are those that minimize $f(x)$ such that $c(x) = 0$, that is, $x'(S', u')$ solves the constrained problem. Stated another way, there are an infinity of values of x that minimize (5.4.89) given sets of s_i and u_i values, $i = 1$ to q. If the two sets of s_i and u_i that correspond to a minimum of (5.4.89), say S' and u', *and* also $x = x'(S', u')$ satisfy $c[x', S', u'] = 0$, then $x'(S', u')$ is the solution to the minimization of $f(x)$ such that $c(x) = 0$.

The difficulty with the exterior penalty function in (5.4.83) was that the multipler s had to approach infinity in order to reach a solution, which is a computational impossibility. In Powell's multiplier penalty method the solution is obtained by satisfying

$$c_i[x'(S, u)] = 0, \qquad i = 1 \text{ to } q, \qquad (5.4.90)$$

for *finite* S. There is an outer and an inner loop to this process: First, finite values of multipliers s_i are selected. Then for a trial set of u_i values, (5.4.89) is minimized by varying x, and that value of x' is tried in (5.4.90). Since (5.4.90) usually will not be satisfied, u is adjusted and the process is repeated. Fortunately, the adjustments to u are not hit or miss; Powell showed that the proper adjustment is

$$u_i^{(k+1)} = u_i^{(k)} - c_i^{(k)}, \qquad i = 1 \text{ to } q, \qquad (5.4.91)$$

where $c_i^{(k)} = c_i[x'(S, u)]$, the values of the constraint functions after the last (kth) minimization over x.

The multiplier penalty process converges for suitably large but finite values of s_i, and it converges faster for larger values of s_i. Also, it is shown that at convergence the Lagrange multipliers associated with the q constraint functions are equal to

$$p_i = s_i u_i, \qquad i = 1 \text{ to } q. \qquad (5.4.92)$$

This process is clarified by performing the multiplier penalty procedure on the previous penalty example.

Example 5.4.6. The multiplier penalty function for the simple problem in (5.4.84) is constructed according to (5.4.89):

$$F(x, s, u) = x^{1/2} + \tfrac{s}{2}[c(x) - u]^2, \qquad (5.4.93)$$

where the constraint function is

$$c(x) = x - 1. \tag{5.4.94}$$

For each minimization with a fixed s and an adjusted value of u, the gradient of the multiplier penalty function is required:

$$g = \nabla F = \tfrac{1}{2}x^{-1/2} + s(x - 1 - u). \tag{5.4.95}$$

Those minima are determined when (5.4.95) is zero. From any value of x, the Newton–Raphson step from (3.2.48) to make (5.4.95) iteratively approach zero is

$$dx = -\frac{g}{g'}, \tag{5.4.96}$$

where g' is the derivative of (5.4.95) with respect to x.

The program in Table 5.4.4 performs the sequence of minimizations, each of which uses values of u adjusted according to (5.4.91), starting from $u = 0$. The iterative Newton–Raphson search is accomplished in defined variable $y = x^{1/2}$ because it is convenient (lines 140–220). Output from that program for a fixed multiplier $s = 2$ and starting from $x = 4$ is shown in Table 5.4.5. The values labeled "objective grad." are $\nabla f(x)$, the gradient of the unconstrained objective function, $x^{1/2}$. Notice that the estimated value of the Lagrange multiplier p computed by (5.4.92) rapidly converges to ∇f in this

Table 5.4.4. A BASIC Program for the Multiplier Penalty Problem in Example 5.4.6

```
10 REM - ONE-VARIABLE MULTPEN EXAMPLE 5.4.4.
20 CLS : PRINT "INPUT MULT. COEF. S="; : INPUT S
30 PRINT "INPUT STARTING VARIABLE X="; : INPUT X
40 U=0 : C=X-1 : REM - INIT OFFSET AND CALC CONSTR FNCN
50 F=SQR(X)+(S/2)*(C-U)*(C-U) : REM - TOTAL PENALTY FNCN
60 PRINT : PRINT "X =";X : PRINT "F =";F : PRINT "CONSTR =";C
70 PRINT "U =";U : PRINT "OBJECTIVE GRAD =";1/2/SQR(X)
80 PRINT "ESTI. LAGRANGE MULT =";S*U
90 PRINT "PRESS <RETURN> TO CONTINUE"; : INPUT S4$
100 IF S4$<>"" THEN GOTO 100
110 GOSUB 140 : REM - FIND X FOR MIN TOTAL PENALTY FNCN
120 C=X-1 : U=U-C : REM - NEW CONSTR & UPDATED OFFSET U
130 GOTO 50 : REM - CLOSE OFFSET U LOOP
140 REM - NEWTON-RAPHSON SUBROUTINE TO FIND MIN F(X)
150 U1=U+1 : Y2=X : Y=SQR(X)
160 G=.5+S*Y*(Y2-U1) : REM - GRAD OF TOTAL PENALTY FNCN WRT X
170 G1=S*(3*Y2-U1) : REM - 2ND DERIV OF TOTAL PENALTY FNCN WRT Y
180 D=-G/G1 : REM - NEWTON-RAPHSON STEP IN Y=SQR(X)
190 IF ABS(D)>.000001 THEN GOTO 210
200 X=Y2 : RETURN : REM - CONVERGED
210 Y=Y+D : Y2=Y*Y : REM - TAKE NEWTON STEP IN Y
220 GOTO 160 : REM - CLOSE NEWTON LOOP
230 END
Ok
```

Table 5.4.5.   Output of the Program in Table 5.4.4 for Example 5.4.6

```
                                        X = .9924326
RUN                                     F = 1.063041
INPUT MULT. COEF. S=? 2                 CONSTR =-7.567466E-03
INPUT STARTING VARIABLE X=? 4           U = .2509514
                                        OBJECTIVE GRAD = .5019027
X = 4                                   ESTI. LAGRANGE MULT = .5019027
F = 11                                  PRESS <RETURN> TO CONTINUE?
CONSTR = 3
U = 0                                   X = 1.001087
OBJECTIVE GRAD = .25                     F = 1.062434
ESTI. LAGRANGE MULT = 0                 CONSTR = 1.08695E-03
PRESS <RETURN> TO CONTINUE?             U = .2498644
                                        OBJECTIVE GRAD = .4997285
X = .7015161                            ESTI. LAGRANGE MULT = .4997288
F = 1.193936                            PRESS <RETURN> TO CONTINUE?
CONSTR =-.298484
U = .298484                             X = .9998462
OBJECTIVE GRAD = .5969682               F = 1.062509
ESTI. LAGRANGE MULT = .596968           CONSTR =-1.5378E-04
PRESS <RETURN> TO CONTINUE?             U = .2500182
                                        OBJECTIVE GRAD = .5000385
X = 1.0551                              ESTI. LAGRANGE MULT = .5000364
F = 1.062632                            PRESS <RETURN> TO CONTINUE?
CONSTR = 5.510009E-02
U = .2433839                            X = 1.000021
OBJECTIVE GRAD = .4867693               F = 1.062499
ESTI. LAGRANGE MULT = .4867678          CONSTR = 2.074242E-05
PRESS <RETURN> TO CONTINUE?             U = .2499974
                                        OBJECTIVE GRAD = .4999948
                                        ESTI. LAGRANGE MULT = .4999949
                                        PRESS <RETURN> TO CONTINUE?
```

case. That is due to (5.4.5):

$$\nabla f(\mathbf{x}', \mathbf{p}') = \mathbf{A}\mathbf{p}'. \tag{5.4.97}$$

In this example, $\mathbf{A} = 1$ according to (5.4.14), since $\mathbf{c}(\mathbf{x}) = x - 1$.

Figure 5.4.6 shows the first three multiplier penalty function curves on which a minimum was determined, as well as the curve for near-optimal $u' = 0.249996$. The corresponding minimum was at $x' = 1.000021$, which gave a constraint function value of 2.1E-5. The most important observation is that the minimum on the converged multiplier penalty function curve was *not* at the solution point and was obtained with a *finite* value of $s = 2$ in this case. Larger values of s, say $s = 4$, only make the convex curves more steep-sided so that convergence is more rapid.

Example 5.4.6 could have been worked in a similar way for inequality constraints by a modification of (5.4.89) that is similar to (5.4.88):

$$F(\mathbf{x}, \mathbf{S}, \mathbf{u}) = f(\mathbf{x}) + \frac{1}{2} \sum_{i=q+1}^{m} s_i \{\min[c_i(\mathbf{x}) - u_i, 0]\}^2. \tag{5.4.98}$$

The locus of $F(\mathbf{x})$ given values of $\mathbf{S}$ and $\mathbf{u}$ would appear as in Figure 5.4.6,

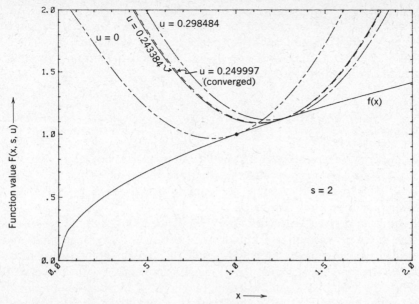

Figure 5.4.6. The first three and the last minimization curves for multiplier penalty function Example 5.4.6.

with the important difference that each curve would be identical to the $f(\mathbf{x})$ locus to the right of the respective points of tangency. Since the second derivative of $F(\mathbf{x})$ is discontinuous at those points, the multiplier penalty function has a clear advantage over the original exterior penalty function (5.4.88) because *the point of tangency does not converge to the solution point in the multiplier penalty method*. Fletcher (1981a) made that advantage clear in his complete analysis of the multiplier penalty method for mixed equality and inequality nonlinear constraints.

With that introduction to multiplier penalty functions, the theoretical basis is described more thoroughly by first developing only the equality constraint case without loss of generality. The reader might be tempted to try solving an equality-constrained problem by minimizing the Lagrangian function (5.4.2), treating both $\mathbf{x}$ and $\mathbf{p}$ as the variables. Of course, that will not work, because (5.4.4) shows that $F(\mathbf{x}, \mathbf{p})$ is a minimum with respect to $\mathbf{x}$ and a maximum with respect to $\mathbf{p}$. However, if one knows the values of the Lagrange multipliers at the solution $\mathbf{p}'$, then it is possible to perform an unconstrained optimization to the constrained minimum over $\mathbf{x}$. The reader may wish to confirm that fact, using the problem posed in Example 5.4.1 with the optimal Lagrange multiplier values given there.

In the spirit of the preceding concepts, Hestenes (1969) suggested what he called the *method of multipliers* based on the *augmented Lagrangian function*

$$P(\mathbf{x}, \mathbf{S}, \mathbf{p}) = f(\mathbf{x}) - \mathbf{p}^T \mathbf{c}(\mathbf{x}) + \tfrac{1}{2}[\mathbf{c}(\mathbf{x})]^T \mathbf{S}[\mathbf{c}(\mathbf{x})]. \qquad (5.4.99)$$

This function is the Lagrangian with an added term, and it is used in the following analysis. The added term can make the Hessian of the augmented function positive definite, as discussed in Section 5.4.5. For the immediate purpose, expansion of (5.4.89), use of the important definition from (5.4.92) and ordinary matrix algebra reveal that

$$F(\mathbf{x}, \mathbf{S}, \mathbf{u}) = P(\mathbf{x}, \mathbf{S}, \mathbf{p}) + \frac{1}{2} \sum_{i=1}^{m} \frac{p_i^2}{s_i}. \tag{5.4.100}$$

The difference between $F[\mathbf{x}(\mathbf{S}, \mathbf{u})]$ and $P[\mathbf{x}(\mathbf{S}, \mathbf{p})]$ is thus not a function of $\mathbf{x}$, so that the trajectory of respective minima are the same. That is, $\mathbf{x}'(\mathbf{S}, \mathbf{u}) = \mathbf{x}'(\mathbf{S}, \mathbf{p})$. This makes it possible to compute using $F(\mathbf{x}, \mathbf{S}, \mathbf{u})$ but make the following analysis using $P(\mathbf{x}, \mathbf{S}, \mathbf{p})$.

In order to derive a correction for offset vector $\mathbf{u}$ in (5.4.89), it is convenient to treat $\mathbf{S}$ as a constant and obtain a correction for Lagrange multiplier $\mathbf{p}$, knowing that adjustment of $\mathbf{p}$ will result in adjustment of $\mathbf{u}$ according to (5.4.92). The point of view that led to (5.4.4) is now taken with respect to P in (5.4.99): the variables are regarded as functions of the Lagrange multipliers, $\mathbf{x} = \mathbf{x}(\mathbf{p})$, so that minimization of $P[\mathbf{x}(\mathbf{p}), \mathbf{p}]$ in (5.4.99) implicitly requires the solution to the nonlinear set of equations

$$\nabla_{\mathbf{p}} P[\mathbf{x}(\mathbf{p}), \mathbf{p}] = \mathbf{0}. \tag{5.4.101}$$

By the implicit function theorem (Section 3.3.1), there exists a neighborhood about a solution point $\mathbf{x}'(\mathbf{p}')$ of (5.4.101) such that

$$P[\mathbf{x}(\mathbf{p}), \mathbf{p}] \le P[\mathbf{x}', \mathbf{p}] = P[\mathbf{x}', \mathbf{p}'], \tag{5.4.102}$$

which holds because $\mathbf{c}(\mathbf{x}') = \mathbf{0}$ in (5.4.99). It has again been shown that $\mathbf{p}'$ is a *maximizer* of $P(\mathbf{p})$ or minimizer of $-P(\mathbf{p})$. A Newton step to approach such a solution in $\mathbf{p}$ requires the first and second derivatives of P with respect to $\mathbf{p}$; from (5.4.99):

$$\nabla_{\mathbf{p}} P(\mathbf{p}) = -\mathbf{c}(\mathbf{p}), \tag{5.4.103}$$

$$\nabla_{\mathbf{p}}^2 P(\mathbf{p}) = -[\mathbf{N}^T \mathbf{W}^{-1} \mathbf{N}], \tag{5.4.104}$$

where the defined matrices $\mathbf{N}$ and $\mathbf{W}$ are

$$\mathbf{N} = [\nabla c_1 \ \nabla c_2 \ \cdots \ \nabla c_m], \tag{5.4.105}$$

$$\mathbf{W} = \nabla_{\mathbf{x}}^2 P[\mathbf{x}(\mathbf{p}), \mathbf{p}]. \tag{5.4.106}$$

To summarize, the process of minimizing $P[\mathbf{x}, \mathbf{p}]$ in (5.4.99) is to estimate a value of $\mathbf{p}$ and then minimize $P[\mathbf{x}, \mathbf{p}]$ with respect to $\mathbf{x}$. Then a correction to $\mathbf{p}$

would be

$$p^{(k+1)} = p^{(k)} + dp, \qquad (5.4.107)$$

where the Newton step **dp** is

$$dp = -[N^T W^{-1} N]^{-1} c[x(p)]. \qquad (5.4.108)$$

There are Newton methods to minimize P with respect to **x** as well as the separate Newton method to adjust the Lagrange multipliers **p** to maximize P. Fletcher (1975) has implemented those, which require both first and second derivatives with respect to **x**, as well as alternative quasi-Newton methods, which require only first derivatives.

Fortunately, both Powell (1969) and Hestenes (1969) employed a more elementary correction **dp** that is valid and performs well if **S** is sufficiently large but finite. Hessian matrix **W** in (5.4.106) is a function of multipliers contained in diagonal matrix **S** as seen by (5.4.99), that is, **W(S)**. Suppose that a large diagonal matrix **D** is added to **S**. Then, from (5.4.99) when $x = x'$,

$$W(S + D) = W(S) + NDN^T, \qquad (5.4.109)$$

where the parentheses serve as functional notation. A first application of the Sherman–Morrison–Woodbury identity in (2.1.36) yields an expression for $[W(S + D)]^{-1}$, and a second application yields

$$\left\{ N^T [W(S + D)]^{-1} N \right\}^{-1} = \left\{ N^T [W(S)]^{-1} N \right\}^{-1} + D. \quad (5.4.110)$$

See Problem 5.25. The left-hand side of (5.4.110) is the main component of the correction in (5.4.108), and (5.4.110) is dominated by the increase in multipliers **S** if **D** is large enough. Therefore, for sufficiently large multipliers in diagonal matrix **S**, the correction to the estimated Lagrange multipliers is

$$dp = -Sc, \qquad (5.4.111)$$

and the corresponding correction to the offset vector **u** according to (5.4.92) is

$$du = -c. \qquad (5.4.112)$$

Regarding the choice of $S = \text{diag}[s_1 \ s_2 \ \cdots \ s_m]$, multipliers s_i are increased from nominal starting values as necessary to force convergence of the constraints c_i at a reasonable linear rate between successive minimizations of $F(x, S, u)$ with respect to **x**. Powell has suggested that each $|c_i^{(k)}|$ should decrease by a factor of at least 4; otherwise, the corresponding s_i multipliers should be increased by a factor of 10. It is important to note that the corresponding u_i is simultaneously decreased by a factor of 10 so that the Lagrange multipliers estimated by (5.4.92) are unchanged.

Table 5.4.6.   Steps in Powell's Multiplier Penalty Algorithm to Minimize (5.4.89)

1. Initialize offset $\mathbf{u} = \mathbf{0}$ and set $s_i = 1$, $i = 1$ to m, or choose s_i such that the penalty terms contribute equally and sum to the magnitude of $f(\mathbf{x})$.
2. Minimize $F(\mathbf{x}, \mathbf{S}, \mathbf{u})$ to find $\mathbf{x}'(\mathbf{S}, \mathbf{u})$ and $\mathbf{c}[\mathbf{x}'(\mathbf{S}, \mathbf{u})]$.
3. Stop if $\|\mathbf{c}(\mathbf{x}')\|_\infty$ is suitable small, but if $\|\mathbf{c}(\mathbf{x}')\|_\infty$ increased, go to step 5.
4. If each $|c_i(\mathbf{x}')|$ decreased by factor 4 or more, set $\mathbf{u} = \mathbf{u} - \mathbf{c}(\mathbf{x}')$ and go to step 2.
5. Corresponding to each $|c_i(\mathbf{x}')|$ not decreasing by at least factor 4, adjust $s_i = 10s_i$ and $u_i = u_i/10$, then go to step 2.

The steps in Powell's algorithm for this most successful penalty method are given in Table 5.4.6; compare this to Example 5.4.6, which did not include the procedure for increasing the multipliers. The algorithm in Table 5.4.6 applies equally well for the multiplier penalty function (5.4.98) for inequality constraints, with some necessary details provided by Fletcher (1975). A precise flow chart and program segment to be merged with QNEWT and BOXMIN are described next.

5.4.4.   Program MULTPEN for Nonlinear Constraints.

Program segment C5-8, MULTPEN, is designed to be merged into the program formed first by quasi-Newton optimizer QNEWT with BOXMIN merged for lower and upper bounds on variables. MULTPEN implements the algorithm described in Table 5.4.6, treating the optimization routine as a subroutine during the penalty iterations that adjust the multipliers in $\mathbf{S}$ and the offset vector $\mathbf{u}$; see multiplier penalty objective functions (5.4.89) and (5.4.98). The quasi-Newton optimizer is started the first time with a Hessian matrix estimate equal to the unit matrix; subsequent minimizations start with the Hessian matrix from the preceding minimization. The result is a longer first minimization followed by noticeably shorter minimizations thereafter. Powell (1969:296) and Fletcher (1975:333) describe a correction to the Hessian matrix to account for any changes made to $\mathbf{S}$ and $\mathbf{u}$ between iterations, but that refinement was not found necessary and has not been included in MULTPEN.

As seen in program listing C5-8 in Appendix C, the MULTPEN algorithm is accomplished by adding lines 4000 to 4990 as well as a few lines that overwrite the original QNEWT code. An important change to QNEWT is provision *in MULTPEN* of a standard objective function and its gradient subroutines (5000 and 7000, respectively), so that the optimizer always addresses the multiplier penalty functions described by (5.4.89) or (5.4.98). Therefore, the *user* must now describe his or her unconstrained objective function $f(\mathbf{x})$ in subroutine *5500*, and the corresponding gradient, $\mathbf{g} = \nabla f(\mathbf{x})$, in subroutine *7500*; these are called by subroutines 5000 and 7000, respectively. Also the *user* must supply a description of the generally nonlinear constraint function(s) in subroutine *8000* and their gradients in a Jacobian array in subroutine *9000*. The details for writing these four subroutines are

Table 5.4.7. Subroutines Included or Required for Code Segment MULTPEN

Name	Lines
Compute Initial Multiplier Values	4700–4795
Find Maximum Constraint Residual Magnitude	4800–4895
Check User's Gradient by Differencing	4900–4990
Standard Multiplier Penalty Function	5000–5140
User's Unconstrained Objective $f(x)$	5500–6999
Standard Gradient for Penalty Function	7000–7100
User's Gradient for Objective $f(x)$	7500–7999
User's (In)equality Constraint Function(s)	8000–8999
User's Jacobian for Constraint Function(s)	9000–9999

given in Example 5.4.7. The various subroutines in MULTPEN are listed in Table 5.4.7. The flow chart for MULTPEN is shown in Figure 5.4.7. Subroutine 4700 assigns initial weights by first determining which constraints are binding; that includes all equality constraints and those inequality constraints that are not satisfied at the starting value of x^0. Then, equal portions of $|f(x^0)|$ are allocated to each squared constraint, thus determining the value of its multiplier. All inequality constraints that are initially satisfied start with unit multipliers. At the end of each minimization (loop), the value of the constraint having the maximum modulus is displayed so that the user may decide to continue or not. Experience with MULTPEN used with optimizer QNEWT suggests that $|c_i|_{\max} < 1\text{E}{-}6$ is a satisfactory stopping criterion.

Flag variable $K7$ plays an important part in the algorithm, indicating when the offsets have been adjusted. The offset adjustment is made every loop unless the required (linear) convergence rate is not obtained. Satisfactory convergence is indicated when the infinity norm of the constraint vector decreases by at least a factor of 4 every loop. Otherwise, the multipliers are increased for the *particular* constraints whose magnitudes failed to decrease by a factor of 4. Multiplier increases by a factor of 10 are accompanied by similar decreases in the corresponding offsets to preserve the Lagrangian multiplier estimates in the fundamental assumption of (5.4.92).

Example 5.4.7.   A problem from Himmelblau (1972:360) is

$$\text{Minimize } f(x) = 4x_1 - x_2^2 - 12$$

such that

$$c_1(x) = 25 - x_1^2 - x_2^2 = 0, \tag{5.4.113}$$

$$c_2(x) = 10x_1 - x_1^2 + 10x_2 - x_2^2 - 34 \geq 0, \qquad x_1 \geq 0, x_2 \geq 0.$$

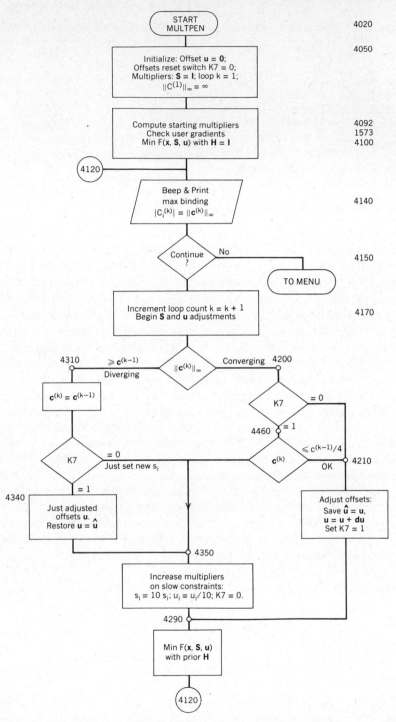

Figure 5.4.7. Flow chart for multiplier penalty program MULTPEN addition to quasi-Newton optimizer QNEWT.

302

Subroutines 5500, 7500, 8000, and 9000 that program the equations (5.4.113) and related gradients are contained in program C5-9, HIM360, in Appendix C; the lower bounds on the two variables are enforced by projection due to BOXMIN.

Subroutines 5500 and 7500 are similar to those normally provided in 5000 and 7000 with QNEWT to describe the unconstrained objective function and its gradient, respectively.

Subroutine 8000 describes constraints $c_1(\mathbf{x})$ and $c_2(\mathbf{x})$ and must contain several important features. In general, there are M constraints in array $C(\ )$, the first $K1$ of these being equality constraints ($K1 = 0$ may occur). Both $K1$ and M *must* be set in subroutine 8000. Inequality constraints are assumed to be in the form $c_i(x) \geq 0$; if given a constraint $y(\mathbf{x}) \leq 0$, then it must be programmed $c_i(\mathbf{x}) = -y(\mathbf{x})$.

Subroutine 9000 contains the Jacobian matrix whose columns are the gradient vectors of the respective constraint functions, that is, $\mathbf{A} = [\nabla c_1 \ \nabla c_2]$. These elements must be programmed in array $A(J, I)$ so that a typical element is the first derivative of the Ith constraint with respect to the Jth variable. Note that these row and column dimensions are reversed from those of the Jacobian matrix in the Gauss–Newton method (4.4.16). The definition employed in this section agrees with most literature on constraints.

To execute this example, LOAD"QNEWT, MERGE"BOXMIN, MERGE"MULTPEN, and MERGE"HIM360. RUN the program, using command 1 to set $x_1 = 3$, $x_2 = 3$; command 2 to print only every 50th iteration; command 5 to set zero lower bounds on x_1 and x_2; and command 6 to begin the sequence of constrained optimizations. The initial output checks the user's subroutine 9000 by comparing computed $\mathbf{g}(\mathbf{x}) = \nabla F(\mathbf{x})$ values by finite differencing the entire penalty function $F(\mathbf{x})$ for $\mathbf{u} = \mathbf{0}$; see (5.4.89) and (5.4.98).

The output from the first and last minimizations are shown in Table 5.4.8. The first of the sequential minimizations required nine iterations and 17 function evaluations and reduced the maximum constraint modulus to 0.9976. Note that $x_1 \geq 0$ came into play twice. The second minimization required four iterations and five function evaluations, and even less work was required for each subsequent minimization. Following the three dots in Table 5.4.8, the last iteration allowed by the user found $\mathbf{x}' = (1.00128 \ 4.89872)^T$ and $\mathbf{p}' = (1.01560 \ 0.75447)^T$ with a maximum constraint modulus of 3.53E–7. A total of 32 function evaluations were required for all six minimizations. It is easy to confirm (5.4.5), $\mathbf{g}(\mathbf{x}') = \mathbf{A}\mathbf{p}'$.

Example 5.4.8. The Rosen–Suzuki test problem was solved by various penalty methods by Gould in Lootsma (1972:356):

$$\text{Minimize } f(\mathbf{x}) = x_1^2 + x_2^2 + 2x_3^2 + x_4^2 - 5x_1 - 5x_2 - 21x_3 + 7x_4$$

Table 5.4.8. Partial Output From Example 5.4.7, Program HIM360

```
AT START OF ITERATION NUMBER 1
    FUNCTION VALUE =-9
  I         X(I)                 G(I)
  1      3.00000000          4.00000000
  2      3.00000000         -6.00000000
              ACTIVATED X( 1 ) LOWER BOUND
    ###### CUT BACK STEP SIZE BY FACTOR OF 4 #####
                STEP-TO-GRADIENT DEGREES=  0.0000
              ACTIVATED X( 1 ) LOWER BOUND
    ###### CUT BACK STEP SIZE BY FACTOR OF 4 #####
    ###### CUT BACK STEP SIZE BY FACTOR OF 4 #####
    ###### CUT BACK STEP SIZE BY FACTOR OF 4 #####
                STEP-TO-GRADIENT DEGREES= 25.6571
              ACTIVATED X( 1 ) LOWER BOUND
    ###### CUT BACK STEP SIZE BY FACTOR OF 4 #####
              ACTIVATED X( 1 ) LOWER BOUND
    ###### CUT BACK STEP SIZE BY FACTOR OF 4 #####
                STEP-TO-GRADIENT DEGREES= 59.1168
    ###### CUT BACK STEP SIZE BY FACTOR OF 4 #####
                STEP-TO-GRADIENT DEGREES= 13.0280
    ###### CUT BACK STEP SIZE BY FACTOR OF 4 #####
                STEP-TO-GRADIENT DEGREES= 10.5119
                STEP-TO-GRADIENT DEGREES= 12.5709
                STEP-TO-GRADIENT DEGREES= 11.2721
CONVERGED; SOLUTION IS:
AT START OF ITERATION NUMBER 9
    FUNCTION VALUE =-32.76677
  I         X(I)                 G(I)
  1      0.91202629          0.00007400
  2      5.01655694         -0.00015321
TOTAL NUMBER OF FUNCTION EVALUATIONS = 17
***********************************************************
ESTIMATED LAGRANGE MULTIPLIERS -
   CONSTRAINT # 1 :      0.00000
   CONSTRAINT # 2 :      0.00000
AFTER   1   PENALTY MINIMIZATIONS,
   THE MAX CONSTRAINT MODULUS # 1  = .9976354
CONTINUE PENALTY MINIMIZATIONS (Y/N)?

                    ● ● ●

AT START OF ITERATION NUMBER 1
    FUNCTION VALUE =-31.19197
  I         X(I)                 G(I)
  1      1.00128162         -0.00005222
  2      4.89871779          0.00000743
                STEP-TO-GRADIENT DEGREES=  9.2431
CONVERGED; SOLUTION IS:
AT START OF ITERATION NUMBER 2
    FUNCTION VALUE =-31.19197
  I         X(I)                 G(I)
  1      1.00128243         -0.00000001
  2      4.89871754          0.00000000
TOTAL NUMBER OF FUNCTION EVALUATIONS = 32
***********************************************************
ESTIMATED LAGRANGE MULTIPLIERS -
   CONSTRAINT # 1 :      1.01560
   CONSTRAINT # 2 :      0.75447
AFTER   6   PENALTY MINIMIZATIONS,
   THE MAX CONSTRAINT MODULUS # 2  = 3.536583E-07
CONTINUE PENALTY MINIMIZATIONS (Y/N)?
```

such that

$$x_1^2 + x_2^2 + x_3^2 + x_4^2 + x_1 - x_2 + x_3 - x_4 \leq 8,$$

$$x_1^2 + 2x_2^2 + x_3^2 + 2x_4^2 - x_1 - x_4 \leq 10, \qquad (5.4.114)$$

$$2x_1^2 + x_2^2 + x_3^2 + 2x_1 - x_2 - x_4 \leq 5.$$

The functions and their gradients in (5.4.114) are coded in program C5-10, LOOT356, in Appendix C. Note that there are three inequality constraints, so $K1 = 0$ and $M = 3$. Also note the required reversal of signs in the inequalities in subroutine 8000.

After accomplishing the LOAD"QNEWT, MERGE"BOXMIN, MERGE"MULTPEN, and MERGE"LOOT356 operations, it is informative to add the following lines that display the multipliers in **S** whenever they are changed:

4430 PRINT "-----------LOOP#";L8;" MULT S(";I;") =";
4432 PRINT USING " ######.####";S(I)
4434 NEXT I

A standard starting point is $\mathbf{x} = (0\ 0\ 0\ 0)^T$, and the constrained optimum is at $\mathbf{x}' = (0\ 1\ 2\ -1)^T$ with Lagrange multipliers $\mathbf{p}' = (1\ 0\ 2)^T$, that is, the second constraint is not binding (active) at the solution. The multipliers all are unity at the start, because none of the constraints were binding. At convergence, the multiplier for the third constraint had been increased to 10,000.

A more interesting starting point is $\mathbf{x} = (1\ 2\ 3\ 4)^T$, from which all the multipliers are increased. Much more accurate results are obtained by changing the optimization stopping criterion (menu command 2) to 0.000001. Note that after 11 penalty loops, the maximum constraint modulus has dropped off the end of the single-precision variable ($P8$) and become zero.

Since it requires 5 minutes to make these runs in IBM interpretive BASICA, IBM BASIC Compiler Version 1.00 was used to generate machine code. That reduced the run time to 48 seconds, a 6.25 : 1 reduction.

This section on program MULTPEN is concluded with a warning to users that the constraints must be appropriately scaled. That amounts to having each equality and inequality function producing values of approximately 1 to 100 in magnitude, and the same must be required of their derivatives. A problem Fletcher (1975:338) classified as difficult and requiring constraint scaling is the classic Colville (1968) test problem 3, which has five variables and 16 constraints, six of them nonlinear. It is recommended for the reader wishing to push MULTPEN to its limit. The solution is obtainable, but the bounds on variables must be included in the penalty function, since BOXMIN fails to perform satisfactorily in that case.

5.4.5. Other Methods for Nonlinear Constraints. Perhaps the most active area of current research in optimization methods concerns problems involving nonlinear constraints. Nonlinear programming algorithms not previously mentioned are usually much more difficult to program, but merit discussion so that the reader may comprehend existing and future articles on these methods. Among those algorithms that have performed well in the past or are currently attractive are the following:

1. Nonlinear elimination methods.
2. Lagrange–Newton method.
3. Exact penalty function method.
4. Sequential quadratic programming (SQP) method.

This section begins with methods that extend the concepts employed to deal with linear constraints, especially projection of the gradient and Hessian. The very nature of projection ignores curvature, which has now been recognized as the main consideration when dealing with nonlinear constraints. Therefore, the second part of this section deals with the Lagrangian function and its first and second derivatives. It turns out that the curvature of the constraint functions is even more important than that of the unconstrained objective function.

The theoretical basis for nonlinearly constrained optimization is the implicit function theorem in the sense that nonlinear constraints implicitly determine a neighborhood (small subspace) about $\mathbf{x}$ in which the necessary and sufficient conditions of unconstrained optimization apply. The neighborhood for which that is valid is the entire linear subspace for *linear* equality constraints, because the subspace is not a function of $\mathbf{x}$, that is, $\mathbf{A} = [\nabla c_1 \; \nabla c_2 \; \cdots \; \nabla c_q]$ is not a function of $\mathbf{x}$, since $\nabla c_i = \mathbf{a}_i$ from $\mathbf{A}^T\mathbf{x} = \mathbf{b}$. Drawing on the concept of a generalized elimination method from Section 5.4.1, note that (5.4.43) can be expressed as

$$[\mathbf{A} \; \mathbf{V}]^{-T} = [\mathbf{S} \; \mathbf{Z}], \qquad (5.4.115)$$

so that a linear transformation of variables may be defined as

$$\begin{bmatrix} \mathbf{z} \\ \mathbf{y} \end{bmatrix} = [\mathbf{A} \; \mathbf{V}]^T\mathbf{x} - \begin{bmatrix} \mathbf{b} \\ \mathbf{0} \end{bmatrix}. \qquad (5.4.116)$$

If $\mathbf{z} = \mathbf{0}$: (5.4.116) yields (5.4.10), $\mathbf{A}^T\mathbf{x} = \mathbf{b}$, and (5.4.115) and (5.4.116) yield (5.4.34), $\mathbf{x} = \mathbf{Sb} + \mathbf{Zy}$. This result is merely the linear transformation of the $\mathbf{x}$ space into the combined $\mathbf{z}, \mathbf{y}$ space, where $\mathbf{z} = \mathbf{0}$ leaves the $\mathbf{y}$ subspace in which the reduced gradient and reduced Hessian were projected for linear constraints.

The analogue of (5.4.116) for the *nonlinear* constraints case is to consider the nonlinear transformations

$$\mathbf{z} = \mathbf{c}(\mathbf{x}), \tag{5.4.117}$$

$$\mathbf{y} = \mathbf{V}^T\mathbf{x}. \tag{5.4.118}$$

The derivatives in the $\mathbf{x}$ space are mapped into a partitioned gradient vector according to

$$\begin{bmatrix} \mathbf{g}_z \\ \mathbf{g}_y \end{bmatrix} = [\mathbf{A} \ \mathbf{V}]^{-1}\mathbf{g}_x. \tag{5.4.119}$$

Gradient $\mathbf{g}_y$ contains the derivatives of $f(\mathbf{x})$ in the subspace of free variables, and $\mathbf{g}_z$ provides a first-order estimate of the Lagrange multipliers as indicated by (5.4.29).

Within the minimization that occurs in subspace $\mathbf{y}$, there must now be an inner Newton iteration that keeps $\mathbf{y}$ constant while maintaining $\mathbf{z} = \mathbf{0}$. The *generalized reduced gradient (GRG) method* by Abadie and Carpentier in 1969 was an early application of these concepts. The interested reader is referred to Fletcher (1981a:147).

To examine the remaining methods to be discussed for optimization with nonlinear constraints $\mathbf{c}(\mathbf{x}) = \mathbf{0}$, the *Lagrangian function*

$$L(\mathbf{x}, \mathbf{p}) = f(\mathbf{x}) - \mathbf{p}^T\mathbf{c} \tag{5.4.120}$$

is differentiated with respect to both variables $\mathbf{x}$ and multipliers $\mathbf{p}$:

$$\nabla_x L = \mathbf{g} - \mathbf{A}\mathbf{p}, \tag{5.4.121}$$

$$\nabla_x^2 L = \mathbf{H} - \sum_{i=1}^{q} p_i \nabla_x^2 c_i, \tag{5.4.122}$$

$$\nabla_p L = -\mathbf{c}, \tag{5.4.123}$$

$$\nabla_p^2 L = \mathbf{0}, \tag{5.4.124}$$

where $\mathbf{g} = \nabla_x f$, $\mathbf{H} = \nabla_x^2 f$, and $\mathbf{A} = [\nabla c_1 \ \nabla c_2 \ \cdots \ \nabla c_q]$. Then the $n + q$ changes in the elements of $\mathbf{x}$ and $\mathbf{p}$ for a Newton step toward a stationary point may be expressed by

$$\begin{bmatrix} \nabla_x^2 L & -\mathbf{A} \\ -\mathbf{A}^T & \mathbf{0} \end{bmatrix} \begin{bmatrix} d\mathbf{x} \\ d\mathbf{p} \end{bmatrix} = \begin{bmatrix} -\mathbf{g} + \mathbf{A}\mathbf{p} \\ \mathbf{c} \end{bmatrix}. \tag{5.4.125}$$

The linear system of equations in (5.4.125) can be solved for $d\mathbf{x} = \mathbf{x}^{(k+1)} - \mathbf{x}^{(k)}$ and $d\mathbf{p} = \mathbf{p}^{(k+1)} - \mathbf{p}^{(k)}$; this iterated process is called the *Lagrange–*

Newton method. It may be viewed as a linearization of the system of nonlinear equations $\nabla L(\mathbf{x}, \mathbf{p}) = \mathbf{0}$ in the same sense that the Newton–Raphson method (3.2.48) solves a set of nonlinear equations.

Example 5.4.9. Conn (1985:11) gave an example of both direct elimination and the Lagrange–Newton method:

$$\text{Minimize } f(\mathbf{x}) = x_1 + x_2^2 + x_3^2$$

$$\text{such that } -x_1 + 3x_2 - x_2 x_3 - 1 = 0. \tag{5.4.126}$$

The solution for this particular problem may be obtained by solving the constraint for x_1 and substituting that expression into $f(\mathbf{x})$:

$$x_1 = 3x_2 - x_2 x_3 - 1. \tag{5.4.127}$$

Therefore, the new unconstrained problem is to minimize

$$F(\mathbf{y}) = 3x_2 + x_2^2 - x_2 x_3 + x_3^2 - 1 \tag{5.4.128}$$

in the subspace $\mathbf{y} = (x_2 \ x_3)^T$. The Newton step in (3.2.48) is exact for a quadratic function:

$$\mathbf{dy} = -\frac{1}{3}\begin{bmatrix} 2 & 1 \\ 1 & 2 \end{bmatrix}\begin{bmatrix} 2x_2 - x_3 + 3 \\ -x_2 + 2x_3 \end{bmatrix}. \tag{5.4.129}$$

For instance, for $\mathbf{y}^{(1)} = (x_2 \ x_3)^T = (2 \ 1)^T$, $\mathbf{dy} = (-4 \ -2)^T$ and $\mathbf{y}^{(2)} = (-2 \ -1)^T$. Therefore, the solution to (5.4.126) is $\mathbf{x}' = (-9 \ -2 \ -1)^T$, using (5.4.127), and it may be verified from (5.4.5) that the Lagrange multiplier is $p' = -1$. Typically, the projected Hessian employed in (5.4.129) is positive-definite, even though (5.4.122), the Hessian of the Lagrangian function, is not; in fact it is singular.

In general, the Lagrange–Newton method requires solution of (5.4.125). Suppose $\mathbf{x} = (4 \ 2 \ 1)^T$ and $p = 3$. Then (5.4.125), which is rank 3, yields $dp = -4$, $dx_1 = 13$, $dx_2 = \frac{12}{5}$, and $dx_3 = -\frac{18}{5}$. Therefore, the Newton point is $\mathbf{x}^{(k+1)} = (17 \ \frac{22}{5} \ -\frac{13}{5})^T$ and $p = -1$. Also, the function value increases from $f^{(k)} = 9$ to $f^{(k+1)} = 43.12$. So even though the Lagrange multiplier is predicted correctly, a sequence of line search iterations would be required.

Exact penalty function methods attempt to estimate the exact Lagrange multipliers as functions of $\mathbf{x}$, $\mathbf{p} = \mathbf{p}(\mathbf{x})$, so that the minimum of $L[\mathbf{x}, \mathbf{p}(\mathbf{x})]$ in (5.4.120) is also the minimizer of $f(\mathbf{x})$. By the chain rule, the gradient of $L(\mathbf{x})$ in (5.4.121) when $\mathbf{p} = \mathbf{p}(\mathbf{x})$ is

$$\nabla_{\mathbf{x}} L(\mathbf{x}) = \mathbf{g}(\mathbf{x}) - \mathbf{A}(\mathbf{x})\mathbf{p}(\mathbf{x}) - \left[\nabla \mathbf{p}(\mathbf{x})^T\right]\mathbf{c}(\mathbf{x}). \tag{5.4.130}$$

At the solution point $\mathbf{x}', \mathbf{p}'$, (5.4.5) requires $\mathbf{g}(\mathbf{x}') = \mathbf{A}'\mathbf{p}'$ and $\mathbf{c}(\mathbf{x}') = \mathbf{0}$; these relationships applied to (5.4.130) meet the necessary conditions for a minimum, $\nabla_x L(\mathbf{x}) = \mathbf{0}$. One way a convergent sequence for $\mathbf{p}(\mathbf{x})$ may be selected is suggested by (5.4.24):

$$\mathbf{p}(\mathbf{x}) = \mathbf{A}^+\mathbf{g}. \tag{5.4.131}$$

There are severe problems with the exact penalty function method. Second derivatives are required, and it is necessary to add a penalty term, say $\mathbf{c}^T\mathbf{S}\mathbf{c}$ to (5.4.120) to make $\nabla_x^2 L(\mathbf{x})$ positive-definite. But the result is similar to $P(\mathbf{x}, \mathbf{S}, \mathbf{p})$ in (5.4.99), which has been successfully applied from a different point of view. Some investigators have defined approximations for (5.4.131) so that least-squares solutions of overdetermined linear equations are not required.

By far the most successful yet complex technique for solving the nonlinear programming problem is the *sequential quadratic programming (SQP) method*. A more convenient form for (5.4.125) is derived by recognizing the iterative nature of the process: $\mathbf{p}^{(k+1)} = \mathbf{p}^{(k)} + d\mathbf{p}$ and $\mathbf{d}^{(k)} = d\mathbf{x}$. Therefore, (5.4.125) may be rewritten as

$$\begin{bmatrix} \nabla_x^2 L & -\mathbf{A} \\ -\mathbf{A}^T & \mathbf{0} \end{bmatrix} \begin{bmatrix} \mathbf{d} \\ \mathbf{p} \end{bmatrix} = \begin{bmatrix} -\mathbf{g} \\ \mathbf{c} \end{bmatrix}. \tag{5.4.132}$$

The first matrix equation in (5.4.132),

$$\nabla_x^2 L\mathbf{d} - \mathbf{A}\mathbf{p} = -\mathbf{g}, \tag{5.4.133}$$

yields

$$\mathbf{d} = \left[\nabla_x^2 L\right]^{-1}(-\mathbf{g} + \mathbf{A}\mathbf{p}) = -\left[\nabla_x^2 L\right]^{-1}\nabla_x L, \tag{5.4.134}$$

using (5.4.121). But (5.4.134) implies that $\mathbf{d}$ is the Newton step for the problem

$$\text{Minimize } L(\mathbf{x} + \mathbf{d}, \mathbf{p}'), \tag{5.4.135}$$

such that

$$\mathbf{A}'^T\mathbf{d} = \mathbf{0}, \tag{5.4.136}$$

where the constraint is the second matrix equation from (5.4.132) with $\mathbf{c}(\mathbf{x}') = \mathbf{0}$.

A typical "major" iteration of the SQP method is to perform the line search

$$\mathbf{x}^{(k+1)} = \mathbf{x}^{(k)} + t\mathbf{d}^{(k)}, \tag{5.4.137}$$

where direction $\mathbf{d}^{(k)}$ solves the quadratic subproblem

$$\text{Minimize } Q(\mathbf{d}) = f + \mathbf{g}^T\mathbf{d} + \tfrac{1}{2}\mathbf{d}^T\left[\nabla_x^2 L\right]\mathbf{d} \tag{5.4.138}$$

such that

$$\mathbf{A}^T\mathbf{d} \geq -\mathbf{c}. \qquad (5.4.139)$$

The inequality constraint has been added for generality. In fact, the key property of the SQP method is that it rapidly identifies the correct set of binding constraints. The efficiency of the SQP method clearly depends on the efficiency of solving the sequence of quadratic subproblems for search direction $\mathbf{d}$. That is clearly possible, based on the analysis of Section 5.4.1, because (5.4.139) is simply a set of linear constraints that amount to a linearization of the nonlinear constraints $\mathbf{c}$. Recall that $\mathbf{A} = [\nabla c_1 \ \nabla c_2 \ \cdots \ \nabla c_q]$.

A practical disadvantage of the quadratic subproblem method just described is that second derivatives are required, and the Hessian in (5.4.138) is often singular or not positive definite. Han (1976) suggested using a Broyden family updated matrix approximation for $\nabla_x^2 L$ in (5.4.138). It is not clear which of several similar updating methods is best, but good practical results have been obtained with these variations of quasi-Newton algorithms involving quadratic subproblems. Reported results indicate that Han's algorithm and similar variations run about twice as fast as the best of alternative methods for nonlinear programming problems. However, there are several ad hoc decisions left to the programmer and the complexity is considerable unless an efficient constrained quadratic minimizer subroutine is available. In fact, Celis (1985) has extended Fletcher's trust radius technique described in Section 4.2.1 to minimize the quadratic subproblem. The method chosen must be able to deal with infeasible or unbounded subproblems, unbounded approximations for the Lagrange multipliers, and nonpositive-definite and singular Hessian approximations.

This section is concluded with the remark that many concepts usually encountered in explanations of nonlinearly constrained optimization have been omitted for the sake of clarity. For example, these discussions usually involve *convexity*, namely, the condition that all points in a subspace having nonlinear boundaries may be reached by linear interpolation between any two boundary points. As Fletcher (1981a:63) remarked, "The subject of convexity is often treated quite extensively in texts on optimization. My experience, however, is that much of this theory contributes little to the development and use of optimization algorithms."

Problems

5.1. Find an equation to compute the nth root of a given positive number e, using the Newton–Raphson method. In other words, obtain a formula for iteratively finding x such that $f(x) = x^n - e = 0$. This approach is programmed in many handheld calculators.

5.2. Use a secant search to find the square root of 80 to six significant figures, starting with a guess of $x = 10$. Compare this performance with that obtained by your equation in the preceding problem. Compute the rates of convergence.

5.3. Verify that $\mathbf{B}_{\text{BFGS}}$ in (5.1.22) satisfies the quasi-Newton condition in (5.1.10). Similarly, verify that $\mathbf{R}_{\text{DFP}}$ in (5.1.25) satisfies the quasi-Newton condition in (5.1.24).

5.4. Modification of a rank n matrix by a term of rank $p \le n$ may be accomplished by the Sherman–Morrison–Woodbury formula (2.1.36). Consider the inverse update formula $\mathbf{R}_{\text{DFP}}$ in (5.1.25). Since $\mathbf{R}_{\text{DFP}}\mathbf{B}_{\text{DFP}} = \mathbf{I}$, use (2.1.36) with $p = 2$ and $\mathbf{Q} = \mathbf{R}$ to obtain $\mathbf{B}_{\text{DFP}}$:

$$\mathbf{B}^*_{\text{DFP}} = \mathbf{B} + \left[1 + \frac{\mathbf{d}^T\mathbf{B}\mathbf{d}}{\mathbf{y}^T\mathbf{d}} \right]\frac{\mathbf{y}\mathbf{y}^T}{\mathbf{y}^T\mathbf{d}} - \left[\frac{\mathbf{y}\mathbf{d}^T\mathbf{B} + \mathbf{B}\,\mathbf{d}\mathbf{y}^T}{\mathbf{y}^T\mathbf{d}} \right]. \quad (5.5.1)$$

5.5. Show that under a linear transformation of the variable space:
   (a) The Secant recursion (5.1.2) is invariant if $B^{(k)}$ is positive.
   (b) Directional derivatives (slopes) are equal in the two spaces according to (5.1.48).
   (c) The steepest descent is not invariant unless under an orthogonal transformation, particularly that the upper triangular form in $\mathbf{U}$ such that $\mathbf{U}^T\mathbf{U} = \mathbf{H}_x$ converts steepest descent into Newton's method.
   (d) The BFGS update formula (5.1.21) is invariant.

5.6. Employ linear transformation and translation of variables, $\tilde{\mathbf{x}} = \mathbf{D}\mathbf{x} + \mathbf{c}$ to transform the interval $400.1234 \le x \le 400.9876$ to $-1 \le \tilde{x} \le +1$. Find $\mathbf{D}$ and $\mathbf{c}$. Indicate any necessary adjustments to first and second derivatives from the $\mathbf{x}$ space to the $\tilde{\mathbf{x}}$ space.

5.7. Linear transformations of variables, $\tilde{\mathbf{x}} = \mathbf{T}\mathbf{x}$, do not provide an invariant metric for Euclidean distance under the two-norm, $\|\tilde{\mathbf{x}}\|_2 = \|\mathbf{x}\|_2$, unless $\mathbf{T}$ is orthonormal. Using (5.1.43), show that the *elliptic norm* (4.1.9) is invariant without such a restriction on $\mathbf{T}$.

5.8. Verify the algebra leading to the conclusion that inequality (5.2.12) involving three sampled function values guarantees that the minimum on a quadratic function occurs at $t < 4$ on the normalized line search t scale.

5.9. Verify the algebra leading to derivative (5.2.24) and location of a minimum (5.2.26) on a cubic function determined by two sampled function values and their first derivatives.

5.10. Use program QNEWT to find the *unconstrained* minimum of the Lagrangian function (5.4.2), with the objective and constraint functions (5.4.26) and the fixed, optimal Lagrange multiplier vector $\mathbf{p}' = (1 \;-\; \frac{1}{3})^T$ found in Example 5.4.1. Observe that the solution $\mathbf{x}'$ agrees with Example 5.4.1, but that trials using nonoptimal multipliers do not converge to the correct solution.

5.11. Find the *maxima* shown in Table 5.4.1 for the camelback function (5.4.75), using QNEWT and starting at various $\mathbf{x}^{(0)}$ in a neighborhood of the origin with radius 3.

5.12. Run program QNEWT with QNEWTGRD merged for Rosenbrock's and Wood's functions; compare the number of iterations and function values with those under "CUTBACK" in Table 5.3.3. Also observe the increase in run time with the larger number of variables.

5.13. Form the Lagrangian function (5.4.2) for the linearly constrained quadratic problem (5.4.12) to (5.4.13). By setting $\nabla_{\mathbf{x}} L = \mathbf{0}$ and $\nabla_{\mathbf{p}} L = \mathbf{0}$, show that the *Lagrangian matrix* is defined by

$$\begin{bmatrix} \mathbf{H} & -\mathbf{A} \\ -\mathbf{A}^T & \mathbf{0} \end{bmatrix} \begin{bmatrix} \mathbf{x} \\ \mathbf{p} \end{bmatrix} = - \begin{bmatrix} \mathbf{d} \\ \mathbf{b} \end{bmatrix}. \qquad (5.5.2)$$

The Lagrangian matrix is symmetric, not positive definite, and may be singular.

5.14. Solve linearly constrained quadratic problem (5.4.26) by computing one Newton step using the projected gradient and Hessian in (5.4.63) and (5.4.64), respectively.

5.15. Run program QNEWT with BOXMIN merged to minimize:

(a) Quadratic function (3.2.11) such that $x_2 \geq 3.9$; start at $\mathbf{x}^{(0)} = (1.9 \; 4.5)^T$. Compare with the results in Section 4.3.3, especially Table 4.3.5 and Figure 4.3.5.

(b) $F(\mathbf{x}) = -x_1 x_2$ such that $x_1 \leq 1$ and $x_2 \leq 2$, starting from $\mathbf{x}^{(0)} = (0.1 \; 0.1)^T$.

5.16. An *inverse barrier function* for inequality constraints was described by Carroll (1961):

$$\text{Minimize } F(\mathbf{x}) = f(\mathbf{x}) + r \sum_{i=1}^{M} \left[c_i(\mathbf{x}) \right]^{-1}, \qquad r \to 0. \quad (5.5.3)$$

The starting and all subsequent values of $\mathbf{x}$ must be feasible, that is, satisfy the constraints. Using this method, minimize $f(\mathbf{x}) = 4x_1 + x_2$ such that $x_1 \geq 0$, $x_2 \geq 0$. First use QNEWT for $r = 1$, and note that an overflow condition results; the BASIC direct-mode statement

"?X(1);X(2)" will show that the search left the first quadrant and proceeded toward negative infinity. Stop that by merging BOXMIN into QNEWT and using BOXMIN to ensure that the minimization remains in the first quadrant, while retaining the inverse terms in (5.5.3). Obtain a sequence of minima $\mathbf{x}'(r)$ for $r = 10, 1, 0.1, 0.001$, and 0.0001. By setting the gradient of (5.5.3) equal to zero, explain the trajectory $\mathbf{x}'(r)$ analytically.

5.17. Expand the multiplier penalty function expressions in (5.4.89) and (5.4.99) for two variables to show that they are equal when $p_1 = u_1 s_1$ and $p_2 = u_2 s_2$, that is, that (5.4.92) is required.

5.18. Find a correction to offset parameter u analytically in (5.4.93) for equality constraint problem (5.4.84). Treating Lagrange multiplier p as the variable, find $\nabla_p P(p)$ and $\nabla_p^2 P(p)$ and form the algebraic expression for the Newton step dp. Show that $du \to (-c)$ as $s \to \infty$, if $p = su$ is maintained. Separately, show that $\nabla_x^2 P(x, p) \to s$ as $s \to \infty$.

5.19. Solve equality constraint problem (5.4.26), using QNEWT with BOXMIN and MULTPEN merged, starting from $\mathbf{x}^{(0)} = (1\ 2\ 3)^T$ and continuing penalty loops until the maximum constraint modulus is less that 10^{-6}. Note that the solution agrees with both the $\mathbf{x}'$ and $\mathbf{p}'$ found by general elimination in Example 5.4.1. Then use finite differences to verify (5.4.8). Separately, perturb each of the two constraint constants (4 and 0) by adding 0.1 and obtaining two new constrained solutions. Evaluate $f(\mathbf{x})$ for the unperturbed and two perturbed solutions for $\mathbf{x}$; use these values in a forward finite difference formula to show that (5.4.8) is true.

5.20. Powell (1969:287) posed a nonlinear programming problem and Celis (1985:79) presented some results:

$$\text{Minimize } f(\mathbf{x}) = x_1 x_2 x_3 x_4 x_5 \text{ such that}$$

$$c_1(\mathbf{x}) = x_1^2 + x_2^2 + x_3^2 + x_4^2 + x_5^2 = 10,$$

$$c_2(\mathbf{x}) = x_2 x_3 - 5x_4 x_5 = 0, \tag{5.5.4}$$

$$c_3(\mathbf{x}) = x_1^3 + x_2^3 = -1.$$

Two solutions are $\mathbf{x}' = (-1.7171\ 1.5957\ \pm 1.8272\ -0.7636$ $\mp 0.7636)^T$. Obtain solutions using QNEWT with BOXMIN and MULTPEN merged and verify (5.4.5). One solution may be obtained using 86 function evaluations for maximum constraint modulus less than 10^{-6} by starting at $\mathbf{x}^{(0)} = (-1\ 1.5\ 2\ -1\ -2)^T$. However, the penalty function method diverges (with the values of $\mathbf{S}$ built into MULTPEN) when $\mathbf{x}^{(0)} = (-1\ -1\ -1\ -1\ -1)^T$ unless BOXMIN variable bounds of $-2 \leq x_j \leq +2$, $j = 1$ to n, are utilized. Other starting points given by Celis (1985:80) will fail for the value of **S** utilized in MULTPEN.

5.21.   From Fletcher (1981a:50), use QNEWT with BOXMIN and MULT-PEN merged to

$$\text{Minimize } f(\mathbf{x}) = -x_1 - x_2 \text{ such that}$$

$$c_1(\mathbf{x}) = x_2 - x_1^2 \geq 0, \qquad\qquad (5.5.5)$$

$$c_2(\mathbf{x}) = 1 - x_1^2 - x_2^2 \geq 0.$$

Sketch the boundaries of the constraints, indicate the feasible region, show several level curves of $f(\mathbf{x})$, and draw the gradient vector $\nabla f(\mathbf{x}')$ at the solution point $\mathbf{x}'$. Discuss the values, signs, and interpretations of the Lagrange multipliers $\mathbf{p}'$.

5.22.   Show how (5.4.125) may be obtained.

5.23.   Discuss the properties of the implicit function theorem listed in Table 3.3.1, as they apply to the Lagrange–Newton stationary point (5.4.125).

5.24.   Obtain the first Lagrange–Newton step according to (5.4.125) for problem (5.4.126) using singular value decomposition (SVD). See (3.1.63) to (3.1.68).

5.25.   Apply the Sherman–Morrison–Woodbury identity twice to verify (5.4.110).

Chapter Six

Network Optimization

Variable parameters of linear physical systems oscillating in the sinusoidal steady state may be adjusted very easily by applying the optimization methods described in Chapters Four and Five. The choice of these variables is often the result of some design process, and it is the redesign or readjustment by optimization that is considered here. This chapter deals with electrical networks, but the mechanical analogues and applications in digital signal processing where the same mathematical principles apply are mentioned.

A brief review of the underlying differential equations leading to simple harmonic motion and the concepts of complex frequency and impedance will establish some common interdisciplinary ground. The fundamental tools of network analysis, constitutive equations and Kirchhoff's laws, will lead to a simple and efficient method for ladder networks and to the nodal matrix method for general network topologies.

It is fortunate that the most useful objective function for systems in steady-state sinusoidal oscillation coincides with the least-pth form for systems sampled over the real frequency domain. Aaron (1956) advocated the use of least squares for electrical networks; the improvements since that time include faster and more robust optimization algorithms and methods for efficiently computing exact gradient vectors. Therefore, the Gauss–Newton optimization method (Chapter Four) employing first derivatives is adapted for constrained adjustment of elements in electrical networks, and projected variable bounds and the multiplier penalty method from Chapter Five are also incorporated. Program TWEAKNET will provide concrete illustrations of the power of these techniques for automatic redesign of ladder networks.

Although it is possible to obtain first partial derivatives by forward finite differences, that time-consuming technique is unnecessary, since exact derivatives may be computed efficiently for all linear systems such as electrical networks and their analogues. The definition and application of Tellegen's remarkable theorem for this purpose will provide insight and computation of exact partial derivatives for any network, especially for lossless networks. The

315

generality of Tellegen's theorem is actually not essential, so that direct differentiation of network matrix equations is shown to yield the same results with greater applicability, even if more obscurely.

The concluding topic concerns robust network response functions, namely, those that are most sensitive yet well behaved for variations in network variables. The dependence of most common network response functions on branch impedances has the form of bilinear or linear fractional transformations. Although the most familiar representation of that form is the electrical engineer's Smith chart, it also is true that such mappings among certain unit disks in the complex variable plane have been studied by mathematicians for centuries, using the Poincaré metric. Bilinear forms are the basis for some of the best components of optimization objective functions. Again, the important issue for optimization of variables that are nonlinearly related to responses of linear systems is that the chosen responses are sensitive yet well conditioned with respect to the variables.

6.1. Network Analysis in the Sinusoidal Steady State

The following sections describe some common mathematical and physical ground among mechanical systems, electrical networks, and sampled functions that are appropriately described by differential equations of simple harmonic motion, Laplace transforms, and z transforms.

The formation and evaluation of network equations involving impedances and admittances (*immittances*) is addressed as network analysis. Based on Kirchhoff's laws, simple recursive equations may be written for ladder networks topologies, and nodal matrix relationships describe linear networks with arbitrary structures.

6.1.1. From Differential Equations to the Frequency Domain. The voltage equilibrium equation for the series resistance, inductance, and capacitance (RLC) circuit shown in Figure 6.1.1 is

$$L\frac{di}{dt} + Ri + \frac{1}{C}\int i\,dt = e_s(t). \tag{6.1.1}$$

The source voltage is e_s, and i is the current through the circuit. If (6.1.1) is differentiated with respect to time t, then it is a linear, second-order differen-

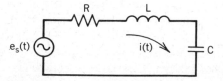

Figure 6.1.1. A series RLC electrical network excited by a sinusoidal voltage source.

tial equation with real, time-invariant coefficients. Equation (6.1.1) models a *linear system* because response $i(t)$ is proportional to scale changes in excitation $e_s(t)$ and because superposition applies. *Superposition* implies that if response $i_1(t)$ is obtained for excitation $e_1(t)$, and similarly for $i_2(t)$ and $e_2(t)$, then $i(t) = i_1(t) + i_2(t)$ when $e(t) = e_1(t) + e_2(t)$. It is especially important to note that various *parameters* or *variables* in linear systems may be nonlinearly related to the response; for example, C in (6.1.1) does not affect response $i(t)$ linearly.

A means for obtaining a solution for the natural frequencies of (6.1.1) is to assume an exponential form for the current given some initial energy stored in the network. Assume that $e_s(t) = 0$; then the response is determined by the *homogeneous equation*

$$i(t) = Ie^{st}. \tag{6.1.2}$$

Since

$$\frac{di}{dt} = si(t) \quad \text{and} \quad \int i(t)\,dt = \frac{i(t)}{s}, \tag{6.1.3}$$

substitution of (6.1.3) into (6.1.1) yields

$$Ls + R + \frac{1}{sC} = 0. \tag{6.1.4}$$

Parameter s is also the frequency in Laplace transform theory. Upon multiplying (6.1.4) by s, the resulting quadratic polynomial will have two roots, s_1 and s_2, and they are generally complex numbers that have real and imaginary parts. Roots s_1 and s_2 are the *characteristic values* and are comparable to the eigenvalues previously discussed in connection with the characteristic equation in (2.2.33). See Maron (1982:398) and Jennings (1977:233). Because the quadratic polynomial in (6.1.4) has real coefficients, the roots must be either real or in conjugate pairs. The latter represent the usual oscillatory case, in which the response decays with time according to (6.1.2) because s has a negative real part. A complete homogeneous solution is the sum of two exponential terms,

$$i(t) = c_1 e^{s_1 t} + c_2 e^{s_2 t}, \tag{6.1.5}$$

where c_1 and c_2 may be viewed as constants of integration that are determined to satisfy the initial condition of the network at $t = 0$.

The *particular solution* of the linear second-order differential equation in (6.1.1) is that obtained with some forcing function, $e_s(t)$. The case of interest in Chapter Six is sinusoidal oscillation, so one choice is

$$e_s(t) = E\cos(\omega t). \tag{6.1.6}$$

However, the usual approach for a particular solution is to impose the exponential forcing function

$$e(t) = Ee^{j\omega t} = E[\cos(\omega t) + j\sin(\omega t)], \qquad (6.1.7)$$

where $j = (-1)^{1/2}$. Then a response function of the form

$$i(t) = Ie^{j\omega t} \qquad (6.1.8)$$

satisfies (6.1.1) and leads to solution of the response magnitude

$$I = \frac{E}{R + sL + 1/sC}, \qquad s = j\omega. \qquad (6.1.9)$$

The key point is that a similar forcing function could have been the complex conjugate of (6.1.7), leading to the complex conjugate of response (6.1.9). Therefore, superposition and the identity $2\cos(\omega t) = e^{j\omega t} + e^{-j\omega t}$ lead to the conclusion that the sum of the two forcing functions is the real sinusoid in (6.1.6). Similarly the response is

$$i(t) = \text{Re}[Ie^{j\omega t}] = |I|\cos(\omega t + \phi), \qquad (6.1.10)$$

where Re means "real part of" and the *phase angle* ϕ is the argument of the current *phasor*

$$I = |I|e^{j\phi}. \qquad (6.1.11)$$

The *general solution* for any forcing function is the sum of the homogeneous and particular solutions; the former may be ignored for the sinusoidal steady-state condition, because it will decay to zero for t sufficiently large. The important quantity in the sinusoidal steady-state solution is the loop *impedance Z* related to Figure 6.1.1:

$$Z = R + sL + \frac{1}{sC}. \qquad (6.1.12)$$

It is the impedance that determines the magnitude and phase angle of the steady-state response according to (6.1.9) through (6.1.12):

$$I = \frac{E}{Z}, \qquad (6.1.13)$$

which is *Ohm's law*. The convention ordinarily employed and used consistently in this chapter is that voltage and current phasors E and I are *root-mean-square* values; therefore, the extreme values of the time function in (6.1.10) have magnitude $2^{1/2}|I|$. Consequently, the real power dissipated in resistance R is

$$P_R = |I|^2 R. \tag{6.1.14}$$

The impedance function in (6.1.12) is a function of *complex frequency s*, say $Z(s)$, where

$$s = \sigma + j\omega. \tag{6.1.15}$$

In the steady state, ω is called the *real frequency* so that the impedance becomes the function $Z(j\omega)$. It is known by the principle of *analytic continuation* that analytic (regular) complex functions of complex variables such as $Z(s)$ are completely determined by their specification over the entire real frequency (ω) axis. Therefore, references to either $Z(s)$ or $Z(j\omega)$ are used as convenient.

The familiar treatment of this section serves as a remainder of the analytical path from linear second-order differential equations representing simple harmonic motion to the complex frequency plane. An alternative explanation may be derived from Laplace transforms with similar results.

6.1.2. Related Technical Disciplines. The most obvious analogues of the preceding analysis of differential equations occur in mechanical systems. Table 6.1.1 compares quantities in several analogous systems. Electrical current (i) is the rate of flow of charge (q) with respect to time, and capacitance (C) is the constant of proportionality between charge and voltage (V): $q = Cv$. Flux linkage is dual to charge, and inductance (L) is the relevant constant of proportionality. An elementary series electrical circuit is illustrated in Figure 6.1.1. A parallel circuit has respective R, L, and C terminals all connected together, in which case a current source that produces a voltage response would be appropriate. As discussed in the next section, practical circuits consist of arbitrary connections of both series and parallel circuits, all of which have mechanical analogues. The mechanical filter field is especially

Table 6.1.1. **Mechanical Analogues of Electrical Quantities**

System	Coordinate	Velocity	Force
Mechanical translation	Position	Velocity	Force
Mechanical rotation	Angular position	Angular velocity	Torque
Series circuit	Charge	Current	Voltage
Parallel circuit	Flux linkage	Voltage	Current

dependent upon electromechanical analogies based on differential equations. The interested reader is referred to Johnson (1973:164) and (1983).

There are several related electrical filter technologies that have important connections to complex functions of complex frequency, such as the impedance function $Z(s)$. It is shown in the next section that such functions are generally rational polynomials. The signal-sampling process based on impulse or delta function modulation leads to the *z transform*,

$$z = e^{sT}, \tag{6.1.16}$$

which is the basis of digital filter technology. Parameter T is the unit sampling interval. One way to design digital filters is by transformation of desirable rational functions of complex frequency s into rational polynomials in z^{-1}. The equivalence in that case is between RLC filters and synchronous sampling switches connected with an arithmetic processor. See Golden (1973).

Other types of filters that are modeled on LC filter prototypes are microwave filters, RC active filters, and the more recent switched-capacitor filters. Interested readers are referred to Temes (1973). The point being emphasized is that optimization of RLC filters illustrates the same potential for optimization of a large number of other technical design problems based on differential equations or rational polynomial functions of complex frequency that define the prototype system.

6.1.3. *Network Analysis.*   A restatement of Ohm's law (6.1.13) is

$$E = IZ \quad \text{or} \quad I = EY, \tag{6.1.17}$$

where Z is an *impedance function* of complex frequency $Z(s)$ and Y is the reciprocal *admittance function*:

$$Z = \frac{1}{Y}. \tag{6.1.18}$$

In the sinusoidal steady state, the complex frequency variable s is purely imaginary, $s = j\omega$. Table 6.1.2 summarizes these functions.

It is customary to associate a parasitic power loss with inductors and capacitors. As indicated by (6.1.14), a current through a resistance results in

Table 6.1.2.   Impedances and Admittances of R, L, and C Network Elements

Element Type	Impedance	Units	Admittance	Units
Resistance	R	Ohms	$G = 1/R$	Mhos
Inductance	sL	Ohms	$1/(sL)$	Mhos
Capacitance	$1/(sC)$	Ohms	sC	Mhos

loss of real power that cannot be recovered, as opposed to stored energy that is returned to the connecting circuit by an L or a C during part of the frequency cycle. Therefore, a dissipative inductor's impedance function and a dissipative capacitor's admittance function of real frequency ω are:

$$Z_L = R_L + jX_L = \omega L(d + j1), \qquad X_L = \omega L, \qquad (6.1.19)$$

$$Y_C = G_C + jB_C = \omega C(d + j1), \qquad B_C = \omega C, \qquad (6.1.20)$$

where the *decrement d* is related to *unloaded quality factor Q* by

$$d = \frac{1}{Q}. \qquad (6.1.21)$$

Quality factor Q relates the power loss element to inductive *reactance* ωL or capacitive *susceptance* ωC:

$$R_L = \frac{\omega L}{Q}, \qquad G_C = \omega C/Q. \qquad (6.1.22)$$

Quality factor Q is different for each kind of element and among elements of the same kind; it is not necessarily independent of frequency (ω), although that is often an acceptable and convenient assumption.

Figure 6.1.2 shows the schematic representation of the R, G, L, and C elements. The loss in conductance G_C is also because of the current, but since it is in parallel with the capacitor, that power dissipation is usually expressed as a function of the voltage:

$$P_G = |V|^2 G_C. \qquad (6.1.23)$$

In Figure 6.1.2 the *reactive power* (average stored energy) is $|I|^2(\omega L)$ in the inductor and $|V|^2(\omega C)$ in the capacitor. Therefore, (6.1.14) and (6.1.23) show

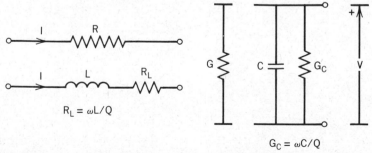

Figure 6.1.2. Schematic representations for resistance, conductance, lossy inductor, and lossy capacitor.

that quality factor Q is the ratio of reactive to real (dissipated) power in an inductor or capacitor.

It is often necessary to invert the immittances in (6.1.19) and (6.1.20) to obtain $Y_L = 1/Z_L$ or $Z_C = 1/Y_C$. A useful identity for that purpose is

$$\frac{1}{d + j1} = \frac{d - j1}{d^2 + 1}. \tag{6.1.24}$$

A *network graph* consists of nodes interconnected by branches that contain elements, as illustrated in Figure 6.1.3. The nodes are numbered from the ground or common node 0 to node n, and the elements in the branches are subscripted with the node numbers at the ends of the branch. There are node voltages with respect to common node 0, and there are branch voltages across each branch. Similarly, there are node currents from the common node (e.g., I_k and I_n) and branch currents (e.g., I_{kn}). A strict polarity convention is necessary: All voltages are potential "rises" to the arrow or $+$ sign, and branch currents always enter the $+$ end of the branch element.

Kirchhoff's current law requires that the sum of all currents leaving a node equals zero; in Figure 6.1.3 $I_{k1} + I_{k2} + I_{kn} + I_{kp} - I_k = 0$. *Kirchhoff's voltage law* requires that the sum of voltages around any loop equals zero; in Figure 6.1.3 $V_2 + V_{k2} - V_{kn} - V_n = 0$. Since the network is assumed to be in the sinusoidal steady state, all voltages and currents are phasors as described in Section 6.1.1. Although the branch elements in Figure 6.1.3 are shown as admittances Y_{ij}, any of them could be described as the equivalent impedance, $Z_{ij} = 1/Y_{ij}$. The choice often is determined by the *constitutive laws* of the elements, for example, Ohm's law for resistors: $V = IR$ or $I = VG$. Another example is a constitutive law for a branch element in Figure 6.1.3, for example, $I_{k2} = V_{k2}Y_{k2}$.

A much more specific graph that is used extensively in this chapter is the *ladder network* illustrated in Figure 6.1.4. It is convenient without loss of

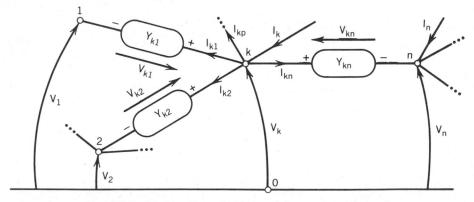

Figure 6.1.3.   Node and branch voltages and currents in a network.

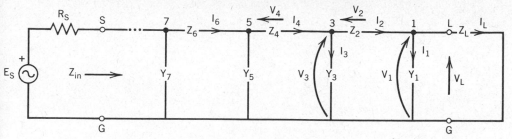

Figure 6.1.4. Convenient notation for ladder network analysis.

generality to be overly restrictive in *topology* (location of nodes and branches), including the source and load (sink), as well as in nomenclature. The *Thevenin source* topology is assumed here, and its impedance is the resistance R_s. All shunt branch elements are admittances, and all series branch elements are impedances, numbered from 1 at load node L and increasing toward the source node S. The nodes have odd numbers except for the common ground node G. Nodes S and G define the *input port*, distinguished by the fact that the net current crossing that two-node interface sums to zero; a similar statement defines the *output port* defined by nodes L and G. Branch currents are defined to flow toward the load or ground; consequently, all branches voltages are defined to rise to the nodes or toward the source. Because these conventions are simple to deduce, ladder networks will usually be marked only with node numbers.

It is quite laborious to assemble rational polynomials that represent responses in complex frequency ($s = \sigma + j\omega$) for any but the most simple networks, such as in Figure 6.1.1 and equation (6.1.9). However, a very simple but efficient real-frequency ($s = j\omega$) analysis scheme is well known for the ladder network in Figure 6.1.4. When the source voltage E_s and all branch immittances are finite and nonzero, then a finite current I_L will flow through load impedance Z_L. Alternatively, it is valid to assume that load current has some arbitrary value and then to apply Kirchhoff's laws to work back to discover the source voltage that is required to produce that load current. If the reader objects that the source voltage thus obtained is not a desired value, then all voltages and currents in the network may be scaled by the particular complex constant that produces the desired source voltage. The suggested step is valid precisely because the network is *linear*.

Figure 6.1.5 displays the load-to-source recursion scheme defined by the recursion formula that has been labeled the *complex linear update* [Cuthbert (1983:71)]:

$$A = BC + D, \tag{6.1.25}$$

where A, B, C, and D are complex variables and are not the so-called chain or *ABCD* parameters often associated with two-port networks. Because computers evaluate trigonometric functions slowly, the cartesian (rectangular)

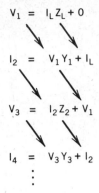

$$V_1 = I_L Z_L + 0$$

$$I_2 = V_1 Y_1 + I_L$$

$$V_3 = I_2 Z_2 + V_1$$

$$I_4 = V_3 Y_3 + I_2$$
$$\vdots$$

Figure 6.1.5. The complex linear update recursion for the ladder network topology defined in Figure 6.1.4.

form for complex quantities is utilized:

$$A = a_r + ja_i, \qquad B = b_r + jb_i, \qquad C = c_r + jc_i, \qquad D = d_r + jd_i. \quad (6.1.26)$$

Therefore, the two real equations defined by complex equation (6.1.25) are

$$a_r = b_r c_r - b_i c_i + d_r, \qquad (6.1.27)$$

$$a_i = b_i c_r + b_r c_i + d_i. \qquad (6.1.28)$$

Figure 6.1.5 also shows how the voltage and current phasors migrate in the recursion. It is shown that only four short BASIC statements are required to accomplish this recursion, no matter how many elements the ladder network contains. Since this method of analysis is incorporated into network optimization program TWEAKNET in Section 6.2, no specific examples are given here.

Real-frequency network analysis requires manipulation of complex numbers, which are merely pairs of real numbers related by specific operations. The reader should note the definitions and identities related to complex numbers that are contained in Table 6.1.3; they are not difficult, but they are pervasive in analysis of networks in the sinusoidal steady state.

This discussion of network analysis concludes with a brief description of the *nodal matrix method*. A network with n terminals and a common terminal such as in Figure 6.1.3 can be described by n equilibrium equations in the admittance form

$$\mathbf{I} = \mathbf{YV}. \qquad (6.1.29)$$

Vector $\mathbf{I}$ contains the complex node currents, I_k, $k = 1$ to n, $\mathbf{Y}$ is the complex $n \times n$ *definite nodal admittance matrix*, and $\mathbf{V}$ is the vector of complex node voltages. To be specific, (6.1.29) has the form

$$
\begin{bmatrix} I_1 \\ I_2 \\ \vdots \\ I_n \end{bmatrix}
=
\begin{bmatrix}
y_{11} & y_{12} & \cdots & y_{1n} \\
y_{21} & y_{22} & \cdots & y_{2n} \\
 & & y_{ij} & \\
y_{n1} & y_{n2} & \cdots & y_{nn}
\end{bmatrix}
\begin{bmatrix} V_1 \\ V_2 \\ \vdots \\ V_n \end{bmatrix}. \qquad (6.1.30)
$$

Table 6.1.3.   Some Useful Identities for Complex Variables[a]

---

1. $Z = R + jX = |Z|e^{j\theta} = |Z| \underline{/\theta}.$

3. $Z^* = R - jX = |Z|e^{-j\theta} = |Z| \underline{/-\theta}.$

3. $e^{j\theta} = \cos\theta + j\sin\theta.$

4. $(Z_1 + Z_2)^* = Z_1^* + Z_2^*.$

5. $(Z_1 Z_2)^* = Z_1^* Z_2^*.$

6. $|Z|^2 = ZZ^* = R^2 + X^2.$

7. $Z + Z^* = 2\,\text{Re}(Z) = 2R.$

8. $\nabla_y Z = \nabla_y R + j\nabla_y X,\ y$ real.

9. $\nabla_y |Z| = \text{Re}[Z^*(\nabla_y Z)]/|Z|.$

10. $\nabla_y(|Z|^2) = 2[R(\nabla_y R) + X(\nabla_y X)].$

11.   If $T(Z) = U + jV,\ \dfrac{dT}{dZ} = \nabla_R U + j\nabla_R V = \nabla_X V - j\nabla_X U.$

---

[a] The asterisk (*) denotes complex conjunction. The del operator (∇_y) denotes the partial derivative with respect to real variable y. $\text{Re}(\cdot)$ denotes the real part of $(\cdot)$. Identity 11 is the *Cauchy–Riemann condition*, where U, V, R, and X are real functions.

The more general problem is to express voltage V_1 through V_n in terms of the currents I_1 through I_n, that is, in the impedance form

$$\mathbf{V} = \mathbf{ZI}. \tag{6.1.31}$$

Comparison of (6.1.29) and (6.1.31) clearly implies that impedance matrix $\mathbf{Z}$ is simply the inverse of admittance matrix $\mathbf{Y}$, $\mathbf{Z} = \mathbf{Y}^{-1}$. So the problem of finding dependent voltages in terms of independent currents is one of constructing the definite admittance matrix and then inverting it. In practice, the LU factorization described in Section 3.1.1 is usually employed for reasons that are discussed in Section 6.3.4.

Example 6.1.1.   Use Kirchhoff's and Ohm's laws to find the nodal admittance matrix for the bridge network in Figure 6.1.6. The equations are

$$\text{Node 1:}\quad I_1 = (V_1 - V_2)Y_2 + (V_1 - V_3)Y_3,$$

$$\text{Node 2:}\quad I_2 = V_2 Y_1 + (V_2 - V_3)Y_5 - (V_1 - V_2)Y_2, \tag{6.1.32}$$

$$\text{Node 3:}\quad I_3 = V_3 Y_4 - (V_2 - V_3)Y_5 - (V_1 - V_3)Y_3.$$

Considering the matrix form in (6.1.29), the definite nodal admittance matrix is

$$\mathbf{Y} = \begin{bmatrix} (Y_2 + Y_3) & (-Y_2) & (-Y_3) \\ (-Y_2) & (Y_1 + Y_2 + Y_5) & (-Y_5) \\ (-Y_3) & (-Y_5) & (Y_3 + Y_4 + Y_5) \end{bmatrix}. \tag{6.1.33}$$

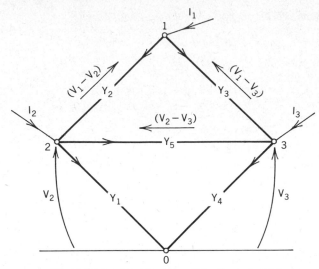

Figure 6.1.6. A bridge network for Example 6.1.1.

Comparing (6.1.33) and (6.1.30), a general rule for forming the $n \times n$ definite admittance matrix for any network composed of uncontrolled branch elements is:

1. Admittance elements y_{ii} are equal to the admittance of all elements connected to node i with all other nodes grounded (connected to node 0).
2. Admittance elements y_{ij} are equal to the negative of the sum of admittances connected between nodes i and j, for i and j chosen from 1 to n, $i \neq j$.

Readers interested in the numerous details involving controlled sources, connection matrices, and programming for general nodal analysis are referred to Staudhammer (1975:97), Adby (1980:151), and Vlach (1983:28, 114).

6.2. Constrained Optimization of Networks

The strategy and significant details of program TWEAKNET are described in this section. TWEAKNET automatically readjusts the values of resistors, inductors, and capacitors in ladder networks to meet or exceed assigned transducer performance targets. The transducer power transfer response of arbitrary ladder networks can be sampled at as many as 40 frequencies, so that as many as 20 variable network elements may be adjusted by a Gauss–Newton optimizer. The target data (desired power transfer values) for those frequency samples supplied by the user are compared in a least-pth summation of residual terms.

The least-pth objective function in TWEAKNET allows the sampled target data to be a mixture of desired goals (by a least-pth criterion), equality constraints, or lower or upper inequality constraint limits. The nonlinear constraints are enforced by the multiplier penalty method described in Sections 5.4.3 and 5.4.4. Also, the variables may be constrained by lower and upper bounds by the projection method described in Sections 5.4.1 and 5.4.2.

The ladder network real-frequency analysis method is based on the complex linear update recursion formula described in Section 6.1.3, especially in Figure 6.1.5. Any of the network's branches may contain a resistor, an inductor, a capacitor, or an inductor in parallel with a capacitor in series branches, or an inductor in series with a capacitor in shunt branches. Each inductor and capacitor may have a unique value of frequency-independent quality factor Q. In addition to the transducer power transfer response (P_{as}/P_L) in decibels (dB), the input impedance is available for all sample values by a menu command before and/or after optimization.

First partial derivatives of the network's transducer response (applicable for the objective function residuals) required for Gauss–Newton optimization are obtained *approximately* by forward finite differences for dissipative (lossy) RLC ladder network variables or *exactly* for lossless LC network variables by a method based on Tellegen's theorem. The first derivative values for any number of lossless-element variables may be obtained efficiently by only one network analysis per frequency sample, as opposed to one additional analysis per frequency for each variable when using finite differences.

Several significant examples are provided to illustrate the power of network optimization in particular and optimization of analogous systems in general. Some readers may wish to bypass technical details by going directly to the summary and examples for TWEAKNET in Section 6.2.4.

6.2.1. Program TWEAKNET Objectives and Structure.

6.2.1. Program TWEAKNET Objectives and Structure. The least-pth objective function minimized by network optimizer TWEAKNET is

$$F(\mathbf{x}, \mathbf{S}, \mathbf{u}) = \frac{1}{p} \sum_{i=1}^{m} \left[r_i(\mathbf{x}, \mathbf{S}, \mathbf{u}) \right]^p. \tag{6.2.1}$$

The TWEAKNET objective function is a combination of (4.4.31) for unconstrained least-pth residuals, (5.4.89) for equality constraints, and (5.4.98) for inequality constraints. The constraints are enforced by the multiplier penalty method described in Sections 5.4.3 and 5.4.4, where matrix $\mathbf{S} = \text{diag}(s_1 \; s_2 \; \cdots \; s_m)$ contains the multipliers, and vector $\mathbf{u}$ contains the constraint offset at each sample. Table 6.2.1 summarizes the four types of residual terms.

Table 6.2.1. Types of Residual Terms in the TWEAKNET Objective Function[a]

Type Residual	$L6()$	Expression	Effect
Unconstrained	0	$1[(L_i - T_i) - 0]$	$\min(L_i - T_i)^p$
Equality constraint	2	$s_i[(L_i - T_i) - u_i]$	$L_i = T_i$
Lower inequality constraint	-1	$s_i \min\{[+(L_i - T_i) - u_i], 0\}$	$L_i \geq T_i$
Upper inequality constraint	$+1$	$s_i \min\{[-(L_i - T_i) - u_i], 0\}$	$L_i \leq T_i$

[a] The constraints are $c_i = L_i - T_i$.

Response L_i at the ith frequency in Table 6.2.1 is the logarithm of the magnitude of the well-known *transducer function*:

$$L_i = 10 \log_{10} \left[\left| \frac{E_s}{V_L} \right|^2 \left(\frac{R_L}{4R_s} \right) \right] \quad \text{in dB}, \tag{6.2.2}$$

where rms voltages E_s and V_L and terminating resistances R_s and R_L are shown in Figure 6.1.4. Note that R_L is the real part (resistance) in load impedance $Z_L = R_L + jX_L$. The strict definition of the transducer function according to (6.2.2) requires that $X_L = 0$, in which case an equivalent definition is

$$L_i = 10 \log_{10} \left(\frac{P_{as}}{P_L} \right) \quad \text{in dB}, \tag{6.2.3}$$

where P_{as} is the maximum power available from the source and P_L is the power delivered to the load (necessarily to R_L). Both (6.2.2) and (6.2.3) provide *dB loss* so that $L_i \geq 0$ for the passive network being considered.

The sample data pairs previously described for program LEASTP in Chapter Four and generated in DATA statements are treated similarly in TWEAKNET. Each pair consists of a sample frequency, ω_i, and the corresponding target data, T_i. The sample data pairs are read into memory when program TWEAKNET is RUN. The four possible interpretations of the target data, T_i, may be changed or reviewed by menu command 7, which sets integer values in array $L6(I)$ according to the column in Table 6.2.1. The user often may wish to change the problem target interpretation during a series of optimization trials.

Figure 6.2.1 illustrates the four possible types of residuals in Table 6.2.1 and includes the unconstrained residual illustrated in Figure 4.4.1 for the least-pth case. In Figure 6.2.1, the lower bound is violated at sample ω_2 for the given set of variables contained in **x**. Residual r_2 is therefore active or binding at this point of the optimization. Samples at ω_4, ω_5, and ω_6 do not represent constraints, so that their residuals are to be minimized in a least-pth sense.

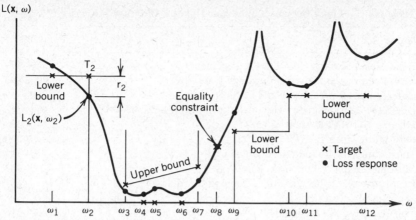

Figure 6.2.1. Sampled network loss response function with targets that are goals, equalities, and lower/upper bounds.

The other bounds are meant to be satisfied if possible, especially the equality constraint at ω_8.

Note that the loss function illustrated in Figure 6.2.1 has two peaks of loss (zeros of transferred power) at two frequencies between ω_9 and ω_{12}; the infinite loss at these two frequencies indicates that either a series branch impedance or a shunt branch admittance was infinite (open or short circuit, respectively). Placement of samples ω_{10}, ω_{11}, and ω_{12} is intended to define a lower bound on the response sag following the peaks.

The listing for program C6-1, TWEAKNET, is contained in Appendix C. The major program segments are shown in Table 6.2.2, and the flow chart is shown in Figure 6.2.2. The format for the network DATA statements in lines 400–890 is described in Section 6.2.2. Command 10 is an alternative to read network data from a disk; that makes the use of compiled BASIC especially convenient. The menu commands for entering values of variables and control parameters are unchanged from LEASTP except that exponent p in the least-pth objective function is now entered with other control parameters (command 2); it defaults to $p = 2$ if not reset. LEASTP in lines 1400–2430 is essentially unchanged from the description in Section 4.5.1. However, the list of residual values is not displayed in TWEAKNET, and no comparison of gradient values is necessary.

Subroutine calls have been inserted into LEASTP for bounding variables by the projection method employed for quasi-Newton optimizer QNEWT in Chapter Five. These additions to LEASTP are indicated by balloons on the LEASTP flow chart in Figure 4.5.1; the actions indicated by the balloons are shown in Table 6.2.3.

The multiplier penalty function algorithm in TWEAKNET is essentially the same as previously employed with QNEWT in Chapter Five according to the

Table 6.2.2.   **Major Segments in Ladder Network Program TWEAKNET, C6-1**

Name	Lines
Read Network Data from DATA Statements	340–379
Sample Pairs and Network Topology Data	400–890
Enter Values for Variables and Control	900–1390
LEASTP Gauss–Newton Unconstrained Optimizer	1400–2430
Display Network Responses and F, x, g	2440–2580
Subroutines for LEASTP Optimizer	2590–3600
Subroutines for Bounding Variables	3610–4190
Multiplier Penalty Function Algorithm	4200–4680
Reconstruct Constraints, Find Maximum Modulus	4690–4810
See/Reset Constraint Sample Number	4820–4965
Calculation of All Residuals	5000–5160
See Network F, L, C Units and Topology	5300–5540
Calculate Partial Derivatives of Residuals	7000–7170
Ladder Network Analysis for $L_i \& Z_{in}$ at ω_i	8000–8800
Calculate Jacobian of Network Response	9000–9700
Read Network Data from Disk	9800–9965

flow chart in Figure 5.4.7. Some differences are that all multipliers start at $s_i = 1$, $i = 1$ to m; no initial estimate of the Hessian matrix is required for the Gauss–Newton method; and the offset (u_i) values for unconstrained samples are maintained at zero as indicated in Table 6.2.1. Variable $K7$ was renamed $K3$ to avoid conflict. Variable $K3$ is also used in a test at line 4266; if there are no constraints, then only one Gauss–Newton minimization is performed. The constraints without offsets, $c_i = L_i - T_i$, are reconstructed in subroutine 4690, so that the treatment of c_i is comparable to that in MULTPEN (Section 5.4.4). Other than the network analysis computations explained in the next two sections, the remaining program segments in Table 6.2.2 are straightforward.

6.2.2.   Ladder Network Analysis.   The analysis method using the complex linear update formula (6.1.25) as illustrated in Figure 6.1.5 for the standard ladder network in Figure 6.1.4 has been implemented in TWEAKNET lines 8000–8800. The flow chart in Figure 6.2.3 describes the algorithm, which is executed for each frequency furnished for analysis. The algorithm operates on data contained in the arrays described in Table 6.2.4.

A specific example will simplify the explanation. Consider the elliptic filter shown in Figure 6.2.4; two equivalent means for entering the descriptive data for real-frequency analysis of that network are shown in Tables 6.2.5 and 6.2.6 The former data set is Appendix C program C6-2, LPTRAP1, to be merged with TWEAKNET. The data set in Table 6.2.6 may be created with the IBM EDLIN line editor or a text editor that can create an ASCII file for storage on magnetic disk. The reader is cautioned that the correct placement of commas

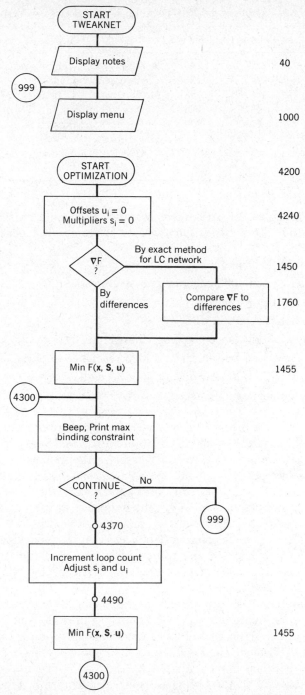

Figure 6.2.2. Flow chart for network optimizer TWEAKNET.

Table 6.2.3.   Additions to LEASTP to Implement Bounds on Variables

Figure	Balloon	Line No.	Purpose
4.5.1	A	1455	Reset/record binding variables
4.5.1	B	1865	Release all bounds
4.5.1	C	1882	Release all K-T bounds
4.5.1	D	1883	Project gradient into subspace
4.5.1	E	2072	Project search direction into subspace
4.5.1	F	2175	Check/set more bounds in linear search

is critical, although extra spaces are allowed. Both data entry methods in Tables 6.2.5 and 6.2.6 illustrate the expected information sequence:

1. Data set title or name.
2. Number of pairs of frequency and target values.
3. Set of frequency samples.
4. Set of target values in dB.
5. Frequency, inductance, and capacitance units.
6. Source and load resistances and load reactance in ohms.
7. Number of lines of network topological data.
8. Set of topological data lines.

From either Table 6.2.5 or 6.2.6 it is seen that there are seven frequency samples. Five frequencies from 0.2 to 2.0 are associated with zero dB targets, and frequencies 1.5 and 2.0 have 40 dB targets. The frequency is assumed to be in hertz (cycles per second), not radian frequency, unless modified by the frequency units constant. That constant might be 1E6 for megahertz (MHz), for instance. In line 600 in Table 6.2.5, the frequency samples are converted to radians per second by the frequency unit factor $(1/2\pi) = 0.159155$. The inductance values are often given in nanohenrys, so the inductance unit factor would be 1E−9; in line 600 the value 1.0 is used, because the values for inductances are furnished in henrys. Similarly, the capacitance values in this example are furnished in farads.

The data in line 610 of Table 6.2.5 indicates that the source resistance is 1 ohm and the load impedance is $1 + j0$ ohms, as illustrated in Figure 6.2.4a. Line 615 shows that seven lines of topological data are required to describe the arrangement of elements in Figure 6.2.4. All branches must be described, starting with the shunt branch adjacent to the load impedance and *including the source resistance*. The first branch is null, as indicated in Figure 6.2.4 and by line 620 in Table 6.2.5. Each topological line contains the list line number, type element (N, R, L, C, or LC), its quality factor Q, and a name.

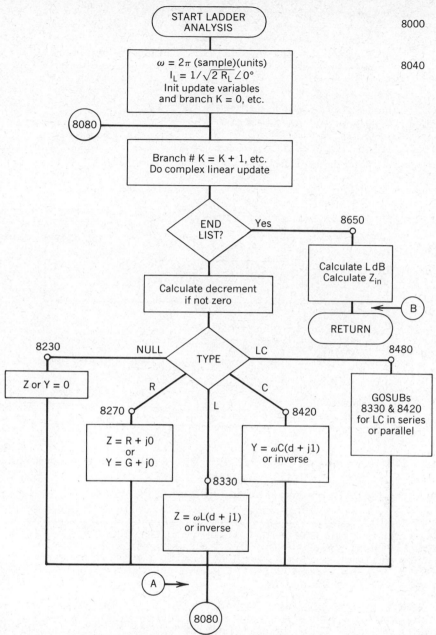

Figure 6.2.3. Flow chart for ladder network analysis in optimizer TWEAKNET. *Note:* Ⓐ and Ⓑ add exact derivatives. See Section 6.2.3.

Table 6.2.4. Ladder Network Element Menu for TWEAKNET

ARRAY:	N$() Name	M$() Symbol	X() Value[a]	Q() Q^a	Odd Branch	Even Branch	M() Code No.
Null	N	—	—	N	N	0
Resistor	R	#	—	R	R	1
Inductor	L	#	0 or #	L	L	2
Capacitor	C	#	0 or #	C	C	3
LC combined	LC	2 #'s	2 #'s	Series LC	Parallel LC	4

[a] # stands for a floating-point number.

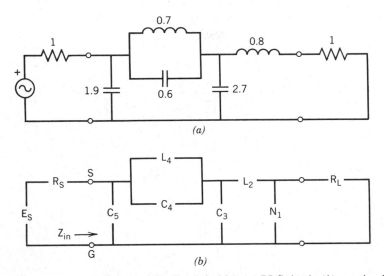

Figure 6.2.4.   Lowpass elliptic network for Example 6.2.1. (*a*) RLC circuit, (*b*) notational circuit.

Table 6.2.5.   The DATA Statement Method for Describing the Network in Figure 6.2.4

```
5 REM - 8511210940. COPYRIGHT  T.R. CUTHBERT.  1985.
7 REM - APPROX. ZVEREV P.201 ELLIPTIC FLTR - C6-2 'LPTRAP1'
400 DATA "LPTRAP1" : REM - NAME DISPLAYED ON FIRST SCREEN
410 DATA 7 : REM - NUMBER OF FOLLOWING FREQ/TARGET DATA PAIRS
420 DATA .2,.4,.6,.8,1,1.5,2
430 DATA  0, 0, 0, 0,0, 40,40
600 DATA .159155,1,1 : REM - FREQ [1/(2PI)], L & C UNITS
610 DATA 1,1,0 : REM - R SOURCE, R LOAD, & X LOAD
615 DATA 7 : REM - NUMBER OF FOLLOWING LADDER TOPOLOGY LINES
620 DATA 1,N,0,NULL1 : REM - LIST#1, NULL BR, DUMMY Q, NAME
630 DATA 2,L,0,L2 H : REM - LIST#2, INDUCTOR, INFINITE Q, NAME
640 DATA 3,C,0,C3 F : REM - LIST#3, CAPACITOR, INFINITE Q, NAME
650 DATA 4,LC,0, L4 H : REM - LIST#4, INDUC (PARALLEL), INF Q, NAME
660 DATA 5,LC,0, C4 F : REM - LIST#5, CAPAC (PARALLEL), INF Q, NAME
670 DATA 6,C,0, C5 F : REM - LIST#6, CAPACITOR, INFINITE Q, NAME
680 DATA 7,R,0,R SOURCE : REM - LIST MUST END WITH SOURCE RESISTOR
Ok
```

Table 6.2.6. The ASCII File Method for Describing the Network in Figure 6.2.4

```
"LPTRAP1 SAMPLE, UNITS, AND TOPOLOGY DATA FOR TWEAKNET"
7
.2,.4,.6,.8,1,1.5,2
0,0,0,0,0,40,40
.159155,1,1
1,1,0
7
1,N,0,NULL1
2,L,0,L2 H
3,C,0,C3 H
4,LC,0,L4 H
5,LC,0,C4 F
6,C,0,C5 F
7,R,0,R SOURCE
```

Infinite Q implies a lossless element. Since $Q = 0$ has no meaning for L's and C's, program TWEAKNET interprets the value $Q = 0$ as infinite Q. Note that lines 650 and 660 in Table 6.2.5 are *both* required to describe branch 4 in Figure 6.2.4. The program assumes that the L and C are in parallel, since these are in a series (even-numbered) branch; the dual case is series-connected L and C in a shunt (odd-numbered) branch.

Table 6.2.4 summarizes the available elements and the BASIC program array names in which data is stored. Note that the element values in the units provided are contained in the $X(\ )$ array of optimizer variables. Because the BASIC computed GOSUB statement operates from numbers and not symbols, array $M(\ )$ contains integers, 0, 1, 2, 3, and 4 corresponding to N, R, L, C, and LC that are made available to the user.

It is now possible to review the flow chart in Figure 6.2.3 in more detail. The radian frequency is computed first. Then the current through the load (caused by a source of yet unknown value) is assumed to have a magnitude such that the power delivered to the load is $\frac{1}{2}$ watt; this arbitrary choice is useful in the scheme for exact partial derivatives to be described in the next section. The load current phase is assumed to be zero; therefore, all other voltages and currents will have a phase angle with respect to this reference. Then the first line of the complex linear update in Figure 6.1.5 is performed, yielding voltage V_1 across branch 1, Figure 6.2.4.

The program determines that branch 1 is a null branch, so admittance $Y_1 = 0 + j0$ is set, and the second line of Figure 6.1.5 is performed to yield current I_2. The program now finds that $M(2) = 2$, which sends the program to subroutine 8330 in Figure 6.2.3 where $Z = \omega L(d + j1)$ is computed according to (6.1.19). That impedance is not inverted, since the branch number is even. The next two updates in Figure 6.1.5 produce I_4 in the branch 4 combination of paralleled L and C. According to the flow chart in Figure 6.2.3, the program finds that $M(4) = 4$ so that it branches to subroutine 8480. That in turn calls subroutines 8330 and 8420 with a branch number alteration that obtains two *admittances* that are added and then inverted to produce the desired even-branch *impedance*.

All the component values are obtained from the $X()$ variable array and adjusted by their unit values so that ωL and ωC are in ohms and mhos, respectively. Obviously, there are some pointers that are adjusted by the algorithm so that the null branches and combined LC branches do not upset the otherwise simple incrementing necessary to obtain desired values at the proper time.

The $M()$ array is initially filled with 9's by program line 337; for the topology shown in Figure 6.2.4, $M()$ contains the integers $0, 2, 3, 4,$ $4, 3, 9, 9, \ldots$. The algorithm detects the "END LIST" condition shown in Figure 6.2.3 when the first value 9 is reached, branching to compute the dB response (6.2.2) and input impedance Z_{in}, Figure 6.1.4.

As previously noted, the power delivered to the load resistance is assumed to be $\frac{1}{2}$ watt; for example, the resulting load current magnitude would be 0.7071 amperes for $R_L = 1$ ohm, according to (6.1.14). For $L_i = 100$ dB, (6.2.2) shows that $|E_s|^2 = 8\text{E}10$, and $|E_s|^2$ increases by 10 for each 10 dB increase in L_i. Since IBM BASIC can only accommodate numbers as large as $1\text{E}+38$, there is the chance that an overflow may occur during some calculations that employ the input voltage or current. That occurrence does not interrupt interpreted BASIC, which simply supplies the machine number for infinity and continues. However, the IBM BASIC compiler halts whenever an overflow occurs. Although overflow does not usually occur, the user should be aware that the easiest way to avoid the possibility in analysis of exceptionally long ladder networks is to employ a BASIC compiler that uses the 8087 math coprocessor chip. That coprocessor can accommodate numbers as large as $1\text{E}+4932$!

Program TWEAKNET calls ladder analysis subroutine 8000 for each of the m frequency samples required in objective function (6.2.1) to form the residuals in Table 6.2.1. Menu command 9 allows the user to compute as many as 40 loss and Z_{in} values over any frequency range specified, using the set of values assigned to the variables (network components). Either linear or logarithmic spacing of frequency samples is available in command 9.

Example 6.2.1. Run program TWEAKNET to obtain values for the transducer loss function and input impedance of a network over a range of frequencies. Consider the network in Figure 6.2.4 and its data in Table 6.2.6, stored on disk by some name, say LPTRAP1.DAT. RUN TWEAKNET and enter the five variable values using command 1: $\mathbf{x}^{(0)} = (.8 \ 2.7 \ .7 \ .6 \ 1.9)^T$. Then use command 10 to read in file LPTRAP1.DAT from the disk. Now the set of seven samples from that file may be reviewed using command 6; the result is shown in Table 6.2.7. Command 8 may then be used to review the topological data corresponding to Figure 6.2.4; the result is in Table 6.2.8. Command 9 may then be used to observe the network response over a frequency range, say, from 0 to 2 radians per second in 11 linearly spaced steps; the results are shown in Table 6.2.9.

For the reader wanting to obtain a more detailed understanding of the BASIC instructions in TWEAKNET lines 8000–8800 for ladder network

Table 6.2.7.   Program TWEAKNET Menu and Command 6 Review of the Sample Data for Examples 6.2.1 and 6.2.2

```
************* COMMAND MENU *************
1. ENTER STARTING VARIABLES (AT LEAST ONCE)
2. REVISE CONTROL PARAMETERS (OPTIONAL)
3. START OPTIMIZATION
4. EXIT (RESUME WITH 'GOTO 999')
5. SEE &/OR RESET LOWER/UPPER BOUNDS ON VARIABLES
6. DISPLAY DATA PAIRS
7. SEE &/OR RESET CONSTRAINT SAMPLE NUMBER(S)
8. SEE FREQUENCY, L, & C UNITS & NETWORK TOPOLOGY
9. SEE NETWORK RESPONSES FOR ALL SAMPLES
10. RECALL SAMPLE, UNITS, & TOPOLOGY DATA FROM DISK
****************************************
INPUT COMMAND NUMBER:? 6
 I      INDEPENDENT     DEPENDENT
 1       0.200000       0.000000
 2       0.400000       0.000000
 3       0.600000       0.000000
 4       0.800000       0.000000
 5       1.000000       0.000000
 6       1.500000      40.000000
 7       2.000000      40.000000
PRESS <RETURN> KEY TO CONTINUE -- READY?
```

analysis, it is useful temporarily to add the following instructions:

8075 PRINT "   BR#   REAL   IMAGINARY"

8105 PRINT "   "; : PRINT K; : PRINT USING S6$;A4,A5

Reloading the variables and topology data using commands 1 and 10 as above, command 9 may be used for just one frequency, say, 1 radian per second; the results are shown in Table 6.2.10. The branch numbers correspond to the

Table 6.2.8.   Topological Data Reviewed by TWEAKNET Command 8 for the Network in Figure 6.2.4

```
UNITS ARE: FREQUENCY =  1.592D-01
           INDUCTANCE =  1.000D+00
           CAPACITANCE =  1.000D+00
R SOURCE, R LOAD, X LOAD =     1.0000     1.0000     0.0000

BRANCH    TYPE    VALUE        Q         NAME
   1       N      0.0000     0.0000     NULL1
   2       L      0.8000     0.0000     L2 H
   3       C      2.7000     0.0000     C3 H
   4       LC     0.7000     0.0000     L4 H
                  0.6000     0.0000     C4 F
   5       C      1.9000     0.0000     C5 F
   6       R      1.0000     0.0000     R SOURCE

PRESS <RETURN> KEY TO CONTINUE -- READY?
```

Table 6.2.9.   Command 9 Frequency Scan from 0 to 2 Radians per Second for Example 6.2.1

```
START FREQUENCY =? 0
STOP FREQUENCY =? 2
NUMBER OF FREQS, MAX 40 (+LIN, -LOG) =? 11
 #     FREQUENCY          RESPONSE dB       Rin        OHMS      Xin
 1      0.00000000        -0.00000058    1.00000000          -0.00000000
 2      0.20000000         0.38934803    0.69498101          -0.40951029
 3      0.40000000         1.24801482    0.38340311          -0.36106296
 4      0.60000000         1.75990001    0.27692510          -0.17500184
 5      0.80000000         0.73275165    0.45733164           0.20426821
 6      1.00000000         3.97692603    0.39480205          -1.41431934
 7      1.20000000        16.71844953    0.00720423          -0.58236194
 8      1.40000000        30.82038474    0.00024101          -0.40497064
 9      1.60000000        43.79171921    0.00001152          -0.32195749
10      1.80000000        34.78645057    0.00008916          -0.27108300
11      2.00000000        33.27204687    0.00012426          -0.23571364
PRESS <RETURN> KEY TO CONTINUE -- READY?
```

standard network configuration in Figure 6.1.4, and the voltages (odd branches) and currents (even branches) are those computed using branch immittances by the complex linear update illustrated in Figure 6.1.5. In Table 6.2.10, $V_1 = I_2$ because of a null branch 1 and a 1-ohm load resistance (with $X_L = 0$). It is seen from Figure 6.2.4 that the rms voltage $V_7 = -1.61122801 - j1.54954716$ from Table 6.2.10 is the source voltage E_s. The last line in Table 6.2.10 is the normal output from TWEAKNET command 9: the sample number, frequency, transducer loss in dB, and input impedance (resistance and reactance) in ohms.

To illustrate the effect of the frequency and component unit factors, the network in Figure 6.2.4 is scaled in both frequency and impedance level. Recall that (6.1.19) and (6.1.20) defined reactance ωL and susceptance ωC, respectively. *Frequency scaling* requires that all reactances and susceptances be held constant for changes in frequency reference, so that all L and C values

Table 6.2.10.   TWEAKNET Output Due to Temporary Lines 8075 and 8105 in Example 6.2.1: Real and Imaginary Parts of Branch Voltages and Currents at 1 Radian per Second

```
START FREQUENCY =? 1
STOP FREQUENCY =? 1
NUMBER OF FREQS, MAX 40 (+LIN, -LOG) =? 1
 #     FREQUENCY          RESPONSE dB       Rin        OHMS      Xin
      BR#          REAL              IMAGINARY
       1        0.70710677         0.00000000
       2        0.70710677         0.00000000
       3        0.70710677         0.56568562
       4       -0.82024494         1.90918896
       5       -1.59708892        -0.42426609
       6       -0.01413909        -1.12528107
       7       -1.61122801        -1.54954716
 1      1.00000000         3.97692603    0.39480205          -1.41431934
PRESS <RETURN> KEY TO CONTINUE -- READY?
```

Table 6.2.11.   Network Topology Data for Figure 6.2.4 Scaled from 1 Radian per Second to 100 MHz and from 1 Ohm to 50 Ohms

```
UNITS ARE: FREQUENCY =  1.000D+06
           INDUCTANCE =  1.000D-09
           CAPACITANCE =  1.000D-12
R SOURCE, R LOAD, X LOAD =   50.0000   50.0000   0.0000

BRANCH    TYPE    VALUE       Q        NAME
   1       N      0.0000    0.0000    NULL1
   2       L     63.6620    0.0000    L2 H
   3       C     85.9440    0.0000    C3 H
   4       LC    55.7040    0.0000    L4 H
                 19.0990    0.0000    C4 F
   5       C     60.4790    0.0000    C5 F
   6       R     50.0000    0.0000    R SOURCE

PRESS <RETURN> KEY TO CONTINUE -- READY?
```

are inversely scaled with frequency. *Impedance scaling* requires that all reactances, ωL and $-1/\omega C$, and resistances be changed by the same amount, so that inductances are increased proportionally and capacitors decreased inversely for increases in impedance level. To change the impedance level from the 1-ohm, 1 radian per second references in Figure 6.2.4 to 50 ohms and $f = 100$ MHz ($\omega = 2\pi f = 6.2832E+8$), the source and load resistances are changed to 50 ohms and $L_2 = 63.662$ nH, $C_3 = 85.944$ pF, $L_4 = 55.704$ nH, $C_4 = 19.099$ pF, and $C_5 = 60.470$ pF. One picofarad (pF) equals 1E−12 farads, and 1 nanohenry (nH) equals 1E−9 henrys.

A new data file was employed so that the scaled network topology is shown in Table 6.2.11 as displayed by command 8. When that data and the new values for the variables given above are used in TWEAKNET, command 9 provides the results shown on Table 6.2.12. These data should be compared with the unscaled network results in Table 6.2.9. Note that the new frequency

Table 6.2.12.   Frequency Response of the Scaled Network in Example 6.2.1 for Comparison with Unscaled Network Data in Table 6.2.9

```
START FREQUENCY =? 0
STOP FREQUENCY =? 200
NUMBER OF FREQS, MAX 40 (+LIN, -LOG) =? 11
 #     FREQUENCY        RESPONSE dB       Rin        OHMS     Xin
 1      0.00000000       0.00000000    50.00000000      -0.00000000
 2     20.00000000       0.38935163    34.74895401     -20.47555717
 3     40.00000000       1.24802544    19.17005335     -18.05312410
 4     60.00000000       1.75991153    13.84618623      -8.75003177
 5     80.00000000       0.73274356    22.86676143      10.21370278
 6    100.00000000       3.97709455    19.73850162     -70.71460116
 7    120.00000000      16.71876638     0.36018283     -29.11779534
 8    140.00000000      30.82114475     0.01204836     -20.24837562
 9    160.00000000      43.78975602     0.00057647     -16.09776810
10    180.00000000      34.78599083     0.00445840     -13.55406670
11    200.00000000      33.27176730     0.00621325     -11.78561340
PRESS <RETURN> KEY TO CONTINUE -- READY?
```

Table 6.2.13. Logarithmic Spacing of Frequency Samples Using Command 9 in TWEAKNET for Example 6.2.1

```
START FREQUENCY =? 2
STOP FREQUENCY =? 200
NUMBER OF FREQS, MAX 40 (+LIN, -LOG) =? -5
   #     FREQUENCY        RESPONSE dB        Rin      OHMS      Xin
   1     2.00000000       0.00417098     49.77821250        -3.08485025
   2     6.32455600       0.04144999     47.86951537        -9.34203119
   3    20.00000000       0.38935163     34.74895401       -20.47555717
   4    63.24556000       1.73054791     13.82977947        -6.76653902
   5   200.00000000      33.27176730      0.00621325       -11.78561230
PRESS <RETURN> KEY TO CONTINUE -- READY?
```

scale has been specified, over which the transducer loss function values are unchanged as expected; see (6.2.2). Of course, the input impedance values have increased by a factor of 50. It is shown later that the gradient values are *different*, because they depend on the component units chosen; there are obvious scaling implications for optimization algorithms.

Finally, logarithmic spacing of frequency samples using command 9 is illustrated in Table 6.2.13. Because geometric frequency mapping is used in translating lowpass filters to their bandpass equivalent filters, logarithmically spaced samples are often more useful than linearly spaced samples. The interested reader is referred to Cuthbert (1983:205, 280). Also, typical quality factor values may be used by replacing the zeros in topology list lines 2 to 6 in Table 6.2.6 with values such as 200 for inductors and 500 for capacitors. Note that Q values do not change with scaling, as seen by (6.1.22).

6.2.3. First Partial Derivatives.

The Gauss–Newton optimization algorithm described in Section 4.4.1 and utilized in programs LEASTP and TWEAK-NET requires only first partial derivatives of the residual terms in the standard objective function (6.2.1). The Hessian matrix of second partial derivatives is estimated by a positive-definite matrix that is a result of the unique form of the objective function. This section describes the two alternative ways in which these first partial derivatives are obtained in program TWEAKNET. A more general analysis of first partial derivatives for electrical networks and analogous linear systems is described in Section 6.3.

The *Jacobian matrix* originally defined by (4.4.15) for the Gauss–Newton method is

$$\mathbf{J} = \left[\nabla_j r_i \right], \tag{6.2.4}$$

where $\mathbf{J}$ is $m \times n$ for m samples of residuals r_i that are functions of n variables. Subroutine 7000 in TWEAKNET assembles the Jacobian in array $A(I, J)$, $I = 1$ to M and $J = 1$ to N. The residual functions are those defined in Table 6.2.1, where it can be seen that the partial derivatives of the several possible residual expressions all are equal to transducer function L_i multiplied by ± 1, multiplier s_i, or zero. Subroutine 7000 receives the first partial derivatives of transducer response L_i with respect to variable x_j in array $A(I, J)$ from subroutine 9000, makes the decisions implied in Table 6.2.1, and places the Jacobian in (6.2.4) into array $A(I, J)$ for later computation of the objective function gradient vector (4.4.32) by subroutine 2750:

$$\nabla F = \sum_{i=1}^{m} r_i^{p-1}(\nabla r_i). \qquad (6.2.5)$$

The remainder of this section describes subroutine 9000, which computes either approximate or exact values of first derivatives of the transducer response function for certain ladder networks. The easiest way to approximate the first partial derivatives of the response function is by forward finite differences: Just perturb each variable in turn and recompute the response. Then

$$\nabla_j L_i \doteq \frac{L_i(\mathbf{x} + 0.0001 x_j \mathbf{e}_j) - L_i(\mathbf{x})}{0.0001 x_j}, \qquad (6.2.6)$$

where $\mathbf{e}_j$ is the $n \times 1$ jth unit vector defined by (2.1.2). An alternative is to perform finite differencing on objective function (6.2.1); this is not implemented in TWEAKNET because of the exact methods that follow. Either way, finite differencing wastes n times as much work to obtain derivatives that are subject to truncation error. The results are accurate to about three to five significant figures, and the effects of those errors usually cause slower final convergence to a nonzero gradient.

Finite differencing is programmed in TWEAKNET lines 9020 to 9160, selected by the user's choice that sets $K6 = 0$ at line 1450. The extra network evaluations required for finite-differenced derivatives are *not* included in the "TOTAL NUMBER OF FUNCTION EVALUATIONS" report provided by TWEAKNET at the conclusion of each minimization, but the total over the sequence of penalty function minimizations is accumulated. It is recommended that the "maximum number of iterations" parameter in command 2 be set to 20 before using finite differences.

Exact first partial derivatives of response functions for *any* electrical network composed of linear elements may be obtained in no more than two analyses at each frequency. In fact, only one analysis is required if the functions being differentiated are not transfer functions. For example, the exact partial derivatives of input impedance Z_{in} with respect to *all* variables in a network may be obtained by just one analysis of that network per frequency. Since the transducer function is a *transfer function* (one that involves quantities at opposite end of the network), two analyses at each frequency are required for unrestricted types of linear elements. Two different ways to arrive at these conclusions are discussed in Section 6.3, one being Tellegen's theorem for systems based on linear constitutive laws.

In order to avoid a substantial increase in BASIC program code, a more limited implementation for exact derivatives is provided in TWEAKNET, namely, that for resistively terminated *lossless LC ladder networks*. In that situation, only one analysis per frequency is required to obtain exact derivatives of transfer functions with respect to all variables. For the transducer function in (6.2.3), assuming $E_s = |E_s| \underline{/0}$ and $P_L = \frac{1}{2}$:

$$\nabla_{X_k} L_i = -(20\log_{10}e)\text{Im}(p_1^* I_k^2), \qquad k \text{ even}, \qquad (6.2.7)$$

$$\nabla_{B_k} L_i = +(20\log_{10}e)\text{Im}(p_1^* V_k^2), \qquad k \text{ odd}, \qquad (6.2.8)$$

where *reflection coefficient* p_1 is defined as

$$p_1 = \frac{Z_{\text{in}} - R_s}{Z_{\text{in}} + R_s}. \qquad (6.2.9)$$

The symbol "Im" in (6.2.7) and (6.2.8) means "imaginary part of," and $(20\log_{10}e) = 8.68589$. Also, I_k and V_k are complex currents and voltages in the kth network branches, respectively.

The even-branch impedances are

$$Z_k = R_k + jX_k, \qquad (6.2.10)$$

where reactance X_k is not to be confused with variable x_j, and the odd-branch admittances are

$$Y_k = G_k + jB_k, \qquad (6.2.11)$$

where B_k is the kth branch susceptance. The partial derivatives in (6.2.7) and (6.2.8) are with respect to the branch reactance, X_k, and the branch susceptance, B_k, respectively. The derivatives are real numbers, even though the currents, voltages, and reflection coefficient are complex. These exact partial derivative formulas for the lossless-element case have been derived by Bandler (1985) using conservation of energy and the Cauchy–Riemann identity (Table 6.1.3) from complex variable theory and by Orchard (1985) using Tellegen's theorem. See Section 6.3.2.

Subroutine 8000 in TWEAKNET for ladder network analysis stores the complex branch voltage or current in $A(I, J)$ for frequency ω_i and variable element x_j associated with that branch. Null branch voltages and currents are skipped over in this storage process. See lines 8150 to 8200 of the BASIC listing C6-1 in Appendix C and balloon A in Figure 6.2.3. Because (6.2.7) and (6.2.8) are not valid unless the phase reference is adjusted to the source voltage (as opposed to the original reference to the load current), it is necessary to subtract the computed source voltage angle from the stored branch voltages and currents. The input impedance is also stored in array $Z(I,)$ at each Ith frequency for later use. These last two complex quantities are not available until the input of the network has been reached by the sequence of complex linear updates. See TWEAKNET lines 8730 to 8790 and balloon B in Figure 6.2.3.

The exact derivatives of the transducer function with respect to all elements of a lossless network are computed in subroutine 9000, lines 9180 to 9700. The reflection coefficient (6.2.9) is computed in lines 9250 to 9290, and the partial derivatives of transducer loss with respect to branch reactance and susceptance, (6.2.7) and (6.2.8), respectively, are computed in lines 9300 to 9380. Programming the preceding calculations involving complex variables is somewhat awkward in the BASIC language, since the complex variable type is not provided.

The chain rule must be applied to the partial derivatives in (6.2.7) and (6.2.8) for them to be properly related to the problem's variables. Thus,

$$\nabla_{x_j} L_i = \left(\nabla_{X_k} L_i \right) \left(\nabla_{x_j} X_k \right), \qquad k \text{ even,} \qquad (6.2.12)$$

$$\nabla_{x_j} L_i = \left(\nabla_{B_k} L_i \right) \left(\nabla_{x_j} B_k \right), \qquad k \text{ odd.} \qquad (6.2.13)$$

If variable x_j is an inductance scaled by unit u_2, then the reactance of lossless branch k is

$$X_k = \omega x_j u_2, \qquad (6.2.14)$$

and the derivative of reactance X_k with respect to variable element x_j is

$$\nabla_{x_j} X_k = \omega u_2, \qquad k \text{ even.} \qquad (6.2.15)$$

Similarly, if variable element x_j is a capacitance scaled by unit u_3, then the derivative of susceptance B_k with respect to variable x_j is

$$\nabla_{x_j} B_k = \omega u_3, \qquad k \text{ odd.} \qquad (6.2.16)$$

When an inductance occurs in an odd branch, then its susceptance is

$$B_k = -\frac{1}{\omega x_j u_2}, \qquad (6.2.17)$$

so that the partial derivative of the inductive susceptance with respect to the scaled inductance value is

$$\nabla_{x_j} B_k = \frac{\omega u_2}{\left(\omega x_j u_2 \right)^2}, \qquad k \text{ odd.} \qquad (6.2.18)$$

Similarly, for a capacitance occurring in an even branch,

$$\nabla_{x_j} X_k = \frac{\omega u_3}{\left(\omega x_j u_3\right)^2}, \qquad k \text{ even.} \qquad (6.2.19)$$

For the parallel LC arrangement in a series branch,

$$X_k = \frac{\omega x_j u_2}{1 - \omega^2 x_j x_{j+1} u_2 u_3}. \qquad (6.2.20)$$

Therefore, the partial derivative of the reactance with respect to the scaled inductance value x_j is

$$\nabla_{x_j} X_k = \frac{\omega u_2}{\left(1 - \omega^2 x_j x_{j+1} u_2 u_3\right)^2}, \qquad k \text{ even,} \qquad (6.2.21)$$

and with respect to the scaled capacitance value x_{j+1} is

$$\nabla_{x_{j+1}} X_k = \frac{\omega u_3 \left(\omega x_j u_2\right)^2}{\left(1 - \omega^2 x_j x_{j+1} u_2 u_3\right)^2}, \qquad k \text{ even.} \qquad (6.2.22)$$

Similarly, for the series LC arrangement occurring in a shunt branch, the partial derivative with respect to the scaled capacitance value x_{j+1} is

$$\nabla_{x_{j+1}} B_k = \frac{\omega u_3}{\left(1 - \omega^2 x_j x_{j+1} u_2 u_3\right)^2}, \qquad k \text{ odd,} \qquad (6.2.23)$$

and the partial derivative with respect to the scaled inductance value x_j is

$$\nabla_{x_j} B_k = \frac{\omega u_2 \left(\omega x_{j+1} u_3\right)^2}{\left(1 - \omega^2 x_j x_{j+1} u_2 u_3\right)^2}, \qquad k \text{ odd.} \qquad (6.2.24)$$

The partial derivatives of reactance X_k or susceptance B_k in (6.2.12) and (6.2.13) are defined in (6.2.15) through (6.2.24). Those factors contain the unit constants u_2 and u_3 for inductors and capacitors, respectively. Therefore, choices of L and C units will affect optimizer conditioning. The symptom to watch is that all elements of the gradient should be within two orders of magnitude of unity at the beginning of the optimization. Otherwise, more appropriate choices for u_2 and u_3 are required.

The chain rule conversions in (6.2.12) to (6.2.24) are accomplished in TWEAKNET lines 9390 to 9700. If this is not programmed efficiently, then the effort wasted in finite differencing may not suffer in comparison, especially for a small number of variables. However, it is shown in the next section that inaccurate derivatives definitely slow the final convergence of the optimization process. Besides, it is shown in Section 6.3 that exact derivatives for any dissipative (lossy) network require no more than twice the work for lossless networks, so computation of exact derivatives is always worthwhile. As is often the case, the price paid is more extensive and sophisticated programming, and this is aggravated by the absence of a complex variable type in many programming languages.

6.2.4. Summary of Program TWEAKNET with Examples.

Program C6-1, TWEAKNET, is listed in Appendix C; remarks are provided throughout the code, which occupies about 32,600 bytes in ASCII storage or 27,700 bytes in compressed binary storage. An additional 19,000 bytes are required for execution because of the defined variables and other BASIC interpreter overhead, leaving about 13,750 bytes free in the standard IBM BASIC environment. TWEAKNET may be compiled into machine code, and compilers that make use of the 8087 math coprocessor chip are highly recommended. This section summarizes the features and use of ladder network constrained optimization program TWEAKNET.

TWEAKNET may be run directly as listed in Appendix C, program C6-1, without merging any additional code. The 10 commands are shown in Table 6.2.7, and the first requirement is to use command 10 to load an ASCII network data file with contents similar to that in Table 6.2.6. Alternatively, the user may wish to first merge DATA statements defining BASIC lines within the range 400 to 889 before running TWEAKNET; see the network data in Table 6.2.5, which is the same as program segment C6-2 in Appendix C. The only other mandatory action before optimization is to use command 1 to furnish the values for the correct number of optimization variables; these are the values of network elements.

Having furnished the network topology data for the network in Figure 6.2.4 by using either Table 6.2.5 or a file containing Table 6.2.6 and command 10, the five variable values shown in Figure 6.2.4 should be entered from the load (right-hand) end toward the source, that is, $\mathbf{x}^{(0)} = (0.8 \ 2.7 \ 0.7 \ 0.6 \ 1.9)^T$. Those values correspond to L_2, C_3, L_4, C_4, and C_5, respectively. Referring to Table 6.2.5, line 600 shows that the frequency scale (normally in Hz) has been rescaled to radians per second by the factor $1/(2\pi) = 0.159155$. Also on line 600, the inductance and capacitance units are scaled by unity and thus remain in henrys and farads, respectively. Elements L_4 and C_4 (in a combined LC branch) always must be furnished in that order, corresponding to lines 650 and 660 in Table 6.2.5. There could have been a sixth variable that would override the fixed R SOURCE value furnished in line 610; in that case the finite differencing option for estimating partial derivatives *must* be used after selecting command 3 (optimization).

Having provided the network topological and variable element data, the user may review the sample data status by using command 6; the result is

Table 6.2.14.   Command 8 Review of Example 6.2.2 Network State Before Optimization

```
UNITS ARE: FREQUENCY =   1.592D-01
           INDUCTANCE =   1.000D+00
           CAPACITANCE =  1.000D+00
  R SOURCE, R LOAD, X LOAD =      1.0000      1.0000      0.0000

  BRANCH    TYPE      VALUE         Q         NAME
     1       N        0.0000      0.0000     NULL1
     2       L        0.8000      0.0000     L2 H
     3       C        2.7000      0.0000     C3 H
     4       LC       0.7000      0.0000     L4 H
                      0.6000      0.0000     C4 F
     5       C        1.9000      0.0000     C5 F
     6       R        1.0000      0.0000     R SOURCE

PRESS <RETURN> KEY TO CONTINUE -- READY?
```

shown in Table 6.2.7. These are the frequencies over which optimization will occur. The corresponding dependent values or targets $(0, 0, 0, 0, 0, 40, 40)$ are the goals to be attained by least-pth minimization. For now, those target values will not be made a part of equality or inequality constraints, requiring the use of command 7.

Command 8 lists the ladder network structure; it may be used before or after optimization. See Table 6.2.14. This allows the user to check the network status; that data is easily recorded by using the IBM-PC's ⟨PrtSc⟩ keyboard command.

Command 2 resets four parameters that already have default values: the exponent p in the least-pth error function (6.2.1) is $p = 2$, each minimization is limited to 50 iterations, the stopping criterion in (1.3.22) for both variables and objective function is $e = 0.0001$, and minimization results are displayed after every iteration. Before selecting finite differencing (after command 3), it is recommended that the maximum number of iterations per minimization be limited to 20, because of the slow final convergence rate. Once the user is familiar with program operation, it is also recommended that the print choice be set to 10 or 50, so that the screen will not be constantly in motion. However, the cutback line search reports are always displayed, so the user will know that the program is running.

Command 9 may be used to display the transducer loss (P_{as}/P_L in dB) and the input impedance Z_{in} (Figure 6.2.4) at one or more frequencies. The frequencies need not correspond to the sampled data, although that may be useful. Command 9 requires the beginning, ending, and number of frequencies to generate the set of responses automatically. A negative number of frequencies produces logarithmic frequency spacing instead of linear spacing. Note that logarithmic spacing requires a nonzero beginning frequency.

Command 3 may now be selected, and the choice of either differences (D) or exact (E) partial derivatives must be made. If all quality factors Q are zero

Table 6.2.15. Unconstrained $p = 2$ Optimization of the Network Described in
Table 6.2.14

```
INPUT COMMAND NUMBER:? 3
*********************************************************
DIFFERENCING OR EXACT LOSSLESS ELEMENT PARTIALS (D/E)? E
AT START OF ITERATION NUMBER 1
   FUNCTION VALUE = 40.7419
 I        X(I)              G(I)
 1     0.80000000      -22.85414195
 2     2.70000000       12.94164591
 3     0.70000000     1141.94880719
 4     0.60000000     1265.14616997
 5     1.90000000        3.65948409
                        LM PARAM V= 1.0D-03
                 STEP-TO-GRADIENT DEGREES= 89.1370
  ###### CUT BACK STEP SIZE BY FACTOR OF 4 ######
AT START OF ITERATION NUMBER 2
   FUNCTION VALUE = 20.37907
 I        X(I)              G(I)
 1     1.08438985      -25.38351211
 2     1.99481314       -4.89851744
 3     0.76848907      324.34956809
 4     0.53401574      475.65427660
 5     2.08282897       -6.25864997
                        LM PARAM V= 1.0D-02
```

●
●
●

```
  ###### CUT BACK STEP SIZE BY FACTOR OF 4 ######
AT START OF ITERATION NUMBER 50
   FUNCTION VALUE = 8.211599
 I        X(I)              G(I)
 1     1.78280955       -0.26355336
 2     1.59812958       -0.08003502
 3     0.92276413       -3.02655326
 4     0.42184804        1.61372724
 5     1.88634557        0.15353740
                        LM PARAM V= 1.0D+04
                 STEP-TO-GRADIENT DEGREES= 84.5001
''''''''''''''''''''''''''''''''''''''
STOPPED AT GIVEN LIMIT OF 50  ITERATIONS; RESULTS ARE:
AT START OF ITERATION NUMBER 51
   FUNCTION VALUE = 8.206866
 I        X(I)              G(I)
 1     1.80429651       -0.00998179
 2     1.60012804        0.23932786
 3     0.92321557       -3.34427939
 4     0.42164131        1.12038790
 5     1.85697765       -0.13935573
TOTAL NUMBER OF FUNCTION EVALUATIONS = 93
EXPONENT P = 2
PRESS <RETURN> KEY TO CONTINUE -- READY?
```

Table 6.2.16.   Frequency Sweep for the Variables at the Bottom of Table 6.2.15

#	FREQUENCY	RESPONSE dB	Rin OHMS	Xin
1	0.00000000	0.00000000	1.00000000	-0.00000000
2	0.10000000	0.01018158	1.03552512	-0.09197870
3	0.20000000	0.08953151	1.08426221	-0.28850966
4	0.30000001	0.35071318	0.97571580	-0.57241620
5	0.40000001	0.86777711	0.68792471	-0.71499266
6	0.50000000	1.56630578	0.43532630	-0.66131558
7	0.60000002	2.21658716	0.29407635	-0.53387052
8	0.69999999	2.49445166	0.23671178	-0.39006792
9	0.80000001	1.98252710	0.25803877	-0.21592986
10	0.89999998	0.39884561	0.54770055	0.07843098
11	1.00000000	1.89357925	0.93110878	-1.42504874
12	1.10000002	9.48817601	0.04989598	-0.81955827
13	1.20000005	16.75686177	0.00718231	-0.58911788
14	1.29999995	23.44359234	0.00139773	-0.48239253
15	1.39999998	30.39448071	0.00026815	-0.41720894
16	1.50000000	39.26755239	0.00003368	-0.37150100
17	1.60000002	73.20648644	0.00000001	-0.33683554
18	1.70000005	44.70595700	0.00000927	-0.30921773
19	1.79999995	40.68735180	0.00002309	-0.28646746
20	1.89999998	39.07531214	0.00003314	-0.26727084
21	2.00000000	38.36082759	0.00003876	-0.25077648
PRESS <RETURN> KEY TO CONTINUE -- READY?				
22	2.09999990	38.09108093	0.00004097	-0.23640155
23	2.20000000	38.07307496	0.00004091	-0.22372965
24	2.30000000	38.20877840	0.00003947	-0.21245314
25	2.40000010	38.44275605	0.00003725	-0.20233828
26	2.50000000	38.74130798	0.00003465	-0.19320357
PRESS <RETURN> KEY TO CONTINUE -- READY?				

(implying infinite Q) and the last variable is not R SOURCE, then exact derivatives are available and are recommended for both speed and accuracy. Compiled BASIC will complete this optimization in a few minutes. Table 6.2.15 shows the first two and last two function evaluations. Command 9 may be used to determine how the optimized network response fits the targets, and the results are shown in Table 6.2.16.

The preceding experiment indicated that the two 40-dB targets were too ambitious, so that the following constrained optimization problem is more appropriate.

Example 6.2.2. Using command 10, the sample data was changed to that shown in Table 6.2.17 as observed by command 6. Command 7 may be used to designate some or all of the samples as lower inequality ($\geq$), equality ($=$), upper inequality ($\leq$) constraints or least-pth goals. Each time command 7 is selected, all designated constraints are displayed; least-pth goals are not. It is a simple matter to select the constraint pattern using command 7 to obtain the results shown in Table 6.2.18.

After 13 iterations with command 3 (using exact partial derivatives) from $\mathbf{x}^{(0)} = (0.8 \ 2.7 \ 0.7 \ 0.6 \ 1.9)^T$, the end of the first sequential optimization (penalty loop) and the subsequent analysis by command 9 are shown in Table 6.2.19.

**Table 6.2.17. A Revised Set of Sample
Data for Example 6.2.2 as Displayed
by Command 6**

I	INDEPENDENT	DEPENDENT
1	0.600000	1.300000
2	0.700000	1.300000
3	0.800000	1.300000
4	0.900000	1.300000
5	1.000000	1.300000
6	1.500000	35.000000
7	2.200000	35.000000

The loaded Q values were then changed to 200 for inductors and 300 for capacitors (with the help of command 10), and optimization was restarted from the last set of variable values. Now the differencing method for estimating partial derivatives *must* be selected, because the network is no longer lossless. Two penalty loops were required to reduce the maximum error constraint modulus to 0.0458. The results are shown by the data from commands 3 and 9; see Table 6.2.20. The response curves for the first starting point, $\mathbf{x}^{(0)} = (0.8\ 2.7\ 0.7\ 0.6\ 1.9)^T$, and the last point, Table 6.2.20, are plotted in Figure 6.2.5.

The preceding example required some unreported experimentation. More important, insight for what is or is not technically possible is most helpful, but it is claimed that the process of nonlinear optimization also contributes to insight. The user will find that repeated experimentation is much easier to accomplish with the aid of certain RAM-resident utility programs for the IBM-PC. Especially useful are *macro commands* that assign a sequence of key strokes to a spare key to avoid retyping and *notepad* capability that enables temporary interrupt of TWEAKNET for modification of an ASCII file that is subsequently acquired using command 10. Two such respective programs are *SuperKey* and *SideKick*, trademarks of Borland International. These programs also provide convenient cut-and-paste features to save noteworthy results.

**Table 6.2.18. The Set of Inequality and Equality Constraints
Imposed on the Sample Data Using Command 7**

CONSTRAINTS NOW SET ARE:				
I	SAMPLE	LOWER	EQUALITY	UPPER
1	0.6000			1.3000
2	0.7000			1.3000
3	0.8000			1.3000
4	0.9000			1.3000
5	1.0000			1.3000
6	1.5000		35.0000	
7	2.2000	35.0000		

Table 6.2.19. **The End of Constrained Optimization and a Frequency Sweep of the Transducer Response Function for the Lossless Network in Figure 6.2.4**

```
    FUNCTION VALUE = 2.256286E-10
  I          X(I)           G(I)
  1        1.25767454      -0.00011281
  2        1.70484432      -0.00001546
  3        0.91281892      -0.00025742
  4        0.41891943      -0.00062080
  5        1.68247745      -0.00005706
                              LM PARAM V= 1.0D-04
                    STEP-TO-GRADIENT DEGREES= 88.2262
CONVERGED; SOLUTION IS:
AT START OF ITERATION NUMBER 13
    FUNCTION VALUE = 2.909471E-12
  I          X(I)           G(I)
  1        1.25768692       0.00001458
  2        1.70484566       0.00001524
  3        0.91281895       0.00017903
  4        0.41891945       0.00033564
  5        1.68246015       0.00001110
TOTAL NUMBER OF FUNCTION EVALUATIONS = 23
EXPONENT P = 2
*********************************************************
AFTER   1   PENALTY MINIMIZATIONS,
   THE MAX CONSTRAINT MODULUS # 6  = 2.412249E-06
CONTINUE PENALTY MINIMIZATIONS (Y/N)?


NUMBER OF FREQS, MAX 40 (+LIN, -LOG) =? 26
 #     FREQUENCY      RESPONSE dB     Rin      OHMS    Xin
 1     0.00000000      0.00000000    1.00000000      -0.00000000
 2     0.10000000      0.01769382    0.99908403      -0.12772648
 3     0.20000000      0.08841475    0.97330122      -0.28170604
 4     0.30000001      0.25169341    0.87698387      -0.44065228
 5     0.40000001      0.53151922    0.71152370      -0.53602045
 6     0.50000000      0.89172274    0.54280863      -0.53465628
 7     0.60000002      1.21311593    0.42423677      -0.46403412
 8     0.69999999      1.29999994    0.37236895      -0.35475487
 9     0.80000001      0.91304890    0.41378490      -0.20882104
10     0.89999998      0.09542350    0.74557252      -0.03897357
11     1.00000000      1.29999931    1.11319530      -1.24138682
12     1.10000002      7.05162603    0.10563878      -0.95952089
13     1.20000005     13.67249314    0.01600889      -0.67778205
14     1.29999995     20.06111275    0.00321637      -0.54620137
15     1.39999998     26.72505746    0.00064836      -0.46772089
16     1.50000000     35.00000241    0.00009261      -0.41378763
17     1.60000002     54.27558940    0.00000106      -0.37349284
18     1.70000005     42.88774168    0.00001436      -0.34174459
19     1.79999995     38.11600553    0.00004243      -0.31581031
20     1.89999998     36.25930254    0.00006428      -0.29406876
21     2.00000000     35.41944921    0.00007724      -0.27548376
PRESS <RETURN> KEY TO CONTINUE -- READY?
22     2.09999990     35.07181375    0.00008300      -0.25935436
23     2.20000000     35.00000060    0.00008382      -0.24518479
24     2.30000000     35.09597058    0.00008153      -0.23261195
25     2.40000010     35.29925108    0.00007742      -0.22136199
26     2.50000000     35.57329942    0.00007238      -0.21122361
PRESS <RETURN> KEY TO CONTINUE -- READY?
```

350

Table 6.2.20. **The End of Constrained Optimization and a Frequency Sweep of the Transducer Response Function for the Dissipative Network in Figure 6.2.4**

```
I          X(I)           G(I)
1       1.58007720      -0.00875166
2       1.43531825      -0.00389084
3       1.05720132      -0.06901293
4       0.36246042      -0.10261966
5       1.37282973      -0.01526351
                     LM PARAM V= 1.0D+02
              STEP-TO-GRADIENT DEGREES= 86.9509
!!!!!!!!!!!!!!!!!!!!!!!!!!!!!!!!!!
STOPPED AT GIVEN LIMIT OF 20  ITERATIONS; RESULTS ARE:
AT START OF ITERATION NUMBER 21
   FUNCTION VALUE = 6.765874E-03
I          X(I)           G(I)
1       1.56613845      -0.00680946
2       1.43370432      -0.00528253
3       1.05818550      -0.02824242
4       0.36211966      -0.00937649
5       1.38812583      -0.00318943
TOTAL NUMBER OF FUNCTION EVALUATIONS = 78
EXPONENT P = 2
****************************************************
AFTER  2  PENALTY MINIMIZATIONS,
   THE MAX CONSTRAINT MODULUS # 2  = 4.448801E-02
CONTINUE PENALTY MINIMIZATIONS (Y/N)? N
```

```
NUMBER OF FREQS, MAX 40 (+LIN, -LOG) =? 26
 #    FREQUENCY       RESPONSE dB      Rin      OHMS      Xin
 1    0.00000000      0.00000000     1.00000000    -0.00000000
 2    0.10000000      0.01238261     1.03684886    -0.03216728
 3    0.20000000      0.05317945     1.12207661    -0.13860064
 4    0.30000001      0.17774808     1.15529263    -0.36599504
 5    0.40000001      0.43261254     1.00864006    -0.61422636
 6    0.50000000      0.80085790     0.74880846    -0.70375131
 7    0.60000002      1.17052571     0.53945426    -0.63789415
 8    0.69999999      1.34448796     0.43148910    -0.50068030
 9    0.80000001      1.08099988     0.43460739    -0.32039924
10    0.89999998      0.37118739     0.69450748    -0.07715498
11    1.00000000      1.31616750     1.66180888    -1.04759004
12    1.10000002      6.79604376     0.18622399    -1.22377406
13    1.20000005     13.44711517     0.03152330    -0.83874336
14    1.29999995     19.90800796     0.00996391    -0.66819049
15    1.39999998     26.64297173     0.00489032    -0.56926513
16    1.50000000     35.00617312     0.00323591    -0.50222420
17    1.60000002     54.35206923     0.00252875    -0.45253043
18    1.70000005     42.63369893     0.00214420    -0.41357129
19    1.79999995     37.99971334     0.00189165    -0.38185512
20    1.89999998     36.19118356     0.00170383    -0.35533248
21    2.00000000     35.37973708     0.00155362    -0.33270352
PRESS <RETURN> KEY TO CONTINUE -- READY?
22    2.09999990     35.05213351     0.00142848    -0.31309378
23    2.20000000     34.99566846     0.00132174    -0.29588750
24    2.30000000     35.10395443     0.00122936    -0.28063527
25    2.40000010     35.31740751     0.00114860    -0.26699914
26    2.50000000     35.50002765     0.00107745    -0.25471897
PRESS <RETURN> KEY TO CONTINUE -- READY?
```

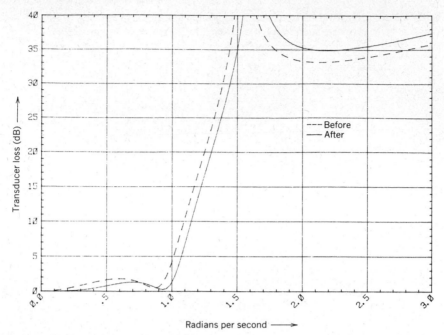

Figure 6.2.5. Beginning and ending response curves for a lossless optimization followed by one for the dissipative network in Figure 6.2.4.

The next example is designed to show how TWEAKNET behaves for larger optimization problems. Although the changes necessary to provide for more than 20 variables are straightforward, many nonlinear optimization problems seem to have fewer variables. The maximum number of variables encountered by the author during many years of network optimization on large computers has been less than 50. For personal computers there is no question of numerical precision, especially when using the math coprocessor for maximum dynamic range, but there is a question of running time. The next example should alleviate that concern when using compiled languages, especially on new personal computers that run many times faster than the original IBM-PC.

Example 6.2.3.   The bandpass filter shown in Figure 6.2.6 was obtained from an ideal lowpass Cauer elliptic filter that was transformed to the standard bandpass configuration and then modified. There were originally three branches in the standard bandpass filter that produced zeros of transmission, each branch originally containing two L's and two C's. The modification replaced each of those branches by two L's and *one* C, eliminating half the transmission zeros and resulting in the three shunt branches at nodes 7 to 9, 13 to 15, and 19 to 21 in Figure 6.2.6. These topics in electrical engineering have been described by Cuthbert (1983:205, 364, 369).

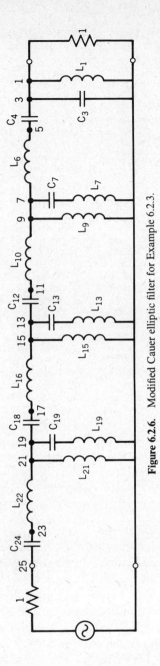

Figure 6.2.6. Modified Cauer elliptic filter for Example 6.2.3.

Table 6.2.21.   Topological Description of the Network in Figure 6.2.6

```
UNITS ARE: FREQUENCY =   1.592D-01
           INDUCTANCE =   1.000D+00
           CAPACITANCE =   1.000D+00
R SOURCE, R LOAD, X LOAD =       1.0000      1.0000      0.0000

BRANCH     TYPE     VALUE        Q          NAME
   1        L        0.0831     0.0000      L1  H
   2        N        0.0000     0.0000      NULL2
   3        C       12.0340     0.0000      C3  F
   4        C        0.0741     0.0000      C4  F
   5        N        0.0000     0.0000      NULL5
   6        L       13.5000     0.0000      L6  H
   7        LC       0.1013     0.0000      L7  H
                     3.4740     0.0000      C7  F
   8        N        0.0000     0.0000      NULL8
   9        L        0.1866     0.0000      L9  H
  10        L       13.9230     0.0000      L10 H
  11        N        0.0000     0.0000      NULL11
  12        C        0.0716     0.0000      C12 F
  13        LC       0.1402     0.0000      L13 H
                     3.2786     0.0000      C13 F
  14        N        0.0000     0.0000      NULL14
  15        L        0.1648     0.0000      L15 H
  16        L       15.8490     0.0000      L16 H
  17        N        0.0000     0.0000      NULL17
  18        C        0.0631     0.0000      C18 F
  19        LC       0.0869     0.0000      L19 H
                     1.8866     0.0000      C19 F
  20        N        0.0000     0.0000      NULL20
  21        L        0.4432     0.0000      L21 H
  22        L       10.9150     0.0000      L22 H
  23        N        0.0000     0.0000      NULL23
  24        C        0.0916     0.0000      C24 F
  25        N        0.0000     0.0000      NULL25
  26        R        1.0000     0.0000      R SOURCE

PRESS <RETURN> KEY TO CONTINUE -- READY?
```

Whether or not the origin of this example is of interest to the reader, the optimization goal is to correct the distortion in the passband from 0.95 to 1.05 radians per second that was caused by the modification and then to compensate for the effect of element dissipation. The TWEAKNET topological description of the network in Figure 6.2.6 including element values is contained in Table 6.6.21. The transducer loss function in the vicinity of the pass band is shown in Figure 6.2.7a for starting values of the lossless elements; the ideal filter had equal ripples in the passband.

The first step is to use TWEAKNET to obtain a 0.5-dB passband for the *lossless* network; Table 6.2.22 also shows that the intention was to maintain at least 55 dB at 0.8 and 1.3 radians per second. The Microway 87BASIC™ compiler was used to convert TWEAKNET to machine language instructions that utilize the 8087 math coprocessor. Only 23 function evaluations were required, using exact derivatives in the first penalty loop to obtain a maximum

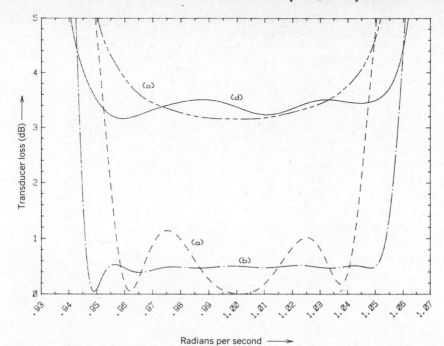

Figure 6.2.7. Transducer loss function for the filter in Figure 6.2.6 for four sets of element values. (*a*) Before optimization with lossless elements; (*b*) after optimization to 0.5 dB with lossless elements; (*c*) effect of then adding dissipative elements; (*d*) after optimization to 3.5 dB with dissipative elements.

constraint error modulus of 2.22E − 6. The new element values and the finite loaded Q values to be introduced at this time are shown in Table 6.2.23. The transducer loss after optimization of the lossless elements is shown in Figure 6.2.7*b*, and the subsequent effect of inductor quality factors of 200 and capacitor quality factors of 500 is shown in Figure 6.2.6*c*.

The second step is to use TWEAKNET to obtain a 3.5-dB passband for the dissipative network. Twenty function evaluations, using derivatives obtained by finite differences (another 380 function evaluations), produced a maximum constraint error modulus of 4.45E − 7. Table 6.2.24 shows that value after the final element values at the end of one penalty loop. The frequency scan confirming the binding 3.5-dB constraints at the passband edges (0.95 and 1.05 radians per second) and at two interior frequencies is also shown at the bottom of Table 6.2.24 and graphed in Figure 6.2.7*d*.

The preceding example will run in interpreted BASIC, but it will take a prohibitive amount of time. There are some OVERFLOW error messages (for Example 6.2.3 only), but IBM BASICA supplies the appropriate number of machine infinity and continues. IBM BASIC compiler Version 1.00 will provide a tenfold increase in speed; unfortunately, the resultant machine code

Table 6.2.22. Constrained Sample Data for Example 6.2.3 as Viewed Using Command 7

```
CONSTRAINTS NOW SET ARE:
 I       SAMPLE       LOWER       EQUALITY       UPPER
 1       0.8000       55.0000
 2       0.9500                                  0.5000
 3       0.9550                                  0.5000
 4       0.9600                                  0.5000
 5       0.9650                                  0.5000
 6       0.9700                                  0.5000
 7       0.9750                                  0.5000
 8       0.9800                                  0.5000
 9       0.9850                                  0.5000
10       0.9900                                  0.5000
11       0.9950                                  0.5000
12       1.0000                                  0.5000
13       1.0050                                  0.5000
14       1.0100                                  0.5000
15       1.0150                                  0.5000
16       1.0200                                  0.5000
17       1.0250                                  0.5000
18       1.0300                                  0.5000
19       1.0350                                  0.5000
20       1.0400                                  0.5000
21       1.0450                                  0.5000
22       1.0500                                  0.5000
23       1.3000       55.0000
SET OR RESET ANY CONSTRAINTS (Y/N)?
```

stops upon overflow. The best resolution of this problem is to use the math coprocessor, which will not underflow or overflow in any practical cases.

Table 6.2.25 summarizes the real time required for the given number of iterations and function evaluations for two examples using three different language modes. Problems that are comparable to Example 6.2.2 (five variables, seven frequency samples) may be run in interpreted BASIC without undue delay. Overflow usually does not occur in problems of that size, and even if it does occur the interpreter will make the intelligent replacement and continue. Overflow in larger problems being run using machine language TWEAKNET from the IBM compiler should be rerun using an equivalent compiler modified to work with the math coprocessor; several such compilers are commercially available. Since the IBM-AT personal computer will decrease the BASICA time in Table 6.2.25 by a factor of 3, much larger problems may be run without compiling TWEAKNET.

Readers who have run the preceding examples should now be able to appreciate the power and potential of constrained nonlinear optimization of networks and analogous physical systems. BASIC is a programming language that most personal computer users can readily understand, so it is a good means for communicating the details of algorithms. As a finished product, TWEAKNET compiled in conjunction with the math coprocessor is an excellent tool for RLC ladder network problems.

Table 6.2.23. Element Values That Resulted from Optimization of the Lossless Network and Finite Q Values to Be Introduced

```
****************************************
INPUT COMMAND NUMBER:? 8

UNITS ARE: FREQUENCY =  1.592D-01
           INDUCTANCE =  1.000D+00
           CAPACITANCE =  1.000D+00
R SOURCE, R LOAD, X LOAD =    1.0000    1.0000    0.0000

BRANCH    TYPE    VALUE      Q         NAME
   1       L      0.1027   200.0000   L1  H
   2       N      0.0000     0.0000   NULL2
   3       C      9.8812   500.0000   C3  F
   4       C      0.0737   500.0000   C4  F
   5       N      0.0000     0.0000   NULL5
   6       L     13.3216   200.0000   L6  H
   7       LC     0.1447   200.0000   L7  H
                  3.4242   500.0000   C7  F
   8       N      0.0000     0.0000   NULL8
   9       L      0.1398   200.0000   L9  H
   9       L      0.1398   200.0000   L9  H
  10       L     10.6267   200.0000   L10 H
  11       N      0.0000     0.0000   NULL11
  12       C      0.0868   500.0000   C12 F
  13       LC     0.1421   200.0000   L13 H
                  3.4242   500.0000   C7  F
   8       N      0.0000     0.0000   NULL8
   9       L      0.1398   200.0000   L9  H
  10       L     10.6267   200.0000   L10 H
  11       N      0.0000     0.0000   NULL11
  12       C      0.0868   500.0000   C12 F
  13       LC     0.1421   200.0000   L13 H
                  3.2692   500.0000   C13 F
  14       N      0.0000     0.0000   NULL14
  15       L      0.1565   200.0000   L15 H
  16       L     15.9345   200.0000   L16 H
  17       N      0.0000     0.0000   NULL17
  18       C      0.0653   500.0000   C18 F
  19       LC     0.0582   200.0000   L19 H
                  2.2811   500.0000   C19 F
  20       N      0.0000     0.0000   NULL20
  21       L      0.5834   200.0000   L21 H
  22       L      8.9024   200.0000   L22 H
  23       N      0.0000     0.0000   NULL23
  24       C      0.1197   500.0000   C24 F
  25       N      0.0000     0.0000   NULL25
  26       R      1.0000     0.0000   R SOURCE

PRESS <RETURN> KEY TO CONTINUE -- READY?
```

Table 6.2.24.   The End of the Optimization with Dissipative Elements and the Resulting Constrained Passband for Example 6.2.6

```
 1       0.10775724       0.00000 ...
 2       9.62782865       0.00000075
 3       0.07748296      -0.00016956
 4      12.75971076      -0.00000102
 5       0.12269792       0.00000531
 6       3.34196177       0.00000033
 7       0.16896779      -0.00000025
 8      10.50612961      -0.00000043
 9       0.08518913      -0.00006046
10       0.11871885       0.00002603
11       3.19510654       0.00000225
12       0.18488478       0.00001771
13      15.87142078      -0.00000015
14       0.06590916      -0.00004858
15       0.03038861      -0.00000532
16       2.23804818      -0.00000099
17       0.61754201      -0.00000208
18       8.47542304      -0.00000026
19       0.12634861      -0.00001923
TOTAL NUMBER OF FUNCTION EVALUATIONS = 20
EXPONENT P = 2
*******************************************************
AFTER   1   PENALTY MINIMIZATIONS,
   THE MAX CONSTRAINT MODULUS # 18   = 4.456584E-07
CONTINUE PENALTY MINIMIZATIONS (Y/N)? N


*************************************
INPUT COMMAND NUMBER:? 9
START FREQUENCY =? .92
STOP FREQUENCY =? 1.08
NUMBER OF FREQS, MAX 40 (+LIN, -LOG) =? 17
 #    FREQUENCY        RESPONSE dB       Rin      OHMS      Xin
 1     0.92000002      22.41810194     0.15077827       -1.51019911
 2     0.93000001      12.51758454     0.24980678       -1.01984439
 3     0.94000000       5.20408380     0.63553912       -0.58705533
 4     0.94999999       3.50000031     1.00105372       -0.47507946
 5     0.95999998       3.16671554     1.38703934       -0.54045038
 6     0.97000003       3.32492903     1.39124909       -0.90003318
 7     0.98000002       3.46381484     1.04797737       -0.96563522
 8     0.99000001       3.49999954     0.73724600       -0.73682705
 9     1.00000000       3.36561215     0.58132184       -0.34894196
10     1.00999999       3.22874116     0.62010047        0.11366616
11     1.01999998       3.34376792     1.03086691        0.58095179
12     1.02999997       3.50000045     1.88762981        0.07682236
13     1.03999996       3.46651651     1.08289018       -0.50851030
14     1.04999995       3.49999953     0.74736096       -0.10705578
15     1.05999994       4.47257345     0.72116641       -0.11917349
16     1.07000005       9.20336718     0.26675704        0.11989430
17     1.08000004      15.66175530     0.14529765        0.50808011
PRESS <RETURN> KEY TO CONTINUE -- READY?
```

There are at least three limitations to further expansion of TWEAKNET in BASIC. Certainly one limitation of IBM BASIC is that variable names are global so that names must be selected with care. Cross-reference programs readily available from computer clubs and commercial sources make it possible to avoid improper reuse of variable names, but it is still not a convenient process. FORTRAN, for example, allows local names in subroutines with

Table 6.2.25. Comparison of Computing Times (in Seconds) for Interpreted, Compiled, and Math Coprocessor BASIC[a]

	Example 6.2.2				Example 6.2.3		
Derivatives	BASICA[b]	Compiled[c]	87BASIC[d]	F#[e]	BASICA[b]	Compiled[c]	87BASIC[d]	F#[e]
Exact	451	45	31	25	—	204	180	6
Differences	800	75	43	22	—	624	398	6

[a] Ten iterations for Example 6.2.2 and three iterations for Example 6.2.3.
[b] IBM Interpreted BASICA Version A2.10.
[c] IBM Compiled BASIC Version 1.00.
[d] Microway 87BASIC Version 3.04.
[e] $F\#$ is the number of function evaluations, not including the N additional evaluations required for differences.

argument lists and named COMMON to pass sets of variable values without worrying about global naming. The interested reader is referred to Wolf (1985).

An important limitation in most versions of BASIC is that the complex variable type is not available. The next section makes it clear that the algebra of complex variables is far too pervasive in the real-frequency analysis of electrical networks to extend the capability of TWEAKNET to more general networks without inconvenience. However, Adby (1980) has written quite general network analysis routines based on the nodal admittance formulation (Section 6.1.3) in a BASIC language that includes matrix (MAT) operations. Since matrix software routines that can be called from BASIC or FORTRAN are commercially available for the IBM-PC, Adby's approach is feasible, and "not a source of much difficulty."

Finally, optimizer program TWEAKNET leaves only 13,750 bytes available for additional code and run-time memory assignment in the standard IBM BASICA environment. Although that is not trivially small, many languages have no such limitation. Additional user-friendly input–output routines, screen graphics, and other valuable features often require large amounts of memory.

6.3. Exact Partial Derivatives for Linear Systems

Gradient optimizers perform substantially better when exact first partial derivatives are available. For networks composed of lossless inductors and capacitors, it has been demonstrated by program TWEAKNET that exact derivatives may be computed with excellent efficiency. This section will show why that result is true for any lossless, reciprocal network, and how it may be extended with only slightly less efficiency to the most general linear networks. The subject is introduced by a brief but essential description of Tellegen's theorem, one of the most simple, powerful, and general tools applicable to all kinds of electrical networks and analogous systems.

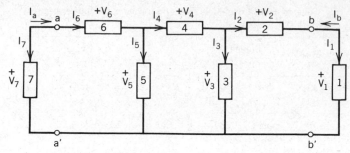

Figure 6.3.1. A network graph having five nodes and seven branches, applicable to the networks in Figures 6.2.4 and Figure 6.3.2.

6.3.1. Tellegen's Theorem.

Several limited forms of Tellegen's theorem for electrical networks will be described before the most general definition is given. Consider the *network graph* shown in Figure 6.3.1. That describes the topology of the network in Figure 6.2.4, for instance, because it specifies the relationships of its nodes and branches. A network having different elements but the same graph is shown in Figure 6.3.2. It is essential that the branch voltages and currents in a network graph be defined in a consistent manner; currents that always enter the plus sign of the voltage are adopted as in Figure 6.3.1.

For two networks having the same graph, let the voltage and current in the kth branch of the first network be called V_k and I_k, and let the corresponding quantities in the second network be called $\hat{V}_k$ and $\hat{I}_k$. One result of Tellegen's theorem is that the sum of all corresponding $V_k \hat{I}_k$ products is zero $(0 + j0$ actually):

$$\sum_{k=1}^{q} V_k \hat{I}_k = 0 = \sum_{k=1}^{q} \hat{V}_k I_k. \qquad (6.3.1)$$

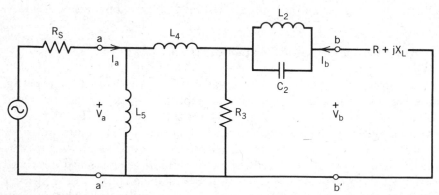

Figure 6.3.2. A second or adjoint network that shares the graph in Figure 6.3.1 with any other network having the same topology.

A better way to write (6.3.1) is in complex vector inner product notation:

$$\mathbf{V}^T \hat{\mathbf{I}} = 0 = \mathbf{I}^T \hat{\mathbf{V}}, \qquad (6.3.2)$$

where vectors $\mathbf{V}$ and $\mathbf{I}$ contain the q complex phasors for the voltages and currents in the q branches of the respective networks. This may seem to be nothing more than conservation of energy, but that is only the special case where the second network is identical to the first except for complex conjugate voltages or currents:

$$\mathbf{V}^T \mathbf{I}^* = 0 = \mathbf{I}^T \mathbf{V}^*. \qquad (6.3.3)$$

The simple addition of two lines to program TWEAKNET was described in Example 6.2.1, Section 6.2.2, to print the branch voltages and currents of ladder networks. The reader would benefit from assigning values to the network in Figure 6.3.2 and using that in conjunction with the network in Figure 6.2.4 to verify (6.3.2) and (6.3.3) numerically for at least one case. Care must be exercised to follow meticulously the branch voltage and current notation in Figure 6.3.1.

The complex conjugation operators distinguish (6.3.3) from (6.3.2). Penfield (1970) has shown that the most general form of *Tellegen's theorem* is

$$(\Phi'\mathbf{V})^T (\Phi''\hat{\mathbf{I}}) = (\Phi'\hat{\mathbf{V}})^T (\Phi''\mathbf{I}). \qquad (6.3.4)$$

The symbol Φ denotes a *Kirchhoff operator* on a set of voltages (currents) that obey Kirchhoff's voltage (current) law such that the result also obeys that law. Complex conjugation and the identity operators were used to produce (6.3.3) from (6.3.4). Other important Kirchhoff operators include differentiation with respect to network variables, frequency, or time. First- or higher-order perturbations are also valid Kirchhoff operators.

Tellegen's theorem applies to networks composed of linear or nonlinear, reciprocal or nonreciprocal, lumped or distributed elements, characterized in frequency or time, discrete or continuous, time-invariant or not, passive or active. There are numerous applications in hydrostatics, mechanical systems, electromagnetic fields, and quantum mechanics. Penfield (1970) has furnished concise proofs of more than 100 important theorems about electrical networks by using Tellegen's theorem.

6.3.2. Derivatives for Lossless Reciprocal Networks.

This section continues the analysis of networks in the sinusoidal steady state, limited to those that consist only of elements that are both lossless and reciprocal, specifically inductors, capacitors, and ideal transformers. In order to derive the simple but exact derivative expressions in (6.2.7) and (6.2.8), both the first-order *differential operator* (Δ) and the complex conjugation operator $(*)$ are employed in Tellegen's theorem (6.3.4).

It is convenient to distinguish between port branches and internal branches of the network. For example, there are two ports, $a-a'$ and $b-b'$, shown in Figures 6.3.1 and 6.3.2. Port currents I_a and I_b are conventionally defined to enter the ports with directions opposite those for internal branches. Therefore, a *difference form* of Tellegen's theorem is

$$I_a \Delta V_a - V_a \Delta I_a = \sum_k (I_k \Delta V_k - V_k \Delta I_k), \qquad (6.3.5)$$

using the identity and differential operators. Similarly, a *sum form* of Tellegen's theorem is

$$I_a^* \Delta V_a + V_a^* \Delta I_a = \sum_k (I_k^* \Delta V_k + V_k^* \Delta I_k), \qquad (6.3.6)$$

using the conjugation and differential operators. The summations in (6.3.5) and (6.3.6) include all branches except the source and the load (ports $a-a'$ and $b-b'$) branches. The load voltage and current, V_b and I_b, do not appear in the left sides of (6.3.5) and (6.3.6) because these may be considered invariant, consistent with the complex linear update network analysis method described in Section 6.1.3.

Figure 6.3.3 shows a resistively terminated two-port network and two typical kinds of branches—one impedance branch, Z_k, and one admittance branch, Y_k. For Tellegen's theorem, assume that the first network is the unperturbed one in Figure 6.3.3 with $\Delta Z_k = 0 = \Delta Y_j$, and the second network is the same network with small but nonzero perturbations.

An interpretation of (6.3.5) is that a small impedance perturbation, ΔZ_k, added in series with the kth branch will change all voltages and currents throughout the network and its terminations. However, it has been demonstrated by the analysis method of Section 6.1.3 that there is a new source

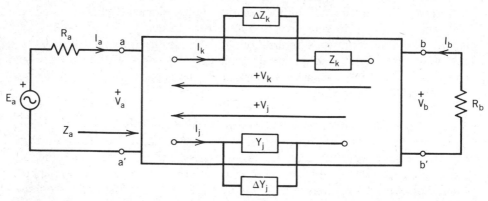

Figure 6.3.3. A resistively terminated two-port network containing perturbed branches Z_k and Y_k.

voltage E_a that will keep load voltage and current V_b and I_b constant. A similar analysis involving perturbation ΔY_j added in parallel with branch Y_j is deferred. The differential voltages and currents, ΔV and ΔI, in (6.3.5) are the changes due to perturbations in all branches of the second network. The same explanation applies for (6.3.6), except that complex conjugate voltages and currents are selected from the first, unperturbed network.

It can now be shown that terms on the right side of (6.3.5) that do not contain ΔZ_k sum to zero. The *constitutive law* for a reciprocal kth impedance branches is

$$V_k = Z_k I_k. \tag{6.3.7}$$

Therefore, neglecting higher-order terms, the differential voltage is

$$\Delta V_k = Z_k \Delta I_k + \Delta Z_k I_k. \tag{6.3.8}$$

Substitution of (6.3.8) and (6.3.7) into (6.3.5) leaves only terms containing ΔZ_k. Terms not containing ΔZ_k on the right side of (6.3.6) also sum to zero,

$$I_k^* \Delta V_k + V_k^* \Delta I_k = I_k^* (Z_k + Z_k^*) \Delta I_k = 0, \tag{6.3.9}$$

because the real part of Z_k is zero.

Orchard (1985) collected the nonzero parts of (6.3.5) and (6.3.6) in the matrix expression

$$\begin{bmatrix} I_a & -V_a \\ I_a^* & V_a^* \end{bmatrix} \begin{bmatrix} \Delta V_a \\ \Delta I_a \end{bmatrix} = I_k \Delta Z_k \begin{bmatrix} I_k \\ I_k^* \end{bmatrix}. \tag{6.3.10}$$

Matrix equation (6.3.10) may be solved for $\Delta V_a / \Delta Z_k$ and $\Delta I_a / \Delta Z_k$; for $\Delta Z_k \to 0$, the partial derivatives of V_a and I_a are obtained with respect to branch impedances Z_k:

$$\nabla_{Z_k} V_a = \frac{I_k(V_a I_k^* + V_a^* I_k)}{2P_b}, \tag{6.3.11}$$

$$\nabla_{Z_k} I_a = \frac{I_k(I_a I_k^* - I_a^* I_k)}{2P_b}. \tag{6.3.12}$$

The determinant of the matrix in (6.3.10) is twice the power delivered to the lossless network; consequently, $2P_b$ is twice the power delivered to the load resistance R_b. As in TWEAKNET line 8050, it is assumed that $P_b = \frac{1}{2}$ watt, so that $2P_b = 1$ in (6.3.11) and (6.3.12).

The exact partial derivatives in (6.3.11) and (6.3.12) are useful for optimization of various ratios of voltages and currents, using the identities in Table 6.1.3 in Section 6.1.3. Also, the partial derivative of input impedance $Z_{\text{in}} =$

V_a/I_a in Figure 6.3.3 may be obtained by using the derivatives of V_a and I_a. By rearranging (6.2.9), it is seen that the corresponding input reflection coefficient is

$$p_a = \frac{V_a - R_a I_a}{V_a + R_a I_a}. \tag{6.3.13}$$

Therefore, partial derivatives of the input reflection coefficient are also available.

The *complex transducer loss function H* is defined:

$$H = \frac{E_a}{(2R_a)^{1/2}}, \qquad P_b = \tfrac{1}{2}. \tag{6.3.14}$$

The transducer loss function in (6.2.2) is

$$L_i = (20 \log_{10} e) \mathrm{Re}[\ln(H)] \ \mathrm{dB}, \tag{6.3.15}$$

where "Re" means a "real part of." Therefore, it is convenient to work with $\ln(H)$:

$$\ln(H) = \ln(E_a) - \tfrac{1}{2}\ln(2R_a), \qquad P_b = \tfrac{1}{2}. \tag{6.3.16}$$

The partial derivative of $\ln(H)$ with respect to a branch impedance is

$$\nabla_{z_k}\ln(H) = \frac{\nabla_{z_k} E_a}{E_a}$$

$$= \frac{\nabla_{z_k} V_a + R_a \nabla_{z_k} I_a}{V_a + R_a I_a}. \tag{6.3.17}$$

Substitution of (6.3.11) and (6.3.12) into (6.3.17) yields

$$\nabla_{z_k}\ln(H) = I_k\left[I_k^* + \left(\frac{V_a^* - R_a I_a^*}{V_a + R_a I_a} \right) I_k \right]. \tag{6.3.18}$$

Assuming that the zero phase reference is the source voltage, that is, $E_a = E_a^*$, then (6.3.13) may be substituted into (6.3.18) to yield

$$\nabla_{z_k}\ln(H) = \left(|I_k|^2 + p_a^* I_k^2 \right). \tag{6.3.19}$$

The partial derivative in (6.3.19) is a complex number, as is

$$\ln(H) = \ln\left[|H| e^{j\theta} \right] = \ln|H| + j\theta. \tag{6.3.20}$$

The Cauchy–Riemann identity, item 11 in Table 6.1.3 in Section 6.1.3, defines the derivative of a complex function of a complex variable. In this case the complex variable is $Z_k = R_k + jX_k$, and $R_k = 0$. The Cauchy–Riemann result of interest is

$$\nabla_{Z_k}\ln(H) = \nabla_{X_k}\theta - j\nabla_{X_k}\ln|H|. \qquad (6.3.21)$$

Equating the imaginary parts of (6.3.21) and (6.3.19):

$$\nabla_{X_k}\ln|H| = -\operatorname{Im}(p_a^*I_k^2). \qquad (6.3.22)$$

Introducing the factor $(20\log_{10}e)$ to (6.3.22) yields the desired partial derivative for lossless, reciprocal networks originally stated in (6.2.7):

$$\nabla_{X_k}L_i = -(20\log_{10}e)\operatorname{Im}(p_a^*I_k^2). \qquad (6.3.23)$$

A similar result may be obtained using branch admittance perturbations, ΔY_k, finding the resulting partial derivatives of V_a and I_a, and finally (6.2.8):

$$\nabla_{B_k}L_i = +(20\log_{10}e)\operatorname{Im}(p_a^*V_k^2), \qquad (6.3.24)$$

where branch admittance $Y_k = G_k + jB_k$, and $G_k = 0$. It should be noted that expressions similar to (6.3.23) and (6.3.24) could be obtained for the transducer phase angle θ. Again, note that (6.3.18) through (6.3.24) are valid only when the power delivered to the load is $P_b = \frac{1}{2}$ watt.

This section is concluded with remarks concerning derivatives evaluated with respect to quantities that have the value zero. That is the present situation with branch resistances in lossless networks. The Cauchy–Riemann identity provides an alternative expression of the partial derivative of $\ln(H)$ with respect to branch resistance R_k:

$$\nabla_{Z_k}\ln(H) = \nabla_{R_k}\ln|H| + j\nabla_{R_k}\theta. \qquad (6.3.25)$$

Therefore, the derivative analogous to (6.3.23) for lossless, reciprocal networks (with $\frac{1}{2}$ watt load power) is:

$$\nabla_{R_k}L_i = +(20\log_{10}e)\left[|I_k|^2 + \operatorname{Re}(p_a^*I_k^2)\right], \qquad (6.3.26)$$

and a similar derivative with respect to branch conductance G_k may be obtained. There is nothing wrong with the fact that the derivative in (6.3.26) occurs at the value $R_k = 0$: It is still a linear prediction in a small neighborhood of $Z_k = 0 + jX_k$ of the expected change in transducer loss L_i with small changes in R_k, typically because of parasitic dissipation (finite quality factor Q). The interested reader is referred to Orchard (1985) for an extended discussion.

Another interesting possibility to consider is the case of derivatives for zero-valued elements, say a shunt capacitor placed in branch 1 of the network in Figure 6.2.4. See problem 6.14. Certainly the partial derivative of L_i with respect to such a capacitance ($C_1 = 0$) is of interest, since a negative derivative value at a passband frequency means that adding some finite amount of C_1 would decrease the transducer loss. This leads to the possibility of "growing" elements into network graphs, a subject explored by Smith (1971). Because of the "Swiss Alps" effect of becoming trapped prematurely in local minima, the technique is unreliable except for determining exact values of circuit parasitic values for resistance, inductance, capacitance, and transmission line (propagation delay) effects in physical networks.

6.3.3. Derivatives for Any Network Using Adjoint Networks.

Tellegen's theorem will now be employed to obtain exact partial derivatives for any linear network, including those that are active, passive, dissipative, or nonreciprocal. Only two analyses of the network are required for each frequency. The response is any voltage or current at any of many ports for voltage or current excitation at any number of ports; it should follow from developments in the last section that exact partial derivatives of other responses, such as the transducer loss, input impedance, reflection coefficient, and so on, are available through algebraic manipulation. The following explanation is restricted to two-port networks as in Figure 6.3.4 without loss of generality.

Tellegen's theorem, (6.3.4), is implemented by defining the differential operator (Δ) on a given network and the identity operator on a second (adjoint) network having the same graph, Figures 6.3.4a and b, respectively. It is convenient to collect the port voltages and currents in vectors containing complex phasors:

$$\mathbf{V}_p = \begin{bmatrix} V_a \\ V_b \end{bmatrix}, \qquad \mathbf{I}_p = \begin{bmatrix} I_a \\ I_b \end{bmatrix}, \qquad \hat{\mathbf{V}}_p = \begin{bmatrix} \hat{V}_a \\ \hat{V}_b \end{bmatrix}, \qquad \hat{\mathbf{I}}_p = \begin{bmatrix} \hat{I}_a \\ \hat{I}_b \end{bmatrix}. \qquad (6.3.27)$$

The $\char`\^$ symbol denotes voltages and currents from the adjoint network. The internal branch voltages and currents are contained in vectors $\mathbf{V}$ and $\mathbf{I}$ that are related by the *branch impedance matrix* from (6.1.31):

$$\mathbf{V} = \mathbf{Z}\mathbf{I}, \qquad (6.3.28)$$

$$\hat{\mathbf{V}} = \hat{\mathbf{Z}}\hat{\mathbf{I}}. \qquad (6.3.29)$$

Then a *difference form* of Tellegen's theorem is

$$\hat{\mathbf{I}}_p^T(\Delta\mathbf{V}_p) - \hat{\mathbf{V}}_p^T(\Delta\mathbf{I}_p) = \hat{\mathbf{I}}^T(\Delta\mathbf{V}) - \hat{\mathbf{V}}^T(\Delta\mathbf{I}). \qquad (6.3.30)$$

The operator (Δ) denotes a differential of every element in the vector or matrix it precedes.

Perturbing the branch impedances (6.3.28) in the given network and ignoring second-order changes,

$$\Delta V = (\Delta Z)I + Z(\Delta I). \tag{6.3.31}$$

Substitution of (6.3.31) into the right side of (6.3.30) yields

$$\hat{I}^T(\Delta Z)I + \hat{I}^T Z\Delta I - \hat{V}^T \Delta I. \tag{6.3.32}$$

Further substitution of $\hat{V}^T = \hat{I}^T \hat{Z}^T$ from (6.3.29) into (6.3.32) shows that the entire right side of (6.3.30) may be reduced to

$$\hat{I}^T(\Delta Z)I \tag{6.3.33}$$

if $\hat{Z} = Z^T$, which defines the *adjoint network*. Reciprocal networks are self-adjoint (e.g., an RLC network).

The adjoint network serves the following purpose in determining exact partial derivatives. A single excitation and resulting response are contemplated for the given and the adjoint network, but they need not respectively agree as to port or kind. Notice that the terminating resistances, if any, are included within the given and adjoint networks in Figure 6.3.4. Whether they are in series or shunt connection is discussed later in this section. A typical excitation pattern might be a unit value of V_a, a zero value of V_b (short-circuit port b–b'), and response I_b. Similarly for the adjoint network, the excitation pattern might be $\hat{V}_b = 1$, $\hat{V}_a = 0$ (short-circuit port $\hat{a}$–$\hat{a}'$), and response $\hat{I}_a$.

For generality, all possible excitation patterns are described by Table 6.3.1, using notation by Bandler (1973a:256). The voltages and currents in Table 6.3.1 are shown as vectors, since there may be multiple sources and multiple responses in a multiport network. The typical excitation pattern just described contemplated scalar V's and I's in Table 6.3.1, but whether they represent port a–a', b–b', $\hat{a}$–$\hat{a}'$, or $\hat{b}$–$\hat{b}'$ voltages or currents is still arbitrary. The "consequences" are due to the assumed independence of the excitation, which is not subject to change as is the response.

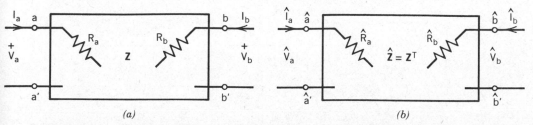

Figure 6.3.4. Two two-port linear networks employed to obtain exact partial derivatives of response due to excitation of the given network. (*a*) Given network, (*b*) adjoint network.

Table 6.3.1.   Excitation Patterns for the Given and Adjoint Networks

	Voltage	Current	Consequences
Excitation	$\mathbf{V}_V$	$\mathbf{I}_I$	$\Delta\mathbf{V}_V = \mathbf{0} = \Delta\mathbf{I}_I$
Response	$\mathbf{I}_V$	$\mathbf{V}_I$	$\Delta\mathbf{I}_V \neq \mathbf{0} \neq \Delta\mathbf{V}_I$

The excitation and response patterns are selected according to the left side of (6.3.30), which simplifies because of the consequences shown in Table 6.3.1:

$$\hat{\mathbf{I}}_I^T(\Delta\mathbf{V}_I) - \hat{\mathbf{V}}_V^T(\Delta\mathbf{I}_V) = \hat{\mathbf{I}}^T(\Delta\mathbf{Z})\mathbf{I}. \tag{6.3.34}$$

Equation (6.3.34) is the final result. Table 6.3.2 summarizes the excitation patterns that satisfy the left side of (6.3.34) for six commonly encountered two-port transfer functions. Compare the first line in Table 6.3.2 with the left side of (6.3.34). The response is a voltage resulting from an exciting current, so $\Delta V_I = \Delta V_a$. The subscript indicates that $\hat{\mathbf{I}}_I = \hat{I}_a = 1$, since it multiplies the desired response. The desired port $b\text{--}b'$ condition, $I_b = 0$, requires that $\Delta I_b = 0$, and its multiplier in (6.3.34) is $\hat{V}_b = 1$.

Notice that the first and third lines in Table 6.3.2 obtain a response at the same port being excited, that is, the response is not a transfer function. It is seen in those cases that the given and adjoint networks are excited and terminated in exactly the same way. Therefore, for reciprocal networks where $\hat{\mathbf{Z}} = \mathbf{Z}^T = \mathbf{Z}$, only one network analysis per frequency is required for exact derivatives; otherwise, two are required, one for the given network and another for the adjoint network.

The right side, $\hat{\mathbf{I}}^T(\Delta\mathbf{Z})\mathbf{I}$, would have been $-\hat{\mathbf{V}}^T(\Delta\mathbf{Y})\mathbf{V}$ had the analysis utilized the *branch admittance matrix* $\mathbf{Y}$ instead of the branch impedance matrix $\mathbf{Z} = \mathbf{Y}^{-1}$. Table 6.3.3 summarizes the right side of (6.3.34) for Z, Y, R, G, L, and C branch elements in the given network.

Table 6.3.2.   Excitation Patterns for Common Two-Port Network Transfer Functions

Response	Given Network		Adjoint Network	
	Source	$b\text{--}b'$	Source	Other Port
V_a/I_a	$I_a = 1$	$I_b = 0$	$\hat{I}_a = 1$	$\hat{I}_b = 0$
V_b/I_a	$I_a = 1$	$I_b = 0$	$\hat{I}_b = 1$	$\hat{I}_a = 0$
I_a/V_a	$V_a = 1$	$V_b = 0$	$\hat{V}_a = 1$	$\hat{V}_b = 0$
I_b/V_a	$V_a = 1$	$V_b = 0$	$\hat{V}_b = 1$	$\hat{V}_a = 0$
V_b/V_a	$V_a = 1$	$I_b = 0$	$\hat{I}_b = 1$	$\hat{V}_a = 0$
I_b/I_a	$I_a = 1$	$V_b = 0$	$\hat{V}_b = 1$	$\hat{I}_a = 0$

Table 6.3.3. Sensitivities for RLC Elements

Branch Element	Sensitivity[a]	Differential
Impedance Z_k	$\hat{I}_k I_k$	ΔZ_k
Admittance Y_k	$-\hat{V}_k V_k$	ΔY_k
Resistance	$\hat{I}_k I_k$	ΔR_k
Conductance	$-\hat{V}_k V_k$	ΔG_k
Inductance	$\omega(d + j1)\hat{I}_k I_k$	ΔL_k
Capacitance	$-\omega(d + j1)\hat{V}_k V_k$	ΔC_k

[a]$d = 1/Q$ for L or C.

Example 6.3.1. To find the exact partial derivative of V_b/V_a with respect to a branch impedance Z_k for a resistively terminated two-port network as shown in Figure 6.3.4a, consult the next-to-last line in Table 6.3.2 to determine the port connections. A unit voltage source is connected at port a–a'; therefore, source resistance R_a is connected in series inside the network (or it would have no effect). Port b–b' is left unterminated so that $I_b = 0$; therefore, load resistance R_b is connected in shunt (or it would have no effect). A unit current source is connected to adjoint network port $\hat{b}$–$\hat{b}'$; therefore, resistance $\hat{R}_b$ is connected in shunt. Adjoint network port $\hat{a}$–$\hat{a}'$ is short-circuited so that $\hat{V}_a = 0$; therefore, resistance $\hat{R}_a$ is connected in series. Both the given and adjoint networks are analyzed at the same frequency, to determine the currents through Z_k: I_k and $\hat{I}_k$, respectively. According to (6.3.34) and Tables 6.3.2 and 6.3.3, the differentials are related by

$$\Delta V_b = \hat{I}_k I_k (\Delta Z_k), \tag{6.3.35}$$

so the exact partial derivative is

$$\nabla_{Z_k} V_b = \hat{I}_k I_k. \tag{6.3.36}$$

Since $V_a = 1$, (6.3.36) is the partial derivative of the transfer function V_b/V_a.

It is not difficult to deal with branch elements that have more than two terminals, such as lengths of transmission line (a two-port branch element) or much larger pieces of subnetworks. Bandler (1970b, 1973a) lists the sensitivity expressions for a large variety of multiport branch elements, especially for microwave networks. Cuthbert (1983:108) provides expressions for the right side of (6.3.34) for scattering, voltage transfer, current transfer, and cascade (chain *ABCD*) network parameter systems that differ from **Z** and **Y** characterization.

6.3.4.   Derivatives Obtained by the Nodal Admittance Matrix.   A conventional way to analyze linear electrical networks is by means of the definite nodal admittance (**Y**) matrix described in Section 6.1.3. It is shown in this section that **LU** factorization may be employed to eliminate much of the work otherwise required for analysis of the adjoint network. Finally, a method is described in which the adjoint network is not involved in efficient calculation of the exact partial derivatives of general network response functions.

The definite nodal admittance matrix was defined as $\mathbf{Y} = [y_{ij}]$ in the system of equations (6.1.30):

$$\mathbf{I} = \mathbf{YV}, \tag{6.3.37}$$

where **V** is the node voltage vector and **I** is the node current vector, both containing phasors related to the sinusoidal steady-state voltages and currents. The adjoint network has the same graph as the given network, and its admittance matrix is $\hat{\mathbf{Y}} = \mathbf{Y}^T$. As Director (1971) has noted, only one nodal admittance matrix need be constructed per frequency, not two, since the node equations of the adjoint network are

$$\mathbf{Y}^T\hat{\mathbf{V}} = \hat{\mathbf{I}}. \tag{6.3.38}$$

Most methods for inverting an $n \times n$ **Y** matrix require on the order of n^3 complex multiplications and divisions, and about n^2 additional operations are required to solve $\mathbf{V} = \mathbf{Y}^{-1}\mathbf{I}$ for a given vector **I**.

The **LU** factorization method was defined for real matrices in (3.1.1) through (3.1.6); exactly the same algebra may be employed by using complex instead of real numbers. Using that method, the nodal admittance matrix may be factored using about $n^3/3$ operations:

$$\mathbf{Y} = \mathbf{LU}. \tag{6.3.39}$$

Then forward and back substitution may be employed to solve (6.3.37) for **V** in about n^2 operations. The solution of (6.3.38) is equivalent to solving

$$\mathbf{U}^T\mathbf{L}^T\hat{\mathbf{V}} = \hat{\mathbf{I}}, \tag{6.3.40}$$

where $\mathbf{U}^T$ is lower triangular and $\mathbf{L}^T$ is upper triangular. Consequently, two forward and back substitution processes using a total of $n^3/3 + 2n^2$ operations are required to solve both the given and adjoint network equations. The **LU** factorization method for obtaining network responses and subsequently their exact derivatives saves about $n^3/3$ operations.

Although Tellegen's theorem and the adjoint method provide useful insight into computation of exact partial derivatives, it turns out that neither of those concepts are necessary. Consider the nodal admittance matrix, $\mathbf{Y} = [y_{ij}]$ and its inverse, $\mathbf{Z} = [z_{ij}]$. Then,

$$\mathbf{ZY} = \mathbf{I}, \tag{6.3.41}$$

where $\mathbf{I}$ is the unit matrix in this case. The object is to obtain the partial derivative of the complex transfer function

$$H = \frac{V_2}{V_1} \quad \text{such that } I_2 = 0. \tag{6.3.42}$$

It is assumed that node 1 is the input terminal and node 2 is the output terminal with respect to a common ground; for example, see Figure 6.1.6 in Section 6.1.3. The elements of $\mathbf{Z}$, z_{ij}, are defined on the basis of zero current at node j, analogous to the definition of the elements of $\mathbf{Y}$, y_{ij}, which are defined on the basis of zero voltage at node j. Since $z_{21} = V_2/I_1$ and $z_{11} = V_1/I_1$,

$$H = z_{21}/z_{11}, \tag{6.3.43}$$

$$\ln(H) = \ln(z_{21}) - \ln(z_{11}). \tag{6.3.44}$$

Taking the derivative of (6.3.44) with respect to some branch admittance Y_k yields

$$\nabla_{Y_k}\ln(H) = (\nabla_{Y_k}z_{21})\left(\frac{1}{z_{21}}\right) - (\nabla_{Y_k}z_{11})\left(\frac{1}{z_{11}}\right). \tag{6.3.45}$$

Therefore, the approach is to find the partial derivatives of z_{21} and z_{11} and then employ (6.3.45) for the desired derivative.

A method by Fidler (1983) is derived by differentiating (6.3.41) with respect to a two-terminal branch element Y_k:

$$\nabla_{Y_k}(\mathbf{ZY}) = (\nabla_{Y_k}\mathbf{Z})\mathbf{Y} + \mathbf{Z}(\nabla_{Y_k}\mathbf{Y}) = \mathbf{0}. \tag{6.3.46}$$

Then postmultiplying (6.3.46) by $\mathbf{Y}^{-1} = \mathbf{Z}$ yields

$$\nabla_{Y_k}\mathbf{Z} = -\mathbf{Z}(\nabla_{Y_k}\mathbf{Y})\mathbf{Z}. \tag{6.3.47}$$

Matrix $\nabla_{Y_k}\mathbf{Y}$ has a simple structure that is not difficult to envision in terms of the two rules given in Section 6.1.3 for formulating $\mathbf{Y}$: diagonal element y_{ii} is the sum of all branch admittances touching node i, and element y_{ij} is the negative sum of all branch admittances connected between nodes i and j. The differentiation is with respect to only one two-terminal branch admittance, Y_k; suppose that it is connected between nodes i and j. Consequently, $\nabla_{Y_k}\mathbf{Y}$ has only four nonzero terms: $+1$ at (i, i) and (j, j) and -1 at (i, j) and (j, i).

Therefore, it is convenient to define a vector

$$
\mathbf{u} = \begin{bmatrix} 0 \\ \vdots \\ +1 \\ \vdots \\ 0 \\ \vdots \\ -1 \\ \vdots \\ 0 \end{bmatrix} \begin{matrix} \\ \\ i \\ \\ \\ \\ j \\ \\ \end{matrix} .
\tag{6.3.48}
$$

Then (6.3.47) may be written

$$
\nabla_{Y_k}\mathbf{Z} = -(\mathbf{Zu})(\mathbf{u}^T\mathbf{Z})
$$

$$
= - \begin{bmatrix} z_{1i} - z_{1j} \\ z_{2i} - z_{2j} \\ \cdots \\ z_{ni} - z_{nj} \end{bmatrix} \begin{bmatrix} z_{i1} - z_{j1} \\ z_{i2} - z_{j2} \\ \cdots \\ z_{in} - z_{jn} \end{bmatrix}^T .
\tag{6.3.49}
$$

The partial derivatives needed in (6.3.45) are available from (6.3.49):

$$
\nabla_{Y_k}z_{21} = -(z_{21} - z_{2j})(z_{i1} - z_{j1}),
\tag{6.3.50}
$$

$$
\nabla_{Y_k}z_{11} = -(z_{1i} - z_{1j})(z_{i1} - z_{j1}).
\tag{6.3.51}
$$

There are slightly more complicated rules for including controlled sources and other more general network components. The only major work required to both analyze and obtain derivatives for any linear network is to form and invert the definite admittance matrix. Branin (1973) has provided a complete treatment of this subject without restriction to the nodal admittance matrix.

6.4.   Robust Response Functions

It is well known that nonlinear programming is much more successfully practiced by those who understand both the optimization methods and the nature of the problems being solved; certainly the more the better. A great deal is known about a large class of useful response functions for optimization of linear electrical networks. Some concepts that have direct bearing on the convergence rate and ultimate success of network function minimization are described. The central theme is the bilinear function, since most network response functions are bilinearly related to branch immittances.

6.4.1. Bilinear Functions and Forms. A *bilinear function* or *linear fractional transformation* may take the form $W(Z)$,

$$W = \frac{bZ + c}{dZ + 1}, \tag{6.4.1}$$

where all quantities are complex numbers, and coefficients $b - cd \neq 0$. Multiplication of both sides of (6.4.1) by the denominator term gives the alternative bilinear form

$$bZ + c - dWZ = W, \tag{6.4.2}$$

which is linear in Z and linear in W, thus the name bilinear. Use of (6.4.2) for any three unique pairs of (W, Z) values enables a solution for coefficients b, c, and d.

A third form of the bilinear function often found in textbooks on complex variables is

$$W = \frac{b}{d} + \frac{c - b/d}{dZ + 1}, \tag{6.4.3}$$

which shows that Z is mapped into W by two linear transformations (e.g., $Q = eF + g$) and one inverse transformation (e.g., $T = 1/Q$). Linear transformations in the rectangular complex plane do not change the shape of sets of points (curves) or angles of intersections of curves. It can be shown that inverse transformations and consequently bilinear transformations map circles and lines in the Z plane into circles or lines in the W plane.

It is important to confirm that "a bilinear function of a bilinear function is bilinear." What is meant by that word twister is that if (6.4.1) is given and a second bilinear function is defined, say

$$Z = \frac{eV + h}{qV + 1}, \tag{6.4.4}$$

then substitution of (6.4.4) into (6.4.1) shows that $W(V)$ also has the bilinear functional form.

There are just two special sets of the three coefficients that are of interest in the present application of bilinear forms, although their general use occurs throughout mathematics and its applications in the physical sciences. Any

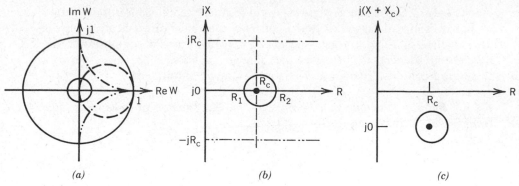

Figure 6.4.1. Mapping of Z into W according to (6.4.5) with $\theta = 0$ and $Z_c = R_c + j0$. (a) Unit circle in the W plane; (b) Z plane for $X_c = 0$; (c) Z plane for $X_c \neq 0$.

bilinear function that maps the right-half complex plane into a unit circle must have the form

$$W = e^{j\theta}\frac{Z - Z_c}{Z + Z_c^*},\qquad(6.4.5)$$

where $Z = R + jX$, $Z_c = R_c + jX_c$, $R_c > 0$, and $Z_c^* = R_c - jX_c$. The mapping of Z into W according to (6.4.5) is illustrated in Figure 6.4.1; the unit circle in the W plane is the image of the entire right-half Z plane. Normally the pure rotation by $e^{j\theta}$ through angle θ is omitted ($\theta = 0$) in (6.4.5).

It is seen from the numerator of (6.4.5) that the origin in the W plane corresponds to the point $Z = Z_c$ in the Z plane. The small circle indicated by the solid line has a *geometric center* located at $Z = Z_c$, that is, $R_c^2 = R_1 R_2$. There is a set of orthogonal circles and intersecting arcs in the W plane that corresponds to the rectangular coordinate grid in the Z plane. Figure 6.4.1a is the Smith chart usually associated with transmission line applications in electrical engineering with $X_c = 0$ as in Figure 6.4.1b. The more general case for $X_c \neq 0$ in Figure 6.4.1c is easily related to it. Lines of constant X in Figure 6.4.1b appear as arcs in the unit circle in Figure 6.4.1a. When $X_c \neq 0$, those arcs represent constant $X + X_c$ loci, and there are no other changes.

The second special bilinear function of interest here is the unique form that maps one unit circle onto another:

$$H = e^{j\phi}\frac{W - w_0}{1 - W w_0}.\qquad(6.4.6)$$

Transformations (6.4.5) and (6.4.6) are illustrated in Figure 6.4.2. It is seen from the numerator of (6.4.6) that w_0 in the W plane maps into the origin of the H plane. The mapping from the Z plane to the H plane must depend on

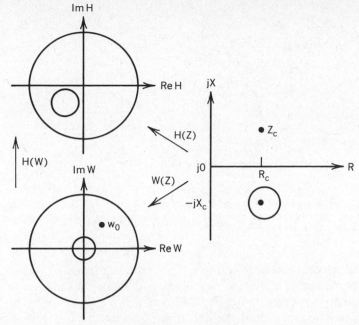

Figure 6.4.2. Mapping Z onto W and H and mapping W onto H using bilinear functions.

some Z-plane constant corresponding to w_0, say $Z = Z_h$, such that

$$H = \frac{Z - Z_h}{Z + Z_h^*}, \tag{6.4.7}$$

$$w_0 = \frac{Z_h - Z_c}{Z_h + Z_c^*}. \tag{6.4.8}$$

One final form of bilinear transformation has been very useful for impedance mapping in electrical engineering:

$$W = T + U\frac{Z - Z_c}{Z + Z_c^*}. \tag{6.4.9}$$

By arranging (6.4.9) in the form of (6.4.1), the constants in (6.4.9) may be identified:

$$Z_c = \frac{1}{d^*}, \tag{6.4.10}$$

$$T = \frac{cd^* + b}{d + d^*}, \tag{6.4.11}$$

$$U = \frac{b}{d} - T. \tag{6.4.12}$$

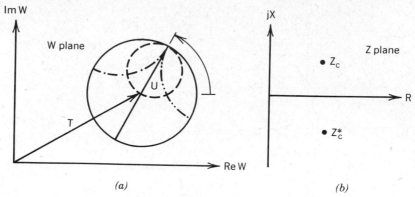

Figure 6.4.3. A bilinear transformation for impedance mapping. (a) The W plane containing a scaled, rotated unit circle; (b) the right-half Z plane that maps into the unit circle.

The complex constants in (6.4.9) have geometric significance: T is a constant displacement, and U scales the magnitude and rotates the unit-circle domain that originally appeared in Figure 6.4.1a as defined by (6.4.5). The bilinear mapping function of (6.4.9) through (6.4.12) is shown in Figure 6.4.3. The circle in Figure 6.4.3a is the image of the entire right-half Z plane in Figure 6.4.3b.

Additional details of these bilinear transformations are available in Cuthbert (1983:242, 370). The relevance to optimization of these mapping transformations in complex planes is discussed next.

6.4.2. The Bilinear Property of Linear Networks.

All voltage and current response ratios such as those listed in Table 6.3.2 are bilinear functions of a linear network's branch immittances, Z_k or Y_k. For instance, input impedance Z_{in} in Figure 6.2.4 is a bilinear function of Z_2, and L_2 is also bilinearly related to Z_2 since $Z_2 = \omega L_2(d + j1)$. Input reflection coefficient p_1 is related to Z_{in} by the bilinear function (6.2.9). Since a bilinear function of a bilinear function is bilinear, p_1 is also a bilinear function of Z_{in}, but with different coefficients b, c, and d in (6.4.1). Other bilinear functions having different coefficients are $p_1(Y_3)$, $p_1(C_4)$, etc. All of the complex constants in these bilinear functions are functions of frequency.

One benefit in knowing that most responses are bilinear functions of branch immittances at a frequency is for *impedance mapping*. The special bilinear form of (6.4.9) as illustrated in Figure 6.4.3 is relevant to the behavior of Z_{in} as a function of Z_2 in Figure 6.2.4, for example. In that case, the W plane represents the Z_{in} plane, and the circular Smith chart is the image of the entire Z_2 right-half plane. Therefore, the extreme values for R_{in} and X_{in} for the worst-case values of L_2 can be determined by inspection, since the rim of the Smith chart represents lossless branch-impedance values.

Another benefit in recognizing the bilinear properties inherent in network behavior is directly applicable to optimization. Suppose that the optimization problem is to adjust branch impedance Z_2 in Figure 6.2.4 to make input impedance Z as close as possible to a given impedance value at a frequency, say $\hat{Z} = \hat{R} + j\hat{X}$. The Euclidean norm (2.1.37) is often used as a metric for the neighborhood shown in Figure 6.4.4a:

$$|\Delta Z|_2 = \left[(R - \hat{R})^2 + (X - \hat{X})^2\right]^{1/2}. \tag{6.4.13}$$

The squared norm also may be used, since a square root is avoided and

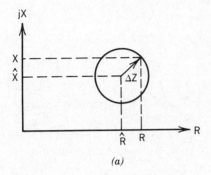

(a)

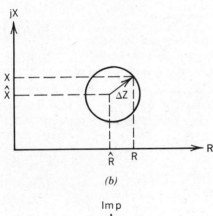

(b)

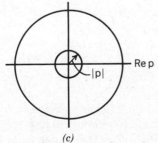

(c)

Figure 6.4.4. Three neighborhoods about $Z = \hat{R} + j\hat{X}$. (*a*) Arithmetic neighborhood; (*b*) geometric neighborhood, (*c*) reflection-plane neighborhood.

squaring is a monotone function that does not introduce extraneous minima. One objection to using the unnormalized sum of squared differences as an error function is that it may not evenly represent the relative sizes of such neighborhoods. For example, $R - \hat{R} = 1.5$ is a large increment if $\hat{R} = 2$ but is quite small if $\hat{R} = 2000$. Another objection to an unnormalized squared error function is that it does not penalize large neighborhoods as severely as others that behave better during optimization.

An alternative definition of a neighborhood in the Z plane is the geometrically centered circle in Figure 6.4.4*b*. The geometric mean approximates the arithmetic mean for small relative errors (less than 10%), but the reason for considering the geometric neighborhood is that (6.4.5) maps that neighborhood into one concentrically located at the origin of the *generalized reflection coefficient* (see Figure 6.4.4*c*):

$$p = \frac{Z - \hat{Z}}{Z + \hat{Z}*}. \tag{6.4.14}$$

The magnitude of the generalized reflection coefficient is a scalar metric that is uniform over the entire right-half Z plane and is normalized to $0 \leq |p| \leq 1$. In order to avoid the square root operation, $|p|^2$ is used, and to increase the penalty for large neighborhoods, a particular monotone *weighting function* is used to map $|p|^2$ to the range zero to infinity:

$$S = \frac{1 + |p|^2}{1 - |p|^2} - 1 = \frac{2|p|^2}{1 - |p|^2}. \tag{6.4.15}$$

The metric in (6.4.15) is a measure over the range $0 \leq S \leq \infty$ for $0 \leq |p| \leq 1$ as ΔZ varies from 0 to infinity, no matter where $\hat{Z}$ occurs in the Z plane.

The resulting normalized metric for the error $\Delta Z = Z - \hat{Z}$ may be expressed in real and imaginary impedance coordinates by substituting (6.4.14) into (6.4.15); after some algebra:

$$S = \frac{(R - \hat{R})^2 + (X - \hat{X})^2}{2R\hat{R}}. \tag{6.4.16}$$

Ordinarily, the 2 in the denominator is ignored when using (6.4.16) to accumulate errors over a number of frequency samples.

6.4.3. Sensitivity of Network Response Functions. Notice that the analysis leading to (6.4.16) mapped a complex error (ΔZ) into a concentric neighborhood in generalized reflection plane p and then imposed an arbitrary weighting function to penalize larger neighborhoods nonlinearly. First, it is shown that nearly all commonly used network response functions are just such scalar functions over the radius of a unit circle, with values that range from zero to infinity as the reflection radius ranges from zero to one. Second, the nature of several of these nonlinear monotone weighting functions, for example (6.4.15), is discussed. A final remark concerns normalized sensitivity.

It is interesting to begin by considering the *transducer function* from (6.2.3):

$$L = 10 \log_{10}\left(\frac{P_{as}}{P_L}\right) \quad \text{in dB.} \tag{6.4.17}$$

If the two-port network is lossless, then the power delivered at the input terminals is the same power that is delivered to the load, $P_L = P_{in}$. In that case it is not difficult to use (6.2.9) as the definition of input reflection coefficient to show that

$$L = -10 \log_{10}(1 - y^2), \tag{6.4.18}$$

where for present and later convenience y is defined to be the reflection magnitude

$$y = |p|. \tag{6.4.19}$$

Note that $0 \le y \le 1$, $1 \ge (1 - y^2) \ge 0$, and $0 \le [-\ln(1 - y^2)] \le \infty$. Therefore, the transducer function is also a weighted metric over the reflection magnitude. (The natural logarithm was used since it differs from base 10 logarithm by just the factor $\log_{10}e$.)

Table 6.4.1 summarizes some weighted reflection magnitude functions and their slopes. Function 1 in Table 6.4.1 is just the reflection magnitude that varies from 0 to 1; it makes very poor use of the dynamic range of digital computers and is not recommended as a component in an optimization

Table 6.4.1. Some Monotone Weighting Functions for the Reflection Magnitude[a]

Name	Response	Modified	Slope
1. Reflection magnitude	y	y	1
2. SWR using $|p|$	$\dfrac{1+y}{1-y}$	$\dfrac{1+y}{1-y} - 1$	$\dfrac{2}{(1-y)^2}$
3. SWR using $|p|^2$	$\dfrac{1+y^2}{1-y^2}$	$\dfrac{1+y^2}{1-y^2} - 1$	$\dfrac{4y}{(1-y^2)^2}$
4. Poincaré metric	$\tanh^{-1} y$	$\ln\dfrac{1+y}{1-y}$	$\dfrac{2}{1-y^2}$
5. Transducer loss	$-10\log(1 - y^2)$	$\ln\dfrac{1}{1-y^2}$	$\dfrac{2y}{1-y^2}$

[a] $y = |p|$. SWR is standing wave ratio function.

objective function. The remaining modified response functions in Table 6.4.1 all range from zero to plus infinity as the reflection magnitude ranges from zero to one. Response function 2 is the normal standing wave ratio (SWR) function that electrical engineers use to express the maximum-to-minimum standing wave ratio on a uniform transmission line (normally one to infinity). It is modified by subtracting unity so that its minimum is zero. Modified response 3 is a function similar to (6.4.15).

Function 4 in Table 6.4.1 is the "classical Poincaré metric... introduced near the turn of the century and is one of the most basic metrics in mathematics besides the Euclidean metric. ...It is amusing to note that a picture of it appearing in most mathematics books looks much like an electrical engineer's Smith chart," according to Helton (1981:1133). The Poincaré metric is defined on reflection magnitude y corresponding to (6.4.6) and is invariant under maps of the unit circle onto itself. The modified form shown in Table 6.4.1 is an identity for $\tanh^{-1} y$ with a factor of $\frac{1}{2}$ omitted. Helton has described the essential role of the Poincaré metric in the design theory for network power transfer over broad frequency ranges.

The clear similarity among the apparently unconnected functions in Table 6.4.1 is remarkable. For example, function 2 in Table 6.4.1 is the conventional SWR function that varies over the range from one to infinity. One way to change the lower bound is to subtract unity from the SWR expression. However, the Poincaré metric achieves the same result by including the logarithmic operator. The major significance of the data in Table 6.4.1 is in the "slope" column, the first derivative of the modified function with respect to y. Table 6.4.2 tabulates some values of interest.

The slope data in Table 6.4.1 describes the effect of a small change in the polar distance of a response in a unit circle and is therefore a *sensitivity*. That unit circle may be centered at the origin of the W plane in Figure 6.4.3, and the possible result of varying some branch impedance occurs in the image circle shown in Figure 6.4.3. It will not be known where or in what direction a branch variable change will cause movement in the response unit circle, but the slope data in Table 6.4.2 does show the response change that will occur, that is, the sensitivity.

Table 6.4.2.   Slope Data for Four Monotone Weighting Functions from Table 6.4.1

| Function | Magnitude of Reflection Coefficient | | | | | | | | |
	0	0.1	0.3	0.5	0.7	0.9	0.99	0.999	0.9999
SWR with $|p|$	2	2.47	4.08	8	22.2	200	20000	2E6	2E8
SWR with $|p|^2$	0	0.408	1.45	3.56	10.8	99.7	1E4	1E6	1E8
Poincaré	2	2.02	2.20	2.67	3.92	10.5	100.5	1000	10000
Transducer	0	0.202	0.659	1.33	2.74	9.47	99.5	1000	10000

The first and third response types in Table 6.4.2 are more sensitive than the others when the reflection magnitude is very small. The Poincaré metric has relatively uniform slope for all reflection magnitudes. Keep in mind that $|p| = 0.9999$ in transducer function (6.4.18) is equivalent to 37 dB, so optimization for reflection magnitudes near unity will occur frequently. The data in Table 6.4.2 also shows that the first two weighting functions would react drastically to changes near the edge of the reflection circle.

It is concluded that certain response functions are more suitable for one purpose than another. Certainly a mixed objective function may be used. Perhaps a good choice would be response function 2 in Table 6.4.1 for network passband performance and response function 5 for stopband selectivity. Or one could recommend the Poincaré metric as a reasonable choice for most situations. In the latter case, the sampled target data, T_i in Table 6.2.1, must be converted to equivalent values in the new domain. That minor inconvenience has proved to be worthwhile in practice.

This section is concluded by mentioning an often touted cure for many problems. It is possible to change the domain of the variables x_j to $\ln(x_j)$, $j = 1$ to n. Suppose that a variable in the new domain is v; then a logarithmic transformation is

$$v = \ln(x),\qquad(6.4.20)$$

or

$$x = e^v.\qquad(6.4.21)$$

At first, (6.4.20) was suggested to keep variables strictly positive, a condition guaranteed since no real value of v will make x negative in (6.4.21). However, a more interesting fact about the nonlinear transformation in (6.4.20) is its effect on the first derivatives in the two domains, using the chain rule:

$$\nabla_x F = (\nabla_v F)(\nabla_x v) = \frac{\nabla_v F}{x}\qquad(6.4.22)$$

Therefore,

$$\nabla_x F = (\nabla_v F)(\nabla_x v)$$
$$= \frac{d(\ln F)}{d(\ln x)} = \frac{(\Delta F/F)}{(\Delta x/x)}.\qquad(6.4.23)$$

The right side of (6.4.23) is the *Bode sensitivity function*, which is a measure of the relative change in the objective function for a relative change in a particular variable.

By optimizing in the logarithmic variable space, it is possible to widen straight but narrow valleys in the functions space and perhaps overcome some of the uneven weighting that occurs as just described. However, it is not difficult to find optimization objective functions that become defective under the logarithmic transformation in (6.4.20); see Wright (1976:73) and problem 1.8. Another difficulty is that all variables and gradients must be transformed by (6.4.20) and (6.4.22) when passing between the optimization algorithm and the objective function and gradient subroutines. Programming details for typical nonlinear transformations have been provided by Cuthbert (1983:160).

There is no uniform agreement concerning the beneficial effects of logarithmic scaling, except that projection of linear constraints for lower bounds on variables is a vastly superior technique.

Problems

6.1. Write the quadratic function in (6.1.4) in terms of parameters $\omega_0^2 = 1/(LC)$ and $\alpha = R/(2L)$. Then find an expression for the roots, using the quadratic formula. How does ω_0 compare to α in order that the solution be oscillatory?

6.2. Derive the rational polynomial $Z(s)$ for the network in Figure 6.5.1, using the branch immittances given in Table 6.1.2, Section 6.1.3. Hint: Form the impedance of the LR branches, invert and add to the admittance of the C branch, and invert once more.

6.3. Find E in the resistive network in Figure 6.5.2. What value of E would make voltage $V_3 = \frac{7}{2}$? What is the value of $R_{in} = V_3/I_4$?

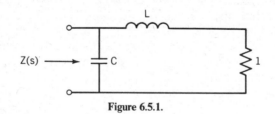

Figure 6.5.1.

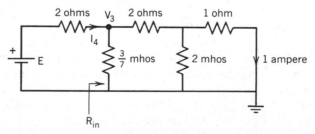

Figure 6.5.2.

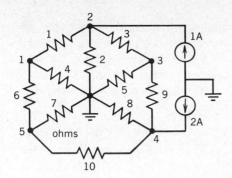

Figure 6.5.3.

6.4. A resistive network having 10 branches connected between five nodes and one common node was described by Ley (1970:292) as shown in Figure 6.5.3. Solve for the node voltages by forming the **Y** node admittance matrix according to the rules in Section 6.1.3 and solve using program MATRIX with LUFAC merged into it as described in Section 3.1.1.

6.5. Prove identity 9 in Table 6.1.3, Section 6.1.3.

6.6. The impedance for the RLC network in Figure 6.1.1 is defined by (6.1.9) and (6.1.13):

$$Z(s) = R + sL + \frac{1}{sC}. \qquad (6.5.1)$$

Write the derivative dZ/ds. Apply the Cauchy–Riemann identity (11 in Table 6.1.3, Section 6.1.3) for $Z = P + jQ$ and $s = \sigma + j\omega$, and evaluate at $s = 0 + j\omega$. What does $\nabla_\sigma P$ equal along the $j\omega$ axis?

6.7. Write the topology description lines similar to lines 620 to 680 in Table 6.2.5 for the network in Figure 6.5.4.

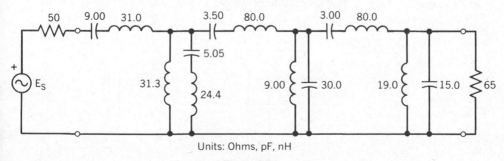

Units: Ohms, pF, nH

Figure 6.5.4.

6.8. Optimize the lossless network in Figure 6.5.4, starting from the 13 element values shown. Sample the passband every 10 MHz from 250 to 400 MHz, placing an upper bound of 0.1 dB on each sample. Sample the stopband at 470 and 490 MHz, using a 40-dB lower bound on those samples. Obtain before and after optimization data. TWEAKNET compiled into machine code is desirable.

6.9. Uhlir (1982) has provided the following equations in BASIC to convert a complex number, $Z = X + jY$, from rectangular to polar form, $Z = R\underline{/A}$:

$$R = \text{SQR}(X*Y* + Y*Y), \tag{6.5.2}$$

$$A = 2*\text{ATN}(Y/(R + X)). \tag{6.5.3}$$

Polar angle A is in radians; for degrees, replace the 2 in (6.5.3) by $(360/\pi)$. The program tests to avoid divide-by-zero errors are $Y = X = 0$ and $Y = 0$ when $X < 0$. Use Uhlir's equations to print the branch voltage or current $A4 + jA5$ in polar form by adding new lines 8075 and 8105 to TWEAKNET, similar to those in Example 6.2.1, Section 6.2.2.

6.10. Prove (6.4.18):

$$L = -10\log_{10}(1 - |p|^2) \quad \text{dB}. \tag{6.5.4}$$

6.11. The standing-wave ratio function is

$$S = \frac{1 + Q}{1 - Q}. \tag{6.5.5}$$

Find an expression for the partial derivative of S with respect to some variable y, $\nabla_y S$, in terms of $\nabla_y Q$.

6.12. Given a value for $\nabla_{Z_k} L$, the partial derivative of transducer loss L with respect to a branch impedance $Z_k = R_k + jX_k$, find an expression for $\nabla_Q L$, where $Q = X_k/R_k$.

6.13. Perform least-pth unconstrained minimization of the transducer loss for the lossless network in Figure 6.5.5. Use samples at $\omega = 0.8$, 1.0, and 1.2 radians per second with target values of 0 dB. What are the optimized L and C values for $p = 2$ and $p = 10$?

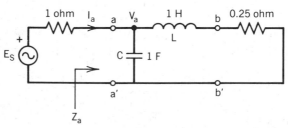

Figure 6.5.5.

6.14. Change the first topology line in Table 6.2.6 for the network in Figure 6.2.4 from a null element to a capacitor in branch 1. Enter the six starting values $\mathbf{x}^{(0)} = (0, 0.8, 2.7, 0.7, 0.6, 1.9)^T$. Define two lower bound constraints on the 40-dB samples at 1.5 and 2.0 radians per second. What is the gradient element $\nabla_{c_1} F[\mathbf{x}^{(0)}]$ for $C_1 = 0$? Perform a least-squares minimization to "grow" a finite capacitor C_1 into branch 1. What are the new element values? Were the constraints satisfied?

6.15. Use the PRINT statements suggested in problem 6.9 to obtain the voltages and currents at one radian per second in the network in Figure 6.5.5 for $Q_L = 100$ and $Q_C = 200$. Compute the exact partial derivatives of $Z_a = V_a/I_a$ with respect to C and L. Check your answers by finite differences, using TWEAKNET command 9.

6.16. Prove that a bilinear function (6.4.1) of a bilinear function is bilinear.

6.17. Solve for $Z(W)$ from (6.4.1).

6.18. Record the gradient in Example 6.2.1, Section 6.2.2, when using the suggested element values in pF and nH. Then compare the gradient after changing to nF (set C unit to 1E–9 and divide capacitances by 1000) and mH (set L unit to 1E–3 and divide inductances by 10^6). Verify that the transducer loss and input impedance do not differ between the two sets of units.

6.19. Consider an $n \times n$ system of *complex* equations

$$(\mathbf{A} + j\mathbf{B})(\mathbf{x} + j\mathbf{y}) = \mathbf{u} + j\mathbf{v}, \qquad (6.5.6)$$

where $\mathbf{A}$ and $\mathbf{B}$ are matrices, the remaining quantities are vectors, and all elements are complex numbers. By writing the two sets of *real* equations resulting from equating real and imaginary parts, show that the $2n \times 2n$ real system of equations also represents the system:

$$\begin{bmatrix} \mathbf{A} & -\mathbf{B} \\ +\mathbf{B} & \mathbf{A} \end{bmatrix} \begin{bmatrix} \mathbf{x} \\ \mathbf{y} \end{bmatrix} = \begin{bmatrix} \mathbf{u} \\ \mathbf{v} \end{bmatrix}. \qquad (6.5.7)$$

This formulation enables use of the real-variable methods described in Chapters Two and Three for solutions of systems of complex linear equations.

6.20. Derive $\nabla_{B_k} L_i$ (6.3.24) and $\nabla_{G_k} L_i$.

6.21. Consider the right side of (6.3.34) for differentials of Z_k and Y_k with respect to small changes in radian frequency, ω. Therefore, show that the *group time delay*, $T_G = -d\theta/d\omega$, is a sum of terms contributed by all L's and C's in the network. Radian angle θ is a part of the transfer function being considered.

6.22. Use the following "triple" of complex data pairs to write three complex equations in three complex unknowns according to (6.4.2):

i	Z_i	W_i
1	$0.1\underline{/\,30°}$	$0.1732\underline{/\,-7.8675}$
2	$0.5\underline{/\,60°}$	$0.4473\underline{/\,129.505}$
3	$1.1\underline{/\,-10°}$	$0.5099\underline{/\,-30.32}$

Solve for complex coefficients b, c, and d using the method in problem 6.19.

6.23. Use the method described by (6.3.45) to find the exact derivatives of the transducer function with respect to each branch *conductance* $(1/R)$ in the network in Figure 6.5.2. Check two of your answers by finite differences.

Appendix A _____

Test Matrices

The purpose of this appendix is to discuss three references where the reader may obtain descriptions and commentary on some well-known and perhaps notorious matrices. Comparison of algorithms is difficult at best, but some standard test problems are helpful. Other resources include reliable sources of FORTRAN programs; four such references are provided.

Knuth (1968:36–37) describes the following matrices and their analytical inverses:

Vandermonde

Combinatorial

Cauchy

Hilbert

Nash (1979:210–211) defines the following matrices and explains why they cause numerical problems:

Hilbert

Ding Dong

Moler

Bordered

Diagonal

Wilkinson W +

Ones

Klema (1980:168, 173–176) lists the following nearly singular matrices and notes why the singular value decomposition algorithm is helpful:

Ostrowski family
Rank 2 3 × 3
Hilbert order 7
Bauer

Dongarra (1979) is the user's guide for the LINPACK collection of FORTRAN matrix programs. These are widely accepted and have been adapted for personal computers.

Smith (1976) and Garbow (1977) document the EISPACK eigensystem package of matrix programs. These subroutines are very well documented.

IBM (1968) is an aging collection of FORTRAN subroutines for numerical analysis in general and matrix computations in particular. These subroutines have been referenced in many books on scientific computation.

Hopper (1981) is the librarian for the Harwell collection of the United Kingdom Atomic Energy Research Establishment. This respected FORTRAN collection includes subroutines for solving eigensystems and many more that are part of gradient nonlinear programming algorithms of all kinds.

Appendix B

Test Problems

The purpose of this appendix is to discuss references where the reader may obtain descriptions and commentary on some well-known nonlinear optimization test problems. These references are listed by importance and accessibility, especially in single sources that have collected articles with problems by different authors.

The foremost set of test problems was assembled by Colville (1968). Colville decided that it was not reasonable to perform a complete analysis on the large variety of algorithms by programming representative problems on a standard computer. Instead, a set of eight standard problems was solved by a number of unrelated organizations, using a variety of computers. Some timing data were obtained. Abadie (1970:532) commented on experiments with Colville's problems, especially as they were solved by the generalized reduced gradient method (Section 5.4.5). Powell (1977:153) commented on the nature of some of the more interesting Colville problems. Colville's summary report is frequently cited, even today.

The book by Himmelblau (1972) contains numerous test problems from a wide variety of sources. Excellent tabular comparisons for various algorithms and computers employed and graphical data where feasible are provided. Pages 69–96, 194–201, 213–216, 366–431, and 475–477 are recommended. This reference is probably the easiest to obtain and utilize for test problems.

Lootsma (1972) edited the proceedings of a conference that were published in an excellent book by a number of contributing authors. Several dozen test problems with tabular results are found on pages 29, 67–68, 99–113, 120, and 162–163. These test problems are less well organized than those in Himmelblau (1972).

Davidon (1977a:17) gives the sources (not necessarily original) for 23 test problems in five categories:

1. Quadratic-like
2. Valleys and helixes

3. Badly scaled, singular, ridges
4. Involving exponentials
5. Variable/large dimension

Also results were tabulated for the performance of Davidon's more recent algorithm (Davidon 1975), a FORTRAN program listing being included (derivative-free version). Previously, Davidon (1976:194) gave three interesting nonlinear least-squares problems, with and without zero residuals.

Nazareth (1978a) discussed quality optimization software, what it should contain, and how it should be tested. He commented on both Colville (1968) and Himmelblau (1972), previously cited. Nazareth (1976:17) provided a list and references for eight test problems, with results from variations on Gauss–Newton algorithms (Levenberg–Marquardt performed well).

Dennis (1979b:15) listed 20 test problems and their original sources. Performance data are given for a Gauss–Newton algorithm that estimates that part of the Hessian matrix normally ignored [see M in equation (4.4.21)].

Finally, there are dozens of articles by recognized researchers with a few test problems that differ from those included in the previous citations. Perhaps the remaining useful comment is that almost all of the numerous articles by Professor Roger Fletcher contain carefully considered test problems, used in full awareness of the contributions of others.

Appendix C _____

Program Listings

CONTENTS

List of Variable Names Used in Program C2-1: MATRIX

A( )	E	K2	M6	P2	S6$
B( )	E1	K3	M8	P3	S7$
B1	H	K4	M9	Q	S8$
B2	I	K5	N	R	S9$
B3	I1	K6	N$	S$	T
C( )	I9	K9	N7	S1$	T1
D( )	J	L8	N8	S2$	T2
D1	J9	L9	N9	S3$	
D2	K	M	P	S4$	
D3	K1	M5	P1	S5$	

```
10 REM - MATRIX MATH UTILITY - C2-1 'MATRIX'
30 REM - USE OF MAJOR VARIABLES AS FOLLOWS -
40 REM     A(I9,J9),B(K9,L9),C(M9,N9),D(M8,N8) VEC OR MAT <= 6x6
50 REM     D1      DETERMINANT
60 REM     I,J     WORKING ROW#,COL#, RESPECTIVELY
70 REM     K       COMMAND # & LOOP INDEX
80 REM     L       LOOP INDEX
90 REM     L8      PRINT FLAG - 0=OFF, 1=LOCAL, 2=GLOBAL
100 REM    M5,M6   TEMPORARY ROW#,COL#
110 REM    N7      HISTORY EVENT COUNTER
120 REM    N$      DATA STATEMENTS FILE NAME
130 REM    S$      FIRST MATRIX NAME
140 REM    S1$     RECALL OR SAVE CHOICE
150 REM    S2$     DISK FILE NAME ON CURRENT DIRECTORY
160 REM    S3$     COMMENT LABEL FOR DISK FILE
170 REM    S4$     MISC STRING
180 REM    S5$     MISC STRING
190 REM    S6$     PRINT-USING IMAGE
200 REM    S7$     MISC NUMBERS IN STRING FORMAT
210 REM    S8$()   HISTORY EVENT NAMES
220 REM    S9$     SECOND MATRIX NAME IF REQUIRED
230 OPTION BASE 1 : REM - NO SUBSCRIPT 0
240 L8=0 : REM - PRINT FLAG OFF
250 N7=0 : REM - HISTORY COUNTER INIT
260 CLS : KEY OFF
270 PRINT "**********ELEMENTARY VECTOR & MATRIX OPERATIONS***********"
280 PRINT : PRINT "NOTES:"
290 PRINT "1. USE ONLY UPPER CASE LETTERS"
300 PRINT "2. MERGE VECTOR AND MATRIX DATA STATEMENTS"
310 PRINT "     INTO RESERVED LINE RANGE 400-620 (OPTIONAL)"
320 PRINT "3. IF 'BREAK' OCCURS, RESTART WITH 'GOTO 999 <RTN>'"
330 DEFDBL A-H,Q-R,T-Z : REM - NOTE THAT P IS SNGL PRECISION
340 DEFINT I-N
350 DEFSTR S
360 DIM S8$(103)
370 REM - DIMENSION FOR 4 MATRICES EACH DIM <= 6:
380 DIM A(6,6),B(6,6),C(6,6),D(6,6)
390 REM **** LINES 400-620 RESERVED FOR MERGING PROBLEM DATA ****
400 REM - THE BYPASS CASE IS SHOWN HERE AS FOLLOWS
410 N$ = "NONE" : REM - A NAME FOR PROBLEM MERGE FILE MUST BE HERE
420 DATA 0,0 : REM - MATRIX A #ROWS,#COLS - MUST HAVE AT LEAST ONE PAIR
630 REM **************************************
640 REM - READ ONE OR MORE MATRICES A,B,C,D
650 PRINT : PRINT "WORKING WITH DATA SET: ";N$
660 READ I9,J9 : REM - DIM OF A
670 IF I9=0 THEN GOTO 1280 : REM - NO A, B, C, OR D
680 PRINT "   A(";I9;",";J9;")"
690 FOR I=1 TO I9
700 FOR J=1 TO J9
```

```
710 READ A(I,J)
720 NEXT J
730 NEXT I
740 READ K9,L9 : REM - DIM OF B (OR ABORT READING)
750 IF K9=0 THEN GOTO 1280 : REM - NO B,C, OR D
760 PRINT "    B(";K9;",";L9;")"
770 FOR I=1 TO K9
780 FOR J=1 TO L9
790 READ B(I,J)
800 NEXT J
810 NEXT I
820 READ M9,N9 : REM - DIM OF C (OR ABORT)
830 IF M9=0 THEN GOTO 1280 : REM - NO C OR D
840 PRINT "    C(";M9;",";N9;")"
850 FOR I=1 TO M9
860 FOR J= 1 TO N9
870 READ C(I,J)
880 NEXT J
890 NEXT I
900 READ M8,N8 : REM - DIM OF D (OR ABORT)
910 IF M8=0 THEN GOTO 1280 : REM - NO D
920 PRINT "    D(";M8;",";N8;")"
930 FOR I=1 TO M8
940 FOR J=1 TO N8
950 READ D(I,J)
960 NEXT J
970 NEXT I
980 GOTO 1280 : REM - TO MENU & SELECTION
990 REM - RE-ENTRY FOR INVALID COMMAND NUMBERS & CONTINUING
999 CLS
1000 PRINT "********* COMMAND MENU **********
1005 PRINT "0. DISPLAY A MATRIX IN FIXED FORMAT"
1010 PRINT "1. SEE COMMAND HISTORY"
1020 PRINT "2. TOGGLE PRINTER ON/OFF"
1030 PRINT "3. EQUATE ONE MATRIX TO ANOTHER"
1040 PRINT "4. TRANSPOSE"
1050 PRINT "5. MATRIX TO/FROM DISK"
1060 PRINT "6. SCALAR * (MATRIX)"
1070 PRINT "7. A = B + C"
1080 PRINT "8. A = D * C"
1090 PRINT "9. D = (invB) & DETERMINANT(B); DESTROYS B!"
1100 PRINT "10. SPARE"
1110 PRINT "11. NORMS OF VECTOR OR MATRIX D"
1120 PRINT "12. EXTREME ELEMENTS OF D"
1130 PRINT "13. SPARE"
1140 PRINT "   FORMAT:    FIXED   SCIENTIFIC   ALL COLUMNS"
1150 PRINT "   PRINT =      14         16           18"
1160 PRINT "   DISPLAY = 0 OR 15       17           19"
1170 PRINT "20. EXIT (RESUME WITH 'GOTO 999')"
1180 REM
1190 PRINT"**************************************
1200 IF N7>100 THEN N7=0 : REM -DON'T RUN OUT OF SUBSCRIPT
1210 PRINT"INPUT COMMAND NUMBER:";:INPUT S$
1220 K=LEN(S$) : IF K=0 THEN GOTO 999 : REM - AVOID <CR>
1230 K=ASC(S$)
1240 IF K<48 OR K>57 THEN GOTO 999 : REM - 1ST CHAR MUST BE 0-9
1250 K=VAL(S$)
1255 IF K=0 THEN K= 15 : REM - ALTERNATIVE DISPLAY NUMBERS
1260 IF K>20 THEN GOTO 999 : REM - CAN'T EXCEED MENU #'S
1270 ON K GOSUB 1330,1410,1470,2470,4580,3870,2890,3040,3220,6370,
              5360,5820,6370,4150,4180,4070,4110,6030,6050,6370
1280 PRINT "PRESS <RETURN> KEY TO CONTINUE -- READY";
1290 INPUT S4$
1295 IF S4$<>"" THEN BEEP : REM - <RETURN> BEFORE NEXT CMD NUMBER
1300 GOTO 999
1310 REM*************************************
1320 REM SEE COMMAND HISTORY - MAX OF 100
1330 IF N7=0 OR N7>100 THEN RETURN
1340 FOR I=1 TO N7
1350 PRINT "   ",S8$(I)
1360 IF L8>0 THEN LPRINT "    ",S8$(I)
```

```
1370 NEXT I
1380 RETURN
1390 REM*****************************************
1400 REM - TOGGLE PRINTER ON/OFF
1410 IF L8=2 THEN GOTO 1430
1420 L8=2 : PRINT "   PRINTER TURNED ON" : GOTO 1440
1430 L8=0 : PRINT "   PRINTER TURNED OFF"
1440 RETURN
1450 REM*****************************************
1460 REM - EQUATE ONE MATRIX TO ANOTHER
1470 PRINT "EQUATE A, B, C, OR D="; : INPUT S$
1480 IF S$="A" OR S$="B" OR S$="C" OR S$="D" THEN GOTO 1500
1490 GOTO 1470
1500 PRINT "EQUATE TO A, B, C, D, OR I (UNIT MATRIX)"; : INPUT S9$
1510 IF S9$="A" OR S9$="B" OR S9$="C" OR S9$="D" OR S9$="I"
        THEN GOTO 1530
1520 GOTO 1500
1530 IF S9$<>"I" THEN GOTO 1560
1540 PRINT "#ROWS=#COLUMNS="; : INPUT S7$
1550 REM ** ENTRY POINT IF PRESET PARAMETERS
1560 N7=N7+1 : S8$(N7)=S$+"="+S9$ : REM STORE OPN NAME
1570 PRINT "   ";S8$(N7)
1580 IF L8>0 THEN LPRINT "   ";S8$(N7)
1590 IF S9$<>"I" THEN GOTO 1640
1600 M5=VAL(S7$)
1610 N7=N7+1 : S8$(N7)="   "+S7$+"X"+S7$+" UNIT MATRIX"
1620 PRINT S8$(N7)
1630 IF L8>0 THEN LPRINT S8$(N7)
1640 IF S$="A" THEN GOTO 1700
1650 IF S$="B" THEN GOTO 1890
1660 IF S$="C" THEN GOTO 2080
1670 IF S$="D" THEN GOTO 2270
1680 GOTO 1470
1690 REM - FOR A=
1700 IF S9$<>"B" THEN GOTO 1720
1710 I9=K9 : J9=L9
1720 IF S9$<>"C" THEN GOTO 1740
1730 I9=M9 : J9=N9
1740 IF S9$<>"D" THEN GOTO 1760
1750 I9=M8 : J9=N8
1760 IF S9$<>"I" THEN GOTO 1780
1770 I9=M5 : J9 = M5
1780 FOR J=1 TO J9
1790 FOR I=1 TO I9
1800 IF S9$="B" THEN A(I,J)=B(I,J)
1810 IF S9$="C" THEN A(I,J)=C(I,J)
1820 IF S9$="D" THEN A(I,J)=D(I,J)
1830 IF S9$="I" THEN A(I,J)=0
1840 IF S9$="I" AND I=J THEN A(I,J)=1
1850 NEXT I
1860 NEXT J
1870 RETURN
1880 REM - FOR B=
1890 IF S9$<>"A" THEN GOTO 1910
1900 K9=I9 : L9=J9
1910 IF S9$<>"C" THEN GOTO 1930
1920 K9=M9 : L9=N9
1930 IF S9$<>"D" THEN GOTO 1950
1940 K9=M8 : L9=N8
1950 IF S9$<>"I" THEN GOTO 1970
1960 K9=M5 : L9 = M5
1970 FOR J=1 TO L9
1980 FOR I=1 TO K9
1990 IF S9$="A" THEN B(I,J)=A(I,J)
2000 IF S9$="C" THEN B(I,J)=C(I,J)
2010 IF S9$="D" THEN B(I,J)=D(I,J)
2020 IF S9$="I" THEN B(I,J)=0
2030 IF S9$="I" AND I=J THEN B(I,J)=1
2040 NEXT I
2050 NEXT J
2060 RETURN
```

```
2070 REM - FOR C=
2080 IF S9$<>"A" THEN GOTO 2100
2090 M9=I9 : N9=J9
2100 IF S9$<>"B" THEN GOTO 2120
2110 M9=K9 :  N9=L9
2120 IF S9$<>"D" THEN GOTO 2140
2130 M9=M8 : N9=N8
2140 IF S9$<>"I" THEN GOTO 2160
2150 M9=M5 : N9 = M5
2160 FOR J=1 TO N9
2170 FOR I=1 TO M9
2180 IF S9$="A" THEN C(I,J)=A(I,J)
2190 IF S9$="B" THEN C(I,J)=B(I,J)
2200 IF S9$="D" THEN C(I,J)=D(I,J)
2210 IF S9$="I" THEN C(I,J)=0
2220 IF S9$="I" AND I=J THEN C(I,J)=1
2230 NEXT I
2240 NEXT J
2250 RETURN
2260 REM - FOR D=
2270 IF S9$<>"A" THEN GOTO 2290
2280 M8=I9 : N8=J9
2290 IF S9$<>"B" THEN GOTO 2310
2300 M8=K9 : N8=L9
2310 IF S9$<>"C" THEN GOTO 2330
2320 M8=M9 : N8=N9
2330 IF S9$<>"I" THEN GOTO 2350
2340 M8=M5 : N8 = M5
2350 FOR J=1 TO N8
2360 FOR I=1 TO M8
2370 IF S9$="A" THEN D(I,J)=A(I,J)
2380 IF S9$="B" THEN D(I,J)=B(I,J)
2390 IF S9$="C" THEN D(I,J)=C(I,J)
2400 IF S9$="I" THEN D(I,J)=0
2410 IF S9$="I" AND I=J THEN D(I,J)=1
2420 NEXT I
2430 NEXT J
2440 RETURN
2450 REM**************************************
2460 REM - TRANSPOSE
2470 PRINT "TRANSPOSE A, B, C, OR D"; : INPUT S$
2480 IF S$="A" OR S$="B" OR S$="C" OR S$="D" THEN GOTO 2510
2490 GOTO 2470
2500 REM ** ENTRY POINT IF PRESET PARAMETERS
2510 N7=N7+1 : S8$(N7)="TRANSPOSE "+S$
2520 PRINT "   ";S8$(N7)
2530 IF L8>0 THEN LPRINT "   ";S8$(N7)
2540 IF S$="A" THEN GOTO 2600
2550 IF S$="B" THEN GOTO 2640
2560 IF S$="C" THEN GOTO 2680
2570 IF S$="D" THEN GOTO 2720
2580 GOTO 2470
2590 REM - TRANSPOSE A
2600 M=I9 : I9=J9 : J9=M
2610 IF I9>M THEN M=I9
2620 GOTO 2740
2630 REM - TRANSPOSE B
2640 M=K9 : K9=L9 : L9=M
2650 IF K9>M THEN M=K9
2660 GOTO 2740
2670 REM - TRANSPOSE C
2680 M=M9 : M9=N9 : N9=M
2690 IF M9>M THEN M=M9
2700 GOTO 2740
2710 REM - TRANSPOSE D
2720 M=M8 : M8=N8 : N8=M
2730 IF M8>M THEN M=M8
2740 FOR I=2 TO M
2750 FOR J=1 TO I-1
2760 IF S$<>"A" THEN GOTO 2780
2770 H=A(I,J) : A(I,J)=A(J,I) : A(J,I)=H
```

```
2780 IF S$<>"B" THEN GOTO 2800
2790 H=B(I,J) : B(I,J)=B(J,I) : B(J,I)=H
2800 IF S$<>"C" THEN GOTO 2820
2810 H=C(I,J) : C(I,J)=C(J,I) : C(J,I)=H
2820 IF S$<>"D" THEN GOTO 2840
2830 H=D(I,J) : D(I,J)=D(J,I) : D(J,I)=H
2840 NEXT J
2850 NEXT I
2860 RETURN
2870 REM****************************************
2880 REM - A=B+C
2890 N7=N7+1 : S8$(N7)="A=B+C"
2900 IF (K9<>M9) OR (L9<>N9) THEN PRINT "B & C DIFFERENT SIZES!"
2910 IF (L8>0) AND (K9<>M9 OR L9<>N9) THEN LPRINT"B & C DIFFERENT SIZES"
2920 PRINT"       ";S8$(N7)
2930 IF L8>0 THEN LPRINT "    ";S8$(N7)
2940 I9=K9 : IF I9<M9 THEN I9=M9
2950 J9=L9 : IF J9<N9 THEN J9=N9
2960 FOR I=1 TO I9
2970 FOR J=1 TO J9
2980 A(I,J) = B(I,J) + C(I,J)
2990 NEXT J
3000 NEXT I
3010 RETURN
3020 REM****************************************
3030 REM - A=D*C
3040 N7=N7+1 : S8$(N7)="A=D*C"
3050 IF N8<>M9 THEN PRINT "WARNING - D & C NOT CONFORMABLE!"
3060 IF L8>0 AND N8<>M9 THEN LPRINT "WARNING - D & C NOT CONFORMABLE!"
3070 PRINT"       ";S8$(N7)
3080 IF L8>0 THEN LPRINT "    ";S8$(N7)
3090 I9=M8 : J9=N9 : REM - RESET MATRIX A DIMENSIONS
3100 FOR I=1 TO I9
3110 FOR J=1 TO J9
3120 H=0
3130 FOR K=1 TO M9
3140 H=H+D(I,K)*C(K,J)
3150 NEXT K
3160 A(I,J)=H
3170 NEXT J
3180 NEXT I
3190 RETURN
3200 REM****************************************
3210 REM - D=inv(B)
3220 IF K9=L9 THEN GOTO 3260
3230 PRINT"MATRIX B IS NOT SQUARE"
3240 IF L8>0 THEN LPRINT "MATRIX B IS NOT SQUARE"
3250 RETURN
3260 N7=N7+1 : S8$(N7)="D=inv(B)"
3270 PRINT"       ";S8$(N7)
3280 IF L8>0 THEN LPRINT "    ";S8$(N7)
3290 M8=K9 : N8=K9
3300 E=.000000000000001#
3310 REM - EQUATE D TO THE UNIT MATRIX
3320 FOR I=1 TO N8
3330 FOR J=1 TO N8
3340 D(I,J)=0
3350 IF I=J THEN D(I,J)=1
3360 NEXT J
3370 NEXT I
3380 REM - FIND MAX MAGNITUDE ON OR BELOW MAIN DIAGONAL
3390 D1=1 : REM - INIT DETERMINANT
3400 FOR K=1 TO N8 : REM - OUTER STAGE LOOP
3410 IF K>=N8 THEN GOTO 3610
3420 I1=K
3430 B1=ABS(B(K,K))
3440 FOR I=K+1 TO N8
3450 IF B1>=ABS(B(I,K)) THEN GOTO 3480
3460 I1=I
3470 B1=ABS(B(I,K)) : REM - NEW MAX PIVOT FOUND
3480 NEXT I
```

```
3490 REM - SWAP ROWS I1 & K IF I1 <> K
3500 IF I1=K THEN GOTO 3610
3510 FOR J=1 TO N8
3520 B2=B(I1,J)
3530 B(I1,J)=B(K,J)
3540 B(K,J)=B2
3550 D2=D(I1,J)
3560 D(I1,J)=D(K,J)
3570 D(K,J)=D2
3580 NEXT J
3590 D1=-D1 : REM - DET SIGN SWAP
3600 REM - TEST FOR SINGULAR MATRIX
3610 IF ABS(B(K,K))<=E THEN GOTO 3820
3620 D1=B(K,K)*D1
3630 REM - DIVIDE PIVOT ROW BY ITS MAIN DIAGONAL ELEMENT
3640 D3=B(K,K)
3650 FOR J=1 TO N8
3660 B(K,J)=B(K,J)/D3
3670 D(K,J)=D(K,J)/D3
3680 NEXT J
3690 REM - REPLACE EACH ROW BY LIN COMBI WITH PIVOT ROW
3700 FOR I=1 TO N8
3710 B3=B(I,K)
3720 IF I=K THEN GOTO 3770
3730 FOR J=1 TO N8
3740 B(I,J)=B(I,J)-B3*B(K,J)
3750 D(I,J)=D(I,J)-B3*D(K,J)
3760 NEXT J
3770 NEXT I
3780 NEXT K
3790 PRINT"    DETERMINANT(B) = ";D1;". NOW B=I."
3800 IF L8>0 THEN LPRINT "    DETERMINANT(B) = ";D1;". NOW B=I."
3810 GOTO 3840
3820 PRINT"B SINGULAR FOR K=";K;". INVERSE ABORTED."
3830 IF L8>0 THEN LPRINT "B SINGULAR FOR K=";K;". INVERSE ABORTED."
3840 RETURN
3850 REM*****************************************
3860 REM - SCALAR * (MATRIX)
3870 PRINT "SCALAR = "; : INPUT S7$
3880 PRINT "MATRIX IS A, B, C, OR D"; : INPUT S$
3890 IF S$="A" OR S$="B" OR S$="C" OR S$="D" THEN GOTO 3910
3900 GOTO 3880
3910 REM ** ENTRY POINT IF PRESET PARAMETERS
3920 N7=N7+1 : S8$(N7)=S7$+"*"+S$
3930 E1=VAL(S7$)
3940 PRINT"   ";S8$(N7)
3950 IF L8>0 THEN LPRINT"   ";S8$(N7)
3960 FOR I=1 TO 6
3970 FOR J=1 TO 6
3980 IF S$="A" THEN A(I,J)=E1*A(I,J)
3990 IF S$="B" THEN B(I,J)=E1*B(I,J)
4000 IF S$="C" THEN C(I,J)=E1*C(I,J)
4010 IF S$="D" THEN D(I,J)=E1*D(I,J)
4020 NEXT J
4030 NEXT I
4040 RETURN
4050 REM*****************************************
4060 REM - PRINT ARRAY IN SCIENTIFIC FORMAT
4070 IF L8=0 THEN L8=1 : REM - TURN ON LOCAL PRINT FLAG
4080 GOTO 4110 : REM - TO STND DISPLAY ROUTINE
4090 REM*****************************************
4100 REM - DISPLAY ARRAY IN SCIENTIFIC FORMAT
4110 S6$=" ##.#####^^^^"
4120 GOTO 4190 : REM - TO STND DISPLAY ROUTINE
4130 REM*****************************************
4140 REM - PRINT ARRAY IN FIXED FORMAT
4150 IF L8=0 THEN L8=1 : GOTO 4180
4160 REM*****************************************
4170 REM - DISPLAY ARRAY IN FIXED FORMAT
4180 S6$=" ######.#####"
4190 PRINT "SEE A, B, C, OR D"; : INPUT S$
```

```
4200 IF S$="A" OR S$="B" OR S$="C" OR S$="D" THEN GOTO 4230
4210 GOTO 4190
4220 REM ** ENTRY POINT IF PRESET PARAMETERS
4230 IF S$<>"A" THEN GOTO 4250
4240 M5=I9 : M6=J9 : GOTO 4310
4250 IF S$<>"B" THEN GOTO 4270
4260 M5=K9 : M6=L9 : GOTO 4310
4270 IF S$<>"C" THEN GOTO 4290
4280 M5=M9 : M6=N9 : GOTO 4310
4290 IF S$<>"D" THEN GOTO 4310
4300 M5=M8 : M6=N8
4310 PRINT "MATRIX ";S$;"(";M5;",";M6;") -"
4320 IF L8>0 THEN LPRINT "MATRIX ";S$;"(";M5;",";M6;") -"
4330 FOR I=1 TO M5
4340 J=M6 : IF J=1 THEN GOTO 4450
4350 FOR J=1 TO M6-1
4360 IF S$="A" THEN PRINT USING S6$; A(I,J);
4370 IF S$="A" AND L8>0 THEN LPRINT USING S6$; A(I,J);
4380 IF S$="B" THEN PRINT USING S6$; B(I,J);
4390 IF S$="B" AND L8>0 THEN LPRINT USING S6$; B(I,J);
4400 IF S$="C" THEN PRINT USING S6$; C(I,J);
4410 IF S$="C" AND L8>0 THEN LPRINT USING S6$; C(I,J);
4420 IF S$="D" THEN PRINT USING S6$; D(I,J);
4430 IF S$="D" AND L8>0 THEN LPRINT USING S6$; D(I,J);
4440 NEXT J
4450 IF S$="A" THEN PRINT USING S6$; A(I,J)
4460 IF S$="A" AND L8>0 THEN LPRINT USING S6$; A(I,J)
4470 IF S$="B" THEN PRINT USING S6$; B(I,J)
4480 IF S$="B" AND L8>0 THEN LPRINT USING S6$; B(I,J)
4490 IF S$="C" THEN PRINT USING S6$; C(I,J)
4500 IF S$="C" AND L8>0 THEN LPRINT USING S6$; C(I,J)
4510 IF S$="D" THEN PRINT USING S6$; D(I,J)
4520 IF S$="D" AND L8>0 THEN LPRINT USING S6$; D(I,J)
4530 NEXT I
4540 IF L8=1 THEN L8=0 : REM - TURN OFF LOCAL PRINT FLAG
4550 RETURN
4560 REM*****************************************
4570 REM - MATRIX TO/FROM DISK
4580 PRINT"SEE DIRECTORY (Y/N)"; : INPUT S4$
4590 IF S4$<>"Y" THEN GOTO 4660
4600 PRINT"FILENAME SPECIFIER (LIKE *.* OR <RETURN>) = "; : INPUT S5$
4610  IF S5$="" THEN S5$="*.*"
4620 FILES S5$
4630 PRINT "SEE DIRECTORY AGAIN (Y/N)"; : INPUT S4$
4640 IF S4$<>"Y" THEN GOTO 4660
4650 GOTO 4600
4660 PRINT "MATRIX INVOLVED IS A, B, C, OR D"; : INPUT S$
4670 IF S$="A" OR S$="B" OR S$="C" OR S$="D" THEN GOTO 4690
4680 GOTO 4660
4690 PRINT "RECALL OR SAVE MATRIX ";S$;" (R/S)"; : INPUT S1$
4700 IF S1$="R" OR S1$="S" THEN GOTO 4720
4710 GOTO 4690
4720 PRINT "FILE NAME IS"; : INPUT S2$
4725 PRINT "     !!! WARNING - DO NOT <CNTRL><BREAK> DURING THIS STEP !!
4730 IF S1$="R" THEN GOTO 5010
4740 REM - SAVE MATRIX TO DISK FILE
4750 PRINT "COMMENT LABEL FOR FILE IS (<96 CHAR): "; : LINE INPUT S3$
4760 REM ** ENTRY POINT IF PRESET PARAMETERS
4770 OPEN S2$ FOR OUTPUT AS #1
4780 WRITE #1,S3$
4790 IF S$<>"A" THEN GOTO 4810
4800 M5=I9 : M6=J9 : GOTO 4870
4810 IF S$<>"B" THEN GOTO 4830
4820 M5=K9 : M6=L9 : GOTO 4870
4830 IF S$<>"C" THEN GOTO 4850
4840 M5=M9 : M6=N9 : GOTO 4870
4850 IF S$<>"D" THEN GOTO 4870
4860 M5=M8 : M6=N8
4870 WRITE #1,M5,M6
4880 FOR I=1 TO M5
4890 FOR J=1 TO M6
```

```
4900 IF S$="A" THEN WRITE #1,A(I,J)
4910 IF S$="B" THEN WRITE #1,B(I,J)
4920 IF S$="C" THEN WRITE #1,C(I,J)
4930 IF S$="D" THEN WRITE #1,D(I,J)
4940 NEXT J
4950 NEXT I
4960 N7=N7+1 : S8$(N7)="SAVED MATRIX "+S$+" IN FILE "+S2$
4970 PRINT "    ";S8$(N7)
4980 IF L8>0 THEN LPRINT "    ";S8$(N7)
4990 CLOSE #1 : RETURN
5000 REM - RECALL MATRIX FROM DISK FILE
5010 OPEN S2$ FOR INPUT AS #1
5020 INPUT #1,S3$
5030 PRINT "READY TO READ FILE ";S2$;" INTO MATRIX ";S$;" TITLED:";
5040 PRINT "    ";S3$
5050 PRINT "PRESS <RETURN> KEY IF OK, ELSE 'ABORT' <RETURN>";
5060 INPUT S4$
5070 IF S4$="" THEN GOTO 5120
5080 PRINT "ABORT RECALL; MATRIX ";S$;" NOT CHANGED" : GOTO 5320
5090 REM ** ENTRY POINT IF PRESET PARAMETERS
5100 OPEN S2$ FOR INPUT AS #1
5110 INPUT #1,S3$
5120 N7=N7+1 : S8$(N7)="READ FILE "+S2$+" INTO MATRIX "+S$
5130 PRINT "    ";S8$(N7)
5140 IF L8>0 THEN LPRINT "    ";S8$(N7)
5150 INPUT #1,M5,M6
5160 FOR I=1 TO M5
5170 FOR J=1 TO M6
5180 IF S$="A" THEN INPUT #1,A(I,J)
5190 IF S$="B" THEN INPUT #1,B(I,J)
5200 IF S$="C" THEN INPUT #1,C(I,J)
5210 IF S$="D" THEN INPUT #1,D(I,J)
5220 NEXT J
5230 NEXT I
5240 IF S$<>"A" THEN GOTO 5260
5250 I9=M5 : J9=M6 : GOTO 5320
5260 IF S$<>"B" THEN GOTO 5280
5270 K9=M5 : L9=M6 : GOTO 5320
5280 IF S$<>"C" THEN GOTO 5300
5290 M9=M5 : N9=M6 : GOTO 5320
5300 IF S$<>"D" THEN GOTO 5320
5310 M8=M5 : N8=M6
5320 CLOSE #1 : RETURN
5330 REM*****************************************
5340 REM
5350 REM - VECTOR AND MATRIX NORMS OF D (EXCEPT SPECTRAL)
5360 IF (M8=1) OR (N8=1) THEN GOTO 5650 : REM -VECTOR
5370 REM - MATRIX CASES
5380 T=0 : K=1 : P=0
5390 FOR I=1 TO M8
5400 H=0
5410 FOR J=1 TO N8
5420 H=H+ABS(D(I,J))
5430 P=P+D(I,J)^2
5440 NEXT J
5450 IF H>T THEN K=I
5460 IF H>T THEN T=H
5470 NEXT I
5480 PRINT "    SQR-ROOT OF SUM SQRS OF D = ";SQR(P)
5490 IF L8>0 THEN LPRINT "    SQR-ROOT OF SUM SQRS OF D = ";SQR(P)
5500 PRINT "    MAX ABSOLUTE ROW SUM OF D = ";T;" FOR ROW";K
5510 IF L8>0 THEN LPRINT"    MAX ABSOLUTE ROW SUM OF D = ";T;" FOR ROW";K
5520 T=0 : K=1
5530 FOR J=1 TO N8
5540 H=0
5550 FOR I=1 TO M8
5560 H=H+ABS(D(I,J))
5570 NEXT I
```

```
5580 IF H>T THEN K=J
5590 IF H>T THEN T=H
5600 NEXT J
5610 PRINT "   MAX ABSOLUTE COL SUM OF D = ";T;" FOR COL";K
5620 IF L8>0 THEN LPRINT "MAX ABSOLUTE COL SUM OF D = ";T;" FOR COL";K
5630 RETURN
5640 REM - VECTOR NORMS
5650 P1=0 : P2=0 : P3=0
5660 FOR I=1 TO M8
5670 FOR J=1 TO N8
5680 P1=P1+ABS(D(I,J))
5690 P2=P2+ABS(D(I,J))^2
5700 IF ABS(D(I,J))>P3 THEN P3=ABS(D(I,J))
5710 NEXT J
5720 NEXT I
5730 PRINT "   P=1 NORM OF D IS";P1
5740 IF L8>0 THEN LPRINT "   P=1 NORM OF D IS";P1
5750 PRINT "   P=2 NORM OF D IS";SQR(P2)
5760 IF L8>0 THEN LPRINT "   P=2 NORM OF D IS";SQR(P2)
5770 PRINT "   INFINITY NORM OF D IS";P3
5780 IF L8>0 THEN LPRINT "   INFINITY NORM OF D IS";P3
5790 RETURN
5800 REM****************************************
5810 REM - EXTREME ELEMENTS OF D
5820 T=0 : T2=-1E+33 : T1=1E+33 :K1=1 : K2=1 :K3=1 : K4=1 :K5=1 : K6=1
5830 FOR I=1 TO M8
5840 FOR J=1 TO N8
5850 IF ABS(D(I,J))<=T THEN GOTO 5870
5860 K1=I : K2=J : T=ABS(D(I,J))
5870 IF D(I,J)>T1 THEN GOTO 5890
5880 K3=I : K4=J : T1=D(I,J)
5890 IF D(I,J)<T2 THEN GOTO 5910
5900 K5=I : K6=J : T2=D(I,J)
5910 NEXT J
5920 NEXT I
5930 S4$=" ELEMENT OF D IS "
5940 PRINT "   MAX ABS"+S4$;T;" @ ROW";K1;", COL";K2
5950 IF L8>0 THEN LPRINT "   MAX ABS"+S4$;T;" @ ROW";K1;", COL";K2
5960 PRINT "   MAX"+S4$;T2;" @ ROW";K5;" COL";K6
5970 IF L8>0 THEN LPRINT "   MAX"+S4$;T2;" @ ROW";K5;" COL";K6
5980 PRINT "   MIN"+S4$;T1;" @ ROW";K3;" COL";K4
5990 IF L8>0 THEN LPRINT "   MIN"+S4$;T1;" @ ROW";K3;" COL";K4
6000 RETURN
6010 REM****************************************
6020 REM - PRINT COLUMNS
6030 IF L8=0 THEN L8=1 : REM - TURN PRINTER ON
6040 REM****************************************
6050 PRINT "DISPLAY COLUMNS OF A, B, C, OR D";:INPUT S$
6060 IF S$="A" OR S$="B" OR S$="C" OR S$="D" THEN GOTO 6080
6070 GOTO 6050
6080 IF S$<>"A" THEN GOTO 6100
6090 M5=I9 : M6=J9 : GOTO 6160
6100 IF S$<>"B" THEN GOTO 6120
6110 M5=K9 : M6=L9 : GOTO 6160
6120 IF S$<>"C" THEN GOTO 6140
6130 M5=M9 : M6=N9 : GOTO 6160
6140 IF S$<>"D" THEN GOTO 6160
6150 M5=M8 : M6=N8
6160 PRINT"****************************************"
6170 PRINT"COLUMNS OF MATRIX ";S$
6175 IF L8>0 THEN LPRINT"COLUMNS OF MATRIX ";S$
6180 FOR J=1 TO M6
6190 PRINT"  COLUMN ";J
6195 IF L8>0 THEN LPRINT"  COLUMN ";J
6200 FOR I=1 TO M5
6210 IF S$="A" THEN PRINT"      ";A(I,J)
6220 IF S$="A" AND L8>0 THEN LPRINT "      ";A(I,J)
6230 IF S$="B" THEN PRINT"      ";B(I,J)
```

```
6240 IF S$="B" AND L8>0 THEN LPRINT "      ";B(I,J)
6250 IF S$="C" THEN PRINT"      ";C(I,J)
6260 IF S$="C" AND L8>0 THEN LPRINT "      ";C(I,J)
6270 IF S$="D" THEN PRINT"      ";D(I,J)
6280 IF S$="D" AND L8>0 THEN LPRINT "      ";D(I,J)
6290 NEXT I
6300 REM - END OF  COLUMN
6310 IF L8 > 0 THEN GOTO 6340
6320 IF J=M6 THEN GOTO 6340
6330 PRINT"PRESS <RETURN> KEY TO CONTINUE -- READY";:INPUT S4$
6340 NEXT J
6350 IF L8=1 THEN L8=0 : REM - TURN PRINTER OFF
6360 RETURN
6370 KEY ON : PRINT "END OF RUN" : END
6380 REM*************************************
6390 REM - ANY NEW CODE GOES HERE
6400 REM*************************************
6410 END



1099 REM - GRAM-SCHMIDT ORTHONORMALIZATION - C2-2 'GSDECOMP'
1100 PRINT "10. G-S DECOMPOSITION OF A=D*C, D ORTHONORMAL"
1270 ON K GOSUB 1330,1410,1470,2470,4580,3870,2890,3040,3220,6400,
          5360,5820,6370,4150,4180,4070,4110,6030,6050,6370
6390 REM - GRAM-SCHMIDT ORTHONORMALIZATION FROM A TO D
6400 N7=N7+1 : S8$(N7)="D=orthodecomp(A), & C triangular"
6410 PRINT"      ";S8$(N7)
6420 IF L8>0 THEN LPRINT"      ";S8$(N7)
6430 M=I9 : N=J9 : M8=M : N8=N : M9=N : N9=N
6440 REM - NULL C
6450 FOR I=1 TO N
6460 FOR J=1 TO N
6470 C(I,J)=0
6480 NEXT J
6490 NEXT I
6500 REM - MAIN LOOP FOR N COLUMNS OF A & D
6510 FOR J=1 TO N
6520 REM - STORE COLUMN J OF A IN D
6530 FOR K=1 TO M
6540 D(K,J) = A(K,J)
6550 NEXT K
6560 I=1
6570 REM - BEGIN I LOOP FOR ROWS OF UPPER TRIANGULAR C
6580 IF I=J THEN GOTO 6720 : REM - GET DIAG ELE & NORM ORTHOG VECTOR
6590 REM - CALC C(I,J) UPPER TRIANG
6600 T=0
6610 FOR K=1 TO M
6620 T=T+D(K,I)*A(K,J)
6630 NEXT K
6640 C(I,J)=T
6650 REM - CONSTRUCT ORTHOG VECTOR IN D
6660 FOR K=1 TO M
6670 D(K,J)=D(K,J)-T*D(K,I)
6680 NEXT K
6690 I=I+1
6700 GOTO 6580 : REM - FOR NEXT ROW OF UPPER TRIANG C
6710 REM - CALCULATE C(I,I)
6720 T=0
6730 FOR K=1 TO M
6740 T=T+D(K,J)*D(K,J)
6750 NEXT K
6760 T=SQR(T)
6770 C(I,I)=T
6780 REM - EUCLIDEAN NORM ORTHOGONAL VECTOR IN COL J OF D
6790 FOR K=1 TO M
6800 D(K,J)=D(K,J)/T
6810 NEXT K
6820 REM - END OF COLUMN J WORK
```

```
6830 NEXT J
6840 REM - CALC ALL COMBI'S OF COL INNER PRODUCTS
6850 FOR I=1 TO N8
6860 FOR J=I TO N8
6870 T=0
6880 FOR K=1 TO M8
6890 T=T+D(K,I)*D(K,J)
6900 NEXT K
6910 PRINT "INNER PRODUCT OF COLUMNS";I;J;T
6920 NEXT J
6930 NEXT I
6940 RETURN
6950 END
```

```
1129 REM - EIGENVALUE BOUNDS FOR MATRIX D. C2-3 'SYMBNDS'
1130 PRINT "13. EIGENVALUE BOUNDS FOR MATRIX D"
1270 ON K GOSUB 1330,1410,1470,2470,4580,3870,2890,3040,3220,6370,
              5360,5820,7030,4150,4180,4070,4110,6030,6050,6370
7000 REM*************************************
7010 REM - GERSCHGORIN BOUNDS ON MAX/MIN EIGENVALUES
7020 REM - ABORT IF MATRIX D NOT SYMMETRIC
7030 IF M8<>N8 THEN GOTO 7270
7040 FOR I=1 TO M8
7050 L=I-1
7060 FOR J=1 TO L
7070 IF D(I,J)<>D(J,I) THEN GOTO 7270
7080 NEXT J
7090 NEXT I
7100 REM - COMPUTE EIGENVALUE BOUNDS FOR MATRIX D
7110 T1=1E+30 : T2=T1 : T3=-1E+30 : T4=T3
7120 FOR I=1 TO M8
7130 R1=0
7140 FOR J=1 TO M8
7150 IF I<>J THEN R1=R1+ABS(D(I,J))
7160 NEXT J
7170 IF D(I,J)>T3 THEN T3=D(I,I)
7180 IF D(I,I)<T2 THEN T2=D(I,I)
7190 R4=D(I,I)+R1
7200 R1=D(I,I)-R1
7210 IF R4>T4 THEN T4=R4
7220 IF R1<T1 THEN T1=R1
7230 NEXT I
7240 PRINT "MAX EIGENVALUE BETWEEN ";T3;" AND ";T4
7250 PRINT "MIN EIGENVALUE BETWEEN ";T1;" AND ";T2
7260 RETURN
7270 PRINT "MATRIX D IS NOT SYMMETRIC!" : RETURN
7280 END
```

```
1129 REM - QR ITERS FOR DIAG FORM OF EIGENVALS - C2-4 'QRITER'
1130 PRINT "13. QR DIAGONALIZATION OF A"
1270 ON K GOSUB 1330,1410,1470,2470,4580,3870,2890,3040,3220,6400,
              5360,5820,6960,4150,4180,4070,4110,6030,6050,6370
6950 REM - QR ITERATIONS FOR DIAGONALIZING MATRIX A
6960 L7=0 : REM - INIT ITER COUNT
6970 S6$=" #####.#####" : S$="A"
6980 PRINT "**********************************************"
6990 GOSUB 4220 : REM - DISPLAY MATRIX A
7000 PRINT "COMPLETED ITERATION # ";L7;" CONTINUE (Y/N)";:INPUT S4$
7010 IF S4$="N" THEN RETURN : REM - GO BACK TO MENU
7020 GOSUB 6430 : REM - DECOMPOSE A=DC BY GRAM-SCHMIDT
7030 S$="B" : S9$="C"
7040 GOSUB 1640 : REM - MOVE C TO B TEMPORARILY
```

```
7050 S$="C" : S9$="D"
7060 GOSUB 1640 : REM - MOVE C=D
7070 S$="D" : S9$="B"
7080 GOSUB 1640 : REM - HAVE NOW SWAPPED D & C
7090 GOSUB 3090 : REM - COMMAND 8 IS A=D*C
7100 L7=L7+1 : REM - INCREMENT ITER COUNTER
7110 GOTO 6970 : REM - START OF ITERATION LOOP
7120 END



1129 REM - SHIFTED INVERSE POWER METHOD FOR EIGENSOLNS - C2-5 'SHINVP'
1130 PRINT "13. EIGENSOLUTION OF B FROM APPROX EIGENVALUE"
1270 ON K GOSUB 1330,1410,1470,2470,4580,3870,2890,3040,3220,6370,
         5360,5820,6960,4150,4180,4070,4110,6030,6050,6370
6950 REM - SHIFTED INVERSE POWER METHOD FOR AN EIGENSOLUTION
6960 L7=0 : REM - INIT ITER COUNT
6970 S6$=" ######.#####" : S$="B"
6980 PRINT "****************************************"
6990 GOSUB 4220 : REM - DISPLAY MATRIX B
7000 PRINT "ESTIMATED EIGENVALUE="; : INPUT T4
7010 S$="C" : S9$="I" : M5=K9 : REM - MAKE C=I DIM LIKE B
7020 GOSUB 1640 : REM - C=I
7030 S$="C" : E1=-T4+.0000000001# : REM - CAN'T BE EXACT
7040 GOSUB 3960 : REM - SHIFT IDENTITY MATRIX BY ESTIMATED EIGENVALUE
7050 GOSUB 2940 : REM - A=B+C SHIFTS MATRIX B
7060 S$="B" : S9$="A"
7070 GOSUB 1640 : REM - B=A. PREPARE TO INVERT SHIFTED MATRIX
7080 GOSUB 3220 : REM - D=inv(B)
7090 S$="B" : S9$="D"
7100 GOSUB 1640 : REM - B=D. HOME OF SHIFTED INVERSE MATRIX
7110 M9=M8 : N9=1 : REM - DIMENSION VECTOR C
7120 FOR I=1 TO M9
7130 C(I,1)=1 : REM - SET C TO ALL 1's
7140 NEXT I
7150 T3=1 : REM - INITIAL VECTOR SCALE FACTOR
7160 REM - START OF LOOP
7165 PRINT
7170 PRINT "ON ITERATION # ";L7;" TRIAL FACTOR=";T3
7180 PRINT " TRIAL TRANSPOSED EIGENVECTOR IS:"
7190 L7=L7+1 : REM - INCREMENT ITERATION COUNTER
7200 FOR I=1 TO M9-1
7210 PRINT USING S6$; C(I,1);
7220 NEXT I
7225 PRINT USING S6$; C(M9,1)
7230 S$="D" : S9$="B" : GOSUB 1640 : REM - D=B
7240 GOSUB 3090 : REM - A=D*C
7250 S$="D" : S9$="A" : GOSUB 1640 : REM - D=A
7260 GOSUB 5820 : REM - MAX ABS ELE IN D
7270 T=D(K1,K2) : REM - RETAIN SIGN
7280 REM - SCALE VECTOR A
7290 FOR I=1 TO I9
7300 A(I,1)=A(I,1)/T
7310 NEXT I
7320 REM - TERMINATION TEST
7330 IF ABS(T3-T)/(1+ABS(T))<.0001 THEN GOTO 7380
7340 T3=T : REM - SAVE T SINCE NOT CONVERGED
7350 S$="C" : S9$="A" : GOSUB 1640 : REM - C=A
7360 GOTO 7165 : REM - GO TO TOP OF LOOP
7370 REM - RENORM FROM INF- TO 2-NORM & PRINT ANSWERS
7380 H=0
7390 FOR I=1 TO I9 : H=H+A(I,1)*A(I,1) : NEXT I
7400 H=SQR(H)
7410 FOR I=1 TO I9 : A(I,1)=A(I,1)/H : NEXT I
7420 PRINT "****************************************"
7430 PRINT " AFTER ";L7;" ITERATIONS:"
```

```
7440 PRINT "  EIGENVALUE NEAREST ";T4;" IS ";1/T+T4
7450 PRINT"   AND ITS EIGENVECTOR IS"
7460 S$="A" : GOSUB 4220 : REM - DISPLAY VECTOR A
7470 RETURN
7480 END


1129 REM - MATRIX D COMPOSITION BY COLUMN VECTORS A - C2-6 'VECTOCOL'
1130 PRINT "13. MATRIX D COMPOSITION BY COLUMN VECTORS A"
1270 ON K GOSUB 1330,1410,1470,2470,4580,3870,2890,3040,3220,6370,
               5360,5820,6960,4150,4180,4070,4110,6030,6050,6370
6950 REM - MATRIX D COMPOSITION BY COLUMN VECTORS A
6960 PRINT "DESTINATION COLUMN NUMBER IN D IS"; : INPUT N8
6970 M8=I9 : REM - HAVE SET BOTH DIMENSIONS OF D
6980 FOR I=1 TO M8
6990 D(I,N8)=A(I,1)
7000 NEXT I
7010 RETURN
7020 END


1129 REM - HOUSEHOLDER TRANS FOR UPPER HESSENBERG - C2-7 'HOUSE'
1130 PRINT "13. TRANSFORMATION OF MATRIX A TO UPPER HESSENBERG FORM
1270 ON K GOSUB 1330,1410,1470,2470,4580,3870,2890,3040,3220,6370,
               5360,5820,6970,4150,4180,4070,4110,6030,6050,6370
6950 REM - HOUSEHOLDER TRANSFORMATION FOR UPPER HESSENBERG
6960 REM - THERE ARE MUCH MORE EFFICIENT WAYS TO IMPLEMENT THIS
6970 FOR K5=1 TO J9-2 : REM - J9 COLUMNS IN MATRIX A
6980 S6$=" ######.#####" : S$="A"
6990 GOSUB 4220 : REM - DISPLAY MATRIX A
7000 PRINT "CONTINUE (Y/N)";:INPUT S4$
7010 PRINT "*****************************************"
7020 IF S4$="N" THEN RETURN : REM - GO BACK TO MENU
7030 S$="B" : S9$="A" : GOSUB 1640 : REM - SAVE B=A
7040 REM - PLACE UNNORM'D VECTOR U INTO D
7050 H=0 : M8=J9 : N8=1 : REM - DIMENSION D
7060 FOR I=K5+1 TO J9
7070 H=H+A(I,K5)*A(I,K5)
7080 NEXT I
7090 R=-SGN(A(K5+1,K5))*SQR(H)
7100 H5=H-R*A(K5+1,K5) : REM - SIGN OF R PREVENTS CANCELLATION
7110 FOR I=1 TO J9
7120 D(I,1)=0
7130 IF I=K5+1 THEN D(I,1)=A(I,K5)-R
7140 IF I<K5+2 THEN GOTO 7160
7150 D(I,1)=A(I,K5)
7160 NEXT I
7170 S$="C" : S9$="D" : GOSUB 1640 : REM - C=D
7180 S$="C" : GOSUB 2540 : REM - TRANSPOSE C
7190 GOSUB 3090 : REM - A=D*C (OUTER PRODUCT OF VECTOR U)
7200 E1=-1/H5 : S$="A" : GOSUB 3960 : REM - SCALE & CHANGE SIGN OF A
7210 S$="C" : S9$="A" : GOSUB 1640 : REM - SETUP FOR SUBTRACTION
7220 S$="D" : S9$="B" : GOSUB 1640: REM - SAVE RESULT
7230 S$="B" : S9$="I" : M5=J9 : GOSUB 1640 : REM - B=I
7240 GOSUB 2940 : REM - A=B+C, A HOUSEHOLDER TRANSFORMATION Q
7250 S$="C" : S9$="A" : GOSUB 1640 : REM - C=A, POSTMULT PLACE
7260 S$="B" : S9$="A" : GOSUB 1640 : REM - SAVE Q FOR PREMULT
7270 GOSUB 3090 : REM - A=D*C, POSTMULT BY Q
7280 S$="C" : S9$="A" : GOSUB 1640 : REM - SETUP PREMULT
7290 S$="D" : S9$="B" : GOSUB 1640 : REM - SETUP PREMULT
7300 GOSUB 3090 : REM -  PREMULT BY Q
7310 PRINT "ANNIHILATED ";J9-K5-1;" ELEMENTS IN COLUMN ";K5;":"
7320 NEXT K5
7330 S$="A" : GOSUB 4220 : REM - DISPLAY MATRIX A
7340 RETURN
7350 END
```

```
1099 REM - FULL-RANK GENERALIZED INVERSE OF MATRIX A - C2-8 'GENINVP'
1100 PRINT "10. FULL-RANK GENERALIZED INVERSE OF MATRIX A"
1129 REM - PROJECTION INTO INTERSECTION OF HYPERPLANES FROM COLS OF A
1130 PRINT "13. PROJECTION MATRIX INTO MANIFOLD FROM COLUMNS OF A"
1270 ON K GOSUB 1330,1410,1470,2470,4580,3870,2890,3040,3220,6960,
               5360,5820,6970,4150,4180,4070,4110,6030,6050,6370
6950 REM - DUAL PURPOSE - COMMANDS 10 AND 13 - ANSWER IN MATRIX A
6960 K5=10 : GOTO 6980 : REM - SET FLAG TO SHOW COMMAND 10 CASE
6970 K5=13 : REM - SET FLAG TO SHOW COMMAND 13 CASE
6980 S6$=" ######.#####" : S$="A"
6990 GOSUB 4220 : REM - DISPLAY MATRIX A
7000 PRINT "CONTINUE (Y/N)";:INPUT S4$
7010 IF S4$="N" THEN RETURN : REM - GO BACK TO MENU
7020 IF K5=10 THEN GOTO 7160 : REM - DON'T NORM GEN INVERSE MATRIX
7030 REM - EUCLIDEAN NORMALIZATION OF COLUMNS OF A (NORMALS)
7040 FOR J=1 TO J9
7050 H=0
7060 FOR I=1 TO I9
7070 H=H+A(I,J)*A(I,J)
7080 NEXT I
7090 IF H<>0 THEN GOTO 7110
7100 PRINT "COLUMN ";J9;" IN MATRIX A WAS NULL. ABORTED." : RETURN
7110 H=SQR(H) : REM - EUCLIDEAN NORM
7120 FOR I=1 TO I9
7130 A(I,J)=A(I,J)/H
7140 NEXT I
7150 NEXT J
7160 S$="D" : S9$="A" : GOSUB 1640 : REM - D=A
7170 S$="D" : GOSUB 2540 : REM - TRANSPOSE D
7180 S$="C" : S9$="A" : GOSUB 1640 : REM - C=A
7190 GOSUB 3090 : REM - A=D*C
7200 S$="B" : S9$="A" : GOSUB 1640 : REM - B=A
7210 GOSUB 3290 : REM - D=inv(B)
7220 S$="C" : GOSUB 2540 : REM - TRANSPOSE C
7230 GOSUB 3090 : REM - A=D*C IS GENERALIZED INVERSE
7240 IF K5=10 THEN RETURN
7250 S$="D" : S9$="C" : GOSUB 1640 : REM - D=C
7260 S$="D" : GOSUB 2540 : REM - TRANSPOSE D
7270 S$="C" : S9$="A" : GOSUB 1640 : REM - C=A
7280 GOSUB 3090 : REM - A=D*C PROJECTION PER RANGE OF GIVEN MATRIX
7290 S$="C" : S9$="A" : GOSUB 1640 : REM - C=A
7300 E1=-1 : S$="C" : GOSUB 3960 : REM - C=-C
7310 S$="B" : S9$="I" : M5=M9 : GOSUB 1640 : REM - B=I
7320 GOSUB 2940 : REM - A=B+C, PROJECTION INTO MANIFOLD
7330 RETURN
7340 END



5 REM - 8503021330. COPYRIGHT T.R. CUTHBERT. 1985.
1099 REM - FORM LU FACTORIZATION OF MATRIX D - C3-1 'LUFAC'
1100 PRINT "10. LU FACTORIZATION OF MATRIX D"
1129 REM - FORWARD AND BACKWARD SUBSTITUTION OF LU IN D FOR VECTOR IN C
1130 PRINT "13. LU SOLUTION OF MATRIX IN D FOR VECTOR IN C"
1270 ON K GOSUB 1330,1410,1470,2470,4580,3870,2890,3040,3220,6960,
               5360,5820,7190,4150,4180,4070,4110,6030,6050,6370
6950 REM - COMMANDS 10 AND 13 - ANSWERS IN D & C, RESPECTIVELY
6960 S6$=" ######.#####" : S$="D"
6970 GOSUB 4220 : REM - DISPLAY MATRIX D
6980 PRINT "CONTINUE (Y/N)";:INPUT S4$
6990 IF S4$="N" THEN RETURN : REM - GO BACK TO MENU
7000 REM - LU SIMILAR TO GAUSSIAN.  INPUT TO D AND OUTPUT TO D
7010 FOR K=1 TO M8-1
7020 K5=K+1
7030 FOR J=K5 TO M8
7040 H=D(K,J)/D(K,K)
7050 D(K,J)=H
7060 FOR I=K5 TO M8
7070 D(I,J)=D(I,J)-H*D(I,K)
```

```
7080 NEXT I
7090 NEXT J
7100 NEXT K
7110 H=1
7120 FOR I=1 TO M8 : REM - CALC DETERMINANT
7130 H=H*D(I,I)
7140 NEXT I
7150 PRINT "D NOW IN LU FACTORED FORM"
7160 PRINT"   DETERMINANT = ";H
7170 RETURN
7180 REM - FWD AND BKWD SUBSTITUTION - INPUT C & OUTPUT C
7190 IF (M9=M8) AND (N9=1) THEN GOTO 7210
7200 PRINT "C DIMENSIONS DON'T MATCH D - ABORT" : RETURN
7210 REM - FORWARD SUBSTITUTION
7220 C(1,1)=C(1,1)/D(1,1)
7230 FOR I=2 TO M9
7240 H=C(I,1)
7250 FOR J=1 TO I-1
7260 H=H-D(I,J)*C(J,1)
7270 NEXT J
7280 C(I,1)=H/D(I,I)
7290 NEXT I
7300 REM - BACKWARD SUBSTITUTION
7310 FOR K=1 TO M9-1
7320 I = M9-K
7330 H=C(I,1)
7340 FOR J=I+1 TO M9
7350 H=H-D(I,J)*C(J,1)
7360 NEXT J
7370 C(I,1)=H
7380 NEXT K
7390 RETURN
7400 END


385 DIM H(21),T(6),E(6) : REM - FOR LDLT FACTORIZATION, UPDATE, & SOLN
1099 REM - LDLT FACTORIZATION, UPDATE, & SOLUTIONS.  C3-2 'LDLTFAC'
1100 PRINT "10. LDLT FACTORIZATION & UPDATE (WITH B) OF MATRIX D"
1129 REM - SOLUTION OF C FOR MATRIX D (in H), ANSWER IN C, BY LDLT
1130 PRINT "13. LDLT SOLN OF C FOR MATRIX D (in H), ANSWER IN C"
1270 ON K GOSUB 1330,1410,1470,2470,4580,3870,2890,3040,3220,6960,
          5360,5820,7810,4150,4180,4070,4110,6030,6050,6370
6950 REM - COMMANDS 10 AND 13 - ANSWERS IN D (&H) & C, RESPECTIVELY
6960 S6$=" ######.#####" : S$="D"
6970 GOSUB 4220 : REM - DISPLAY MATRIX D
6980 PRINT "CONTINUE (Y/N)";:INPUT S4$
6990 IF S4$="N" THEN RETURN : REM - GO BACK TO MENU
7000 REM - LDLT PERFORMED IN VECTOR H BUT CAN BE VIEWED IN D
7010 IF M8=N8 THEN GOTO 7040 : REM - MATRIX D MUST BE SQUARE
7020 PRINT "MATRIX D NOT SQUARE - ABORTED." : RETURN
7030 REM - PLACE SYM LWR TRIANG MATRIX D IN VECTOR H
7040 K=0
7050 FOR J=1 TO N8
7060 FOR I=1 TO M8
7070 IF I<J THEN GOTO 7100
7080 K=K+1
7090 H(K)=D(I,J)
7100 NEXT I
7110 NEXT J
7120 REM - LDLT FACTORIZATION OF MATRIX IN SITU IN VECTOR H
7130 K5=1
7140 FOR I=2 TO N8
7150 Z=H(K5)
7160 IF Z<=0 THEN GOTO 7340
7170 K5=K5+1
7180 I1=K5
7190 FOR J=I TO M8
7200 Z5=H(K5)
```

```
7210 H(K5)=H(K5)/Z
7220 J5=K5
7230 I5=I1
7240 FOR K=I TO J
7250 J5=J5+N8+1-K
7260 H(J5)=H(J5)-H(I5)*Z5
7270 I5=I5+1
7280 NEXT K
7290 K5=K5+1
7300 NEXT J
7310 NEXT I
7320 IF H(K5)<=0 THEN GOTO 7340
7330 GOTO 7350
7340 PRINT"H(";K5;") IS NEGATIVE - ABORT" : RETURN
7350 PRINT "NEED RANK-1 UPDATE OF D USING VECTOR B (Y/N)"; : INPUT S4$
7360 IF S4$<>"Y" THEN GOTO 7690
7370 PRINT"SCALAR Q = "; : INPUT Q
7380 REM - SOLN OF LV=Z FOR V BY FWD SUBSTITUTION
7390 T(1)=B(1,1)
7400 FOR I= 2 TO N8
7410 I4=I
7420 Z=B(I,1)
7430 FOR J=1 TO I-1
7440 Z=Z-H(I4)*T(J)
7450 I4=I4+N8-J
7460 NEXT J
7470 T(I)=Z
7480 NEXT I
7490 REM - UPDATE dii IN H DIAGONAL & FILL E(.)
7500 I4=1
7510 FOR I=1 TO N8
7520 Z=H(I4)+Q*T(I)*T(I)
7530 IF Z<=0 THEN GOTO 7790 : REM - dii NEGATIVE
7540 H(I4)=Z
7550 E(I)=T(I)*Q/Z
7560 Q=Q-E(I)*E(I)*Z
7570 I4=I4+N8+1-I
7580 NEXT I
7590 REM - UPDATE L* = LLhat
7600 I4=1
7610 FOR I=1 TO N8-1
7620 I4=I4+1
7630 FOR J=I+1 TO N8
7640 B(J,1)=B(J,1)-H(I4)*T(I)
7650 H(I4)=H(I4)+E(I)*B(J,1)
7660 I4=I4+1
7670 NEXT J
7680 NEXT I
7690 K=0 : REM - RETURN DECOMPOSITION TO MATRIX D
7700 FOR J=1 TO N8
7710 FOR I=1 TO N8
7720 D(I,J)=0
7730 IF I<J THEN GOTO 7760
7740 K=K+1
7750 D(I,J)=H(K)
7760 NEXT I
7770 NEXT J
7780 RETURN
7790 PRINT"H(";I;") IS NEGATIVE - ABORT" : RETURN
7800 REM - SOLUTION OF DX=C GIVEN C WITH ANSWER X IN C
7810 FOR I=2 TO N8
7820 I4=I
7830 V=C(I,1)
7840 FOR J=1 TO I-1
7850 V=V-H(I4)*C(J,1)
7860 I4=I4+N8-J
7870 NEXT J
7880 C(I,1)=V
```

```
7890 NEXT I
7900 C(N8,1)=C(N8,1)/H(I4)
7910 FOR K=2 TO N8
7920 I=N8+1-K
7930 I1=I4-K
7940 V=C(I,1)/H(I1)
7950 I4=I1
7960 FOR J=I+1 TO N8
7970 I1=I1+1
7980 V=V-H(I1)*C(J,1)
7990 NEXT J
8000 C(I,1)=V
8010 NEXT K
8020 RETURN
8030 END




10 REM - SINGULAR VALUE DECOMPOSITION - | C3-3 | 'SVD'
20 OPTION BASE 1 : REM - FORTRAN-TYPE SUBSCRIPTS
30 CLS : KEY OFF : L8=0 : REM - INIT PRINT FLAG
40 PRINT "*********** SINGULAR VALUE DECOMPOSITION *************"
50 PRINT "NOTES:"
60 PRINT "1. USE ONLY UPPER CASE LETTERS"
70 PRINT "2. MERGE MATRIX DATA STATEMENTS"
80 PRINT "      INTO RESERVED LINE RANGE 400-620 (OPTIONAL)"
90 PRINT "3. IF 'BREAK' OCCURS, RESTART WITH 'GOTO 999 <RTN>'"
100 DEFDBL A-H,Q-R,T-Z
110 DEFINT I-N
120 REM - DIMENSIONED FOR MAXIMUM OF 6x6 SYSTEM
130 DIM A(6,6),U(6,6),V(6,6),S(6,6),W(6),R(6)
390 REM **** LINES 400-620 RESERVED FOR MERGING PROBLEM DATA ****
400 REM - THE BYPASS CASE IS SHOWN HERE AS FOLLOWS
410 N$ = "NONE" : REM - A NAME FOR PROBLEM MERGE FILE MUST BE HERE
420 DATA 0,0 : REM - MATRIX A #ROWS,#COLS - HAVE AT LEAST ONE PAIR
630 REM *********************************
640 REM - READ MATRIX A
650 PRINT : PRINT "WORKING WITH DATA SET: ";N$
660 READ I9,J9 : REM - DIM OF A
670 IF I9=0 THEN GOTO 1160 : REM - MATRIX A NOT LOADED THIS WAY
680 PRINT "   A(";I9;",";J9;")"
690 FOR I=1 TO I9
700 FOR J=1 TO J9
710 READ A(I,J)
720 NEXT J
730 NEXT I
740 GOTO 1160 : REM - TO MENU AND SELECTION
990 REM - RE-ENTRY FOR INVALID COMMAND NUMBERS & CONTINUING
999 CLS
1000 PRINT "*********** COMMAND MENU ***********"
1010 PRINT "0. DISPLAY A MATRIX IN FIXED FORMAT"
1020 PRINT "1. MATRIX TO/FROM DISK"
1030 PRINT "2. SINGULAR VALUE DECOMPOSITION OF MATRIX A"
1040 PRINT "3. PRINT A MATRIX IN FIXED FORMAT"
1050 PRINT "4. EXIT"
1060 PRINT"*********************************
1070 REM  - SOLICIT INPUT COMMAND NUMBER AND CHECK VALIDITY
1080 PRINT"INPUT COMMAND NUMBER:";:INPUT S$
1090 K=LEN(S$) : IF K=0 THEN GOTO 999 : REM - AVOID <CR>
1100 K=ASC(S$)
1110 IF K<48 OR K>52 THEN GOTO 999 : REM - 1ST CHAR MUST BE 0-4
1120 K=VAL(S$)
1130 IF K=0 THEN K=5 : REM - GOSUB CAN'T USE 0
1140 IF K>5 THEN GOTO 999 : REM - CAN'T EXCEED MENU #'S
1150 ON K GOSUB 3970,1220,3530,4730,3570
1160 PRINT "PRESS <RETURN> KEY TO CONTINUE -- READY";
1170 INPUT S4$
1180 IF S4$<>"" THEN BEEP : REM - <RETURN> BEFORE NEXT CMD NUMBER
```

```
1190 GOTO 999
1200 REM*******************************************
1210 REM - PERFORM SVD ON MATRIX A
1220 I3=0 : K9=I9 : L9=J9 : M9=J9 : N9=J9 : M8=J9 : N8 =J9
1230 PRINT "    SVD WORKING - PLEASE WAIT"
1240 FOR I=1 TO I9
1250 FOR J=1 TO J9
1260 U(I,J)=A(I,J)
1270 NEXT J
1280 NEXT I
1290 REM - BEGIN HOUSEHOLDER TRANSFORMATION TO BIDIAGONAL FORM
1300 G=0 : S1=0 : A1=0
1310 FOR I=1 TO J9
1320 L=I+1
1330 R(I)=S1*G
1340 G=0 : S=0 : S1=0
1350 IF I>I9 THEN GOTO 1620
1360 FOR K=I TO I9
1370 S1=S1+ABS(U(K,I))
1380 NEXT K
1390 IF S1=0 THEN GOTO 1620
1400 FOR K=I TO I9
1410 U(K,I)=U(K,I)/S1
1420 S=S+U(K,I)*U(K,I)
1430 NEXT K
1440 F=U(I,I)
1450 G=-SGN(F)*SQR(S)
1460 H=F*G-S
1470 U(I,I)=F-G
1480 IF I=J9 THEN GOTO 1590
1490 FOR J=L TO J9
1500 S=0
1510 FOR K=I TO I9
1520 S=S+U(K,I)*U(K,J)
1530 NEXT K
1540 F=S/H
1550 FOR K=I TO I9
1560 U(K,J)=U(K,J)+F*U(K,I)
1570 NEXT K
1580 NEXT J
1590 FOR K=I TO I9
1600 U(K,I)=S1*U(K,I)
1610 NEXT K
1620 W(I)=S1*G
1630 G=0 : S=0 : S1=0
1640 IF (I>I9) OR (I=J9) THEN GOTO 1930
1650 FOR K=L TO J9
1660 S1=S1+ABS(U(I,K))
1670 NEXT K
1680 IF S1=0 THEN GOTO 1930
1690 FOR K=L TO J9
1700 U(I,K)=U(I,K)/S1
1710 S=S+U(I,K)*U(I,K)
1720 NEXT K
1730 F=U(I,L)
1740 G=-SGN(F)*SQR(S)
1750 H=F*G-S
1760 U(I,L)=F-G
1770 FOR K=L TO J9
1780 R(K)=U(I,K)/H
1790 NEXT K
1800 IF I=I9 THEN GOTO 1900
1810 FOR J=L TO I9
1820 S=0
1830 FOR K=L TO J9
1840 S=S+U(J,K)*U(I,K)
1850 NEXT K
1860 FOR K=L TO J9
```

```
1870 U(J,K)=U(J,K)+S*R(K)
1880 NEXT K
1890 NEXT J
1900 FOR K=L TO J9
1910 U(I,K)=S1*U(I,K)
1920 NEXT K
1930 T=ABS(W(I))+ABS(R(I))
1940 IF T>A1 THEN A1=T
1950 NEXT I
1960 REM - PRODUCT OF RIGHT-HAND TRANSFORMATIONS
1970 FOR I2=1 TO J9
1980 I=J9+1-I2
1990 IF I=J9 THEN GOTO 2160
2000 IF G=0 THEN GOTO 2130
2010 FOR J=L TO J9
2020 V(J,I)=(U(I,J)/U(I,L))/G
2030 NEXT J
2040 FOR J=L TO J9
2050 S=0
2060 FOR K=L TO J9
2070 S=S+U(I,K)*V(K,J)
2080 NEXT K
2090 FOR K=L TO J9
2100 V(K,J)=V(K,J)+S*V(K,I)
2110 NEXT K
2120 NEXT J
2130 FOR J=L TO J9
2140 V(I,J)=0 : V(J,I)=0
2150 NEXT J
2160 V(I,I)=1
2170 G=R(I)
2180 L=I
2190 NEXT I2
2200 REM - PRODUCT OF LEFT-HAND TRANSFORMATIONS
2210 M1=J9
2220 IF I9<J9 THEN M1=I9
2230 FOR I2=1 TO M1
2240 I=M1+1-I2
2250 L=I+1
2260 G=W(I)
2270 IF I=J9 THEN GOTO 2310
2280 FOR J=L TO J9
2290 U(I,J)=0
2300 NEXT J
2310 IF G=0 THEN GOTO 2470
2320 IF I=M1 THEN GOTO 2430
2330 FOR J=L TO J9
2340 S=0
2350 FOR K=L TO I9
2360 S=S+U(K,I)*U(K,J)
2370 NEXT K
2380 F=(S/U(I,I))/G
2390 FOR K=I TO I9
2400 U(K,J)=U(K,J)+F*U(K,I)
2410 NEXT K
2420 NEXT J
2430 FOR J=I TO I9
2440 U(J,I)=U(J,I)/G
2450 NEXT J
2460 GOTO 2500
2470 FOR J=I TO I9
2480 U(J,I)=0
2490 NEXT J
2500 U(I,I)=U(I,I)+1
2510 NEXT I2
2520 REM - DIAGONALIZE THE BIDIAGONAL FORM
2530 FOR K2=1 TO J9
2540 K1=J9-K2
```

```
2550 K=K1+1
2560 I4=0
2570 REM - SPLITTING TEST
2580 FOR L2=1 TO K
2590 L1=K-L2
2600 L=L1+1
2610 IF ABS(R(L))+A1=A1 THEN GOTO 2830
2620 REM - NOTE THAT R(1)=0 ALWAYS
2630 IF (ABS(W(L1))+A1)=A1 THEN GOTO 2650
2640 NEXT L2
2650 C=0 : S=1
2660 FOR I=L TO K
2670 F=S*R(I)
2680 R(I)=C*R(I)
2690 IF (ABS(F)+A1)=A1 THEN GOTO 2830
2700 G=W(I)
2710 H=SQR(F*F+G*G)
2720 W(I)=H
2730 C=G/H
2740 S=-F/H
2750 FOR J=1 TO I9
2760 Y=U(J,L1)
2770 Z=U(J,I)
2780 U(J,L1)=Y*C+Z*S
2790 U(J,I)=-Y*S+Z*C
2800 NEXT J
2810 NEXT I
2820 REM - CONVERGENCE TEST
2830 Z=W(K)
2840 IF (L=K) THEN GOTO 3360
2850 REM - SHIFT FROM THE BOTTOM DIAGONAL 2x2 SUBMATRIX
2860 IF I4=30 THEN GOTO 3500 : REM - 30 ITERATION LIMIT
2870 I4=I4+1
2880 X=W(L)
2890 Y=W(K1)
2900 G=R(K1)
2910 H=R(K)
2920 F=((Y-Z)*(Y+Z)+(G-H)*(G+H))/(2*H*Y)
2930 G=SQR(F*F+1)
2940 F=((X-Z)*(X+Z)+H*(Y/(F+SGN(F)*ABS(G))-H))/X
2950 REM - NEXT QR TRANSFORMATION
2960 C=1 : S=1
2970 FOR I1=L TO K1
2980 I=I1+1
2990 G=R(I)
3000 Y=W(I)
3010 H=S*G
3020 G=C*G
3030 Z=SQR(F*F+H*H)
3040 R(I1)=Z
3050 C=F/Z
3060 S=H/Z
3070 F=X*C+G*S
3080 G=-X*S+G*C
3090 H=Y*S
3100 Y=Y*C
3110 FOR J=1 TO J9
3120 X=V(J,I1)
3130 Z=V(J,I)
3140 V(J,I1)=X*C+Z*S
3150 V(J,I)=-X*S+Z*C
3160 NEXT J
3170 Z=SQR(F*F+H*H)
3180 W(I1)=Z
3190 IF Z=0 THEN GOTO 3220
3200 C=F/Z
3210 S=H/Z
3220 F=C*G+S*Y
```

```
3230 X=-S*G+C*Y
3240 FOR J=1 TO I9
3250 Y=U(J,I1)
3260 Z=U(J,I)
3270 U(J,I1)=Y*C+Z*S
3280 U(J,I)=-Y*S+Z*C
3290 NEXT J
3300 NEXT I1
3310 R(L)=0
3320 R(K)=F
3330 W(K)=X
3340 GOTO 2580
3350 REM - CONVERGENCE OBTAINED
3360 IF Z>=0 THEN GOTO 3410
3370 W(K)=-Z
3380 FOR J=1 TO J9
3390 V(J,K)=-V(J,K)
3400 NEXT J
3410 NEXT K2
3420 FOR I=1 TO J9 : REM - SIGMA'S FROM VECTOR TO MATRIX
3430 FOR J=1 TO J9
3440 S(I,J)=0
3450 IF I=J THEN S(I,J)=W(I)
3460 NEXT J
3470 NEXT I
3480 PRINT "END OF SINGULAR VALUE DECOMPOSITION"
3490 RETURN
3500 PRINT " SINGULAR VALUE DECOMPOSITION FAILED AFTER 30 ITERATIONS!"
3510 RETURN
3520 REM*****************************************
3530 REM - PRINT ARRAY IN FIXED FORMAT
3540 IF L8=0 THEN L8=1 : GOTO 3570
3550 REM*****************************************
3560 REM - DISPLAY ARRAY IN FIXED FORMAT
3570 S6$=" ######.#####"
3580 PRINT "SEE A, U, V, OR S"; : INPUT S$
3590 IF S$="A" OR S$="U" OR S$="V" OR S$="S" THEN GOTO 3620
3600 GOTO 3580
3610 REM ** ENTRY POINT IF PRESET PARAMETERS
3620 IF S$<>"A" THEN GOTO 3640
3630 M5=I9 : M6=J9 : GOTO 3700
3640 IF S$<>"U" THEN GOTO 3660
3650 M5=K9 : M6=L9 : GOTO 3700
3660 IF S$<>"V" THEN GOTO 3680
3670 M5=M9 : M6=N9 : GOTO 3700
3680 IF S$<>"S" THEN GOTO 3700
3690 M5=M8 : M6=N8
3700 PRINT "MATRIX ";S$;"(";M5;",";M6;") -"
3710 IF L8>0 THEN LPRINT "MATRIX ";S$;"(";M5;",";M6;") -"
3720 FOR I=1 TO M5
3730 J=M6 : IF J=1 THEN GOTO 3840
3740 FOR J=1 TO M6-1
3750 IF S$="A" THEN PRINT USING S6$; A(I,J);
3760 IF S$="A" AND L8>0 THEN LPRINT USING S6$; A(I,J);
3770 IF S$="U" THEN PRINT USING S6$; U(I,J);
3780 IF S$="U" AND L8>0 THEN LPRINT USING S6$; U(I,J);
3790 IF S$="V" THEN PRINT USING S6$; V(I,J);
3800 IF S$="V" AND L8>0 THEN LPRINT USING S6$; V(I,J);
3810 IF S$="S" THEN PRINT USING S6$; S(I,J);
3820 IF S$="S" AND L8>0 THEN LPRINT USING S6$; S(I,J);
3830 NEXT J
3840 IF S$="A" THEN PRINT USING S6$; A(I,J)
3850 IF S$="A" AND L8>0 THEN LPRINT USING S6$; A(I,J)
3860 IF S$="U" THEN PRINT USING S6$; U(I,J)
3870 IF S$="U" AND L8>0 THEN LPRINT USING S6$; U(I,J)
3880 IF S$="V" THEN PRINT USING S6$; V(I,J)
3890 IF S$="V" AND L8>0 THEN LPRINT USING S6$; V(I,J)
3900 IF S$="S" THEN PRINT USING S6$; S(I,J)
```

```
3910 IF S$="S" AND L8>0 THEN LPRINT USING S6$; S(I,J)
3920 NEXT I
3930 IF L8=1 THEN L8=0 : REM - TURN OFF LOCAL PRINT FLAG
3940 RETURN
3950 REM****************************************
3960 REM - MATRIX TO/FROM DISK
3970 PRINT"SEE DIRECTORY (Y/N)"; : INPUT S4$
3980 IF S4$<>"Y" THEN GOTO 4050
3990 PRINT"FILENAME SPECIFIER (LIKE *.* OR <RETURN>) = "; : INPUT S5$
4000   IF S5$="" THEN S5$="*.*"
4010 FILES S5$
4020 PRINT "SEE DIRECTORY AGAIN (Y/N)"; : INPUT S4$
4030 IF S4$<>"Y" THEN GOTO 4050
4040 GOTO 3990
4050 PRINT "MATRIX INVOLVED IS A, U, V, OR S"; : INPUT S$
4060 IF S$="A" OR S$="U" OR S$="V" OR S$="S" THEN GOTO 4080
4070 GOTO 4050
4080 PRINT "RECALL OR SAVE MATRIX ";S$;" (R/S)"; : INPUT S1$
4090 IF S1$="R" OR S1$="S" THEN GOTO 4110
4100 GOTO 4080
4110 PRINT "FILE NAME IS"; : INPUT S2$
4120 PRINT "  !!! WARNING - DO NOT <CNTRL><BREAK> DURING THIS STEP !!!"
4130 IF S1$="R" THEN GOTO 4410
4140 REM - SAVE MATRIX TO DISK FILE
4150 PRINT "COMMENT LABEL FOR FILE IS (<96 CHAR): "; : LINE INPUT S3$
4160 REM ** ENTRY POINT IF PRESET PARAMETERS
4170 OPEN S2$ FOR OUTPUT AS #1
4180 WRITE #1,S3$
4190 IF S$<>"A" THEN GOTO 4210
4200 M5=I9 : M6=J9 : GOTO 4270
4210 IF S$<>"U" THEN GOTO 4230
4220 M5=K9 : M6=L9 : GOTO 4270
4230 IF S$<>"V" THEN GOTO 4250
4240 M5=M9 : M6=N9 : GOTO 4270
4250 IF S$<>"S" THEN GOTO 4270
4260 M5=M8 : M6=N8
4270 WRITE #1,M5,M6
4280 FOR I=1 TO M5
4290 FOR J=1 TO M6
4300 IF S$="A" THEN WRITE #1,A(I,J)
4310 IF S$="U" THEN WRITE #1,U(I,J)
4320 IF S$="V" THEN WRITE #1,V(I,J)
4330 IF S$="S" THEN WRITE #1,S(I,J)
4340 NEXT J
4350 NEXT I
4360 N7=N7+1 : S8$(N7)="SAVED MATRIX "+S$+" IN FILE "+S2$
4370 PRINT "   ";S8$(N7)
4380 IF L8>0 THEN LPRINT "   ";S8$(N7)
4390 CLOSE #1 : RETURN
4400 REM - RECALL MATRIX FROM DISK FILE
4410 OPEN S2$ FOR INPUT AS #1
4420 INPUT #1,S3$
4430 PRINT "READY TO READ FILE ";S2$;" INTO MATRIX ";S$;" TITLED:";
4440 PRINT "   ";S3$
4450 PRINT "PRESS <RETURN> KEY IF OK, ELSE 'ABORT' <RETURN>";
4460 INPUT S4$
4470 IF S4$="" THEN GOTO 4520
4480 PRINT "ABORT RECALL; MATRIX ";S$;" NOT CHANGED" : GOTO 4720
4490 REM ** ENTRY POINT IF PRESET PARAMETERS
4500 OPEN S2$ FOR INPUT AS #1
4510 INPUT #1,S3$
4520 N7=N7+1 : S8$(N7)="READ FILE "+S2$+" INTO MATRIX "+S$
4530 PRINT "   ";S8$(N7)
4540 IF L8>0 THEN LPRINT "   ";S8$(N7)
4550 INPUT #1,M5,M6
4560 FOR I=1 TO M5
4570 FOR J=1 TO M6
4580 IF S$="A" THEN INPUT #1,A(I,J)
```

```
4590 IF S$="U" THEN INPUT #1,U(I,J)
4600 IF S$="V" THEN INPUT #1,V(I,J)
4610 IF S$="S" THEN INPUT #1,S(I,J)
4620 NEXT J
4630 NEXT I
4640 IF S$<>"A" THEN GOTO 4660
4650 I9=M5 : J9=M6 : GOTO 4720
4660 IF S$<>"U" THEN GOTO 4680
4670 K9=M5 : L9=M6 : GOTO 4720
4680 IF S$<>"V" THEN GOTO 4700
4690 M9=M5 : N9=M6 : GOTO 4720
4700 IF S$<>"S" THEN GOTO 4720
4710 M8=M5 : N8=M6
4720 CLOSE #1 : RETURN
4730 KEY ON : PRINT "END OF RUN" : END
```

```
1129 REM - EXAMPLES 3.2.12 & 3.3.4; NR ITERS - C3-4 'LAGRANGE'
1130 PRINT "13. LAGRANGE EXAMPLE USING NEWTON ITERATIONS"
1270 ON K GOSUB 1330,1410,1470,2470,4580,3870,2890,3040,3220,6370,
               5360,5820,6960,4150,4180,4070,4110,6030,6050,6370
6950 REM - INPUT STARTING X1,X2,X3 VECTOR
6960 PRINT"INPUT STARTING X1,X2,X3:" : INPUT X1,X2,X3
6970 L7=0 : REM - INIT ITER# & PUT -F VECTOR INTO C
6980 IF X1>0 THEN GOTO 7000
6990 PRINT "X1<=0, SO LOG(X1) IMPOSSIBLE" : RETURN
7000 M9=3 : N9=1 : L7=L7+1 : REM - SET DIM'S AND INCR ITER#
7010 C(1,1)=-(-X2+X3*(-4+2*X1+1/X1))
7020 C(2,1)=-(-X1+X3)
7030 C(3,1)=-(2-X1*(4-X1)+LOG(X1)+X2)
7040 REM - PUT JACOBIAN INTO B
7050 K9=3 : L9=3
7060 B(1,1)=X3*(2-1/(X1*X1)) : B(1,2)=-1 : B(1,3)=-4+2*X1+1/X1
7070 B(2,1)=-1 : B(2,2)=0 : B(2,3)=1
7080 B(3,1)=B(1,3) : B(3,2)=1 : B(3,3)=0
7090 GOSUB 3290 : REM - INVERT JACOBIAN
7100 GOSUB 3090 : REM - A=D*C TO CALC DX
7110 REM - UPDATE X VECTOR BY DX COMPONENTS
7120 X1=X1+A(1,1) : X2=X2+A(2,1) : X3=X3+A(3,1)
7130 PRINT
7140 PRINT "AFTER ITERATION #";L7;" TRANSPOSED X VECTOR IS:"
7150 PRINT USING " ######.#####";X1;X2;X3
7160 PRINT"CONTINUE (Y/N)" : INPUT S4$
7170 IF S4$="N" THEN RETURN
7180 GOTO 6980 : REM - NEXT ITERATION
7190 END
```

List of Variable Names Used in Program C4-1: NEWTON

C1	F1	I5	L7	S6$
C2	FNACS()	I7	M	T()
C3	G()	J	M1	T1
C4	G1	J5	N	V
D5	G2	K	P1	X()
D6	H()	K1	Q	Z
E()	I	K2	S$	Z5
E1	I1	K5	S4$	
F	I4	L	S5$	

```
10 REM - NEWTON OPTIMIZER - PROGRAM C4-1 'NEWTON'
20 OPTION BASE 1 : REM - NO SUBSCRIPT 0
30 CLS : KEY OFF
40 PRINT "********* NEWTON OPTIMIZER *************"
50 PRINT "NOTES:"
60 PRINT "1.   USE ONLY UPPER CASE LETTERS"
70 PRINT "2.   IF 'BREAK' OCCURS, RESTART WITH 'GOTO 999'"
80 PRINT "3.   USER MUST PROVIDE SUBROUTINE 5000 FOR FUNCTION EVALUATION"
90 PRINT "          AND SUBROUTINE 7000 FOR GRADIENT EVALUATION"
100 PRINT "4.   ENTER DEFAULT ANSWERS TO QUESTIONS BY <RETURN>."
130 REM - USE OF MAJOR VARIABLES AS FOLLOWS -
140 REM     E()      SEARCH STEP VECTOR
150 REM     F        FUNCTION VALUE RETURNED BY USER SUBROUTINE 5000
160 REM     G()      GRADIENT VECTOR RETURNED BY USER SUBROUTINE 7000
170 REM     H()      VECTOR CONTAINING SYMMETRIC HESSIAN - SEE (4.1.14)
180 REM     I,J      WORKING ROW#, COL#, RESPECTIVELY
190 REM     K        COMMAND # & LOOP INDEX
200 REM     L7       ITERATION #
210 REM     N        NUMBER OF VARIABLES IN VECTOR X()
220 REM     X()      VECTOR OF VARIABLES RELATED TO PARTICULAR PROBLEM
230 REM              IN USER-SUPPLIED SUBROUTINES 5000 AND 7000.
240 DEFDBL A-H,Q-R,T-Z : REM - NOTE THAT P IS SNGL PRECISION
250 DEFINT I-N
260 DEFSTR S
270 S6$=" ######.######"
280 DEF FNACS(X)=1.570796-ATN(X/SQR(1-X*X)) : REM - ARC COS
290 M1=50 : E1=.0001 : D6=.0001 : I7=1 : REM - SET DEFAULT PARAMETERS
300 REM - FOLLOWING DIMENSIONS ARE FOR N<=30.  THE HESSIAN VECTOR
310 REM      H() MUST BE DIMENSIONED N*(N+1)/2.
320 DIM X(30),G(30),H(465),T(30),E(30)
330 REM - HESSIAN H(.) STORED AS AS VECTOR; SEE EQUATION (4.1.14)
340 GOTO 1150 : REM - TO MENU & SELECTION
350 REM - RE-ENTRY FOR INVALID COMMAND NUMBERS & CONTINUING
999 CLS : K2=0 : REM - INIT FUNCTION EVALUATION COUNTER
1000 PRINT "************* COMMAND MENU *************"
1010 PRINT "1. ENTER STARTING VARIABLES (AT LEAST ONCE)"
1020 PRINT "2. REVISE CONTROL PARAMETERS (OPTIONAL)'
1030 PRINT "3. START OPTIMIZATION"
1040 PRINT "4. EXIT (RESUME WITH 'GOTO 999')"
1050 REM
1060 PRINT"******************************************
1070 PRINT"INPUT COMMAND NUMBER:";:INPUT S$
1080 K=LEN(S$) : IF K=0 THEN GOTO 999 : REM - AVOID <CR>
1090 K=ASC(S$)
1100 IF K<48 OR K>57 THEN GOTO 999 : REM - 1ST CHAR MUST BE 0-9
1110 K=VAL(S$)
1120 IF K=0 THEN K= 15 : REM - ALTERNATIVE DISPLAY NUMBERS
1130 IF K>20 THEN GOTO 999 : REM - CAN'T EXCEED MENU #'S
1140 ON K GOSUB 1210,1300,1450,1420
1150 PRINT "PRESS <RETURN> KEY TO CONTINUE -- READY";
1160 INPUT S4$
1170 IF S4$<>"" THEN BEEP : REM - <RETURN> BEFORE NEXT CMD NUMBER
1180 GOTO 999
1190 REM**********************************
1200 REM - ENTER VARIABLES
1210 PRINT"NUMBER OF VARIABLES = "; : INPUT N
1220 PRINT "ENTER STARTING VARIABLES X(I):"
1230 FOR I=1 TO N
1240 PRINT "    X(";I;")="; : INPUT X(I)
1250 NEXT I
1260 PRINT "TRUST REGION RADIUS ="; : INPUT T1
1270 IF T1=0 THEN T1=1000000! : REM - DEFAULT TO UNBOUNDED NEWTON
1280 RETURN
1290 REM ********************************
1300 REM - REVISE CONTROL PARAMETERS
1310 PRINT "MAXIMUM # OF ITERATIONS (DEFAULT=50):"; : INPUT S4$
1320 M1=50 : IF S4$<>"" THEN M1=VAL(S4$)
1330 PRINT "STOPPING CRITERION (DEFAULT=.0001):"; : INPUT S4$
```

```
1340 E1=.0001 : IF S4$<>"" THEN E1=VAL(S4$)
1350 PRINT "ENTER FINITE DIFF FACTOR (DEFAULT=.0001):"; : INPUT S4$
1360 D6=.0001 : IF S4$<>"" THEN D6=VAL(S4$)
1370 PRINT "PRINT EVERY Ith ITERATION (DEFAULT=1):"; : INPUT S4$
1380 I7=1 : IF S4$<>"" THEN I7=VAL(S4$)
1390 RETURN
1400 REM *********************************
1410 REM - NORMAL STOP
1420 KEY ON : PRINT "END OF RUN" : END
1430 REM *********************************
1440 REM - MAIN OPTIMIZATION ALGORITHM - SEE CHAPTER FOUR
1450 GOSUB 5000 : REM - GET INITIAL FUNCTION VALUE
1455 K2=K2+1 : REM - INCREMENT FNCN EVAL COUNT
1460 L7=0 : REM - INITIAL ITERATION COUNT
1470 REM - RE-ENTRY POINT FOR NEW ITERATION
1480 L7=L7+1 : M=0 : REM - INCREMENT ITER # & INIT CUTBACK COUNT
1490 F1=F : REM -SAVE FOR DOWNHILL COMPARISON
1500 GOSUB 7000 : REM - CALC GRADIENT
1510 IF((L7-1) MOD I7)=0 THEN GOSUB 2320 : REM - RPT F, X, AND G
1520 GOSUB 2410 : REM - CALC HESSIAN IN VECTOR H(.) BY DIFF'G G
1530 GOSUB 2710 : REM - FACTOR H=LDLT IN SITU IN H(.)
1540 FOR I=1 TO N : E(I)=-G(I) : NEXT I : REM - NEG GRADIENT
1550 GOSUB 2100 : REM - CALC Q DENOM FOR CAUCHY POINT
1560 REM - CALC LENGTH OF GRADIENT
1570 G1=0 : FOR I=1 TO N : G1=G1+G(I)*G(I) : NEXT I : G1=SQR(G1)
1580 IF G1<>0 THEN GOTO 1600
1590 PRINT "***** GRADIENT IS ZERO *****" : GOTO 2050
1600 C4=G1 : REM - SAVE GRADIENT LENGTH FOR ANGLE CALC
1610 G2=G1*G1*G1/Q : REM - LENGTH OF CAUCHY STEP
1620 S4$="CAUCHY"
1630 S5$=" (BOUNDED)"
1640 IF G2>T1 THEN GOTO 1700 : REM - CP IS OUTSIDE TRUST RADIUS
1650 GOSUB 2990 : REM - CALC NEWTON STEP IN E()
1660 S4$="NEWTON"
1670 G1=0 : FOR I=1 TO N : G1=G1+E(I)*E(I) : NEXT I : G1=SQR(G1)
1680 IF G1<T1 THEN S5$=" (UNBOUNDED)"
1690 IF G1<T1 THEN GOTO 1710 : REM - NEWTON STEP INSIDE TRUST RADIUS
1700 FOR I=1 TO N : E(I)=E(I)*T1/G1 : NEXT I : REM - STEP LENGTH=RADIUS
1710 PRINT "***************************** ";S4$+S5$
1720 IF S4$="CAUCHY" THEN GOTO 1820
1730 REM - CALC NEWTON-TO-GRADIENT DEGREES
1740 C2=0 : C3=0
1750 FOR I=1 TO N : C2=C2+G(I)*E(I) : C3=C3+E(I)*E(I) : NEXT I
1760 P1=-C2/C4/SQR(C3) : IF P1<1 THEN GOTO 1780
1770 P1=0 : GOTO 1790 : REM - AVOID /0 IN ACS
1780 P1=57.29578*FNACS(P1)
1790 PRINT"                 NEWTON-TO-GRADIENT DEGREES=";
1800 PRINT USING " ##.#";P1
1810  REM - TAKE STEP WITH INCREMENT IN E(.)
1820 FOR I=1 TO N : X(I)=X(I)+E(I) : NEXT I
1830 GOSUB 5000 : REM - CALC F(X+dX)
1835 K2=K2+1 : REM - INCREMENT FNCN EVAL COUNT
1840 IF F<F1 THEN GOTO 1900
1850 REM - GET BACK TO LAST TURNING POINT & CUTBACK dX
1860 FOR I=1 TO N : X(I)=X(I)-E(I) : E(I)=E(I)/4 : NEXT I
1870 PRINT "    ###### CUT BACK STEP SIZE BY FACTOR OF 4 #####"
1880 M=M+1 : IF M<11 THEN GOTO 1820
1890 PRINT "STEP SIZE TOO SMALL - TERMINATED" : GOTO 2070
1900 IF S4$="CAUCHY" THEN GOTO 1960 : REM - BYPASS QUAD FACTOR REPORT
1910 REM - CALC QUADRATIC BEHAVIOR FACTOR R
1920 GOSUB 2100 : REM - CALC QUAD DELTA F USING NEWTON STEP
1930 IF Q<.000001 THEN GOTO 1960 : REM - PROBABLY RESET HESSIAN
1940 P1=(F1-F)/(Q/2) : REM - SEE CHAPTER FOUR
1950 PRINT "              QUADRATIC BEHAVIOR FACTOR R=";P1
1960 IF L7<M1 THEN GOTO 2010
1970 PRINT "!!!!!!!!!!!!!!!!!!!!!!!!!!!!!!"
1980 PRINT "STOPPED AT GIVEN LIMIT OF";M1;" ITERATIONS; RESULTS ARE:"
1990 L7=L7+1 : GOTO 2060
```

```
2000 REM - TEST CONVERGENCE OF BOTH F AND EACH X(I)
2010 IF ABS(F1-F)/(1+ABS(F1))>E1 THEN GOTO 1480
2020 FOR I=1 TO N
2030 IF ABS(E(I))/(1+ABS(X(I)))>E1 THEN GOTO 1480
2040 NEXT I
2050 L7=L7+1 : PRINT "CONVERGED; SOLUTION IS:"
2060 GOSUB 7000 : REM - GET GRADIENT AT SOLUTION POINT
2070 GOSUB 2320 : REM - REPORT CONDITIONS AT STOPPING POINT
2075 PRINT "TOTAL NUMBER OF FUNCTION EVALUATIONS =";K2
2080 RETURN : REM - RETURN TO MENU
2090 REM *************************************
2100 REM - CALC QUAD FORM Q FOR E USING LDLT IN H(.)
2110 REM - CALC T(I)
2120 FOR I=1 TO N : T(I)=0 : NEXT I : K=0 : REM - INIT
2130 FOR J=1 TO N : REM - COLUMN LOOP
2140 FOR I=1 TO N : REM - WORK DOWN COLUMN J OF MATRIX L (3.1.12)
2150 IF I<J THEN GOTO 2200
2160 K=K+1
2170 C1=H(K)
2180 IF I=J THEN C1=1
2190 T(J)=T(J)+C1*E(I)
2200 NEXT I
2210 NEXT J
2220 REM - SUM DxT^2 TERMS FOR Q
2230 K=0 : Q=0 : REM - INIT
2240 FOR J=1 TO N
2250 FOR I=1 TO N
2260 IF I<J THEN GOTO 2290
2270 K=K+1
2280 IF I=J THEN Q=Q+H(K)*T(I)*T(I)
2290 NEXT I
2300 NEXT J
2310 RETURN
2320 REM *************************************
2330 REM - PRINT FUNCTION, VARIABLES, AND GRADIENT
2340 PRINT "AT START OF ITERATION NUMBER";L7
2350 P1=F : PRINT "   FUNCTION VALUE =";P1
2360 PRINT " I          X(I)          G(I)"
2370 FOR I=1 TO N
2380 PRINT I; : PRINT USING S6$;X(I),G(I)
2390 NEXT I
2400 RETURN
2410 REM *************************************
2420 REM - OBTAIN HESSIAN IN H(.) BY DIFFERENCING GRADIENT
2430 FOR I=1 TO N : E(I)=G(I) : NEXT I : REM - SAVE NOMINAL GRADIENT
2440 K=0 : REM - INDEX FOR H(K)
2450 FOR J=1 TO N : REM -COLUMNS OF HESSIAN
2460 D5=D6*ABS(X(J)) : REM - INCREMENT FOR X(J) USING PARAMETER D6
2470 IF D5<.000001, THEN D5=.000001 : REM - MINIMUM PERTURBATION
2480 X(J)=X(J)+D5 : REM - PERTURB X(J)
2490 GOSUB 7000 : REM - GET ALL G(I) FOR PERTURBED X(J)
2500 FOR I=1 TO N : REM -ROWS OF Jth HESSIAN COLUMN
2510 T(I)=(G(I)-E(I))/D5 : REM - DELj(DELiF) APPROXIMATE DERIVATIVE
2520 IF I<J THEN GOTO 2550 : REM - ABOVE HESSIAN MAIN DIAGONAL
2530 K=K+1 : H(K)=T(I) : REM - STORE UNAVERAGED VALUE IN H VECTOR
2540 GOTO 2640 : REM - NEXT I
2550 K1=0 : REM - COMPUTE SYMMETRIC ELEMENT INDEX
2560 FOR L=1 TO I
2570 FOR M=1 TO N
2580 IF M<L THEN GOTO 2610
2590 K1=K1+1
2600 IF M=J AND L=I THEN GOTO 2630
2610 NEXT M
2620 NEXT L
2630 H(K1)=(H(K1)+T(I))/2 : REM - AVERAGE SYMMETRIC ELEMENTS
2640 NEXT I
2650 X(J)=X(J)-D5 : REM - RESTORE TO NOMINAL
2660 NEXT J
```

```
2670 FOR I=1 TO N : G(I)=E(I) : NEXT I : REM - RESTORE NOMINAL
2680 RETURN
2690 REM ********************************
2700 REM - LDLT FACTORIZATION OF MATRIX IN SITU IN VECTOR H
2710 K5=1
2720 FOR I=2 TO N
2730 IF H(K5)>0 THEN GOTO 2760
2740 PRINT "                        HESSIAN MADE P.D."
2750 H(K5)=.000001 : REM - FORCE POSITIVE DEFINITENESS
2760 Z=H(K5)
2770 K5=K5+1
2780 I1=K5
2790 FOR J=I TO N
2800 Z5=H(K5)
2810 H(K5)=H(K5)/Z
2820 J5=K5
2830 I5=I1
2840 FOR K=I TO J
2850 J5=J5+N+1-K
2860 H(J5)=H(J5)-H(I5)*Z5
2870 I5=I5+1
2880 NEXT K
2890 K5=K5+1
2900 NEXT J
2910 NEXT I
2920 IF H(K5)<=0 THEN GOTO 2940
2930 RETURN
2940 PRINT "                        HESSIAN MADE P.D."
2950 H(K5)=.000001 : REM - FORCE POSITIVE DEFINITENESS
2960 RETURN
2970 REM ****************************************
2980 REM - SOLUTION E=Inv(H)E FOR NEWTON STEP
2990 FOR I=2 TO N
3000 I4=I
3010 V=E(I)
3020 FOR J=1 TO I-1
3030 V=V-H(I4)*E(J)
3040 I4=I4+N-J
3050 NEXT J
3060 E(I)=V
3070 NEXT I
3080 E(N)=E(N)/H(I4)
3090 FOR K=2 TO N
3100 I=N+1-K
3110 I1=I4-K
3120 V=E(I)/H(I1)
3130 I4=I1
3140 FOR J=I+1 TO N
3150 I1=I1+1
3160 V=V-H(I1)*E(J)
3170 NEXT J
3180 E(I)=V
3190 NEXT K
3200 RETURN
3210 END



7 REM - ROSENBROCKS FUNCTION IN TWO-SPACE - [C4-2] 'ROSEN'
5000 REM ***********************************
5010 REM - ROSENBROCK BANANA FUNCTION - OBJECTIVE
5020 F=100*(X(2)-X(1)*X(1))^2+(1-X(1))^2
5030 RETURN
7000 REM ***********************************
7010 REM - ROSENBROCK BANANA FUNCTION - GRADIENT
7020 G(1)=-400*(X(1)*X(2)-X(1)^3)-2*(1-X(1))
7030 G(2)=200*(X(2)-X(1)*X(1))
7040 RETURN
```

```
7 REM - WOODS FUNCTION IN FOUR-SPACE - C4-3 'WOODS'
5000 REM ***************************************
5010 REM - WOOD'S FUNCTION
5020 F=100*(X(2)-X(1)^2)^2+(1-X(1))^2+90*(X(4)-X(3)^2)^2
      +(1-X(3))^2+10.1*((X(2)-1)^2+(X(4)-1)^2)+19.8*(X(2)-1)*(X(4)-1)
5030 RETURN
7000 REM **************************************
7010 REM - GRADIENT OF WOOD'S FUNCTION
7020 G(1)=-400*X(1)*(X(2)-X(1)^2)-2*(1-X(1))
7030 G(2)=200*(X(2)-X(1)^2)+20.2*(X(2)-1)+19.8*(X(4)-1)
7040 G(3)=-360*X(3)*(X(4)-X(3)^2)-2*(1-X(3))
7050 G(4)=180*(X(4)-X(3)^2)+20.2*(X(4)-1)+19.8*(X(2)-1)
7060 RETURN
```

```
5 REM - 8512100909.  COPYRIGHT T.R.CUTHBERT.  1985.
7 REM - ADDS TO NEWTON - SIMPLE BOUNDS ON VARIABLES - PRGM C4-4 'NBOUNDS

325 DIM L5(30),P5(30,2) : REM - THE SET ON/OFF AND LOWER/UPPER ARRAYS
335 N=30 : GOSUB 3210 : REM - UNBOUND ALL POSSIBLE VARIABLES
999 CLS : K2=0 : L6=0 : REM - INIT FNCN COUNT & BINDING BOUND(S) FLAG
1050 PRINT "5. SEE &/OR RESET LOWER/UPPER BOUNDS ON VARIABLES"
1140 ON K GOSUB 1210,1300,1450,1420,3320
1635 IF L6<>0 THEN GOTO 1700 : REM - FORCE CAUCHY IF BOUNDS ARE BINDING
1820 GOSUB 3515 : REM - CHECK/SET BINDING BOUNDS IN STEPS IN E()
1825 FOR I=1 TO N : X(I)=X(I)+E(I) : NEXT I
1880 M=M+1 : IF M<11 THEN GOTO 1825 : REM - TRY FEASIBLE CUTBACK STEP
3210 REM************************************
3220 REM - INIT FLAGS AND LOWER/UPPER BOUNDS
3225 L6=0 : REM - CLEAR THE 'BINDING BOUND(S)' FLAG
3230   FOR I=1 TO N : L5(I)=0
3240 P5(I,1)=-10000 : P5(I,2)=+10000
3250 NEXT I
3260 RETURN
3300 REM***********************************************
3310 REM - SEE OR RESET LOWER/UPPER BOUND ON VARIABLES
3320 S4$="NONE.   " : PRINT "BOUNDS NOW SET ARE:"
3330 PRINT " I          LOWER          UPPER"
3340 FOR I=1 TO 30
3350 IF L5(I)=0 THEN GOTO 3370
3360 S4$="" : PRINT I;"   "; : PRINT USING S6$;P5(I,1);P5(I,2)
3370 NEXT I
3380 PRINT S4$;"SET OR RESET ANY BOUNDS (Y/N)"; : INPUT S4$
3390 IF S4$<>"Y" THEN RETURN
3400 REM - RE-ENTRY FOR MORE BOUND SETTING
3410 PRINT "ENTER 0 TO RETURN TO MENU, ELSE ENTER VARIABLE # =";
3420 INPUT I : IF I=0 THEN RETURN
3430 PRINT "PRESS <RETURN> IF NO BOUND DESIRED"
3440 PRINT "  LOWER BOUND ="; : INPUT S4$
3450 P5(I,1)=-10000 : IF S4$<>"" THEN P5(I,1)=VAL(S4$)
3455 IF S4$<>"" THEN L5(I)=1
3460 PRINT "  UPPER BOUND ="; : INPUT S4$
3470 P5(I,2)=+10000 : IF S4$<>"" THEN P5(I,2)=VAL(S4$)
3475 IF S4$<>"" THEN L5(I)=1
3480 GOTO 3410
3500 REM*****************************************
3510 REM - CHECK BOUNDS AND RESET STEP IN E() IF BINDING
3515 L6=0 : REM - CLEAR THE 'BINDING BOUND(S)' FLAG
3520 FOR I=1 TO N
3530 IF L5(I)=0 THEN GOTO 3560
3540 IF (X(I)+E(I))<P5(I,1) THEN L6=1
3545 IF (X(I)+E(I))<P5(I,1) THEN E(I)=P5(I,1)-X(I)
3550 IF (X(I)+E(I))>P5(I,2) THEN L6=1
3555 IF (X(I)+E(I))>P5(I,2) THEN E(I)=P5(I,2)-X(I)
3560 NEXT I
3570 RETURN
3580 END
```

List of Variable Names Used in Program C4-5: LEASTP

A( )	F1	I7	L7	S$
C2	F2	J	M	S( )
C3	FNACS( )	J1	M1	S4$
D( )	G( )	J5	M3	S6$
D1	G1	K	M7	V
D2	H( )	K2	N	V5
D5	I	K5	N$	X( )
E( )	I1	K7	N5	Z
E1	I4	L	P1	Z5
F	I5	L1	R( )	

```
10 REM - LEASTP OPTIMIZER - PROGRAM C4-5 'LEASTP'
20 OPTION BASE 1 : REM - NO SUBSCRIPT O
30 CLS : KEY OFF : M7=0 : REM - DEFAULT NUMBER OF SAMPLE DATA
40 PRINT "********* LEASTP OPTIMIZER *************" : PRINT
50 PRINT "NOTES:"
60 PRINT "1.   USE ONLY UPPER CASE LETTERS"
70 PRINT "2.   IF 'BREAK' OCCURS, RESTART WITH 'GOTO 999'"
80 PRINT "3.   USER MUST PROVIDE SUBROUTINE 5000 FOR RESIDUALS,"
90 PRINT "          SUBROUTINE 7000 FOR THE JACOBIAN MATRIX, AND"
95 PRINT "          SAMPLE DATA IN LINES 400-600 IF REQUIRED"
100 PRINT "4.   ENTER DEFAULT ANSWERS TO QUESTIONS BY <RETURN>."
110 PRINT
130 REM - USE OF MAJOR VARIABLES AS FOLLOWS -
140 REM      A(,)   JACOBIAN MATRIX.  A(K,J) IS DERIV OF Kth RESIDUAL
142 REM              WITH RESPECT TO Jth VARIABLE. IS DIM MxN.
144 REM      D()    VECTOR FOR LM DIAGONAL SCALING MATRIX.
146 REM      D1     DETERMINANT OF LDLT FACTORIZATION.
148 REM      E()    SEARCH STEP VECTOR.
150 REM      F      HALF THE SUM OF Pth POWER RESIDUALS.
152 REM      F1     SAVED VALUE OF F FOR DOWNHILL COMPARISON.
154 REM      FNACS  INVERSE COSINE FUNCTION
156 REM      G()    GRADIENT OF F.
158 REM      G1     LENGTH OF GRADIENT.
160 REM      H()    VECTOR STORAGE OF APPROXIMATE HESSIAN MATRIX.
162 REM      K2     COUNT OF NUMBER OF F EVALUATIONS.
164 REM      K7     EXPONENT P - POWER TO WHICH RESIDUALS RAISED.
166 REM      L7     ITERATION COUNTER.
168 REM      M      NUMBER OF DATA SAMPLES.
170 REM      M3     NUMBER OF DATA SAMPLES READ IN FROM DATA STATEMENTS.
172 REM      N      NUMBNER OF VARIABLES.
174 REM      R()    RESIDUALS, DIM M.
176 REM      S(,)   SAMPLES S(I,1) IS INDEPENDENT & S(I,2) IS DEPENDENT.
178 REM      V      LEVENBERG-MARQUARDT (LM) PARAMETER.
180 REM      X()    VARIABLES VECTOR, DIM N.
240 DEFDBL A-H,O-Z : REM - NOTE THAT P IS SNGL PRECISION
250 DEFINT I-N
270 S6$=" ######.########"
280 DEF FNACS(X)=1.570796-ATN(X/SQR(1-X*X)) : REM - ARC COS
290 M1=50 : E1=.0001 : I7=1 : L1=0 : M3=0 : V=.001# : REM - INIT PARAMS
300 REM - FOLLOWING DIMENSIONS ARE FOR N<=20.  THE HESSIAN VECTOR
310 REM       H() MUST BE DIMENSIONED N*(N+1)/2.  # SAMPLES M<=40.
320 DIM X(20),G(20),H(210),E(20),A(40,20),S(40,2),D(20),R(40)
330 REM - HESSIAN H(.) STORED AS AS VECTOR; SEE EQUATION (4.1.14)
340 READ N$ : PRINT "WORKING WITH DATA SET ";N$ : PRINT
350 READ M7 : REM - M7 SHOULD EQUAL M SET BY USER IN SUBROUTINE 5000
360 FOR K=1 TO M7 : READ S(K,1) : NEXT K
```

```
370 FOR K=1 TO M7 : READ S(K,2) : NEXT K
375 GOTO 1160 : REM - TO MENU & SELECTION
380 REM - SAMPLE DATA WILL BE EMPLOYED BY USER IN SUBROUTINE 5000
390 REM - ENTER FOUR DATA STATEMENTS FOR SAMPLE PAIRS AS FOLLOWS
400 DATA "DUMMY" : REM - THE NAME OF THE DATA SET
410 DATA 5 : REM - THE NUMBER OF SAMPLE PAIRS <=40
420 DATA  1 , 2 , 3 , 4 , 5  : REM - THE INDEPENDENT DATA VALUES
430 DATA 1.1,2.2,3.3,4.4,5.5 : REM - THE DEPENDENT DATA VALUES
990 REM - RE-ENTRY FOR INVALID COMMAND NUMBERS & CONTINUING
999 CLS : K2=0 : REM - INIT FUNCTION EVALUATION COUNTER
1000 PRINT "*********** COMMAND MENU ***********"
1010 PRINT "1. ENTER STARTING VARIABLES (AT LEAST ONCE)"
1020 PRINT "2. REVISE CONTROL PARAMETERS (OPTIONAL)"
1030 PRINT "3. START OPTIMIZATION"
1040 PRINT "4. EXIT (RESUME WITH 'GOTO 999')"
1050 PRINT "5. SPARE"
1060 PRINT "6. DISPLAY DATA PAIRS"
1070 PRINT"*************************************
1080 PRINT"INPUT COMMAND NUMBER:";:INPUT S$
1090 K=LEN(S$) : IF K=0 THEN GOTO 999 : REM - AVOID <CR>
1100 K=ASC(S$)
1110 IF K<48 OR K>57 THEN GOTO 999 : REM - 1ST CHAR MUST BE 0-9
1120 K=VAL(S$)
1130 IF K=0 THEN K= 15 : REM - ALTERNATIVE DISPLAY NUMBERS
1140 IF K>20 THEN GOTO 999 : REM - CAN'T EXCEED MENU #'S
1150 ON K GOSUB 1220,1290,1420,1390, 999,3000
1160 PRINT "PRESS <RETURN> KEY TO CONTINUE -- READY";
1170 INPUT S4$
1180 IF S4$<>"" THEN BEEP : REM - <RETURN> BEFORE NEXT CMD NUMBER
1190 GOTO 999
1200 REM*************************************
1210 REM - ENTER VARIABLES
1220 PRINT"NUMBER OF VARIABLES = "; : INPUT N
1230 PRINT "ENTER STARTING VARIABLES X(I):"
1240 FOR I=1 TO N
1250 PRINT "   X(";I;")=";  : INPUT X(I)
1260 NEXT I
1270 RETURN
1280 REM *********************************
1290 REM - REVISE CONTROL PARAMETERS
1300 PRINT "MAXIMUM # OF ITERATIONS (DEFAULT=50):"; : INPUT S4$
1310 M1=50 : IF S4$<>"" THEN M1=VAL(S4$)
1320 PRINT "STOPPING CRITERION (DEFAULT=.0001):"; : INPUT S4$
1330 E1=.0001 : IF S4$<>"" THEN E1=VAL(S4$)
1340 PRINT "PRINT EVERY Ith ITERATION (DEFAULT=1):"; : INPUT S4$
1350 I7=1 : IF S4$<>"" THEN I7=VAL(S4$)
1360 RETURN
1370 REM *********************************
1380 REM - NORMAL STOP
1390 KEY ON : PRINT "END OF RUN" : END
1400 REM *********************************
1410 REM - MAIN OPTIMIZATION ALGORITHM - SEE CHAPTER FOUR
1420 IF N>0 THEN GOTO 1450
1430 PRINT "----- NUMBER OF VARIABLES N NOT SET; USE COMMAND #1 -----"
1440 RETURN
1450 PRINT "EXPONENT P (2,4,6,8, OR 10) ="; : INPUT K7
1460 FOR I=1 TO M : REM - NULL A(I,J) JACOBIAN MATRIX
1470 FOR J=1 TO N
1480 A(I,J)=0
1490 NEXT J
1500 NEXT I
1510 GOSUB 5000 : REM - FIRST CALC OF RESIDUALS
1520 PRINT "USER SET NUMBER OF SAMPLES M =";M;" IN SUBROUTINE 5000"
1530 PRINT "          IS THIS CONSISTENT WITH THIS PROBLEM (Y/N)";
1540 INPUT S4$ : IF S4$="N" THEN RETURN
1550 K2=K2+1 : REM - INCRE F EVAL COUNT
1560 IF L1=0 THEN GOTO 1580 : REM - NO FAILURE IN RESIDUALS SUBROUTINE
1570 PRINT "SUBROUTINE 5000 UNABLE TO COMPUTE" : RETURN
```

```
1580 F=0 : REM - CALC FIRST SUM Pth RESIDUALS
1590 FOR K=1 TO M : F=F+R(K)^K7 : NEXT K : F=F/K7
1600 GOSUB 7000 : REM - CALC FIRST JACOBIAN
1610 GOSUB 2600 : REM - CALC/STORE NORMAL MATRIX IN H()
1620 REM - PUT NORMALIZED SCALING FACTORS INTO D()
1630 L=0 : D2=0
1640 FOR J=1 TO N
1650 FOR I=1 TO N
1660 IF I<J THEN GOTO 1690
1670 L=L+1
1680 IF I=J THEN D(J)=H(L)
1690 NEXT I
1700 IF D(J)<=0 THEN D(J)=1
1710 D2=D2+D(J)*D(J)
1720 NEXT J
1730 D2=SQR(D2)
1740 FOR J=1 TO N : D(J)=D(J)/D2 : NEXT J : REM - NORMALIZE
1750 GOSUB 2750 : REM - CALC GRADIENT G() AND LENGTH G1
1760 REM - COMPARE GRADIENT TO FINITE DIFFERENCES
1770 PRINT "GRADIENT     VIA SUB7000        VIA DIFFERENCES"
1780 FOR J1=1 TO N
1790 D5=.0001#*ABS(X(J1)) : IF D5<.000001# THEN D5=.000001#
1800 X(J1)=X(J1)+D5 : GOSUB 5000 : REM - PERTURBED RESIDUALS
1810 F2=0 : FOR I=1 TO M : F2=F2+R(I)^K7 : NEXT I : F2=F2/K7
1820 PRINT USING "      ######.########";G(J1),(F2-F)/D5
1830 X(J1)=X(J1)-D5
1840 NEXT J1
1850 PRINT "PRESS <RETURN> KEY TO CONTINUE -- READY"; : INPUT S4$
1860 L7=0 : REM - INIT ITERATION COUNT
1870 REM***************************************
1880 REM - RE-ENTRY POINT FOR NEW ITERATION
1890 L7=L7+1 : F1=F : REM - INCRE ITER COUNT & SAVE LAST F VALUE
1900 IF M3=0 THEN V=V/10 : REM - LAST STEP WAS A GOOD ONE SO REDUCE V
1910 IF V<1D-20        THEN V=1D-20       : REM - V=0 NOT ALLOWED
1920 IF M3<>0 THEN V=10*V : REM - LAST STEP REQ'D CUTBACK, SO INCREASE V
1930 M3=0 : REM - CLEAR CUTBACK COUNTER
1940 IF L7=1 THEN GOTO 1980 : REM - ELSE CALC GRADIENT
1950 GOSUB 7000 : REM - GET JACOBIAN
1960 GOSUB 2750 : REM - CALC GRADIENT
1970 GOSUB 2600 : REM - CALC/STORE NORMAL MATRIX INTO H()
1980 IF ((L7-1) MOD I7)=0 THEN GOSUB 2500 : REM - RPT F,X, & G
1990 GOSUB 2870 : REM - ADD LM PARAM TO H()
2000 GOSUB 3110 : REM - FACTOR (H+vD)=LDLT IN SITU IN H()
2010 IF N5=0 THEN GOTO 2060 : REM - FACTORIZATION OK
2020 V=100*V : REM - INCREASE LM PARAM V
2030 GOSUB 2600 : REM - CALC/STORE NORMAL MATRIX INTO H()
2040 GOTO 1990 : REM - REVISE NORMAL MATRIX AND RE-FACTOR
2050 REM - SET RIGHTHAND SIDE = -G()
2060 FOR I=1 TO N : E(I)=-G(I) : NEXT I
2070 GOSUB 3390 : REM - CALC STEP dx IN E()
2080 REM - CALC STEP-TO-GRADIENT DEGREES
2090 C2=0 : C3=0
2100 FOR I=1 TO N : C2=C2+G(I)*E(I) : C3=C3+E(I)*E(I) : NEXT I
2110 P1=-C2/G1/SQR(C3) : IF P1<1 THEN GOTO 2130
2120 P1=0 : GOTO 2140 : REM - AVOID /0 IN ACS
2130 P1=57.29578*FNACS(P1)
2140 PRINT "              STEP-TO-GRADIENT DEGREES=";
2150 PRINT USING " ##.####";P1
2160 REM - TAKE STEP WITH INCREMENT IN E()
2170 FOR I=1 TO N : X(I)=X(I)+E(I) : NEXT I
2180 GOSUB 5000 : REM - CALC RESIDUALS
2190 K2=K2+1 : F=0 : REM - INCRE F EVAL COUNT & CALC SUM Pth RESIDUALS
2200 FOR K=1 TO M : F=F+R(K)^K7 : NEXT K : F=F/K7
2210 IF F<F1 THEN GOTO 2270
2220 REM - GET BACK TO LAST TURNING POINT & CUTBACK dx
2230 FOR I=1 TO N : X(I)=X(I)-E(I) : E(I)=E(I)/4 : NEXT I
2240 PRINT "  ###### CUT BACK STEP SIZE BY FACTOR OF 4 ######"
2250 M3=M3+1 : IF M3<11 THEN GOTO 2170 : REM - TRY CUTBACK STEP
```

```
2260 PRINT "STEP SIZE TOO SMALL - TERMINATED" : GOTO 2390
2270 IF L7<M1 THEN GOTO 2320 : REM - NOT AT MAX ITERATIONS
2280 PRINT "!!!!!!!!!!!!!!!!!!!!!!!!!!!!!!!!!"
2290 PRINT "STOPPED AT GIVEN LIMIT OF";M1;" ITERATIONS; RESULTS ARE:"
2300 L7=L7+1 : GOTO 2370
2310 REM - TEST CONVERGENCE OF BOTH F AND EACH X(I)
2320 IF ABS(F1-F)/(1+ABS(F1))>E1 THEN GOTO 1890
2330 FOR I=1 TO N
2340 IF ABS(E(I))/(1+ABS(X(I)))>E1 THEN GOTO 1890
2350 NEXT I
2360 L7=L7+1 : PRINT "CONVERGED; SOLUTION IS:"
2370 GOSUB 7000 : REM - GET JACOBIAN
2380 GOSUB 2750 : REM - GET GRADIANT
2390 GOSUB 2500 : REM - REPORT F,X, & G AT STOPPING POINT
2400 PRINT "PRESS <RETURN> KEY TO CONTINUE -- READY"; : INPUT S4$
2410 K=0 : PRINT "RESIDUALS ARE:"
2420 FOR I=1 TO M : PRINT I,R(I) : REM - PRINT RESIDUALS
2430 IF I<21 OR K=1 THEN GOTO 2460
2440 PRINT "PRESS <RETURN> KEY TO CONTINUE -- READY"; : INPUT S4$
2450 K=1 : REM - DON'T PAUSE FOR 2ND HALF OF DISPLAY
2460 NEXT I
2470 PRINT "TOTAL NUMBER OF FUNCTION EVALUATIONS =";K2
2480 PRINT "EXPONENT P =";K7
2490 RETURN
2500 REM *********************************
2510 REM - PRINT FUNCTION, VARIABLES, AND GRADIENT
2520 PRINT "AT START OF ITERATION NUMBER";L7
2530 P1=F : PRINT "   FUNCTION VALUE =";P1
2540 PRINT " I          X(I)            G(I)"
2550 FOR I=1 TO N
2560 PRINT I; : PRINT USING S6$;X(I),G(I)
2570 NEXT I
2580 RETURN
2590 REM *********************************
2600 REM - CALC/STORE NORMAL MATRIX IN H()
2610 FOR I=1 TO N*(N+1)/2 : H(I)=0 : NEXT I
2620 FOR K=1 TO M
2630 L=0
2640 FOR J=1 TO N
2650 FOR I=1 TO N
2660 IF I<J THEN GOTO 2690
2670 L=L+1
2680 H(L)=H(L)+A(K,I)*A(K,J)*R(K)^(K7-2)
2690 NEXT I
2700 NEXT J
2710 NEXT K
2720 FOR L=1 TO N*(N+1)/2 : H(L)=(K7-1)*H(L) : NEXT L
2730 RETURN
2740 REM*********************************
2750 REM - CALC GRADIENT AND ITS LENGTH
2760 G1=0
2770 FOR I=1 TO N
2780 G(I)=0
2790 FOR K=1 TO M
2800 G(I)=G(I)+A(K,I)*R(K)^(K7-1)
2810 NEXT K
2820 G1=G1+G(I)*G(I)
2830 NEXT I
2840 G1=SQR(G1)
2850 RETURN
2860 REM*********************************
2870 REM - ADD LM PARAM TO NORMAL MATRIX
2880 L=0 : PRINT "                          LM PARAM V=";
2890 PRINT USING "##.#^^^^";V
2900 FOR J=1 TO N
2910 FOR I=1 TO N
2920 IF I<J THEN GOTO 2950
2930 L=L+1
```

```
2940 IF I=J THEN H(L)=H(L)+V*D(J)
2950 NEXT I
2960 NEXT J
2970 RETURN
2980 REM*******************************************
2990 REM - DISPLAY SAMPLE DATA FROM LINES 400 ...
3000 PRINT " I       INDEPENDENT     DEPENDENT"
3010 K=0
3020 FOR I=1 TO M7
3030 PRINT I; : PRINT USING " ######.######";S(I,1);S(I,2)
3040 IF I<21 OR K=1 THEN GOTO 3070
3050 PRINT "PRESS <RETURN> KEY TO CONTINUE -- READY"; : INPUT S4$
3060 K=1 : REM - DON'T PAUSE FOR 2ND HALF OF DISPLAY
3070 NEXT I
3080 RETURN
3090 REM ********************************
3100 REM - LDLT FACTORIZATION OF MATRIX IN SITU IN VECTOR H
3110 K5=1 : N5=1 : D1=1 : REM - N5=1 NOT PD OR DET=D1<1D-6
3120 FOR I=2 TO N
3130 IF H(K5)>0 THEN GOTO 3150
3140 GOTO 3340
3150 Z=H(K5) : D1=D1*H(K5)
3160 K5=K5+1
3170 I1=K5
3180 FOR J=I TO N
3190 Z5=H(K5)
3200 H(K5)=H(K5)/Z
3210 J5=K5
3220 I5=I1
3230 FOR K=I TO J
3240 J5=J5+N+1-K
3250 H(J5)=H(J5)-H(I5)*Z5
3260 I5=I5+1
3270 NEXT K
3280 K5=K5+1
3290 NEXT J
3300 NEXT I
3310 D1=D1*H(K5) : IF D1>.0000000001# THEN N5=0
3320 IF N5=1 THEN GOTO 3340
3330 RETURN
3340 PRINT "HESSIAN NOT PD OR TOO SMALL DETERMINANT =";
3350 PRINT USING "##.####^^^^";D1
3360 RETURN
3370 REM ****************************************
3380 REM - SOLUTION E=Inv(H)E FOR SEARCH STEP
3390 FOR I=2 TO N
3400 I4=I
3410 V5=E(I)
3420 FOR J=1 TO I-1
3430 V5=V5-H(I4)*E(J)
3440 I4=I4+N-J
3450 NEXT J
3460 E(I)=V5
3470 NEXT I
3480 E(N)=E(N)/H(I4)
3490 FOR K=2 TO N
3500 I=N+1-K
3510 I1=I4-K
3520 V5=E(I)/H(I1)
3530 I4=I1
3540 FOR J=I+1 TO N
3550 I1=I1+1
3560 V5=V5-H(I1)*E(J)
3570 NEXT J
3580 E(I)=V5
3590 NEXT K
3600 RETURN
```

```
8 REM - ROSENBROCK IN RESIDUAL FORM (1/2)*F - PRGM  C4-6  'ROSENPTH'
5000 REM********************************
5010 REM - ROSENBROCK BANANA FUNCTION RESIDUALS
5020 M=2 : REM - THE NUMBER OF SAMPLES - NO DATA REQUIRED IN THIS CASE
5030 L1=0 : REM - NO LOGS OR SQR INVOLVED HERE SO ALWAYS OK
5040 R(1)=10*(X(2)-X(1)*X(1))
5050 R(2)=1-X(1)
5060 RETURN
7000 REM********************************
7010 REM ROSENBROCK BANANA FUNCTION JACOBIAN
7020 A(1,1)=-20*X(1)
7030 A(2,1)=-1
7040 A(1,2)=10
7050 A(2,2)=0
7060 RETURN
7070 END
```

```
8 REM - GENERAL GAUSS QUADRATURE COEFF'S VIA LEASTP.  C4-7  'GAUSS'
5000 REM********************************
5010 REM - RESIDUALS FOR GENERAL GAUSS
5020 M=N : REM - SAME NUMBER OF SAMPLES AS VARIABLES
5025 K1=M\2 : REM - NUMBER OF PRODUCT TERMS
5030 FOR K=1 TO M
5040 R(K)=0
5060 FOR J=1 TO K1
5070 R(K)=R(K)+2*X(2*J-1)*X(2*J)^(2*(K-1))
5080 NEXT J
5090 R(K)=R(K)-2/(2*K-1)
5100 IF M=2*(M\2) THEN GOTO 5120
5105 IF K<>1 THEN GOTO 5120
5110 R(K)=R(K)+X(M)
5120 NEXT K
5130 RETURN
7000 REM********************************
7010 REM - JACOBIAN FOR GENERAL GAUSS
7015 K1=M\2
7020 FOR K=1 TO M
7040 FOR J=1 TO K1
7050 A(K,2*J-1)=2*X(2*J)^(2*(K-1))
7060 A(K,2*J)=2*(2*(K-1))*X(2*J-1)*X(2*J)^(2*(K-1)-1)
7070 NEXT J
7075 IF M=2*(M\2) THEN GOTO 7095 : REM - M IS EVEN
7080  A(K,M)=0
7085  IF K=1 THEN A(K,M)=1
7090  GOTO 7110
7095 IF K=1 THEN A(K,M)=0
7110 NEXT K
7120 RETURN
7130 END
```

```
8 REM-USE WITH LEASTP OPTIMIZER FOR SARGESON PROBLEM.  C4-8  - 'SARGESON'
325 DIM Y4(33),Y5(33) : REM-SAVE FROM SUBR5000 FOR SUBR7000-EFFICIENCY
400 DATA "SARGESON"
410 DATA 33
420 DATA 0,10,20,30,40,50,60,70,80,90,100
430 DATA 110,120,130,140,150,160,170,180,190,200
440 DATA 210,220,230,240,250,260,270,280,290,300
450 DATA 310,320
460 DATA .844,.908,.932,.936,.925,.908,.881,.850,.818,.784,.751
470 DATA .718,.685,.658,.628,.603,.580,.558,.538,.522,.506,.490,.478
480 DATA .467,.457,.448,.438,.431,.424,.420,.414,.411,.406
5000 REM********************************
5010 REM-RESIDUALS FOR SARGESON IN LOOTSMA 1972:185. N=5 VARIABLES
5020 M=33 : REM - NUMBER OF SAMPLE POINTS
5030 FOR I=1 TO M
5040 Y4(I)=EXP(-X(4)*S(I,1))
5050 Y5(I)=EXP(-X(5)*S(I,1))
```

```
5060 R(I)=X(1)+X(2)*Y4(I)+X(3)*Y5(I)-S(I,2)
5070 NEXT I
5080 L1=0 : REM - NOTHING TO BLOW UP HERE
5090 RETURN
7000 REM*****************************************
7010 REM-JACOBIAN FOR SARGESON PROBLEM-USES Y4() & Y5() FROM SUBR5000
7020 FOR I=1 TO M
7030 A(I,1)=1
7040 A(I,2)=Y4(I)
7050 A(I,3)=Y5(I)
7060 A(I,4)=-X(2)*S(I,1)*Y4(I)
7070 A(I,5)=-X(3)*S(I,1)*Y5(I)
7080 NEXT I
7090 RETURN
7100 END



8 REM - USE WITH LEASTP OPTIMIZER FOR CHEBYSHEV PROBLEM C4-9 - 'CHEBY'
400 DATA "CHEBY"
410 DATA 11
420 DATA 0,.1,.2,.3,.4,.5,.6,.7,.8,.9,1
430 DATA -1,0,0,0,0,0,0,0,0,0,0
5000 REM*****************************************
5010 REM - RESIDUALS FOR CHEBYSHEV APPROXIMATION (LEAST Pth). N=3.
5020 M=11 : REM - NUMBER OF SAMPLE POINTS
5030 FOR I=1 TO M
5040 Y=S(I,1)
5050 R(I)=-1+Y*Y*(X(1)+Y*Y*(X(2)+Y*Y*X(3))) - S(I,2)
5060 NEXT I
5070 L1=0 : REM - NOTHING TO BLOW UP HERE
5080 RETURN
7000 REM*****************************************
7010 REM - JACOBIAN FOR CHEBYSHEV APPROXIMATION
7020 FOR I=1 TO M
7030 Y=S(I,1)
7040 A(I,1)=Y*Y
7050 A(I,2)=Y*Y*A(I,1)
7060 A(I,3)=Y*Y*A(I,2)
7070 NEXT I
7080 RETURN
7090 END
```

List of Variable Names Used in Program C5-1: QNEWT

B()	F	G5()	I7	M1	T
C2	F1	G6	J	M3	T()
C3	F5	H()	J5	N	V
D2	F6	I	K	P1	V5
D3	FNACS()	I1	K2	Q	W()
D4	G()	I2	K5	S$	X()
E()	G1	I4	L	S4$	Z
E1	G3	I5	L7	S6$	Z2
					Z5

```
10 REM - QUASI-NEWTON OPTIMIZER - PROGRAM C5-1 'QNEWT'
20 OPTION BASE 1 : REM - NO SUBSCRIPT O
30 CLS : KEY OFF
40 PRINT "********* QNEWT OPTIMIZER *************" : PRINT
50 PRINT "NOTES:"
60 PRINT "1.  USE ONLY UPPER CASE LETTERS"
```

```
70 PRINT "2.  IF 'BREAK' OCCURS, RESTART WITH 'GOTO 999'"
80 PRINT "3.  USER MUST PROVIDE SUBROUTINE 5000 FOR FUNCTION VALUE"
90 PRINT "       AND SUBROUTINE 7000 FOR THE GRADIENT VECTOR."
100 PRINT "4.  ENTER DEFAULT ANSWERS TO QUESTIONS BY <RETURN>."
110 PRINT
120 REM - USE OF MAJOR VARIABLES AS FOLLOWS -
130 REM     B()     VECTOR FOR RANK-1 UPDATE OF HESSIAN MATRIX
140 REM     D3      DELTA FNCN VALUE USED WITH LINQUAD & LINCUBIC
150 REM     D4      DIFFERENCE BETWEEN CURRENT AND TURNING-POINT SLOPES
160 REM     E()     SEARCH DIRECTION VECTOR.
170 REM     F       OBJECTIVE FUNCTION VALUE USER COMPUTES IN SUB5000
180 REM     F5      OBJECTIVE FUNCTION VALUE USED INTERNALLY
190 REM     FNACS   INVERSE COSINE FUNCTION
200 REM     G()     GRADIENT OF F USER COMPUTES IN SUB7000
210 REM     G2      SLOPE AT CURRENT X POINT
220 REM     G3      SLOPE AT TURNING POINT
230 REM     G5()    GRADIENT OF F USED INTERNALLY
240 REM     H()     VECTOR STORAGE OF APPROXIMATE HESSIAN MATRIX.
250 REM     K2      COUNT OF NUMBER OF F EVALUATIONS.
260 REM     L7      ITERATION COUNTER.
270 REM     M3      NUMBER OF DATA SAMPLES READ IN FROM DATA STATEMENTS.
280 REM     N       NUMBER OF VARIABLES.
290 REM     Q       SCALAR COEFFICIENT IN RANK-1 UPDATE OF HESSIAN
300 REM     T       LINE SEARCH METRIC (VARIABLE)
310 REM     T()     WORKING VECTOR USED IN VARIOUS WAYS
320 REM     W()     WORKING VECTOR USED IN VARIOUS WAYS
330 REM     X()     VARIABLES VECTOR, DIM N.
340 DEFDBL A-H,Q-Z : REM - NOTE THAT P IS SNGL PRECISION
350 DEFINT I-N
360 S6$=" ######.########"
370 DEF FNACS(X)=1.570796-ATN(X/SQR(1-X*X)) : REM - ARC COS
380 M1=50 : E1=.0001 : I7=1 : V=.001# : REM - INIT PARAMS
390 REM - FOLLOWING DIMENSIONS ARE FOR N<=20.  THE HESSIAN VECTOR
400 REM     H() MUST BE DIMENSIONED N*(N+1)/2.
410 DIM X(20),G(20),H(210),E(20),G5(20),W(20),T(20),B(20)
420 REM - HESSIAN H(.) STORED AS AS VECTOR; SEE EQUATION (3.1.14)
430 GOTO 1150 : REM - TO MENU & SELECTION
990 REM - RE-ENTRY FOR INVALID COMMAND NUMBERS & CONTINUING
999 CLS : K2=0 : REM - INIT FUNCTION EVALUATION COUNTER
1000 PRINT "************ COMMAND MENU ************"
1010 PRINT "1. ENTER STARTING VARIABLES (AT LEAST ONCE)"
1020 PRINT "2. REVISE CONTROL PARAMETERS (OPTIONAL)"
1030 PRINT "3. START OPTIMIZATION"
1040 PRINT "4. EXIT (RESUME WITH 'GOTO 999')"
1050 PRINT "5. SPARE"
1060 PRINT"*************************************"
1070 PRINT"INPUT COMMAND NUMBER:";:INPUT S$
1080 K=LEN(S$) : IF K=0 THEN GOTO 999 : REM - AVOID <CR>
1090 K=ASC(S$)
1100 IF K<48 OR K>57 THEN GOTO 999 : REM - 1ST CHAR MUST BE 0-9
1110 K=VAL(S$)
1120 IF K=0 THEN K= 15 : REM - ALTERNATIVE DISPLAY NUMBERS
1130 IF K>20 THEN GOTO 999 : REM - CAN'T EXCEED MENU #'S
1140 ON K GOSUB 1210,1280,1410,1380, 999
1150 PRINT "PRESS <RETURN> KEY TO CONTINUE -- READY";
1160 INPUT S4$
1170 IF S4$<>"" THEN BEEP : REM - <RETURN> BEFORE NEXT CMD NUMBER
1180 GOTO 999
1190 REM***************************************
1200 REM - ENTER VARIABLES
1210 PRINT"NUMBER OF VARIABLES = "; : INPUT N
1220 PRINT "ENTER STARTING VARIABLES X(I):"
1230 FOR I=1 TO N
1240 PRINT "   X(";I;")="; : INPUT X(I)
1250 NEXT I
1260 RETURN
1270 REM ******************************
1280 REM - REVISE CONTROL PARAMETERS
1290 PRINT "MAXIMUM # OF ITERATIONS (DEFAULT=50):"; : INPUT S4$
```

```
1300 M1=50 : IF S4$<>"" THEN M1=VAL(S4$)
1310 PRINT "STOPPING CRITERION (DEFAULT=.0001):"; : INPUT S4$
1320 E1=.0001 : IF S4$<>"" THEN E1=VAL(S4$)
1330 PRINT "PRINT EVERY Ith ITERATION (DEFAULT=1):"; : INPUT S4$
1340 I7=1 : IF S4$<>"" THEN I7=VAL(S4$)
1350 RETURN
1360 REM ***********************************
1370 REM - NORMAL STOP
1380 KEY ON : PRINT "END OF RUN" : END
1390 REM ***********************************
1400 REM - MAIN OPTIMIZATION ALGORITHM - SEE CHAPTER FIVE
1410 IF N>0 THEN GOTO 1433
1420 PRINT "----- NUMBER OF VARIABLES N NOT SET; USE COMMAND #1 --"
1430 RETURN
1433 REM - INITIAL HESSIAN COULD BE MADE EXACT OR APPROXIMATE
1440 GOSUB 2860 : REM - INITIALIZE HESSIAN IN H(.)
1450 GOSUB 2980 : REM - LDLT FACTORIZATION OF HESSIAN
1460 REM - FIND & SAVE MIN POSITIVE DIAG ELEMENT IN H
1470 I2=N+1 : D2=H(1)
1480 FOR I=2 TO N
1490 IF H(I2)>=D2 THEN GOTO 1510
1500 D2=H(I2) : REM - THE MIN POSITIVE DIAGONAL ELEMENT
1510 I2=I2+N+1-I
1520 NEXT I
1530 IF D2>0 THEN GOTO 1550
1540 PRINT "STARTING HESSIAN NOT POSITIVE DEFINITE." : RETURN
1550 GOSUB 5000 : K2=K2+1 : F5=F : REM - STARTING FUNCTION VALUE
1560 D3=.1*ABS(F) : REM - PREDICT 10% FNCN REDUCTION ON ITER #1
1570 L7=0 : REM - INITIALIZE ITERATION COUNTER
1572 GOSUB 7000 : REM - CALC GRADIENT VECTOR
1574 FOR I=1 TO N : G5(I)=G(I) : NEXT I
1580 L7=L7+1 : REM - RE-ENTRY POINT FOR ITERATION LOOP
1600 IF ((L7-1) MOD I7)=0 THEN GOSUB 2770 : REM - REPORT F, X AND G
1610 FOR I=1 TO N : E(I)=-G5(I) : NEXT I : REM - RHS OF NEWTON EQUATION
1620 GOSUB 3260 : REM - SOLVE FOR SEARCH DIRECTION VECTOR
1630 GOSUB 2480 : REM - PRINT STEP-TO-GRADIENT DEGREES
1640 REM***********************************
1650 REM - BEGIN LINE SEARCH USING ONLY CUTBACKS
1660 REM - CALC SLOPE AT TURNING POINT
1670 G3=0
1680 FOR I=1 TO N
1690 G3=G3+G5(I)*E(I)
1700 T(I)=G5(I) : REM - SAVE GRADIENT AT TURNING POINT
1710 W(I)=X(I) : REM - SAVE TURNING POINT
1720 NEXT I
1730 IF G3<0 THEN GOTO 1750
1740 PRINT "POS SLOPE @ TURNING POINT STARTING ITER #";L7 : GOTO 2700
1750 M3=0 : T=1 : REM - INITIAL COUNTER & STEP METRIC
1760 REM - RE-ENTRY IN LINE SEARCH USING ONLY CUTBACKS
1770 FOR I=1 TO N : X(I)=W(I)+T*E(I) : NEXT I : REM - STEP
1780 GOSUB 5000 : K2=K2+1 : F1=F : REM - TRIAL FNCN
1790 IF F1>=F5 THEN GOTO 1810 : REM - CUT BACK STEP SIZE
1800 F5=F1 : GOTO 2610 : REM - TEST FOR TERMINATION
1810 M3=M3+1 : REM - INCREMENT CUTBACK COUNT
1820 IF M3<11 THEN GOTO 1850
1830 FOR I=1 TO N : X(I)=W(I) : NEXT I : REM - SET X AT TURNING POINT
1840 PRINT " STEP SIZE TOO SMALL - TERMINATED" : GOTO 2710
1850 T=T/4
1860 PRINT"   ###### CUT BACK STEP SIZE BY FACTOR OF 4 #####"
1870 GOTO 1770 : REM - TRY REDUCED STEP
2300 REM***********************************
2310  REM - BEGIN TWO RANK-ONE UPDATES USING BFGS FORMULA
2320 GOSUB 7000 : REM - NEW GRADIENT
2330 G6=0 : REM - CALC CURRENT SLOPE
2340 FOR I=1 TO N : REM - LAST GRADIENT WAS SAVED IN T(I)
2350 W(I)=T(I) : G5(I)=G(I) : G6=G6+G5(I)*E(I)
2360 NEXT I
2370 D3=F6-F5 : REM - NEW DIFFERENCE FOR INIT STEP CALC
2380 D4=G6-G3 : REM - CURRENT SLOPE MINUS OLD SLOPE
```

```
2390 IF D4<0 THEN GOTO 1580 : REM - START NEXT ITERATION
2400 FOR I=1 TO N : B(I)=G5(I)-W(I) : NEXT I : REM - GRADIANT DIFF
2410 Q=1/(T*D4) : REM = SCALAR MULTIPLIER IN FIRST RANK-1 UPDATE
2420 GOSUB 3520 : REM - PERFORM FIRST BFGS RANK-1 UPDATE
2430 FOR I=1 TO N : B(I)=W(I) : NEXT I : REM - VECTOR FOR 2ND UPDATE
2440 Q=1/G3 : REM - SCALAR MULTIPLIER IN 2ND RANK-1 UPDATE
2450 GOSUB 3520 : REM - PERFORM SECOND BFGS RANK-1 UPDATE
2460 GOTO 1580 : REM - START NEXT ITERATION
2470 REM******************************************
2480 REM - CALC STEP-TO-GRADIENT DEGREES
2490 C2=0 : C3=0 : G1=0
2500 FOR I=1 TO N
2510 C2=C2+G(I)*E(I) : C3=C3+E(I)*E(I) : G1=G1+G(I)*G(I)
2520 NEXT I
2525 IF C3=0 OR G3=0 THEN RETURN : REM - AVOID DIVISION BY ZERO
2530 P1=-C2/SQR(C3*G1) : IF P1<1 THEN GOTO 2550
2540 P1=0 : GOTO 2560 : REM - AVOID /0 IN ACS
2550 P1=57.29578*FNACS(P1)
2560 PRINT "                    STEP-TO-GRADIENT DEGREES=";
2570 PRINT USING " ##.####";P1
2580 RETURN
2590 REM**********************************************
2600 REM - TEST FOR TERMINATION
2610 IF L7<M1 THEN GOTO 2660 : REM - NOT AT MAX ITERATIONS
2620 PRINT "!!!!!!!!!!!!!!!!!!!!!!!!!!!!!!!!!"
2630 PRINT "STOPPED AT GIVEN LIMIT OF";M1;" ITERATIONS; RESULTS ARE:"
2640 L7=L7+1 : GOTO 2710
2650 REM - TEST CONVERGENCE OF BOTH F AND EACH X(I)
2660 IF ABS(F1-F)/(1+ABS(F1))>E1 THEN GOTO 2320
2670 FOR I=1 TO N
2680 IF ABS(T*E(I))/(1+ABS(X(I)))>E1 THEN GOTO 2320
2690 NEXT I
2700 L7=L7+1 : PRINT "CONVERGED; SOLUTION IS:"
2710 GOSUB 5000 : F5=F : REM - GET FUNCTION VALUE
2720 GOSUB 7000 : REM - GET GRADIANT
2730 GOSUB 2780 : REM - REPORT F,X, & G AT STOPPING POINT
2740 PRINT "TOTAL NUMBER OF FUNCTION EVALUATIONS =";K2
2750 RETURN
2760 REM *********************************
2770 REM - PRINT FUNCTION, VARIABLES, AND GRADIENT
2780 PRINT "AT START OF ITERATION NUMBER";L7
2790 P1=F5 : PRINT "   FUNCTION VALUE =";P1
2800 PRINT " I          X(I)              G(I)"
2810 FOR I=1 TO N
2820 PRINT I; : PRINT USING S6$;X(I),G(I)
2830 NEXT I
2840 RETURN
2850 REM *************************************
2860 REM - STORE UNIT MATRIX IN H()
2870 L=0
2880 FOR J=1 TO N
2890 FOR I=1 TO N
2900 IF I<J THEN GOTO 2940
2910 L=L+1
2920 H(L)=0
2930 IF I=J THEN H(L)=1
2940 NEXT I
2950 NEXT J
2960 RETURN
2970 REM **********************************
2980 REM - LDLT FACTORIZATION OF MATRIX IN SITU IN VECTOR H
2990 K5=1
3000 FOR I=2 TO N
3010 IF H(K5)>0 THEN GOTO 3040
3020 PRINT "                 HESSIAN MADE P.D."
3030 H(K)=.000001 : REM - FORCE POSITVE DEFINITENESS
3040 Z2=H(K5)
3050 K5=K5+1
```

```
3060 I1=K5
3070 FOR J=I TO N
3080 Z5=H(K5)
3090 H(K5)=H(K5)/Z2
3100 J5=K5
3110 I5=I1
3120 FOR K=I TO J
3130 J5=J5+N+1-K
3140 H(J5)=H(J5)-H(I5)*Z5
3150 I5=I5+1
3160 NEXT K
3170 K5=K5+1
3180 NEXT J
3190 NEXT I
3200 IF H(K5)<=0 THEN GOTO 3220
3210 RETURN
3220 PRINT "                    HESSIAN MADE P.D."
3230 H(K5)=.000001 : REM - FORCE POSITIVE DEFINITENESS
3240 RETURN
3250 REM *****************************************
3260 REM - SOLUTION E=Inv(H)E FOR SEARCH STEP
3270 FOR I=2 TO N
3280 I4=I
3290 V5=E(I)
3300 FOR J=1 TO I-1
3310 V5=V5-H(I4)*E(J)
3320 I4=I4+N-J
3330 NEXT J
3340 E(I)=V5
3350 NEXT I
3360 E(N)=E(N)/H(I4)
3370 FOR K=2 TO N
3380 I=N+1-K
3390 I1=I4-K
3400 V5=E(I)/H(I1)
3410 I4=I1
3420 FOR J=I+1 TO N
3430 I1=I1+1
3440 V5=V5-H(I1)*E(J)
3450 NEXT J
3460 E(I)=V5
3470 NEXT K
3480 RETURN
3490 REM******************************************
3500 REM - RANK 1 UPDATE OF H WITH QBBT
3510 REM - SOLN OF LV=Z FOR V BY FWD SUBSTITUTION
3520 T(1)=B(1)
3530 FOR I= 2 TO N
3540 I4=I
3550 Z=B(I)
3560 FOR J=1 TO I-1
3570 Z=Z-H(I4)*T(J)
3580 I4=I4+N-J
3590 NEXT J
3600 T(I)=Z
3610 NEXT I
3620 REM - UPDATE dii IN H DIAGONAL & FILL E()
3630 I4=1
3640 FOR I=1 TO N
3650 Z=H(I4)+Q*T(I)*T(I)
3660 IF Z<=0 THEN GOTO 3830 : REM - dii NEGATIVE
3670 H(I4)=Z
3680 E(I)=T(I)*Q/Z
3690 Q=Q-E(I)*E(I)*Z
3700 I4=I4+N+1-I
3710 NEXT I
3720 REM - UPDATE L* = LLhat
3730 I4=1
```

```
3740 FOR I=1 TO N-1
3750 I4=I4+1
3760 FOR J=I+1 TO N
3770 B(J)=B(J)-H(I4)*T(I)
3780 H(I4)=H(I4)+E(I)*B(J)
3790 I4=I4+1
3800 NEXT J
3810 NEXT I
3820 RETURN
3830 PRINT"H(";I;") IS NEGATIVE - ABORT" : RETURN
3840 REM - END OF QUASI-NEWTON MINIMIZATION MAIN PROGRAM
3850 REM *********************************
```

```
1660 REM - LINE SEARCH, NO DERIVS & QUAD INTERP - C5-2 'LINQUAD'
1670 REM - CALC SLOPE AT TURNING POINT
1680 G3=0
1690 FOR I=1 TO N
1700 G3=G3+G5(I)*E(I)
1710 T(I)=G5(I) : REM - SAVE GRADIENT AT TURNING POINT
1720 W(I)=X(I) : REM - SAVE REFERENCE POINT
1730 NEXT I
1740 IF G3<0 THEN GOTO 1760
1750 PRINT "POS SLOPE @ TURNING POINT STARTING ITER #";L7 : RETURN
1760 T=-2*D3/G3 : REM - CALC INITIAL STEP METRIC
1770 IF T>1 THEN T=1
1780 F6=F5 : T5=0 : I3=0 : REM - SAVE FNCN VALUE & INIT ACCUM
1790 REM - RE-ENTRY IN LINE SEARCH WITHOUT DERIVATIVES
1800 FOR I=1 TO N : X(I)=W(I)+T*E(I) : NEXT I : REM - STEP
1810 GOSUB 5000 : K2=K2+1 : F1=F : REM - TRIAL FNCN AT UNIT METRIC STEP
1820 IF I3>2 THEN GOTO 1840 : REM - ALLOW ONLY 3 CONSECUTIVE INTERPOLATIONS
1830 IF F1>=F5 THEN GOTO 2010
1840 F2=F5
1850 T5=T5+T : REM - ACCUMULATE STEP METRIC
1860 REM - RE-RENTRY AFTER STEP METRIC DOUBLED
1870 FOR I=1 TO N : W(I)=X(I) : NEXT I : REM - ACCEPT STEP
1880 F5=F1
1890 IF I3>=1 THEN GOTO 2180 : REM - END LINE SEARCH
1900 FOR I=1 TO N : X(I)=W(I)+T*E(I) : NEXT I : REM - STEP
1910 PRINT "                    REPEAT STEP"
1920 GOSUB 5000 : K2=K2+1 : F1=F : REM - TRIAL FNCN
1930 IF F1>=F5 THEN GOTO 2170 : REM - EXIT EXTRAPOLATION
1940 REM - SET I3=2 IF NEXT EXTRAPOLATION MIGHT BOUND MINIMUM
1950 IF ((F1+F2)>=(2*F) AND (7*F1+5*F2)>(12*F5)) THEN I3=2
1960 T5=T5+T : REM - ACCUMULATE LINE METRIC
1970 T=2*T : REM - DOUBLE STEP SIZE (EXTRAPOLATE)
1980 PRINT "                    DOUBLE STEP"
1990 GOTO 1870
2000 REM - ENTRY FOR INTERPOLATION AFTER UNIT STEP FAILED
2010 T=T/2 : REM - CUT STEP SIZE IN HALF
2020 PRINT "                    HALVE STEP"
2030 FOR I=1 TO N : X(I)=W(I)+T*E(I) : NEXT I : REM - STEP
2040 GOSUB 5000 : K2=K2+1 : F2=F : REM - FNCN EVALUATION
2050 IF F2 >= F5 THEN GOTO 2100 : REM - TO QUADRATIC INTERPOLATION
2060 T5=T5+T : REM - ACCUMULATE STEP METRIC
2070 F5=F2
2080 GOTO 2180 : REM - EXIT LINE SEARCH
2090 REM - CALCULATE QUADRATIC INTERPOLATION WITH LOWER BOUND
2100 Z=.1
2110 IF ((F1+F5)>(2*F2)) THEN Z=1+(F5-F1)/(F5+F1-2*F2)/2
2120 IF Z<.1 THEN Z=.1 : REM - LOWER BOUND
2130 PRINT "                    INTERPOLATE"
2140 T=Z*T : REM - INTERPOLATE ON STEP SIZE
2150 I3=I3+1 : REM - LIMIT TO 1 EXTRAPOLATION & 3 INTERPOLATIONS
2160 GOTO 1800 : REM - END INTERPOLATION
2170 FOR I=1 TO N : X(I)=W(I) : NEXT I : REM - MOVE BACK
2180 T=T5 : REM - FINAL LINE-SEARCH METRIC
2190 PRINT "                    STEP=";
2200 PRINT USING " ####.######";T
2210 GOTO 2610 : REM - TEST FOR TERMINATION
```

```
1660 REM - LINE SEARCH USING DERIVS & CUBIC INTERP - C5-3 'LINCUBIC'
1670 REM - CALC SLOPE AT TURNING POINT
1680 G3=0
1690 FOR I=1 TO N
1700 G3=G3+G5(I)*E(I)
1710 T(I)=G5(I) : REM - SAVE GRADIENT AT TURNING POINT
1720 W(I)=X(I) : REM - SAVE TURNING POINT
1730 NEXT I
1740 IF G3<0 THEN GOTO 1760
1750 PRINT "POS SLOPE @ TURNING POINT STARTING ITER #";L7 : RETURN
1760 T=ABS(2*D3/G3) : REM - CALC INITIAL STEP METRIC
1770 IF T>1 THEN T=1
1780 G2=G3 : REM - SAVE SLOPE
1790 F6=F5 : T5=0 : I3=0 : REM - SAVE FNCN VAL, INIT ACCUM, SET FLAG
1800 REM - RE-ENTRY IN LINE SEARCH WITH DERIVATIVES
1810 FOR I=1 TO N : X(I)=X(I)+T*E(I) : NEXT I : REM - STEP
1820 GOSUB 5000 : K2=K2+1 : F1=F : REM - TRIAL FNCN
1830 GOSUB 7000 : REM - GET GRADIENT
1840 G6=0 : REM - GET SLOPE
1850 FOR I=1 TO N : G6=G6+G(I)*E(I) : G5(I)=G(I) : NEXT I
1860 IF I3>2 THEN GOTO 2130 : REM - ALLOW ONLY 3 INTERPOLATIONS
1870 IF F1>=F5 THEN PRINT "                         FNCN FAILED"
1880 IF F1>=F5 THEN GOTO 2010
1890 IF ABS(G6/G3)<=.9 THEN GOTO 2140
1900 IF G6>0 THEN PRINT "                         SLOPE FAILED"
1910 IF G6>0 THEN GOTO 2010
1920 PRINT "                         EXTRAPOLATE"
1930 T5=T5+T
1940 Z=10
1950 IF (G2<G6) THEN Z=G6/(G2-G6) : REM - LINEAR EXTRAP ON SLOPE
1960 IF Z>10 THEN Z=10
1970 T=Z*T
1980 F5=F1
1990 G2=G6 : REM - UPDATE REFERENCE SLOPE
2000 GOTO 1800
2010 FOR I=1 TO N : X(I)=X(I)-T*E(I) : NEXT I : REM - BACK UP
2020 IF I3>2 THEN GOTO 1800 : REM - GET FNCN VALUE & EXIT
2030 PRINT "                         INTERPOLATE"
2040 Z=3*(F5-F1)/T+G6+G2 : REM - CUBIC INTERPOLATION
2050 Q=Z*Z-G2*G6
2060 IF Q>0 THEN GOTO 2080 : REM  - IF NOT, EXIT LINE SEARCH
2070 I3=3 : GOTO 1800
2080 Z3=SQR(Q)
2090 Z=1-(G6+Z3-Z)/(2*Z3+G6-G2)
2100 T=Z*T
2110 I3=I3+1 : REM - COUNT INTERPOLATIONS
2120 GOTO 1800
2130 FOR I=1 TO N : X(I)=W(I) : NEXT I : REM - SET X TO TURNING POINT
2140 T=T5+T
2150 F5=F1
2160 PRINT "                         STEP=";
2170 PRINT USING " ####.######";T
2180 GOTO 2610 : REM - TEST FOR TERMINATION




7 REM - FINITE DIFF GRADIENT FOR QNEWT - C5-4 'QNEWTGRD'
7000 REM ***********************************
7010 REM - FINITE DIFFERENCES FOR GRADIENT FOR QNEWT
7020 F9=F : REM - SAVE NOMINAL FUNCTION VALUE
7030 FOR II=1 TO N : REM - CALC POS PERTURBATIONS
7040 DX=ABS(X(II))/10000
7050 IF DX<.000001 THEN DX = .000001 : REM - IF X NEARLY ZERO
7060 X(II)=X(II)+DX
7070 GOSUB 5000 : K2=K2+1 : REM - GET PERTURBED FUNCTION VALUE
```

```
7080 G(II)=(F-F9)/DX : REM - FIRST-ORDER DIFFERENCE
7090 X(II)=X(II)-DX : REM - RESTORE X(II) VALUE
7100 NEXT II
7110 F=F9 : REM - RESTORE NOMINAL FNCN VALUE
7120 RETURN


7 REM - CAMELBACK FNCN, BRANNIN 1972 - PRGM [C5-5] 'CAMEL
5000 REM********************************************
5010 REM - CAMELBACK FUNCTION WITH PARAMETERS A-E
5020 A=-4:B=+2.1:C=-1/3:D=+4:E=-4
5030 X1=X(1)*X(1) : X2=X(2)*X(2)
5040 F=X1*(A+X1*(B+X1*C))-X(1)*X(2)+X2*(D+X2*E)
5050 RETURN
7000 REM********************************************
7010 REM - CAMELBACK GRADIENT WITH PARAMETERS A-E
7020 G(1)=X(1)*(2*A+X1*(4*B+X1*6*C))-X(2)
7030 G(2)=-X(1)+X(2)*(2*D+X2*4*E)
7040 RETURN


7 REM - ADDS TO QNEWT - SIMPLE BOUNDS ON VARIABLES - PRGM [C5-6] 'BOXMIN'
415 DIM L4(20),L5(20),P5(20,2) : REM - CONSTRAINT AND BOUNDS ARRAYS
425 GOSUB 2030 : REM - UNBOUND ALL POSSIBLE VARIABLES
1050 PRINT "5. SEE &/OR RESET LOWER/UPPER BOUNDS ON VARIABLES"
1140 ON K GOSUB 1210,1280,1410,1380,2090
1435 GOSUB 3850 : REM - RESET & RECORD BINDING VARIABLES
1576 FOR I=1 TO N : L5(I)=0 : NEXT I : REM - UNBIND ALL CONSTRAINTS
1602 REM - RELEASE NON-K-T CONSTRAINTS
1604 FOR I=1 TO N
1606 IF L5(I)*G(I)>0 THEN L5(I)=0
1608 NEXT I
1610 FOR I=1 TO N : REM - PROJECT GRADIENT INTO FIXED SUBSPACE
1612 G5(I)=G(I)*(1-ABS(L5(I)))
1614 E(I)=-G5(I) : REM - RIGHTHAND SIDE OF NEWTON LINEAR EQUATIONS
1616 NEXT I
1622 FOR I=1 TO N : REM - PROJECT SEARCH DIRECTION INTO SUBSPACE
1624 E(I)=E(I)*(1-ABS(L5(I)))
1626 NEXT I
1775 GOSUB 1880 : REM - CHECK/SET ANY ADDITIONAL BOUNDS
1875 REM********************************************
1880 REM - CHECK FOR MORE BOUNDS AND RESET STEP SIZE T IF BINDING
1900 FOR I=1 TO N
1905 IF E(I)=0 THEN GOTO 1990 : REM - TEST ONLY SUBSPACE BOUNDS
1907 L5(I)=0 : REM - CANCEL BOUNDS, THEN RETEST THEM
1910 REM - PROCESS LOWER BOUNDS
1920 IF (W(I)+T*E(I))>(P5(I,1)+P2) THEN GOTO 1960
1940 X(I)=P5(I,1)
1945 L5(I)=-1 : REM - NOW AN ACTIVE CONSTRAINT
1948 PRINT "              ACTIVATED X(";I;") LOWER BOUND"
1950 GOTO 1990 : REM - NO NEED TO PROCESS UPPER BOUNDS
1960 IF (W(I)+T*E(I))<P5(I,2) THEN GOTO 1990
1980 X(I)=P5(I,2)
1985 L5(I)=+1 : REM - NOW AN ACTIVE CONSTRAINT
1988 PRINT "              ACTIVATED X(";I;") UPPER BOUND"
1990 NEXT I
2000 RETURN
2010 REM********************************************
2020 REM - INIT FLAGS AND LOWER/UPPER BOUNDS
2030 REM - CLEAR THE 'BINDING BOUND(S)' & SET DEFAULT LIMITS
2040 FOR I=1 TO 20 : L4(I)=0 : L5(I)=0
2050 P5(I,1)=-10000 : P5(I,2)=+10000
2060 NEXT I
2070 RETURN
2080 REM********************************************
2090 REM - SEE OR RESET LOWER/UPPER BOUND ON VARIABLES
```

```
2100 S4$="NONE.   " : PRINT "BOUNDS NOW SET ARE:"
2110 PRINT " I          LOWER            UPPER"
2120 FOR I=1 TO 20
2130 IF L4(I)=0 THEN GOTO 2150
2140 S4$="" : PRINT I;"    "; : PRINT USING S6$;P5(I,1);P5(I,2)
2150 NEXT I
2160 PRINT S4$;"SET OR RESET ANY BOUNDS (Y/N)"; : INPUT S4$
2170 IF S4$<>"Y" THEN RETURN
2180 REM - RE-ENTRY FOR MORE BOUND SETTING
2190 PRINT "ENTER 0 TO RETURN TO MENU, ELSE ENTER VARIABLE # =";
2200 INPUT I : IF I=0 THEN RETURN
2210 PRINT "PRESS <RETURN> IF NO BOUND DESIRED"
2220 PRINT " LOWER BOUND ="; : INPUT S4$
2230 P5(I,1)=-10000 : IF S4$<>"" THEN P5(I,1)=VAL(S4$)
2240 IF S4$<>"" THEN L4(I)=1
2250 PRINT " UPPER BOUND ="; : INPUT S4$
2260 P5(I,2)=+10000 : IF S4$<>"" THEN P5(I,2)=VAL(S4$)
2270 IF S4$<>"" THEN L4(I)=1
2280 GOTO 2190
2350 W(I)=T(I) : G5(I)=G(I)*(1-ABS(L5(I))) : G6=G6+G5(I)*E(I)
3840 REM************************************
3850 REM - RESET & RECORD BINDING VARIABLES
3860 FOR I=1 TO N
3870 REM - PROCESS LOWER BOUNDS
3880 IF X(I)>P5(I,1) THEN GOTO 3920
3890 X(I)=P5(I,1) : L5(I)=-1
3900 PRINT "SET X(";I;")=";X(I);" (LOWER BOUND)"
3910 GOTO 3950 : REM - NO NEED TO PROCESS UPPER BOUNDS
3920 IF X(I)<P5(I,2) THEN GOTO 3950
3930 X(I)=P5(I,2) : L5(I)=+1
3940 PRINT "SET X(";I;")=";X(I);" (UPPER BOUND)"
3950 NEXT I
3960 RETURN
3970 REM - END OF QUASI-NEWTON OPTIMIZER WITH BOXMIN CONSTRAINTS


7 REM - FNCN FOR QNEWT+BOXMIN+QNEWTGRD - PRGRM  C5-7   'PAV17'
5000 REM - FNCN SUBROUTINE FROM QNEWT - FROM HIMMELBLAU P.416
5010 F=0 : A3=1 : REM - INITIALIZE
5020 FOR I9=1 TO 10
5030 A1=LOG(X(I9)-2) : A2=LOG(10-X(I9)) : A3=A3*X(I9)
5040 F=F+A1*A1+A2*A2
5050 NEXT I9
5060 F=F-A3^(.2)
5070 RETURN
7000 REM *************************************
7010 REM - GRADIENT FOR PAV17 FUNCTION
7020 FOR I9=1 TO N
7030 G(I9)=2*(LOG(X(I9)-2)/(X(I9)-2)-LOG(10-X(I9))/(10-X(I9)))
          -.2*A3^(.2)/X(I9)
7040 NEXT I9
7050 RETURN


7 REM - ADDS TO QNEWT WITH BOXMIN - MULTIPLIER PENALTIES - C5-8 'MULTPEN
40 PRINT "**********QNEWT WITH BOXMIN AND MULTPEN************"
80 PRINT "3.  USER MUST PROVIDE SUBROUTINE 5500 FOR OBJECTIVE FNCN"
90 PRINT "          AND SUBROUTINE 7500 FOR ITS GRADIENT VECTOR."
105 PRINT "5.  USER MUST SUPPLY SUBROUTINE 8000 FOR CONSTRAINT FNCNS"
107 PRINT "          AND SUBROUTINE 9000 FOR CONSTRAINTS GRADIENTS"
```

```
417 DIM C(30),A(20,30),U(30),U9(30),S(30) : REM - MAX OF 30 CONSTRAINTS
1030 PRINT "3. START BOUNDED OPTIMIZATION"
1052 PRINT "6. START CONSTRAINED OPTIMIZATION"
1140 ON K GOSUB 1210, 1280, 1410, 1380, 2090, 4000
1573 IF L8=1 THEN GOSUB 4900 : REM - CHECK USER'S GRADIENT BY DIFF'G
4000 REM*********************************
4010 REM - START MULTIPLIER PENALTY FUNCTION METHOD
4020 PRINT "*******************************************************"
4040 M=0 : REM - INIT TOTAL NUMBER OF CONSTRAINTS TO CHECK USER
4045 C6=1E+20 : K7=0 : REM - CONSTR CONVERGENCE CONSTANTS
4050 FOR I=1 TO 30
4055 U(I)=0 : REM - INIT CONSTRAINT RESIDUAL OFFSET
4057 S(I)=1 : REM - FOR CALCS AT LINE 4092
4060 FOR J=1 TO N
4070 A(J,I)=0 : REM - INIT CONSTRAINT JACOBIAN
4080 NEXT J
4090 NEXT I
4092 GOSUB 4700 : REM - SET INITIAL CONSTR RESIDUAL WEIGHTS
4095 L8=1 : REM - INIT PENALTY LOOP COUNT
4100 GOSUB 1400 : REM - MIN F(X) FROM COLD START
4110 REM - RE-ENTRY FOR OUTER PENALTY LOOP
4120 PRINT "*******************************************************"
4122 BEEP
4125 GOSUB 4800 : REM - FIND MAX PNLTY MODULUS & ESTI LAGR MULTIPLIERS
4130 PRINT "AFTER ";L8;" PENALTY MINIMIZATIONS,"
4140 PRINT "  THE MAX CONSTRAINT MODULUS #";K8;" =";P8
4150 PRINT "CONTINUE PENALTY MINIMIZATIONS (Y/N)"; : INPUT S4$
4160 IF S4$="N" THEN RETURN
4170 L8=L8+1 : REM - INCREMENT PENALTY LOOP COUNT
4180 REM - POWELL'S PARAMETERS ADJUST SCHEME
4190 C7=C6 : C6=P8 : IF C6>=C7 THEN GOTO 4310
4200 IF K7=1 THEN GOTO 4460
4210 FOR I=1 TO M : REM - ADJUST ALL OFFSETS
4220 U9(I)=U(I) : REM - SAVE OFFSETS
4230 C8=C(I)
4240 IF I<=K1 THEN GOTO 4260 : REM - EQUALITY CONSTR CASE
4250 IF U(I)<C(I) THEN C8=U(I) : REM - CHOOSE MIN
4260 U(I)=U(I)-C8 : REM - NEW CONSTR RESIDUAL OFFSET ESTIMATE
4270 NEXT I
4280 K7=1 : REM - JUST RESET ALL OFFSETS
4290 GOSUB 1460 : REM - MIN F(X) START'G WITH CURRENT HESSIAN
4300 GOTO 4120 : REM - CLOSE OUTER PENALTY LOOP
4310 REM - DIVERGING CASE
4320 C6=C7 : REM - USE PRIOR MAX C() NORM
4330 IF K7=0 THEN GOTO 4350
4340 FOR I=1 TO M : U(I)=U9(I) : NEXT I : REM - USE PRIOR OFFSETS
4350 REM - SELECTIVELY INCREASE WEIGHTS ON CONSTR RESIDUALS
4360 FOR I=1 TO M
4370 C9=C(I)
4380 IF I<=K1 THEN GOTO 4410 : REM - EQUALITY CONSTR
4390 IF C9<0 THEN GOTO 4410 : REM - BIND'G INEQUALITY CONSTR
4400 GOTO 4430 : REM - UNBINDING INEQUALITY CONSTRAINT
4410 IF ABS(C9)<(C7/4) THEN GOTO 4430 : REM - WEIGHT IS OK
4420 S(I)=S(I)*10 : U(I)=U(I)/10 : REM - FORCE CONVERGENCE
4430 NEXT I
4440 K7=0 : GOTO 4290 : REM - TO START OF PENALTY LOOP
4450 REM - TEST FOR MIN CONVERGENCE RATE
4460 IF C6>(C7/4) THEN GOTO 4350 : REM - FORCE GREATER CONVERGENCE RATE
4470 GOTO 4210  : REM - IS OK - ADJUST ALL OFFSETS
4700 REM****************************
4705 REM - INIT PENALTY WEIGHTS S()
4710 GOSUB 5000 :  REM - CALC F(X) WITH S(I)=1
4715 REM - C8=SUM C(I)^2 FROM SUB5000
4720 REM - F9= UNCONSTR'D OBJECTIVE FNCN VALUE
4725 REM - U(I)=0 NOW
4730 REM - CHANGE WEIGHTS ON BINDING CONSTRAINTS - LEAVE REST =1
4735 K9=0
4740 FOR I= K1+1 TO M : REM - TEST FOR # BINDING INEQUALITIES
```

```
4745 IF C(I)<0 THEN K9=K9+1
4750 NEXT I
4755 IF (K1+K9)=0 THEN RETURN : REM - NONE BINDING SO S()=1
4757 C8=2*ABS(F9)/(K1+K9) : REM - AVERAGE ALLOWED EACH C(I)^2 TERM
4760 FOR I=1 TO M
4770 IF I<=K1 THEN GOTO 4785 : REM - IS AN EQUALITY CONSTRAINT
4775 IF C(I)>0 THEN GOTO 4790 : REM - UNBIND'G INEQUALITY CONSTR
4785 S(I)=C8/(C(I)*C(I)+.001#) : IF S(I)>1000 THEN S(I)=1000
4790 NEXT I
4795 RETURN
4800 REM**********************************
4810 REM - FIND MAX CONSTR RESID MAGNITUDE P8=ABS(C(K8))
4820 P8=0 : K8=0 : REM - INIT
4825 PRINT "ESTIMATED LAGRANGE MULTIPLIERS -"
4830 FOR I=1 TO M
4840 C9=C(I)
4850 IF I<=K1 THEN GOTO 4870 : REM - IS EQUALITY CONSTRAINT
4860 IF C9>0 THEN C9=0
4870 C9=ABS(C9)
4880 IF C9<=P8 THEN GOTO 4887
4885 P8=C9 : K8=I : REM - NEW MAX MODULUS
4887 PRINT "   CONSTRAINT #";I;":";
4888 PRINT USING "#####.#####";U(I)*S(I)
4890 NEXT I
4895 RETURN
4900 REM*******************************************
4910 REM - COMPARE USER'S GRADIENT WITH FINITE DIFFERENCES
4920 PRINT "GRADIENT    VIA SUB9000        VIA DIFFERENCES"
4930 FOR J1=1 TO N
4940 D5=.0001#*ABS(X(J1)) : IF D5<.000001# THEN D5=.000001#
4950 X(J1)=X(J1)+D5 : GOSUB 5000 : REM - PERTURBED FUNCTION
4960 PRINT USING "     #####.#########";G(J1),(F-F5)/D5
4970 X(J1)=X(J1)-D5 : REM - RESTORE NOMINAL X(J1)
4980 NEXT J1
4982 F = F5 : REM - RESTORE NOMINAL FNCN
4985 PRINT "PRESS <RETURN> KEY TO CONTINUE -- READY"; : INPUT S4$
4990 RETURN
5000 REM*****************************
5010 REM - STND MULTIPLIER PENALTY FUNCTION
5020 C8=0 : REM - INIT SUM OF PENALTIES
5030 GOSUB 8000 : REM - CALC CONSTRAINTS C()
5040 IF M>0 THEN GOTO 5050
5045 PRINT "WARNING - USER FAILED TO ASSIGN M & K1 VALUES IN SUB8000!"
5050 FOR I9=1 TO M
5060 C9=C(I9)-U(I9) : REM - OFFSET CONSTRAINT RESIDUAL
5070 IF I9<=K1 THEN GOTO 5100 : REM - IS EQUALITY CONSTRAINT
5080 IF C9<0 THEN GOTO 5100 : REM - PENALIZE
5090 GOTO 5110
5100 C8=C8+S(I9)*C9*C9 : REM - ACCUMULATE PENALTIES
5110 NEXT I9
5120 GOSUB 5500 : REM - CALC UNCONSTR'D OBJECTIVE FNCN
5125 F9=F : REM - SAVE OBJECTIVE VALUE TO INIT S() IN SUB4700
5130 F=F+C8/2
5140 RETURN
7000 REM*****************************
7010 REM - STND MULT PENALTY GRADIENT
7020 GOSUB 7500 : REM - CALC UNCONSTR'D F GRADIENT
7030 GOSUB 9000 : REM - CALC CONSTRAINTS GRADIENTS
7040 FOR J9=1 TO N : REM - VARIABLES LOOP
7050 G9=0
7060 FOR I9=1 TO M : REM - CONSTRAINTS LOOP
7062 C9=C(I9)-U(I9) : REM - OFFSET CONSTR RESIDUAL
7064 IF I9<=K1 THEN GOTO 7070 : REM - IS EQUALITY CONSTR
7066 IF C9<0 THEN GOTO 7070 : REM - IS BIND'G INEQUALITY CONSTR
7068 GOTO 7080
7070 G9=G9+S(I9)*C9*A(J9,I9)
7080 NEXT I9
7090 G(J9)=G(J9)+G9
```

```
7095 NEXT J9
7100 RETURN




7 REM - HIMMELBLAU P360 OBJ&CONSTR F&GRADS - [C5-9] 'HIM360'
5500 REM******************************
5510 REM - HIMMELBLAU P.360 OBJECTIVE FUNCTION
5520 F=4*X(1)-X(2)*X(2)-12
5530 RETURN
7500 REM*******************************
7510 REM - HIMMELBLAU P.360 GRADIENT OF OBJECTIVE FNCN
7520 G(1)=4 : G(2)=-2*X(2)
7530 RETURN
8000 REM*******************************
8010 REM - HIMMELBLAU P.360 CONSTRAINT FUNCTIONS C()
8020 REM - K=# OF EQUALITY CONSTR'S (K<=N)
8030 REM - M=TOTAL # OF ALL CONSTRAINTS
8040 REM - DEFINE EQUALITY CONSTR'S FIRST
8060 K=1 : M=2 : REM - USER MUST SET THESE TWO VALUES HERE
8070 C(1)=25-X(1)*X(1)-X(2)*X(2)
8080 C(2)=10*X(1)-X(1)*X(1)+10*X(2)-X(2)*X(2)-34
8090 RETURN
9000 REM*******************************
9010 REM - HIMMELBLAU P.360 CONSTRAINTS GRADIENTS (JACOBIAN)
9020 REM -  COL J OF A(J,I) IS GRADIENT VECTOR OF C(I)
9025 REM - MAIN PGRM HAS SET ALL A(,)=0
9030 A(1,1)=-2*X(1)
9040 A(2,1)=-2*X(2)
9050 A(1,2)=10-2*X(1)
9060 A(2,2)=10-2*X(2)
9070 RETURN




5 REM 8510071334. COPYRIGHT T.R. CUTHBERT 1985.
7 REM ROSEN SUZUKI, LOOTSMA BOOK.  [C5-10] 'LOOT356'
5500 REM******************************
5510 REM ROSEN-SUZUKI OBJECTIVE FUNCTION
5520 X1=X(1)*X(1) : X2=X(2)*X(2) : X3=X(3)*X(3) : X4=X(4)*X(4)
5530 F=X1+X2+2*X3+X4-5*X(1)-5*X(2)-21*X(3)+7*X(4)
5540 RETURN
7500 REM******************************
7510 REM ROSEN-SUZUKI OBJECTIVE FUNCTION GRADIENT
7520 G(1)=2*X(1)-5
7530 G(2)=2*X(2)-5
7540 G(3)=4*X(3)-21
7550 G(4)=2*X(4)+7
7560 RETURN
8000 REM******************************
8010 REM ROSEN-SUZUKI CONSTRAINT FUNCTIONS
8015 M=3 : K1=0
8017 X1=X(1)*X(1) : X2=X(2)*X(2) : X3=X(3)*X(3) : X4=X(4)*X(4)
8020 C(1)=-X1-X2-X3-X4-X(1)+X(2)-X(3)+X(4)+8
8030 C(2)=-X1-2*X2-X3-2*X4+X(1)+X(4)+10
8040 C(3)=-2*X1-X2-X3-2*X(1)+X(2)+X(4)+5
8050 RETURN
9000 REM******************************
9010 REM ROSEN-SUZUKI CONSTRAINTS GRADIENTS (JACOBIAN)
9030 A(1,1)=-2*X(1)-1:A(2,1)=-2*X(2)+1:A(3,1)=-2*X(3)-1:A(4,1)=-2*X(4)+1
9040 A(1,2)=-2*X(1)+1:A(2,2)=-4*X(2):A(3,2)=-2*X(3):A(4,2)=-4*X(4)+1
9050 A(1,3)=-4*X(1)-2:A(2,3)=-2*X(2)+1:A(3,3)=-2*X(3):A(4,3)=1
9060 RETURN
```

Basic Variable Names Used in Program C6-1, TWEAKNET

A( )	D5	J2	M	R5	U1
A4	D6	J5	M$( )	R6	U2
A5	D9	J9	M( )	S$	U3
B( )	E( )	K	M1	S( )	U9( )
B4	E1	K1	M2	S1( )	V
B5	F	K2	M3	S2$	V5
C( )	F1	K3	N	S3$	W
C2	G( )	K4	N$	S4	W( )
C3	G1	K5	N$( )	S4$	X( )
C4	G9	K6	N1	S5	X4
C5	H	K7	N2	S5$	X5
C6	H( )	K8	N3( )	S6$	Z
C7	I	K9	N5	S7$	Z( )
C8	I1	L	P1	S8$	Z5
C9	I2	L1	P5( )	T	
D( )	I4	L4( )	P8	T1	
D1	I5	L5( )	Q	T2	
D2	I7	L6( )	Q( )	T3	
D3	I9	L7	R( )	T4	
D4	J	L8	R4	U( )	

```
7 REM - GAUSS-NEWTON WITH BNDD VARS AND CNSTRS -| PGM C6-1 | 'TWEAKNET'
10 REM - LADDER NETWORK OPTIMIZER BASE PROGRAM 'TWEAKNET'
20 OPTION BASE 1 : REM - NO SUBSCRIPT 0
30 CLS : KEY OFF : M=0 : REM - DEFAULT NUMBER OF SAMPLE DATA
40 PRINT "********* NETWORK OPTIMIZER ************" : PRINT
50 PRINT "NOTES:"
60 PRINT "1.   USE ONLY UPPER CASE LETTERS"
70 PRINT "2.   IF 'BREAK' OCCURS, RESTART WITH 'GOTO 999'"
80 PRINT "3.   USER MUST PROVIDE SAMPLE DATA AND UNITS (FREQ,"
83 PRINT "          L,C) AND TOPOLOGY DATA IN LINES 400-889,"
85 PRINT "          OR RECALL THAT FROM DISK FILE USING CMD 10."
90 PRINT "          AT LEAST CMD 1 MUST BE USED TO SET VARIABLES."
100 PRINT "4.  ENTER DEFAULT ANSWERS TO QUESTIONS BY <RETURN>."
110 PRINT
130 REM - USE OF MAJOR VARIABLES AS FOLLOWS -
140 REM      A(,)   JACOBIAN MATRIX.  A(K,J) IS DERIV OF Kth RESIDUAL
142 REM             WITH RESPECT TO Jth VARIABLE. IS DIM MxN.
144 REM      D()    VECTOR FOR LM DIAGONAL SCALING MATRIX.
146 REM      D1     DETERMINANT OF LDLT FACTORIZATION.
148 REM      E()    SEARCH STEP VECTOR.
150 REM      F      HALF THE SUM OF Pth POWER RESIDUALS.
152 REM      F1     SAVED VALUE OF F FOR DOWNHILL COMPARISON.
154 REM      FNACS  INVERSE COSINE FUNCTION
156 REM      G()    GRADIENT OF F.
158 REM      G1     LENGTH OF GRADIENT.
160 REM      H()    VECTOR STORAGE OF APPROXIMATE HESSIAN MATRIX.
162 REM      K2     COUNT OF NUMBER OF F EVALUATIONS.
164 REM      K7     EXPONENT P - POWER TO WHICH RESIDUALS RAISED.
166 REM      L7     ITERATION COUNTER.
168 REM      M      NUMBER OF DATA SAMPLES.
170 REM      M3     NUMBER OF DATA SAMPLES READ IN FROM DATA STATEMENTS.
```

```
172 REM    N      NUMBNER OF VARIABLES.
174 REM    R()    RESIDUALS, DIM M.
176 REM    S(,)   SAMPLES S(I,1) IS INDEPENDENT & S(I,2) IS DEPENDENT.
178 REM    V      LEVENBERG-MARQUARDT (LM) PARAMETER.
180 REM    X()    VARIABLES VECTOR, DIM N.
240 DEFDBL A-H,Q-R,T-Z : REM - NOTE THAT P IS SNGL PRECISION
250 DEFINT I-N
270 S5$=" ####.####" : S6$=" ######.########" : S7$=" ##.###^^^^"
280 DEF FNACS(X)=1.570796-ATN(X/SQR(1-X*X)) : REM - ARC COS
290 M1=50 : E1=.0001 : I7=1 : K7=2 : M3=0 : V=.001# : REM - INIT PARAMS
300 REM - FOLLOWING DIMENSIONS ARE FOR N<=20.  THE HESSIAN VECTOR
310 REM    H() MUST BE DIMENSIONED N*(N+1)/2.  # SAMPLES M<=40.
320 DIM X(20),G(20),H(210),E(20),A(40,20),S(40,2),D(20),R(40)
323 DIM L4(20),L5(20),P5(20,2),W(20) : REM - CONSTR, BND, & SAVE ARRAYS
325 DIM C(40),L6(40),U(40),U9(40),S1(40) : REM - MAX OF 40 CONSTRAINTS
327 DIM M(35),M$(35),Q(35),N$(35) : REM - MAX 20 NON-NULL BRANCHES
328 DIM N3(20,2),Z(40,2),B(40,20) : REM - STORAGE PER TELLEGEN DERIVS
330 REM - HESSIAN H(.) STORED AS AS VECTOR; SEE EQUATION (4.1.14)
335 GOSUB 3800 : REM - UNBOUND ALL POSSIBLE VARIABLES
337 FOR I=1 TO 35 : M(I)=9 : NEXT I : REM - TOPOLOGY 'END' TYPE
340 READ N$ : PRINT "WORKING WITH DATA SET ";N$ : PRINT
350 READ M : REM - M IS NUMBER OF SAMPLES
355 IF M=0 THEN GOTO 1160 : REM - NO SAMPLES & TOPOLOGY FROM DATA STMNTS
360 FOR K=1 TO M : READ S(K,1) : NEXT K
365 FOR K=1 TO M : READ S(K,2) : NEXT K
370 READ U1,U2,U3 : REM-FREQ,L,C UNITS
371 READ R6,R4,X4,K9 : REM - R SOURCE, R LOAD, X LOAD, # TOPOL LINES
372 FOR I=1 TO K9 : REM - READ TOPOLOGY LINES
373 READ J,M$(I),Q(I),N$(I)
374 IF M$(I)="N" THEN M(I)=0 : REM - NULL BRANCH
375 IF M$(I)="R" THEN M(I)=1 : REM - RESISTOR
376 IF M$(I)="L" THEN M(I)=2 : REM - INDUCTOR
377 IF M$(I)="C" THEN M(I)=3 : REM - CAPACITOR
378 IF M$(I)="LC" THEN M(I)=4 : REM - SERIES LC IN SHUNT, OR DUAL
379 NEXT I
380 GOTO 1160 : REM - TO MENU & SELECTION
385 REM - SAMPLE DATA WILL BE EMPLOYED IN SUBROUTINE 5000 & OTHERS
390 REM - ENTER DATA STATEMENTS IN LINES 400-888 FOR SAMPLE PAIRS
395 REM -    AND FREQ,L,C UNITS & TOPOLOGY
400 DATA "DUMMY"
410 DATA 0 : REM - M=0. PROVIDE THESE LAST TWO LINES IF NO DATA FOLLOWS
990 REM - RE-ENTRY FOR INVALID COMMAND NUMBERS & CONTINUING
999 CLS : K2=0 : REM - INIT FUNCTION EVALUATION COUNTER
1000 PRINT "************* COMMAND MENU ************"
1010 PRINT "1. ENTER STARTING VARIABLES (AT LEAST ONCE)"
1020 PRINT "2. REVISE CONTROL PARAMETERS (OPTIONAL)"
1030 PRINT "3. START OPTIMIZATION"
1040 PRINT "4. EXIT (RESUME WITH 'GOTO 999')"
1050 PRINT "5. SEE &/OR RESET LOWER/UPPER BOUNDS ON VARIABLES"
1060 PRINT "6. DISPLAY DATA PAIRS"
1062 PRINT "7. SEE &/OR RESET CONSTRAINT SAMPLE NUMBER(S)"
1064 PRINT "8. SEE FREQUENCY, L, & C UNITS & NETWORK TOPOLOGY"
1066 PRINT "9. SEE NETWORK RESPONSES FOR ALL SAMPLES"
1068 PRINT "10. RECALL SAMPLE, UNITS, & TOPOLOGY DATA FROM DISK"
1070 PRINT"************************************"
1080 PRINT"INPUT COMMAND NUMBER:";:INPUT S$
1090 K=LEN(S$) : IF K=0 THEN GOTO 999 : REM - AVOID <CR>
1100 K=ASC(S$)
1110 IF K<48 OR K>57 THEN GOTO 999 : REM - 1ST CHAR MUST BE 0-9
1120 K=VAL(S$)
1130 IF K=0 THEN K= 15 : REM - ALTERNATIVE DISPLAY NUMBERS
1140 IF K>20 THEN GOTO 999 : REM - CAN'T EXCEED MENU #'S
1150 ON K GOSUB 1220,1290,4200,1390,3860,3000,4830,5310,2442,9800
1160 PRINT "PRESS <RETURN> KEY TO CONTINUE -- READY";
1170 INPUT S4$
1180 IF S4$<>"" THEN BEEP : REM - <RETURN> BEFORE NEXT CMD NUMBER
1190 GOTO 999
1200 REM*****************************************
```

```
1210 REM - ENTER VARIABLES
1220 PRINT"NUMBER OF VARIABLES = "; : INPUT N
1230 PRINT "ENTER STARTING VARIABLES X(I):"
1240 FOR I=1 TO N
1250 PRINT "   X(";I;")="; : INPUT X(I)
1260 NEXT I
1270 RETURN
1280 REM *******************************
1290 REM - REVISE CONTROL PARAMETERS
1292 PRINT "EXPONENT P (2,4,6,8, OR 10) ="; : INPUT S4$
1294 K7=2 : IF S4$<>"" THEN K7=VAL(S4$)
1300 PRINT "MAXIMUM # OF ITERATIONS (DEFAULT=50):"; : INPUT S4$
1310 M1=50 : IF S4$<>"" THEN M1=VAL(S4$)
1320 PRINT "STOPPING CRITERION (DEFAULT=.0001):"; : INPUT S4$
1330 E1=.0001 : IF S4$<>"" THEN E1=VAL(S4$)
1340 PRINT "PRINT EVERY Ith ITERATION (DEFAULT=1):"; : INPUT S4$
1350 I7=1 : IF S4$<>"" THEN I7=VAL(S4$)
1360 RETURN
1370 REM ***********************************
1380 REM - NORMAL STOP
1390 KEY ON : PRINT "END OF RUN" : END
1400 REM ***********************************
1410 REM - MAIN LEASTP OPTIMIZATION ALGORITHM - SEE CHAPTERS 4 & 5
1420 IF N>0 THEN GOTO 1445
1430 PRINT "----- NUMBER OF VARIABLES N NOT SET; USE COMMAND #1 -----"
1440 RETURN
1445 K6=0 : REM - DEFAULT TO FINITE DIFFERENCES
1450 PRINT "DIFFERENCING OR EXACT LOSSLESS ELEMENT PARTIALS (D/E)";
1452 INPUT S4$ : IF S4$="E" THEN K6=1
1455 GOSUB 4080 : REM - RESET & RECORD BINDING VARIABLES
1460 FOR I=1 TO M : REM - NULL A(I,J) JACOBIAN MATRIX
1470 FOR J=1 TO N
1480 A(I,J)=0
1490 NEXT J
1500 NEXT I
1510 GOSUB 5000 : REM - FIRST CALC OF RESIDUALS
1550 K2=K2+1 : REM - INCRE F EVAL COUNT
1580 F=0 : REM - CALC FIRST SUM Pth RESIDUALS
1590 FOR K=1 TO M : F=F+R(K)^K7 : NEXT K : F=F/K7
1600 GOSUB 7000 : REM - CALC FIRST JACOBIAN
1610 GOSUB 2600 : REM - CALC/STORE NORMAL MATRIX IN H()
1620 REM - PUT NORMALIZED SCALING FACTORS INTO D()
1630 L=0 : D2=0
1640 FOR J=1 TO N
1650 FOR I=1 TO N
1660 IF I<J THEN GOTO 1690
1670 L=L+1
1680 IF I=J THEN D(J)=H(L)
1690 NEXT I
1700 IF D(J)<=0 THEN D(J)=1
1710 D2=D2+D(J)*D(J)
1720 NEXT J
1730 D2=SQR(D2)
1740 FOR J=1 TO N : D(J)=D(J)/D2 : NEXT J : REM - NORMALIZE
1750 GOSUB 2750 : REM - CALC GRADIENT G() AND LENGTH G1
1760 IF G1<>0 THEN GOTO 1860
1770 PRINT "GRADIENT IS ZERO; CONVERGED.":RETURN:REM - MAYBE FEASIBLE XO
1860 L7=0 : V = .01# : REM - INIT ITERATION COUNT & LM PARAM
1865 FOR I=1 TO N : L5(I)=0 : NEXT I : REM - UNBIND ALL CONSTRAINTS
1870 REM***********************************
1880 REM - RE-ENTRY POINT FOR NEW ITERATION
1890 L7=L7+1 : F1=F : REM - INCRE ITER COUNT & SAVE LAST F VALUE
1900 IF M3=0 THEN V=V/10 : REM - LAST STEP WAS A GOOD ONE SO REDUCE V
1910 IF V<1D-20        THEN V=1D-20        : REM - V=0 NOT ALLOWED
1920 IF M3<>0 THEN V=10*V : REM - LAST STEP REQ'D CUTBACK, SO INCREASE V
1930 M3=0 : REM - CLEAR CUTBACK COUNTER
1940 IF L7=1 THEN GOTO 1980 : REM - ELSE CALC GRADIENT
1950 GOSUB 7000 : REM - GET JACOBIAN
```

```
1960 GOSUB 2750 : REM - CALC GRADIENT
1970 GOSUB 2600 : REM - CALC/STORE NORMAL MATRIX INTO H()
1980 IF ((L7-1) MOD I7)=0 THEN GOSUB 2500 : REM - RPT F,X, & G
1981 REM - RELEASE NON-K-T CONSTRAINTS
1982 FOR I=1 TO N : IF L5(I)*G(I)>0 THEN L5(I)=0 : NEXT I
1983 FOR I=1 TO N : REM - PROJECT GRADIENT INTO FIXED SUBSPACE
1984 G(I)=G(I)*(1-ABS(L5(I)))
1986 NEXT I
1990 GOSUB 2870 : REM - ADD LM PARAM TO H()
2000 GOSUB 3110 : REM - FACTOR (H+vD)=LDLT IN SITU IN H()
2010 IF N5=0 THEN GOTO 2060 : REM - FACTORIZATION OK
2020 V=100*V : REM - INCREASE LM PARAM V
2030 GOSUB 2600 : REM - CALC/STORE NORMAL MATRIX INTO H()
2040 GOTO 1990 : REM - REVISE NORMAL MATRIX AND RE-FACTOR
2050 REM - SET RIGHTHAND SIDE = -G()
2060 FOR I=1 TO N : E(I)=-G(I) : NEXT I
2070 GOSUB 3390 : REM - CALC STEP dx IN E()
2072 FOR I=1 TO N : REM - PROJECT SEARCH DIRECTION INTO SUBSPACE
2074 E(I)=E(I)*(1-ABS(L5(I)))
2076 NEXT I
2080 REM - CALC STEP-TO-GRADIENT DEGREES
2090 C2=0 : C3=0
2100 FOR I=1 TO N : C2=C2+G(I)*E(I) : C3=C3+E(I)*E(I) : NEXT I
2105 IF G1=0 OR C3<=0 THEN GOTO 2160 : REM - CAN'T CALC ANGLE
2110 P1=-C2/G1/SQR(C3) : IF P1<1 THEN GOTO 2130
2120 P1=0 : GOTO 2140 : REM - AVOID /0 IN ACS
2130 P1=57.29578*FNACS(P1)
2140 PRINT "                    STEP-TO-GRADIENT DEGREES=";
2150 PRINT USING " ##.####";P1
2160 REM - TAKE STEP WITH INCREMENT IN E()
2165 FOR I=1 TO N : W(I)=X(I) : NEXT I : REM - SAVE BASE POINT
2170 FOR I=1 TO N : X(I)=W(I)+E(I) : NEXT I
2175 GOSUB 3620 : REM - CHECK/SET ANY ADDITIONAL BOUNDS
2180 GOSUB 5000 : REM - CALC RESIDUALS
2190 K2=K2+1 : F=0 : REM - INCRE F EVAL COUNT & CALC SUM Pth RESIDUALS
2200 FOR K=1 TO M : F=F+R(K)^K7 : NEXT K : F=F/K7
2210 IF F<F1 THEN GOTO 2270
2220 REM - GET BACK TO LAST TURNING POINT & CUTBACK dx
2230 FOR I=1 TO N : X(I)=W(I) : E(I)=E(I)/4 : NEXT I
2240 PRINT "  ###### CUT BACK STEP SIZE BY FACTOR OF 4 ######"
2250 M3=M3+1 : IF M3<11 THEN GOTO 2170 : REM - TRY CUTBACK STEP
2260 PRINT "STEP SIZE TOO SMALL - TERMINATED" : GOTO 2390
2270 IF L7<M1 THEN GOTO 2320 : REM - NOT AT MAX ITERATIONS
2280 PRINT "!!!!!!!!!!!!!!!!!!!!!!!!!!!!!!!!!!!"
2290 PRINT "STOPPED AT GIVEN LIMIT OF";M1;" ITERATIONS; RESULTS ARE:"
2300 L7=L7+1 : GOTO 2370
2310 REM - TEST CONVERGENCE OF BOTH F AND EACH X(I)
2320 IF ABS(F1-F)/(1+ABS(F1))>E1 THEN GOTO 1890
2330 FOR I=1 TO N
2340 IF ABS(E(I))/(1+ABS(X(I)))>E1 THEN GOTO 1890
2350 NEXT I
2360 L7=L7+1 : PRINT "CONVERGED; SOLUTION IS:"
2370 GOSUB 7000 : REM - GET JACOBIAN
2380 GOSUB 2750 : REM - GET GRADIANT
2390 GOSUB 2500 : REM - REPORT F,X, & G AT STOPPING POINT
2410 PRINT "TOTAL NUMBER OF FUNCTION EVALUATIONS =";K2
2420 PRINT "EXPONENT P =";K7
2430 RETURN
2440 REM**********************************
2442 REM - DISPLAY dB AND Zin OVER A FREQ RANGE
2444 IF N=0 THEN GOTO 2494 : REM - MUST HAVE VARIABLES SET
2446 K1=0 : K6=0 : REM - PAGING AND SUB8000 FLAGS
2448 PRINT "START FREQUENCY =";  : INPUT T1
2450 PRINT "STOP FREQUENCY =";  : INPUT T2
2452 PRINT "NUMBER OF FREQS, MAX 40 (+LIN, -LOG) =";  : INPUT L1
2453 IF L1<0 AND T1=0 THEN T1=.001 : REM - DUE TO LOG CASE
2454 T3=T1 : D6=1 : IF ABS(L1)=1 THEN GOTO 2468
2456 IF ABS(L1)>40 THEN L1=SGN(L1)*40
```

```
2458 D6=ABS(L1)-1 : X5=T2-T1
2460 IF L1<0 THEN X5=LOG(T2/T1)
2462 X5=X5/D6
2464 IF L1<0 THEN X5=EXP(X5)
2466 D6=X5
2468 PRINT " #       FREQUENCY          RESPONSE dB        Rin      OHMS      Xin"
2470 FOR I2=1 TO ABS(L1)
2472 T4=S(I2,1) : S(I2,1)=T3 : REM - SAVE SAMPLE DATA BASE
2474 GOSUB 8000 : REM - GET dB & Zin
2476 PRINT I2;:PRINT USING S6$;S(I2,1),C(I2),A4,A5
2478 IF I2<21 OR I2=ABS(L1) OR K1=1 THEN GOTO 2484
2480 PRINT "PRESS <RETURN> KEY TO CONTINUE -- READY"; : INPUT S4$
2482 K1=1 : REM - DON'T PAUSE FOR 2ND HALF OF DISPLAY
2484 S(I2,1)=T4 : REM - RESTORE SAMPLE DATA BASE
2486 IF L1>0 THEN T3=T3+D6 : REM - INCREMENT FREQUENCY
2488 IF L1<0 THEN T3=T3*D6
2490 NEXT I2
2492 RETURN
2494 PRINT "*****  MUST ASSIGN VALUES TO VARIABLES FIRST  *****"
2496 RETURN
2500 REM **********************************
2510 REM - PRINT FUNCTION, VARIABLES, AND GRADIENT
2520 PRINT "AT START OF ITERATION NUMBER";L7
2530 P1=F : PRINT "   FUNCTION VALUE =";P1
2540 PRINT " I          X(I)            G(I)"
2550 FOR I=1 TO N
2560 PRINT I; : PRINT USING S6$;X(I),G(I)
2570 NEXT I
2580 RETURN
2590 REM **************************************
2600 REM - CALC/STORE NORMAL MATRIX IN H()
2610 FOR I=1 TO N*(N+1)/2 : H(I)=0 : NEXT I
2620 FOR K=1 TO M
2630 L=0
2640 FOR J=1 TO N
2650 FOR I=1 TO N
2660 IF I<J THEN GOTO 2690
2670 L=L+1
2680 H(L)=H(L)+A(K,I)*A(K,J)*R(K)^(K7-2)
2690 NEXT I
2700 NEXT J
2710 NEXT K
2720 FOR L=1 TO N*(N+1)/2 : H(L)=(K7-1)*H(L) : NEXT L
2730 RETURN
2740 REM***************************************
2750 REM - CALC GRADIENT AND ITS LENGTH
2760 G1=0
2770 FOR I=1 TO N
2780 G(I)=0
2790 FOR K=1 TO M
2800 G(I)=G(I)+A(K,I)*R(K)^(K7-1)
2810 NEXT K
2820 G1=G1+G(I)*G(I)
2830 NEXT I
2840 G1=SQR(G1)
2850 RETURN
2860 REM*****************************************
2870 REM - ADD LM PARAM TO NORMAL MATRIX
2880 L=0 : PRINT "                    LM PARAM V=";
2890 PRINT USING "##.#^^^^";V
2900 FOR J=1 TO N
2910 FOR I=1 TO N
2920 IF I<J THEN GOTO 2950
2930 L=L+1
2940 IF I=J THEN H(L)=H(L)+V*D(J)
2950 NEXT I
2960 NEXT J
2970 RETURN
```

```
2980 REM*******************************************
2990 REM - DISPLAY SAMPLE DATA FROM LINES 400 ...
3000 PRINT " I      INDEPENDENT      DEPENDENT"
3010 K=0
3020 FOR I=1 TO M
3030 PRINT I; : PRINT USING " ######.######";S(I,1);S(I,2)
3040 IF I<21 OR K=1 THEN GOTO 3070
3050 PRINT "PRESS <RETURN> KEY TO CONTINUE -- READY"; : INPUT S4$
3060 K=1 : REM - DON'T PAUSE FOR 2ND HALF OF DISPLAY
3070 NEXT I
3080 RETURN
3090 REM ********************************
3100 REM - LDLT FACTORIZATION OF MATRIX IN SITU IN VECTOR H
3110 K5=1 : N5=1 : D1=1 : REM - N5=1 NOT PD OR DET=D1<1D-6
3120 FOR I=2 TO N
3130 IF H(K5)>0 THEN GOTO 3150
3140 GOTO 3340
3150 Z=H(K5) : D1=D1*H(K5)
3160 K5=K5+1
3170 I1=K5
3180 FOR J=I TO N
3190 Z5=H(K5)
3200 H(K5)=H(K5)/Z
3210 J5=K5
3220 I5=I1
3230 FOR K=I TO J
3240 J5=J5+N+1-K
3250 H(J5)=H(J5)-H(I5)*Z5
3260 I5=I5+1
3270 NEXT K
3280 K5=K5+1
3290 NEXT J
3300 NEXT I
3310 D1=D1*H(K5) : IF D1>.0000000001# THEN N5=0
3320 IF N5=1 THEN GOTO 3340
3330 RETURN
3340 PRINT "HESSIAN NOT PD OR TOO SMALL DETERMINANT =";
3350 PRINT USING "##.####^^^^";D1
3360 RETURN
3370 REM *****************************************
3380 REM - SOLUTION E=Inv(H)E FOR SEARCH STEP
3390 FOR I=2 TO N
3400 I4=I
3410 V5=E(I)
3420 FOR J=1 TO I-1
3430 V5=V5-H(I4)*E(J)
3440 I4=I4+N-J
3450 NEXT J
3460 E(I)=V5
3470 NEXT I
3480 E(N)=E(N)/H(I4)
3490 FOR K=2 TO N
3500 I=N+1-K
3510 I1=I4-K
3520 V5=E(I)/H(I1)
3530 I4=I1
3540 FOR J=I+1 TO N
3550 I1=I1+1
3560 V5=V5-H(I1)*E(J)
3570 NEXT J
3580 E(I)=V5
3590 NEXT K
3600 RETURN
3610 REM*********************************************
3620 REM - CHECK FOR MORE BOUNDS AND RESET VARIABLE IF BINDING
3630 FOR I=1 TO N
3640 IF E(I)=0 THEN GOTO 3760 : REM - TEST ONLY SUBSPACE BOUNDS
3650 L5(I)=0 : REM - CANCEL BOUNDS, THEN RETEST THEM
```

```
3660 REM - PROCESS LOWER BOUNDS
3670 IF X(I)>P5(I,1) THEN GOTO 3720
3680 X(I)=P5(I,1)
3690 L5(I)=-1 : REM - NOW AN ACTIVE CONSTRAINT
3700 PRINT "            ACTIVATED X(";I;") LOWER BOUND"
3710 GOTO 3760 : REM - NO NEED TO PROCESS UPPER BOUNDS
3720 IF X(I)<P5(I,2) THEN GOTO 3760
3730 X(I)=P5(I,2)
3740 L5(I)=+1 : REM - NOW AN ACTIVE CONSTRAINT
3750 PRINT "            ACTIVATED X(";I;") UPPER BOUND"
3760 NEXT I
3770 RETURN
3780 REM*****************************************
3790 REM - INIT FLAGS AND LOWER/UPPER BOUNDS
3800 REM - CLEAR THE 'BINDING BOUND(S)' & SET DEFAULT LIMITS
3810 FOR I=1 TO 20 : L4(I)=0 : L5(I)=0
3820 P5(I,1)=-10000 : P5(I,2)=+10000
3830 NEXT I
3840 RETURN
3850 REM*****************************************
3860 REM - SEE OR RESET LOWER/UPPER BOUND ON VARIABLES
3870 S4$="NONE.  " : PRINT "BOUNDS NOW SET ARE:"
3880 PRINT " I        LOWER          UPPER"
3890 FOR I=1 TO 20
3900 IF L4(I)=0 THEN GOTO 3920
3910 S4$="" : PRINT I;"   "; : PRINT USING S6$;P5(I,1);P5(I,2)
3920 NEXT I
3930 PRINT S4$;"SET OR RESET ANY BOUNDS (Y/N)"; : INPUT S4$
3940 IF S4$<>"Y" THEN RETURN
3950 REM - RE-ENTRY FOR MORE BOUND SETTING
3960 PRINT "ENTER 0 TO RETURN TO MENU, ELSE ENTER VARIABLE # =";
3970 INPUT I : IF I=0 THEN RETURN
3980 PRINT "PRESS <RETURN> IF NO BOUND DESIRED"
3990 PRINT "  LOWER BOUND ="; : INPUT S4$
4000 P5(I,1)=-10000 : IF S4$<>"" THEN P5(I,1)=VAL(S4$)
4010 IF S4$<>"" THEN L4(I)=1
4020 PRINT "  UPPER BOUND ="; : INPUT S4$
4030 P5(I,2)=+10000 : IF S4$<>"" THEN P5(I,2)=VAL(S4$)
4040 IF S4$<>"" THEN L4(I)=1
4050 GOTO 3960
4070 REM*****************************************
4080 REM - RESET & RECORD BINDING VARIABLES
4090 FOR I=1 TO N
4100 REM - PROCESS LOWER BOUNDS
4110 IF X(I)>P5(I,1) THEN GOTO 4150
4120 X(I)=P5(I,1) : L5(I)=-1
4130 PRINT "SET X(";I;")=";X(I);" (LOWER BOUND)"
4140 GOTO 4180 : REM - NO NEED TO PROCESS UPPER BOUNDS
4150 IF X(I)<P5(I,2) THEN GOTO 4180
4160 X(I)=P5(I,2) : L5(I)=+1
4170 PRINT "SET X(";I;")=";X(I);" (UPPER BOUND)"
4180 NEXT I
4190 RETURN
4200 REM********************************
4210 REM - START MULTIPLIER PENALTY FUNCTION METHOD
4220 PRINT "*********************************************************"
4230 C6=1E+20 : K3=1 : REM - CONSTR CONVERGENCE CONSTANT & FLAG
4240 FOR I=1 TO 40
4250 U(I)=0 : REM - INIT CONSTRAINT RESIDUAL OFFSET
4260 S1(I)=1 : REM - INIT PENALTY MULTIPLIERS
4262 IF L6(I)<>0 THEN K3=0 : REM - CHECKING FOR ANY CONSTRAINTS
4264 NEXT I
4266 IF K3=1 THEN GOTO 1400 : REM - IS UNCONSTRAINED PROBLEM
4270 L8=1 : REM - INIT PENALTY LOOP COUNT
4280 GOSUB 1400 : REM - MIN F(X,S,U) BY LEASTP
4290 REM - RE-ENTRY FOR OUTER PENALTY LOOP
4300 PRINT "*********************************************************"
4310 BEEP
```

```
4320 GOSUB 4690 : REM - FIND MAX PENALTY MODULUS
4330 PRINT "AFTER ";L8;" PENALTY MINIMIZATIONS,"
4340 PRINT "  THE MAX CONSTRAINT MODULUS #";K8;" =";P8
4350 PRINT "CONTINUE PENALTY MINIMIZATIONS (Y/N)"; : INPUT S4$
4360 IF S4$="N" THEN RETURN : REM - GO TO COMMAND MENU
4370 L8=L8+1 : REM - INCREMENT PENALTY LOOP COUNT
4380 REM - POWELL'S PARAMETERS ADJUST SCHEME
4390 C7=C6 : C6=P8 : IF C6>=C7 THEN GOTO 4510
4400 IF K3=1 THEN GOTO 4670
4410 FOR I=1 TO M : REM - ADJUST ALL CONSTRAINT OFFSETS
4415 IF L6(I)=0 THEN GOTO 4470 : REM - UNCONSTRAINED SAMPLE
4420 U9(I)=U(I) : REM - SAVE OFFSETS
4430 C8=C(I)
4440 IF L6(I)=2 THEN GOTO 4460 : REM - EQUALITY CONSTR CASE
4450 IF U(I)<C(I) THEN C8=U(I) : REM - CHOOSE MIN
4460 U(I)=U(I)-C8 : REM - NEW CONSTR RESIDUAL OFFSET ESTIMATE
4470 NEXT I
4480 K3=1 : REM - JUST RESET ALL OFFSETS
4490 K6=0 : GOSUB 1455 : REM - MIN F(X,S,U) BY LEASTP
4500 GOTO 4300 : REM - CLOSE OUTER PENALTY LOOP
4510 REM - DIVERGING CASE
4520 C6=C7 : REM - USE PRIOR MAX C() NORM
4530 IF K3=0 THEN GOTO 4550
4540 FOR I=1 TO M : U(I)=U9(I) : NEXT I : REM - USE PRIOR OFFSETS
4550 REM - SELECTIVELY INCREASE WEIGHTS ON CONSTR RESIDUALS
4560 FOR I=1 TO M
4570 IF L6(I)=0 THEN GOTO 4640 : REM - NOT A CONSTRAINED RESIDUAL
4580 C9=C(I)
4590 IF L6(I)=2 THEN GOTO 4620 : REM - EQUALITY CONSTR
4600 IF C9<0 THEN GOTO 4620 : REM - BIND'G INEQUALITY CONSTR
4610 GOTO 4640 : REM - UNBINDING INEQUALITY CONSTRAINT
4620 IF ABS(C9)<(C7/4) THEN GOTO 4640 : REM - WEIGHT IS OK
4630 S1(I)=S1(I)*3 : U(I)=U(I)/3 : REM - FORCE CONVERGENCE
4640 NEXT I
4650 K3=0 : GOTO 4490 : REM - TO START OF PENALTY LOOP
4660 REM - TEST FOR MIN CONVERGENCE RATE
4670 IF C6>(C7/4) THEN GOTO 4550 : REM - FORCE GREATER CONVERGENCE RATE
4680 GOTO 4410  : REM - IS OK - ADJUST CONSTRAINT OFFSETS
4690 REM*********************************
4700 REM - RECONSTRUCT CONSTRAINTS & FIND MAX MAGNITUDE P8=ABS(C(K8))
4710 P8=0 : K8=0 : REM - INIT
4720 FOR I=1 TO M
4730 IF L6(I)=0 THEN GOTO 4800 : REM - NOT A CONSTRAINED SAMPLE
4740 C9=C(I)-S(I,2) : IF L6(I)=1 THEN C9=-C9 : REM - UNBIASED CONSTR'S
4750 IF L6(I)=2 THEN GOTO 4770 : REM - IS EQUALITY CONSTRAINT
4760 IF C9>0 THEN C9=0
4770 C(I)=C9 : C9=ABS(C9) : REM - C(I)=CONSTRAINT; SEE TABLE 6.2.1.
4780 IF C9<=P8 THEN GOTO 4800
4790 P8=C9 : K8=I : REM - NEW MAX MODULUS
4800 NEXT I
4810 RETURN
4820 REM*****************************
4825 REM - SEE/RESET CONSTRAINT SAMPLE NUMBERS
4830 S4$="NONE.  " : PRINT "CONSTRAINTS NOW SET ARE:"
4835 PRINT " I    SAMPLE     LOWER     EQUALITY    UPPER"
4845 FOR I=1 TO M
4850 IF L6(I)=0 THEN GOTO 4895
4855 S4$=""
4860 IF L6(I)>0 THEN GOTO 4875
4865 PRINT I;:PRINT USING S5$;S(I,1);:PRINT TAB(15);
4867 PRINT USING S5$;S(I,2)
4870 GOTO 4895
4875 IF L6(I)>1 THEN GOTO 4890
4880 PRINT I;:PRINT USING S5$;S(I,1);:PRINT TAB(37);
4882 PRINT USING S5$;S(I,2)
4885 GOTO 4895
4890 PRINT I;:PRINT USING S5$;S(I,1);:PRINT TAB(26);
4892 PRINT USING S5$;S(I,2)
```

```
4895 NEXT I
4900 PRINT S4$;"SET OR RESET ANY CONSTRAINTS (Y/N)"; : INPUT S4$
4905 IF S4$<>"Y" THEN RETURN
4910 REM - RE-ENTRY FOR MORE CONSTRAINTS
4915 PRINT "ENTER 0 TO RETURN TO MENU, ELSE ENTER SAMPLE # =";
4920 INPUT I : IF I=0 THEN RETURN
4925 IF I<=M THEN GOTO 4935
4930 PRINT "*** YOU PROVIDED ONLY";M;" SAMPLES ***" : GOTO 4915
4935 PRINT"SAMPLE PAIR #";I;" ="; : PRINT USING S5$;S(I,1);S(I,2)
4940 PRINT "LOWER, EQUALITY, UPPER, OR GOAL (L,E,U,G):"; : INPUT S4$
4945 IF S4$="L" THEN L6(I)=-1
4950 IF S4$="E" THEN L6(I)=+2
4955 IF S4$="U" THEN L6(I)=+1
4960 IF S4$="G" OR S4$="" THEN L6(I)=0 : REM - CLEAR CONSTRAINT
4965 GOTO 4910 : REM - LOOP BACK
5000 REM******************************
5010 REM - CALCULATION OF ALL RESIDUALS
5020 IF M>0 THEN GOTO 5040
5030 PRINT "WARNING - USER FAILED TO ASSIGN M IN DATA STMNT, LINE 410"
5040 FOR I2=1 TO M
5050 GOSUB 8000 : REM - GETS NETWORK RESPONSE IN C(I2)
5060 C9=C(I2)-S(I2,2) : REM - (RESPONSE-GOAL)
5070 IF L6(I2)=+1 THEN C9=-C9 : REM - IS INEQUALITY UPPER BOUND
5080 C9=C9-U(I2) : REM - OFFSET CONSTRAINT RESIDUAL
5090 IF L6(I2)=0 THEN GOTO 5150 : REM - NOT A CONSTRAINED RESIDUAL
5100 IF L6(I2)=2 THEN GOTO 5140 : REM - IS EQUALITY CONSTRAINT
5110 IF C9>0 THEN C9=0 : REM - INEQUALITY IS SATISFIED
5120 IF C9<0 THEN GOTO 5140 : REM - PENALIZE
5130 GOTO 5150
5140 C9=S1(I2)*C9 : REM - PENALTY MULTIPLIER
5150 R(I2) = C9
5155 NEXT I2
5160 RETURN
5300 REM******************************
5310 REM - SEE UNITS AND NETWORK TOPOLOGY
5320 PRINT : PRINT "UNITS ARE: FREQUENCY ="; : PRINT USING S7$;U1
5330 PRINT "          INDUCTANCE ="; : PRINT USING S7$;U2
5340 PRINT "          CAPACITANCE ="; : PRINT USING S7$;U3
5350 R5=R6:IF N=K9 THEN R5=X(N):REM-R SOURCE MAY BE VARIABLE
5360 PRINT "R SOURCE, R LOAD, X LOAD =";:PRINT USING S5$;R5,R4,X4
5370 PRINT
5380 S4$="NONE.  USER MUST PROVIDE DATA IN LINES 600-888 OR BY CMD#10."
5390 PRINT "BRANCH    TYPE    VALUE        Q        NAME"
5400 N1=0 : N2=0 : REM - LIST INDEX & NULL BRANCH COUNT
5410 FOR I=1 TO 35 : REM - BRANCH LOOP
5420 N1=N1+1 : REM - INCREMENT LIST INDEX
5430 IF M(N1)=9 THEN GOTO 5540 : REM - REACHED NTWK INPUT END
5440 IF M(N1)=0 THEN N2=N2+1 : REM - NULL BRANCH COUNT
5450 S4$="" : IF N1>N2 THEN T=X(N1-N2)
5455 IF M(N1)=0 THEN T=0 : REM - NULL BRANCH
5460 IF (N1-N2)=N+1 AND M$(N1)="R" THEN T=R6
5470 PRINT " ";I;"      ";M$(N1),:PRINT USING S5$;T,Q(N1);
5480 PRINT "     ";N$(N1)
5490 IF M(N1)<4 THEN GOTO 5530 : REM - ELE TYPE OCCUPIES ONLY 1 LINE
5500 N1=N1+1 : REM - POINT TO C IN LC BRANCH
5510 PRINT TAB(15); : PRINT USING S5$;X(N1-N2),Q(N1);
5520 PRINT "     ";N$(N1)
5530 NEXT I
5540 PRINT S4$ : RETURN
7000 REM******************************
7010 REM - CALCULATION OF PARTIAL DERIVATIVES OF ALL RESIDUALS
7020 GOSUB 9000 : REM - GETS NETWORK RESPONSE PARTIALS IN A(,)
7030 REM - PARTIAL OF Ith RESPONSE WRT Jth VARIABLE IN A(I,J)
7040 FOR I9=1 TO M : REM - CONSTRAINTS LOOP
7050 IF L6(I9)=0 THEN GOTO 7170 : REM - NOT A CONSTRAINED RESIDUAL
7060 G9=1
7070 C9=C(I9)-S(I9,2) : REM - (RESPONSE-GOAL)
7080 IF L6(I9)<>1 THEN GOTO 7100 : REM - IS NOT INEQUALITY UPPER BOUND
```

```
7090 C9=-C9 : G9=-G9 : REM - REVERSE SIGN FOR INEQUALITY UPPER BOUND
7100 C9=C9-U(I9) : REM - OFFSET CONSTRAINT RESIDUAL
7110 IF L6(I9)=2 THEN GOTO 7150 : REM - IS EQUALITY CONSTRAINT
7120 IF C9>0 THEN G9=0 : REM - INEQUALITY IS SATISFIED
7130 IF C9<0 THEN GOTO 7150 : REM - PENALIZE
7140 GOTO 7160
7150 G9=S1(I9)*G9 : REM - PENALTY MULTIPLIER
7160 FOR J9=1 TO N : A(I9,J9)=G9*A(I9,J9) : NEXT J9 : REM - VARS LOOP
7170 NEXT I9 : RETURN
8000 REM******************************
8010 REM - LADDER NETWORK ANALYSIS FOR TRANSDUCER TRANSFER FUNCTION
8020 REM - USER MUST SUPPLY SERIES SOURCE RESISTANCE IN TOPOLOGY
8030 REM - RESPONSE AT SAMPLE # I2 IS H IN C(I2). ALSO Zin=A4+jA5
8040 W=6.28318530718#*S(I2,1)*U1 : IF W=0 THEN W=.00000000001#
8050 B4=1/SQR(2*R4) :B5=0 : D4=0 :D5=0 : REM - LOAD POWER = 1/2 WATT
8060 C4=R4 :C5=X4 : REM - LOAD IMPEDANCE
8070 K=0:N1=0:N2=0:K4=0:REM - K=BR#,N1=LIST#,N2=# NULL BR'S,K4=VAR#
8080 REM - RE-ENTRY TO PROCESS NEXT BRANCH
8090 K=K+1 : N1=N1+1 : M2=M(N1) : IF M2=0 THEN M2=10 : REM - NULL BRANCH
8100 GOSUB 8590 : REM - COMPLEX LINEAR UPDATE
8110 IF M2=9 THEN GOTO 8650 : REM - AT SOURCE; COMPUTE RESPONSE
8130 ON M2 GOSUB 8270,8330,8420,8480,8230,8230,8230,8230,8230,8230
8140 IF K6=0 THEN GOTO 8080 : REM - USING FINITE DIFF PARTIALS
8150 IF M2=10 OR M2=1 THEN GOTO 8080 : REM - WAS NULL BRANCH OR RESISTOR
8160 K4=K4+1 : N3(K4,2)=K : N3(K4,2)=N1 : REM - VAR#K4's BR# & LIST#
8170 A(I2,K4)=A4 : B(I2,K4)=A5 : REM - SAVE VAR#K4's V OR I
8180 IF M2<>4 THEN GOTO 8080 : REM - NOT LC BRANCH
8190 N3(K4,2)=N1-1:K4=K4+1:N3(K4,1)=K:N3(K4,2)=N1: REM - SPECIAL LC CASE
8200 A(I2,K4)=A4 : B(I2,K4)=A5 : REM - REPEAT SAVE V OR I FOR LC BRANCH
8210 GOTO 8080 : REM - LOOP TO PROCESS NEXT BRANCH
8220 REM******************************
8230 REM - NULL BRANCH
8240 C4=0 : C5=0 : N2=N2+1 : REM - NULL BRANCH HAS NO X()
8250 RETURN
8260 REM******************************
8270 REM - RESISTOR
8280 C4=X(N1-N2) : C5=0 :IF (N1-N2)<=N THEN GOTO 8300 : REM - VARIABLE R
8290 C4=R6 : REM - R SOURCE IS FIXED
8300 IF K=INT(K/2)*2 THEN RETURN : REM - K IS EVEN (SERIES) BRANCH
8310 C4=1/C4 : RETURN
8320 REM******************************
8330 REM -INDUCTOR
8335 Q=Q(N1) : IF Q<>0 THEN Q=1/Q : REM - NOW DECREMENT=1/Q
8340 C5=W*X(N1-N2)*U2 : REM - REACTANCE
8350 C4=C5*Q : REM - SERIES RESISTANCE
8360 IF K=INT(K/2)*2 THEN RETURN
8370 D3=Q*Q+1 : REM - INVERT THE IMMITTANCE
8380 C4=Q/D3/C5
8390 C5=-1/D3/C5
8400 RETURN
8410 REM******************************
8420 REM - CAPACITOR
8425 Q=Q(N1) : IF Q<>0 THEN Q=1/Q : REM - NOW DECREMENT=1/Q
8430 C5=W*X(N1-N2)*U3 : REM - SUSCEPTANCE
8440 C4=C5*Q : REM - SHUNT CONDUCTANCE
8450 IF K=INT(K/2)*2 THEN GOTO 8370 : REM - INVERT ADMITTANCE
8460 RETURN
8470 REM******************************
8480 REM - SERIES-LC-IN-SHUNT OR PARALLEL-LC-IN-SERIES BRANCH
8490 K=K+1 : REM - MAKE BRANCH LOOK ODD IF EVEN OR VICE VERSA
8500 GOSUB 8330 : REM - INDUCTOR
8510 S4=C4 : S5=C5 : REM - SAVE PARTS
8520 N1=N1+1 : REM - SET LIST INDEX TO CAPACITOR
8530 GOSUB 8420 : REM - CAPACITOR
8540 K=K-1 : REM - RESTORE CORRECT BRANCH #
8550 C5=C5+S5 : C4=C4+S4 : REM - COMBINE IMMITTANCES
8560 S4=C4*C4+C5*C5 : C5=-C5/S4 : C4=C4/S4 : REM - INVERT
8570 RETURN
```

```
8580 REM******************************
8590 REM - COMPLEX LINEAR UPDATE
8600 A4=B4*C4-B5*C5+D4
8610 A5=B5*C4+B4*C5+D5
8620 D4=B4 : D5=B5 :B4=A4 : B5=A5
8630 RETURN
8640 REM******************************
8650 REM - PLACE RESPONSE DECIBELS LOSS IN C(I) & SCALE V & I
8660 IF K=INT(K/2)*2 GOTO 8680
8670 A4=B4 : A5=B5 : B4=D4 : B5=D5 : D4=A4 : D5=A5 : REM - SWAP
8680 C5=D4*D4+D5*D5 : REM - E SOURCE = D4+jD5
8685 C(I2)=C5/2/C4 : REM - C4 = R SOURCE
8690 C(I2)=4.34294481904#*LOG(C(I2))
8700 A5=B4*B4+B5*B5 : REM - INPUT I MAG SGD
8710 A4=(D4*B4+D5*B5)/A5 : A4=A4-C4 : REM - Rin TO RIGHT OF R SOURCE
8720 A5=(D5*B4-D4*B5)/A5 : REM - Xin
8725 IF K6=0 THEN RETURN : REM - USING FINITE DIFF PARTIALS
8730 Z(I2,1)=A4 : Z(I2,2)=A5 : REM - SAVE Zin FOR TELLEGEN DERIVS
8740 REM - SUBTRACT Es ANGLE & THEN SQUARE Ik OR Vk
8750 FOR K4=1 TO N : REM - VARIABLES LOOP
8760 A4=A(I2,K4) : A5=B(I2,K4) : REM - BRANCH Ik OR Vk
8770 B4=D4*A4+D5*A5 : B5=D4*A5-D5*A4
8780 A(I2,K4)=B4/C5*B4-B5/C5*B5 : B(I2,K4)=2*B4/C5*B5:REM-AVOID OVRFLW
8790 NEXT K4
8800 RETURN
9000 REM******************************
9010 IF K6=1 THEN GOTO 9180 : REM - EXACT PARTIALS FOR LOSSLESS NTWK
9020 REM - JACOBIAN OF RESPONSE PARTIALS BY DIFFERENCING
9030 REM - PARTIAL OF Ith SAMPLE WRT Jth VARIABLE IN A(I,J)
9040 FOR I2=1 TO M : Z(I2,1)=C(I2) : NEXT I2 : REM - SAVE UNPERTURBED
9050 FOR J2=1 TO N : REM - VARIABLES LOOP
9060 D9=ABS(X(J2))/10000#
9070 IF D9<.000001# THEN D9=.000001#
9080 X(J2)=X(J2)+D9 : REM  - PERTURB VARIABLE
9090 FOR I2=1 TO M : REM - FREQUENCY SAMPLES LOOP
9100 GOSUB 8000 : REM - NETWORK ANALYSIS
9110 A(I2,J2) = (C(I2)-Z(I2,1))/D9
9120 NEXT I2
9130 X(J2)=X(J2)-D9 : REM - RESTORE NOMINAL VARIABLE VALUE
9140 NEXT J2
9150 FOR I2=1 TO M : C(I2)=Z(I2,1) : NEXT I2 : REM - RESTORE UNPERTURBED
9160 RETURN
9170 REM******************************
9180 REM - EXACT LOSSLESS NETWORK DERIVATIVES
9190 FOR I2=1 TO M : REM - FREQ SAMPLES LOOP
9200 FOR K4=1 TO N : REM - VARIABLES LOOP
9210 K=N3(K4,1) : N1=N3(K4,2) : REM - BRANCH & LIST NUMBERS
9220 IF M(N1)<>1 THEN GOTO 9250
9230 PRINT "NO VARIABLE RESISTORS IN LOSSLESS MODE"
9240 A(I2,K4)=0 : GOTO 9370
9250 REM - RHO = A4+jA5
9260 B4=Z(I2,1) : B5=Z(I2,2) : REM - Zin(I2)
9270 A5=(B4+R6)*(B4+R6)+B5*B5
9280 A4=(B4*B4-R6*R6+B5*B5)/A5
9290 A5=2*R6*B5/A5
9300 B4=A(I2,K4) : B5=B(I2,K4) : REM - ROTATED, SQUARED BR V or I
9320 D5=8.68588963806#*(A4*B5-A5*B4): REM - IMAG (RHO CONJG * Ik SQRD)
9330 IF K=INT(K/2)*2 THEN GOTO 9360 : REM - EVEN BRANCH
9340 A(I2,K4)=+D5 : REM - ODD BR d(dB)/d(BK)
9350 GOTO 9370
9360 A(I2,K4)=-D5 : REM - EVEN BR d(dB)/d(XK)
9370 NEXT K4
9380 NEXT I2
9390 REM - APPLY CHAIN RULE PER BRANCH & VARIABLE
9400 FOR I2=1 TO M : REM - SAMPLE LOOP
9410 W=6.28318530718#*S(I2,1)#*U1 : IF W=0 THEN W=.00000000001#
9420 FOR K4=1 TO N : REM - VARIABLES LOOP
9430 K=N3(K4,1) : N1=N3(K4,2) : REM - BR# & LIST#
```

```
9440 A4=A(I2,K4) : A(I2,K4)=0
9450 M2=M(N1) : REM - ELEMENT TYPE#
9460 IF M2<2 THEN GOTO 9480 : REM - NO NULL OR RESISTOR BRANCHES ALLOWED
9470 ON M2 GOSUB 9500,9510,9560,9610
9480 NEXT K4
9490 NEXT I2
9500 RETURN
9510 REM - INDUCTOR
9520 A(I2,K4) = A4*W*U2
9530 IF K=INT(K/2)*2 THEN RETURN : REM - EVEN BRANCH
9540 A(I2,K4)=A(I2,K4)/(W*X(K4)*U2)^2
9550 RETURN
9560 REM - CAPACITOR
9570 A(I2,K4) = A4*W*U3
9580 IF K<>INT(K/2)*2 THEN RETURN : REM - ODD BRANCH
9590 A(I2,K4) = A(I2,K4)/(W*X(K4)*U3)^2
9600 RETURN
9610 REM - LC BRANCH
9620 A5=(1-W*W*X(K4)*X(K4+1)*U2*U3)^2
9630 IF K<>INT(K/2)*2 THEN GOTO 9670 : REM - ODD BRANCH
9640 A(I2,K4) = A4*W*U2/A5
9650 A(I2,K4+1)=A(I2,K4+1)*W*U3*(W*X(K4)*U2)^2/A5
9660 GOTO 9690
9670 A(I2,K4+1)=A(I2,K4+1)*W*U3/A5
9680 A(I2,K4)=A4*W*U2*(W*X(K4+1)*U3)^2/A5
9690 K4=K4+1 : REM = SET VAR# TO CAPACITOR
9700 RETURN
9800 REM*****************************************
9805 REM - READ SAMPLES, UNITS AND NETWORK TOPOLOGY FROM DISK
9810 PRINT "SEE DIRECTORY (Y/N)"; : INPUT S4$
9815 IF S4$<>"Y" THEN GOTO 9850
9820 PRINT "FILENAME SPECIFIER (LIKE *.* OR <RETURN>) = "; : INPUT S8$
9825 IF S8$="" THEN S8$="*.*"
9830 FILES S8$
9835 PRINT "SEE DIRECTORY AGAIN (Y/N)"; : INPUT S4$
9840 IF S4$<>"Y" THEN GOTO 9850
9845 GOTO 9820
9850 PRINT "FILE NAME IS"; : INPUT S2$
9855 PRINT "   !!! WARNING - DO NOT <CNTRL><BREAK> DURING THIS STEP !!!"
9860 OPEN S2$ FOR INPUT AS #1
9865 INPUT #1,S3$
9870 PRINT "READY TO READ FILE ";S2$;" TITLED:";S3$
9875 PRINT "PRESS <RETURN> KEY IF OK, ELSE 'ABORT' <RETURN>";
9880 INPUT S4$
9885 IF S4$="" THEN GOTO 9895
9890 PRINT "ABORT DISK READ; NETWORK DATA NOT CHANGED" : GOTO 9965
9895 PRINT "READ FILE";S2$;" INTO MEMORY."
9900 INPUT #1,M : REM - M IS # SAMPLES
9905 FOR K=1 TO M : INPUT#1,S(K,1) : NEXT K : REM - FREQUENCIES
9910 FOR K=1 TO M : INPUT#1,S(K,2) : NEXT K : REM - TARGETS (dB)
9915 INPUT#1,U1,U2,U3 : REM - FREQ, L, C UNITS
9920 INPUT#1,R6,R4,X4,K9 : REM - R SOURCE, R LOAD, X LOAD, # TOPOL LINES
9925 FOR I=1 TO K9 : REM - READ TOPOLOGY LINES
9930 INPUT#1,J,M$(I),Q(I),N$(I)
9935 IF M$(I)="N" THEN M(I)=0 : REM - NULL BRANCH
9940 IF M$(I)="R" THEN M(I)=1 : REM - RESISTOR
9945 IF M$(I)="L" THEN M(I)=2 : REM - INDUCTOR
9950 IF M$(I)="C" THEN M(I)=3 : REM - CAPACITOR
9955 IF M$(I)="LC" THEN M(I)=4 : REM - SERIES LC IN SHUNT OR DUAL
9960 NEXT I
9965 CLOSE #1 : RETURN
9970 END


7 REM - APPROX. ZVEREV P.201 ELLIPTIC FLTR - [C6-2] 'LPTRAP1'
400 DATA "LPTRAP1" : REM - NAME DISPLAYED ON FIRST SCREEN
410 DATA 7 : REM - NUMBER OF FOLLOWING FREQ/TARGET DATA PAIRS
```

```
420 DATA .2,.4,.6,.8,1,1.5,2
430 DATA  0, 0, 0, 0,0, 40,40
600 DATA .159155,1,1 : REM - FREQ [1/(2PI)], L & C UNITS
610 DATA 1,1,0 : REM - R SOURCE, R LOAD, & X LOAD
615 DATA 7 : REM - NUMBER OF FOLLOWING LADDER TOPOLOGY LINES
620 DATA 1,N,0,NULL1 : REM - LIST#1, NULL BR, DUMMY Q, NAME
630 DATA 2,L,0,L2 H : REM - LIST#2, INDUCTOR, INFINITE Q, NAME
640 DATA 3,C,0,C3 F : REM - LIST#3, CAPACITOR, INFINITE Q, NAME
650 DATA 4,LC,0, L4 H : REM - LIST#4, INDUC (PARALLEL), INF Q, NAME
660 DATA 5,LC,0, C5 F : REM - LIST#5, CAPAC (PARALLEL), INF Q, NAME
670 DATA 6,C,0, C5 F : REM - LIST#6, CAPACITOR, INFINITE Q, NAME
680 DATA 7,R,0,R SOURCE : REM - LIST MUST END WITH SOURCE RESISTOR
```

References

Aaron, M. R. (1956). The use of least squares in system design. *IEEE Trans. Circuits Syst.*, December, pp. 224–231.

Abadie, J. (1970). *Integer and Nonlinear Programming*. Amsterdam: North Holland.

Abramowitz, M., and I. A. Stegun (1972). *Handbook of Mathematical Functions with Formulas, Graphs, and Mathematical Tables*. Washington, D.C.: U.S. Government Printing Office.

Acton, F. S. (1970). *Numerical Methods That Work*. New York: Harper & Row.

Adby, P. R. (1980). *Applied Circuit Theory: Matrix and Computer Methods*. New York: Wiley.

Adey, R. A. (1983). *Software for Engineering Problems*. London: Gulf Publishing Co.

Agnew, D. G. (1978). Efficient use of the Hessian matrix for circuit optimization. *IEEE Trans. Circuits Syst.*, August, pp. 600–608.

―――― (1979). Minimax optimization techniques for electronic circuits. *IEE International Conf. Computer Aided Des. Manuf.*, pp. 12–14.

Aleksandrov, A. D., A. N. Kolmogorov, and M. A. Lavrent'ev (1956). *Mathematics*, Vol. 1. Cambridge, MA: M.I.T Press.

Antreich, K. T., and S. A. Huss (1984). An interactive technique for the nominal design of integrated circuits. *IEEE Circuits Syst.*, February, pp. 203–212.

Aoki, M. (1971). *Introduction to Optimization Techniques*. New York: Macmillan.

Avriel, M. (1976). *Nonlinear Programming Analysis and Methods*. Englewood Cliffs, NJ: Prentice-Hall.

Ayers, F. (1962). *Theory and Problems of Matrices*. New York: Schaum.

Bandler, J. W. (1970a). Conditions for a minimax optimum. *Proc. 1970 Allerton Conf. IEEE Circuits Syst.*, October, pp. 23–30.

―――― and R. E. Seviora (1970b). Current trends in network optimization. *IEEE Trans. Microwave Theory Tech.*, December, pp. 1159–1170.

―――― and C. Charalambous (1971a). Practical least Pth approximation with extremely large values of P. *Proc. 1971 Asilomar Conf. IEEE Circuits Syst.*, November, pp. 66–70.

―――― N. D. Markettos, and N. K. Sinha (1971b). Optimum approximation of high-order systems by low-order models using recent gradient methods. *Proc. 1971 Allerton Conf. IEEE Circuits Syst.*, October, pp. 170–179.

―――― (1973a). Computer-aided circuit optimization. *Modern Filter Theory and Design* (G. C. Temes and S. J. Mitra, eds.). New York: Wiley, pp. 211–271.

―――― N. D. Markettos, and T. V. Srinivasan (1973b). Gradient minimax techniques for system modelling. *Int. J. Syst. Sci.*, pp. 317–331.

_____ and C. Charalambous (1974a). Nonlinear programming using minimax techniques. *J. Opt. Theory Appl.*, **13** (6), pp. 607–619.

_____ and W. Y. Chu (1974b). Computational merits of extrapolation in least pTH approximation and nonlinear programming. *Proc. 12th Allerton Conf. IEEE Circuits Syst.*, October, pp. 912–921.

_____ and W. Y. Chu (1975). *Nonlinear Programming Using Least Pth Optimization with Extrapolation*. Hamilton, Ontario, Canada: McMaster University, Report. No. SOC-78.

_____ and M. R. M. Rizk (1979). Optimization of electrical circuits. *Math. Programming*, pp. 1–64.

_____ S. H. Chen, and S. Daijavad (1984). Proof and extension of general sensitivity formulas for lossless two-ports. *Elec. Lett.*, **20** (11), pp. 481–482.

_____ S. H. Chen, and S. Daijavad (1985). Simple derivation of a general sensitivity formula for lossless two-ports. *Proc. IEEE*, 73 (1), January, pp. 165–166.

Bard, Y. (1968). On a numerical instability of Davidon-like methods. *Math. Comput.*, pp. 665–666.

Bauer, F. L. (1963). Optimally scaled matrices. *Numerische Mathematik*, pp. 73–87.

Beckman, F. S. (1960). The solution of linear equations by the conjugate gradient method. *Mathematical Methods for Digital Computers* (A. Ralston and H. S. Wilf, eds.). New York: Wiley.

Bellman, R. (1960). *Introduction to Matrix Analysis*. New York: McGraw-Hill.

Best, M. J. (1975). A method to accelerate the rate of convergence of a class of optimization algorithms. *Math. Programming*, pp. 139–160.

Beveridge, G. S., and R. S. Schecter (1970). *Optimization: Theory and Practice*. New York: McGraw-Hill.

Bowdler, H. J., R. S. Martin, G. Peters, and J. H. Wilkinson (1966). Solution of real and complex systems of linear equations. *Numerische Mathematik*, pp. 217–234.

Box, M. J. (1965). A new method of constrained optimization and a comparison with other methods. *Comput. J.*, pp. 42–52.

_____ D. Davies, and W. H. Swann (1969). *Nonlinear Optimization Techniques*. Edinburgh: Oliver and Boyd.

Bracken, J., and G. P. McCormick (1968). *Selected Applications of Nonlinear Programming*. New York: Wiley.

Branin, F. H. (1972). Widely convergent method for finding multiple solutions of simultaneous nonlinear equations. *IBM J. Res. Develop.*, pp. 504–522.

_____ (1973). Network sensitivity and noise analysis simplified. *IEEE Trans. Circuit Theory*, May, pp. 285–288.

Brayton, R. K., and J. Cullum (1977). An algorithm for minimizing a differentiable function subject to box constraints and errors. IBM Res. Report RC6429. Yorktown Heights, NY: IBM Thomas J. Watson Research Center.

_____ and R. Spence (1980). *Sensitivity and Optimization*. Amsterdam: Elsevier.

Breen, R. H., and G. C. Temes (1973). Applications of Golub's algorithm in circuit optimization and analysis. *IEEE Trans. Circuit Theory*, November, pp. 687–690.

Broyden, G. C. (1965). A class of methods for solving nonlinear simultaneous equations. *Math. Comput.*, October, pp. 577–593.

_____ (1973). Quasi-Newton or modification methods. *Numerical Solution of Systems of Nonlinear Equations* (G. D. Byrne and C. A. Hall, eds.). New York: Academic, pp. 241–280.

Buckley, A. (1973). An Alternate Implementation of Goldfarb's Minimization Algorithm. Harwell, Berkshire, England: Atomic Energy Research Establishment, Report T.P.44.

Caceci, M. S., and W. P. Cacheris (1984). Fitting curves to data. *Byte*, May, pp. 340–362.

Calahan, D. A. (1968). *Modern Network Synthesis*. New York: Hayden.

———— (1972). *Computer-Aided Network Design* (rev. ed.). New York: McGraw-Hill.

Carroll, C. W. (1961). The created response surface technique for optimizing nonlinear restrained systems. *Opns. Res.*, **9** (2), pp. 169–184.

Celis, M. R., J. E. Dennis, and R. A. Tapia (1985). A trust region strategy for nonlinearly constrained optimization. *Numerical Optimization 1984* (P. T. Boggs, R. H. Byrd, and R. B. Schnabel, eds.). Philadelphia, PA: SIAM, pp. 71–82.

Charalambous, C., and A. R. Conn (1978). An efficient method to solve the minimax problem directly. *SIAM J. Num. Anal.*, **15**, pp. 278–290.

Cline, A. K., C. B. Moler, G. W. Stewart, and J. H. Wilkinson (1979). An estimate for the condition number of a matrix. *SIAM J. Numer. Anal.*, April, pp. 368–375.

Colville, A. R. (1968). A comparative study on nonlinear programming codes, IBM New York Sci. Center, Tech. Report 320–2949, June.

Conn, A. R. (1985). Nonlinear programming, exact penalty functions, and projection techniques for non-smooth functions. *Numerical Optimization 1984* (P. T. Boggs, R. H. Byrd, and R. B. Schnabel, eds.). Philadelphia, PA: SIAM, pp. 3–25.

Courant, R. (1936). *Differential and Integral Calculus*. New York: Wiley, p. 91.

———— (1943). Variational methods for the solution of problems of equilibrium and vibration. *Bull. Amer. Math. Soc.*, **49**, pp. 1–23.

Crowder, H., and P. Wolfe (1972). Linear convergence of the conjugate gradient method. *IBM J. Res. Develop.*, July, pp. 431–433.

Cullum, J. (1972). An algorithm for minimizing a differentiable function that uses only function values. *Techniques of Optimization* (A. V. Balakrishnan, ed.). New York: Academic.

Cuthbert, T. R. (1967). Optimization—making the best of it by computer redesign. Richardson, TX: Collins Radio Company Report WP-5713.

———— (1983). *Circuit Design Using Personal Computers*. New York: Wiley.

Cutteridge, O. P. D. (1974). Powerful 2-part program for solution of nonlinear simultaneous equations. *Electron. Lett.*, **10** (10), pp. 182–184.

Davidon, W. C. (1959). Variable metric method for minimization. Argonne, IL.: Argonne National Laboratory, ANL-5900 Revised.

———— (1969). Variance algorithm for minimization. *Comput. J.*, February, pp. 406–410.

———— (1975). Optimally conditioned optimization algorithms without line searches. *Math. Programming*, pp. 1–9.

———— (1976). New least-square algorithm. *J. Optimiz. Theory Appl.*, **18** (2), pp. 187–197.

———— and L. Nazareth (1977a). OCOPTR—A derivative free FORTRAN implementation of Davidon's optimally conditioned method. Argonne, IL: Argonne National Laboratory, ANL-AMD-TM-303.

———— and L. Nazareth (1977b). DRVOCR—A FORTRAN implementation of Davidon's optimally conditioned method. Argonne, IL: Argonne National Laboratory, ANL-AMD-TM-306.

Davies, D. (1968). The use of Davidon's method in non-linear programming. Imperial Chemical Industries, Ltd., August.

Dejka, W. J., and D. C. McCall (1978). Mathematical programming. San Diego, CA: USNELC, Tech. Note TN-1487.

Dembo, R. S., S. C. Eisenstat, and T. Steinhaug (1982). Inexact Newton methods. *SIAM J. Num. Anal.*, **19** (2), pp. 400–408.

Dennis, J. E., and H. H. W. Mei (1979a). Two new unconstrained optimization algorithms which use function and gradient values. *J. Opt. Theory Appl.*, **28** (4), pp. 453–482.

_____ D. M. Gay, and R. E. Welsch (1979b). An adaptive nonlinear least-squares algorithm. Madison, WI: University Wisconsin Mathematics Research Center, Report. No. 2010, ASTIA AD A079 716.

_____ and R. B. Schnabel (1983). *Numerical Methods for Unconstrained Optimization and Nonlinear Equations*. Englewood Cliffs, NJ: Prentice-Hall.

_____ (1984). A user's guide to nonlinear optimization algorithms. *Proc. IEEE.*, December, pp. 1765–1776.

DiMambro, P. H. (1983). Calculating transfer function and its first- and second-order sensitivities using one network analysis. *Electron Lett.*, May 26, pp. 421–423.

Director, S. W. (1971). LU factorization in network sensitivity calculations. *IEEE Trans. Circuit Theory*, January, pp. 184–185.

Dixon, L. C. W. (1972a). Quasi-Newton algorithms generate identical points. *Math. Prog.*, pp. 383–387.

_____ (1972b). *Nonlinear Optimisation*. New York: Crane, Russak.

_____ (1977). *Optimization in Action*. London: Academic.

Dongarra, J. J., C. B. Moler, J. R. Bunch, and G. W. Stewart (1979). *LINPACK User's Guide*. Philadelphia: SIAM.

Emery, F. E., and M. O'Hagan (1966). Optimal design of matching networks for microwave transistor amplifiers. *IEEE Trans. Microwave Theory Tech.*, December, pp. 696–698.

Faddeev, D. K., and V. N. Faddeeva (1963). *Computational Methods of Linear Algebra*. San Francisco: Freeman.

Ferrero di Roccaferrera, G. M. (1964). *Operations Research Models for Business and Industry*. Cincinnati: South-Western.

Fiacco, A. V. and G. P. McCormick (1963). Programming under nonlinear constraints by unconstrained minimization: A primal-dual method. McLean, VA: Research Analysis Corp., Tech. Paper RAC-TP-96, ASTIA AD 423 903.

_____ and G. P. McCormick (1964a). Computational algorithm for the sequential unconstrained minimization technique for nonlinear programming. *Management Sci.*, July, pp. 601–617.

_____ and G. P. McCormick (1964b). The sequential unconstrained minimization technique for nonlinear programming, Algorithm II, Optimum gradients by Fibonacci search. McLean, VA: Research Analysis Corp., Tech. Paper RAC-TP-123, ASTIA AD 450 546L. Also in *Management Sci.*, January, pp. 360–366.

_____ and G. P. McCormick (1965). The sequential unconstrained minimization technique for convex programming with equality constraints. McLean, VA: Research Analysis Corp., Tech Paper RAC-TP-155, ASTIA AD 623 093.

_____ and G. P. McCormick (1966a). Extensions of SUMT for nonlinear programming: Equality constraint and extrapolation. *Management Sci.*, July, pp. 816–828.

_____ and G. P. McCormick (1966b). The slacked unconstrained minimization technique for convex programming. McLean, VA: Research Analysis Corp., Tech. Paper RAC-TP-227, ASTIA AD 643 817. Also, *SIAM J. Applied Math.*, May, 1967, pp. 505–515.

_____ and G. P. McCormick (1966c). The sequential unconstrained minimization technique without parameters. McLean, VA: Research Analysis Corp., Tech. Paper RAC-TP-228, ASTIA AD 643 818. Also, *Opns. Res.*, 1967, pp. 820–827.

_____ and G. P. McCormick (1966d). Extensions of SUMT for nonlinear programming: equality constraints and extrapolation. *Management Sci.*, July, pp. 816–828.

_____ (1967a). Second order sufficient conditions for weak and strict constrained minima. Evanston, IL: Northwestern University Systems Research, Memo No. 175, April.

_____ (1967b). Mathematical programming by generalized sequential unconstrained methods. Evanston, IL: Northwestern University Systems Research, Memo No. 178, ASTIA AD 656 904.

_____ and G. P. McCormick (1968). *Nonlinear Programming: Sequential Unconstrained Minimization Techniques*. New York: Wiley.

_____ and A. P. Jones (1969). Generalized penalty methods in topological spaces. *SIAM J. Appl. Math.*, September, pp. 996–1000.

Fidler, J. K. (1983). Comment—Calculating transfer function and its first- and second-order sensitivities using one network analysis. *Electron. Lett.*, October 27, pp. 914–916.

_____ and C. Nightingale (1978). *Computer-Aided Circuit Design*. New York: Wiley.

Finkbeiner, D. T. (1966). *Introduction to Matrices and Linear Transformations*. San Francisco: Freeman.

Fletcher, R., and M. J. D. Powell (1963). A rapidly convergent descent method for minimization. *Comput. J.*, pp. 163–168.

_____ and C. M. Reeves (1964). Function minimization by conjugate gradients. *Comput. J.*, pp. 149–154.

_____ (1965). Function minimization without evaluating derivatives—a review. *Comput. J.*, **8**, pp. 33–44.

_____ (1968a). Programming under linear equality and inequality constraints. Imperial Chemical Industries, Ltd., January.

_____ (1968b). Generalized inverse methods for the best least squares solution of systems of non-linear equations. *Comput. J.*, pp. 392–399.

_____ (1969). *Optimization*. London: Academic.

_____ (1970a). Generalized inverses for nonlinear equations and optimization. *Numerical Methods for Non-linear Algebraic Equations* (P. Rabinowitz, ed.). London: Gordon and Breach.

_____ (1970b). A new approach to variable metric algorithms. *Comput. J.*, August, pp. 317–322.

_____ (1971a). A modified Marquardt subroutine for non-linear least squares. Harwell, Berkshire, England: Atomic Energy Research Establishment, Report No. AERE-R.6799.

_____ J. A. Grant, and M. D. Hebden (1971b). The calculation of linear best L_p approximations. *Comput. J.*, August, pp. 276–279.

_____ (1972). FORTRAN subroutines for minimization by quasi-Newton methods. Harwell, Berkshire, England: Atomic Energy Research Establishment, Report. No. AERE-R7125.

_____ (1973). An exact penalty function for nonlinear programming with inequalities. *Math. Programming*, pp. 129–150.

_____ J. A. Grant, and M. D. Hebden (1974a). Linear minimax approximation as the limit of best L_p-approximation. *SIAM J. Numer. Anal.*, March, pp. 123–136.

_____ and M. P. Jackson (1974b). Minimization of a quadratic function of many variables subject only to lower and upper bounds. *J. Inst. Maths. Appl.*, pp. 159–174.

_____ and M. J. D. Powell (1974c). On the modification of LDL^T factorizations. *Math. Comput.*, October, pp. 1067–1087.

_____ (1975). An ideal penalty function for constrained optimization. *J. Inst. Math. Appl.*, pp. 319–342.

_____ and T. L. Freeman (1977). A modified method for minimization. *J. Opt. Theory Appl.*, November, pp. 357–372.

_____ (1980). *Practical Methods of Optimization: Volume 1, Unconstrained Optimization*. Chichester, England: Wiley.

_____ (1981a). *Practical Methods of Optimization: Volume 2, Constrained Optimization*. Chichester, England: Wiley.

_____ and J. W. Sinclair (1981b). Degenerate values for Broyden methods. *J. Opt. Theory Appl.*, March, pp. 331–324.

Forsythe, G. E. (1970). Pitfalls in computation, or why a math book isn't enough. *Amer. Math. Monthly*, November, pp. 931–956.

_____ M. A. Malcolm, and C. B. Moler (1977). *Computer Methods for Mathematical Computation*. Englewood Cliffs, NJ: Prentice-Hall.

Fox, R. L., and E. L. Stanton (1968). Developments in structural analysis by direct energy minimization. *AIAA J.*, June, pp. 1036–1042.

Francis, J. G. F. (1961, 1962). The QR transformation, I & II. *Computer. J.*, **4**, pp. 265–271, 332–345.

Fried, S. S. (1984). Evaluating 8087 performance on the IBM PC. *Byte Guide to the IBM PC*, Fall, pp. 197–208.

Garbow, B. S., J. M. Boyle, J. J. Dongarra, and C. B. Moler (1977). *Matrix Eigensystem Routines*, New York: Springer-Verlag.

Gentleman, W. M. (1973). Least squares computations by Givens transformations without square roots. *J. Inst. Math. Appl.*, pp. 329–336.

Gill, P. E., and W. Murray (1972). Quasi-Newton methods for unconstrained optimization. *J. Inst. Math. Appl.*, pp. 91–108.

_____ G. H. Golub, W. Murray, and M. A. Saunders (1974a). Methods for modifying matrix factorizations. *Math. Comput.*, April, pp. 505–535.

_____ and W. Murray (1974b). *Numerical Methods for Constrained Optimization*. London: Academic.

_____ and W. Murray (1974c). Newton-type methods for unconstrained and linearly-constrained optimization. *Math. Programming*, pp. 311–350.

_____ W. Murray, and M. A. Saunders (1974d). Methods for computing and modifying the LDV factors of a matrix. Teddington, Middlesex, England: National Physical Laboratory, Report. No. NAC56.

_____ and W. Murray (1978). Algorithms for the solution of the nonlinear least-squares problem. *SIAM J. Numer. Anal.*, October, pp. 977–992.

_____ W. Murray, S. M. Pickle, and M. H. Wright (1979a). The design and structure of a Fortran program library for optimization. *ACM Trans. Math. Software*, **5** (3), pp. 259–283.

_____ and W. Murray (1979b). Conjugate-gradient methods for large-scale nonlinear optimization. Systems Opt. Lab, Dept. Opt. Res., Tech. Report SOL 79-15. Stanford, CA: Stanford U., October.

_____ W. Murray, and M. H. Wright (1981). *Practical Optimization*. London: Academic.

Goddard, P. J., and R. Spence (1969). Efficient method for the calculation of first- and second-order network sensitivities. *Electron. Lett.*, August, pp. 351–352.

_____ P. A. Villalaz, and R. Spence (1971). Method for the efficient computation of the large-change sensitivity of linear nonreciprocal networks. *Electron. Lett.*, February 25, pp. 112–113.

Golden, R. M. (1973). Digital filters, in *Modern Filter Theory and Design* (G. C. Temes and S. K. Mitra, eds.). New York: Wiley, pp. 505–557.

Goldfarb, D., and L. Lapidus (1968). Conjugate gradient method for nonlinear programming problems with linear constraints. *I & EC Fundamentals*, February, pp. 142–151.

_____ (1980). Curvilinear path steplength algorithms for minimization which use directions of negative curvature. *Math. Programming*, pp. 31–40.

Goldstein, A. A., and J. F. Price (1967). An effective algorithm for minimization. *Numerische Mathematik*, pp. 184–189.

Golub, G. H., and M. A. Saunders (1970). Linear least squares and quadratic programming. *Integer and Nonlinear Programming* (J. A. Abadie, ed.). Amsterdam: North Holland, pp. 229–256.

_____ and C. F. Van Loan (1980). An analysis of the total least squares problem. *SIAM J. Numer. Anal.*, **17** (6), pp. 883–893.

_____ and C. F. Van Loan (1983). *Matrix Computations*. Baltimore: Johns Hopkins University Press.

Greville, T. N. E. (1959). The pseudoinverse of a rectangular or singular matrix and its applications to the solution of systems of linear equations. *SIAM Rev.*, January, pp. 38–43.

_____ (1960). Some applications of the pseudoinverse of a matrix. *SIAM Rev.*, January, pp. 15–22.

_____ (1966). Note on the generalized inverse of a matrix product. *SIAM Rev.*, October, pp. 518–521.

Gyurcsik, R. S., K. Mayanum, T. Yee, F. Ma, and D. O. Pedersen (1984). Language comparison for circuit simulation on desktop computers. *Digest 1984 IEEE ISCAS*, pp. 527–529.

Haarhoff, P. C., and J. D. Buys (1970). A new method for the optimization of a nonlinear function subject to nonlinear constraints. *Comput. J.*, May, pp. 178–184.

Hachtel, G. D., T. R. Scott, and R. P. Zug (1980). An interactive linear programming approach to model parameter fitting and worst case circuit analysis. *IEEE Trans. Circuits Syst.*, October, pp. 871–881.

Hadley, G. (1963). *Linear Programming*. Reading, MA: Addison-Wesley.

_____ (1964). *Nonlinear and Dynamic Programming*. Reading, MA: Addison-Wesley.

Hald, J., and K. Madsen (1981). Combined LP and quasi-Newton methods for minimax optimization. *Math. Programming*, pp. 49–62.

Han, S. P. (1976). Superlinearly convergent variable metric algorithms for general nonlinear programming problems. *Math. Programming*, pp. 263–282.

Hebden, M. D. (1973). An algorithm for minimization using exact second derivatives. Atomic Energy Research Establishment, Report TP515. Harwell, England.

Helton, J. W. (1981). Broadbanding: gain equalization directly from data. *IEEE Trans. Circuits Syst.*, December, pp. 1125–1137.

Herskowitz, G. J., and M. Sankaran (1969). Application of NASAP to the design of communication circuits and extension of NASAP routines to large scale circuits. Hoboken, NJ: Stevens Inst. Tech., Report. NASA N69-31005.

Hestenes, M. R. (1969). Multiplier and gradient methods. *J. Opt. Theory Appl.*, **4**, pp. 3030–3320.

Hewlett-Packard Co. (1982). *HP-15C Advanced Functions Handbook*. Part No. 00015-90011.

Himmelblau, D. M. (1972). *Applied Nonlinear Programming*. New York: McGraw-Hill.

Holmes, R., and T. A. Jeeves (1981). Practical aspects of nonlinear programming. MIT Industrial Liaison Program, Report 10-36-81.

Hopper, M. J. (1981). Harwell Subroutine Library. United Kingdom Atomic Energy Authority, Oxfordshire, England.

Hooke, R., and T. A. Jeeves (1961). "Direct search" solution of numerical and statistical problems. *J. ACM*, April, pp. 212–221.

Huang, H. Y. (1970). Unified approach to quadratically convergent algorithms for function minimization. *J. Opt. Theory Appl.*, **5**, pp. 405–423.

IBM (1968). *System 360 Scientific Subroutine Package (360A-CM-03X), Version III: Programmer's Manual*. White Plains, NY, Part No. H20-0205-3.

Jacobs, D. (1977). *The State of the Art in Numerical Analysis*. London: Academic.

Jacobson, D. H. and W. Oksman (1970). An algorithm that minimizes homogeneous functions of N variables in $N + 2$ iterations and rapidly minimizes general functions. Cambridge, MA: Harvard Univ. Div. Eng. Appl. Physics, Tech. Report. No. 618 (Revised).

Jennings, A. (1977). *Matrix Computation for Engineers and Scientists*. Chichester, England: Wiley.

Johnson, R. A. (1973). Mechanical bandpass filters, in *Modern Filter Theory and Design* (G. C. Temes and S. K. Mitra, eds.). New York: Wiley, pp. 157–210.

―――― (1983). *Mechanical Filters in Electronics*. New York: Wiley.

Jones, A. (1970). SPIRAL―A new algorithm for non-linear parameter estimation using least squares. *Comput. J.*, August, pp. 301–308.

Kamiel, M. S., and A. Dax (1979). A modified Newton's method for unconstrained minimization. *SIAM J. Numer. Anal.*, **16** (2), pp. 324–331.

―――― (1984). Computing the singular value decomposition in image processing. *Proc. 1984 Conf. Info. Systems and Sciences*. Princeton University, pp. 91–93.

Kaplan, W. (1959). *Advanced Calculus*. Reading, MA: Addison-Wesley.

Klema, V. C., and A. J. Laub (1980). The singular value decomposition: its computation and some applications. *IEEE Trans. Auto. Control*, AC-25, N.2, April, pp. 164–176.

Knuth, D. E. (1968). *The Art of Computer Programming*. Reading, MA: Addison-Wesley.

Krebs, M. G. (1973). The design of a least-Pth electric circuit optimization program. Los Angeles, CA: University of California.

Kuester, J. L., and J. H. Mize (1973). *Optimization Techniques with Fortran*. New York: McGraw-Hill.

Kuhn, H. W., and A. W. Tucker (1951). Nonlinear programming, in *Proc. Second Berkeley Symp. on Math. Stat. Probability* (J. Neyman, ed.). Berkeley, CA: University of California Press, pp. 481–492.

Lasdon, L. S., and A. D. Waren (1966). Optimal design of filters with bounded, lossy elements. *IEEE Trans. Circuit Theory*, June, pp. 175–187.

―――― and A. D. Waren (1967). Mathematical programming for optimal design. *Electro-Technology*, November, pp. 55–70.

Lawson, C. L., and R. J. Hanson (1974). *Solving Least Squares Problems*. Englewood Cliffs, NJ: Prentice-Hall.

Lee, H. B., P. Carvey, R. Grabowski, and D. Evans (1970). Program refines circuit from rough design data. *Electronics*, November 23, pp. 58–65.

Lenard, M. L. (1978). Accelerated conjugate direction methods for unconstrained optimization. *J. Opt. Theory Appl.*, May, pp. 11–31.

Levenberg, K. (1944). A method for the solution of certain nonlinear problems in least squares. *Quart. Appl. Math.*, **2**, pp. 164–168.

Levy, A. V., and S. Gomez (1984). The tunneling method applied to global optimization. *Numerical Optimization 1984* (P. T. Boggs, R. H. Byrd, and R. B. Schnabel, eds.). Philadelphia, PA: SIAM, pp. 213–244.

Ley, B. J. (1970). *Computer Aided Analysis and Design for Electrical Engineers*. New York: Holt, Rinehart and Winston.

Li, S. T., J. W. Rockway, J. C. Logan, and D. W. S. Tan (1983). *Microcomputer Tools for Communications Engineering*. Dedham, MA: Artech.

Lootsma, F. A. (1972). *Numerical Methods for Nonlinear Optimization*. London: Academic.

Luns, R., and T. H. I. Jaakola (1973). Optimization by direct search and systematic reduction of the size of the search region. *AIChE J.*, **19** (4), pp. 760–766.

Lyness, J. N. (1976). An interface problem in numerical software. *Proc. 6th Manitoba Conf. Num. Math.*, pp. 251–263.

Maffioli, F. (1970). Constrained variable metric optimization of layered electromagnetic absorbers. *Alta Frequenza*, February, pp. 154–164.

Marjkowski, G. (1984). *A Comprehensive Guide to the IBM Personal Computer*. Englewood Cliffs, NJ: Prentice-Hall.

Maron, M. J. (1982). *Numerical Analysis: A Practical Approach*. New York: Macmillan.

Marquart, D. W. (1963). An algorithm for least-squares estimation of nonlinear parameters. *J. SIAM*. June, pp. 431–441.

Martin, R. S., G. Peters, and J. H. Wilkinson (1965). Symmetric decomposition of a positive definite matrix. *Numerische Mathematik*, pp. 362–383.

Massara, R. E., and J. K. Fidler (1975). Efficient damping method for least-squares algorithms. *Electron. Lett.*, January 23, pp. 33–34.

McCalla, T. R. (1967). *Introduction to Numerical Methods and FORTRAN Programming*. New York: Wiley.

McCormick, G. P., W. C. Mylander, and A. V. Fiacco (1965). Computer program implementing the sequential unconstrained minimization technique for nonlinear programming. McLean, VA: Research Analysis Corp., Tech. Paper RAC-TP-151.

———— (1967). Minimizing structured unconstrained functions. McLean, VA: Research Analysis Corp., Tech Paper RAC-TP-277.

———— (1969). Antizigzagging by bending. *Management Sci.*, **15**, pp. 315–320.

Miller, A. R. (1981). *BASIC Programs for Scientists and Engineers*. Berkeley, CA: Sybex.

Moore, P. G., and S. D. Hodges (1970). *Programming for Optimal Decisions*. Middlesex, England: Penguin.

Moré, J. J., and D. C. Sorensen (1984). Newton's method. *Studies in Numerical Analysis*, Vol. 24 (G. H. Golub, ed.). Washington, DC: Mathematical Association of America.

Morris, J. (1983). *Computational Methods in Elementary Numerical Analysis*. New York: Wiley.

Mosteller, H. W. (1978). Heuristic direct-search minimization. *IEEE Trans. Auto. Control*. June, pp. 493–494.

Murray, W. (1972). *Numerical Methods for Unconstrained Optimization*. London: Academic.

Murtaugh, B. A., and R. W. H. Sargent (1970). Computational experience with quadratically convergent minimisation methods. *Comput. J.*, pp. 185–194.

Musson, J. T. B., B. Nicholson, and M. Sadler (1970). The application of optimisation to the design of LCR networks and microwave components. *Marconi Rev.*, pp. 202–224.

Mylander, W. C., R. L. Hodges, and G. P. McCormick (1971). A guide to SUMT—Version 4. McLean, VA: Research Analysis Corp., Tech. Paper RAC-P-63.

Nash, J. C. (1979). *Compact Numerical Methods for Computers: Linear Algebra and Function Minimisation*. New York: Wiley.

Nazareth, L. (1973). Unified approach to unconstrained minimization (I). Generation of conjugate directions for unconstrained minimization without derivatives (II). Berkeley, CA: Univ. California, Tech. Report. ASTIA AD-770 616.

———— (1976). A hybrid least squares method. Argonne, IL: Argonne National Laboratory, Report ANL-AMD-TM-254 (Revised).

———— (1977). A relationship between the BFGS and conjugate gradient algorithms. Argonne, IL: Argonne National Laboratory, Report ANL-AMD-TM-282 (Revised).

———— (1978a). Software for Optimization. Menlo Park, CA: Stanford Systems Optimization Laboratory, Tech. Report SOL 78-32. ASTIA AD-A066 343.

———— and J. Nocedal (1978b). A study of conjugate gradient methods. Menlo Park, CA: Stanford Systems Optimization Laboratory, Tech. Report SOL-78-29. ASTIA AD-A066 391.

———— and J. Nocedal (1978c). Properties of conjugate gradient methods with inexact line searches. Menlo Park, CA: Stanford Systems Optimization Laboratory, Tech. Report SOL-78-1.

Noble, B. (1969). *Applied Linear Algebra*. Englewood Cliffs, NJ: Prentice-Hall.

Norton, P. (1984). *MS-DOS and PC-DOS User's Guide*. Bowie, MD: Robert J. Brady Co.

Ogata, K. (1967). *State Space Analysis of Control Systems*. Englewood Cliffs, NJ: Prentice-Hall.

O'Leary, D. P. (1982). A discrete Newton algorithm for minimizing a function of many variables. *Math. Programming*, **23**, pp. 20–33.

Orchard, H. J., G. C. Temes, and T. Cataltepe (1985). Sensitivity formulas for terminated lossless two-ports. *IEEE Trans. Circuits Systs.*, CAS-32, N.5, May, pp. 459–466.

Oren, S. S. (1973). Self-scaling variable metric algorithms without line search for unconstrained minimization. *1973 Joint Auto. Controls Conf.*, June.

Osborne, M. R., and G. A. Watson (1969). An algorithm for minimax approximation in the nonlinear case. *Comput. J.*, pp. 63–68.

Penfield, P., R. Spence, and S. Duinker (1970). *Tellegen's Theorem and Electrical Networks*. Cambridge, MA: M.I.T. Press.

Penrose, R. (1955). A generalized inverse for matrices. *Proc. Cambridge Philos. Soc.*, pp. 406–413.

Peters, G., and J. H. Wilkinson (1970). The least squares problem and pseudo-inverses. *Comput. J.*, pp. 309–316.

Popovic, J. R., J. W. Bandler, and C. Charalambous (1974). General programs for least pth and near minimax approximation. *Int. J. Systems Sci.*, pp. 907–932.

Powell, M. J. D. (1964). An efficient way for finding the minimum of a function of several variables without calculating derivatives. *Comput. J.*, pp. 155–162.

_____ (1965). A method for minimizing a sum of squares of non-linear functions without calculating derivatives. *Comput. J.*, pp. 303–307.

_____ (1967). Minimization of functions of several variables. *Numerical Analysis* (J. Walsh, ed.). Washington, DC: Thompson Book Co., pp. 143–157.

_____ (1969). A method for nonlinear constraints in optimization problems. *Optimization* (R. Fletcher, ed.). London: Academic, pp. 283–297.

_____ (1970). A hybrid method for nonlinear equations. *Numerical Methods for Nonlinear Algebraic Equations* (P. Rabinowitz, ed.). London: Gordon and Breach.

_____ (1977). Quadratic termination properties of Davidon's new variable metric algorithm. *Math. Programming*, pp. 141–147.

_____ (1978). A fast algorithm for nonlinearly constrained optimization calculations. *Numerical Analysis*, Dundee 1977 (G. A. Watson, ed.). Berlin: Springer-Verlag, pp. 144–157.

Ralston, A. (1965). *A First Course in Numerical Analysis*. New York: McGraw-Hill.

Rheinboldt, W. C. (1974). Methods for solving systems of nonlinear equations. Philadelphia, PA: Society for Industrial and Applied Mathematics.

Rosen, J. B. (1960). The gradient projection method for nonlinear programming. Part I. Linear constraints. *J. SIAM.*, March, pp. 181–217.

Rosenbrock, H. H. (1960). An automatic method for finding the greatest or least value of a function. *Comput. J.*, pp. 175–184.

Rust, B. (1966). A simple algorithm for computing the generalized inverse of a matrix. *Comm. ACM*, May, pp. 381–385.

Schrack, G., and N. Borowski (1972). An experimental comparison of three random searches. *Numerical Methods for Nonlinear Optimization* (F. A. Lootsma, ed.). New York: Academic, pp. 137–147.

Shanno, D. F. (1970). An accelerated gradient projection method for linearly constrained nonlinear estimation. *SIAM J. Appl. Math.*, March, pp. 322–334.

Sibul, L. H., and A. L. Fogelsanger (1984). Application of coordinate rotation algorithm to singular value decomposition. *Digest ISCAS '84*, pp. 821–824.

Smith, B. T., J. M. Boyle, J. J. Dongarra, B. S. Garbow, Y. Ikebe, V. C. Klema, and C. B. Moler (1976). *Matrix Eigensystem Routines*. New York: Springer-Verlag.

Smith, R. C. (1971). Considerations in automated design using topological alterations. Southern Methodist University, University Microfilm No. AAD72-00662.

Spence, R. (1970). *Linear Active Networks*. London: Wiley.

Staudhammer, J. (1975). *Circuit Analysis by Digital Computer*. Englewood Cliffs, NJ: Prentice-Hall.

Steihaug, T. (1983). The conjugate gradient method and trust regions in large scale optimization. *SIAM J. Numer. Anal.*, **20** (3), pp. 626–637.

Stewart, G. W. (1967). A modification of Davidon's minimization method to accept difference approximations of derivatives. *J. ACM*, January, pp. 72–83.

Strang, G. (1980). *Linear Algebra and Its Applications*, 2nd Ed. New York: Academic.

Talisa, S. H. (1985). Application of Davidenko's method to the solution of dispersion relations in lossy waveguide systems. *IEEE Trans. Microwave Theory Tech.*, MTT-33, No. 10, October, pp. 967–971.

Temes, G. C., and D. A. Calahan (1967). Computer-aided network optimization, the state-of-the-art. *Proc. IEEE*, November, pp. 1832–1863.

_____ and D. Y. F. Zai (1969). Least pth approximation. *IEEE Trans. Circuit Theory*, May, pp. 235–237.

_____ and S. K. Mitra (1973). *Modern Filter Design*. New York: Wiley.

Tesler, L. G. (1984). Programming languages. *Sci. Amer.*, September, pp. 70–78.

Traub, J. F. (1964). *Iterative Methods for the Solution of Equations*. New York: Chelsea.

Tufts, D. W., and R. Kumarsan (1982). Singular value decomposition and improved frequency estimation using linear prediction. *IEEE Trans. Acoustics, Speech, Sig. Proc.*, ASSP-30, No. 4, August, pp. 671–675.

Uhlir, A. (1982). Rectangular-to-polar conversion in BASIC. *Microwave Jour.*, March, p. 18.

Van Huffel, S., J. Vanderwalle, and J. Starr (1983). The total linear least squares problem: formulation, algorithm, and applications. *International Circuits Systs. Conf. 1984*, pp. 328–331.

Vlach, J., and K. Singhal (1983). *Computer Methods for Circuit Analysis and Design*. New York: Van Nostrand Reinhold.

Voith, R. P., W. G. Vogt, and M. H. Mickle (1969). A direct computational procedure for the generalized inverse. *Conference Record, 12th Annual Midwest Symp. on Circuit Theory*, April, pp. VI.7.1–VI.7.8.

Walsh, G. R. (1975). *Methods of Optimization*. Chichester, England: Wiley.

Walster, G. W., E. R. Hansen, and S. Sengupta (1984). Test results for a global optimization algorithm. *Numerical Optimization 1984* (P. T. Boggs, R. H. Byrd, and R. B. Schnabel, eds.). Philadelphia: SIAM, pp. 272–287.

Waren, A. D., and L. S. Lasdon (1979). The status of nonlinear programming software. *Operations Res.* May–June, pp. 431–456.

Wilde, D. J., and C. S. Beightler (1967). *Foundations of Optimization*. Englewood Cliffs, NJ: Prentice-Hall.

Wilf, H. S. (1962). *Mathematics for the Physical Sciences*. New York: Wiley.

Wilkinson, J. H. (1963). *Rounding Errors in Algebraic Processes*. Englewood Cliffs, NJ: Prentice-Hall.

_____ (1965). *The Algebraic Eigenvalue Problem*. London: Oxford University Press.

Willoughby, J. K., and B. L. Pierson (1973). The projection operator applied to gradient methods for solving optimal control problems with terminal state constraints. *Int. J. Systems Sci.*, pp. 45–57.

Wisner, D. A. and R. Chattergy (1978). *Introduction to Nonlinear Optimization: A Problem Solving Approach*. New York: North-Holland.

Wolf, C. (1985). Serious FORTRAN. *PC Magazine*, December 24, pp. 161–171.

Wolfe, P. (1961). A duality theorem for non-linear programming. *Quart. Appl. Math.*, pp. 239–244.

Wright, D. J., and O. P. D. Cutteridge (1976). Applied optimization and circuit design. *Computer Aided Design*. April, pp. 70–76.

Wylie, C. R. (1951). *Advanced Engineering Mathematics*. New York: McGraw-Hill.

Zirilli, F. (1982). The solution of nonlinear systems of equations by second order systems of o.d.e. and linearly implicit A-stable techniques. *SIAM J. Numer. Anal.*, **19** (4), pp. 800–814.

Index